Fourier and Fourier-Stieltjes Algebras on Locally Compact Groups

Mathematical
Surveys
and
Monographs

Volume 231

Fourier and Fourier-Stieltjes Algebras on Locally Compact Groups

Eberhard Kaniuth
Anthony To-Ming Lau

AMERICAN MATHEMATICAL SOCIETY
Providence, Rhode Island

2010 *Mathematics Subject Classification.* Primary 43-02, 43A10, 43A20, 43A30, 43A25, 46-02, 22-02.

For additional information and updates on this book, visit
www.ams.org/bookpages/surv-231

Library of Congress Cataloging-in-Publication Data

Names: Kaniuth, Eberhard, author. | Lau, Anthony To-Ming, author.
Title: Fourier and Fourier-Stieltjes algebras on locally compact groups / Eberhard Kaniuth, Anthony To-Ming Lau.
Description: Providence, Rhode Island: American Mathematical Society, [2018] | Series: Mathematical surveys and monographs; volume 231 | Includes bibliographical references and index.
Identifiers: LCCN 2017052436 | ISBN 9780821853658 (alk. paper)
Subjects: LCSH: Topological groups. | Group algebras. | Fourier analysis. | Stieltjes transform. | Locally compact groups. | AMS: Abstract harmonic analysis – Research exposition (monographs, survey articles). msc | Abstract harmonic analysis – Abstract harmonic analysis – Measure algebras on groups, semigroups, etc. msc | Abstract harmonic analysis – Abstract harmonic analysis – L^1-algebras on groups, semigroups, etc. msc | Abstract harmonic analysis – Abstract harmonic analysis – Fourier and Fourier-Stieltjes transforms on nonabelian groups and on semigroups, etc. msc | Abstract harmonic analysis – Abstract harmonic analysis – Fourier and Fourier-Stieltjes transforms on locally compact and other abelian groups. msc | Functional analysis – Research exposition (monographs, survey articles). msc | Topological groups, Lie groups – Research exposition (monographs, survey articles). msc
Classification: LCC QA387 .K354 2018 | DDC 515/.2433–dc23
LC record available at https://lccn.loc.gov/2017052436

∞ The paper used in this book is acid-free and falls within the guidelines
established to ensure permanence and durability.
Visit the AMS home page at `http://www.ams.org/`

10 9 8 7 6 5 4 3 2 1 23 22 21 20 19 18

We dedicate this book to our wives, Ulla and Alice, for their lifetime of support and exceptional patience during the preparation of the manuscript.

Contents

Preface

Let G be a locally compact group. Let $C^b(G)$ be the C^*-algebra of bounded continuous complex-valued functions on G with the supremum norm, and let $C_0(G)$ be the closed $*$-subalgebra of $C^b(G)$ that consists of functions vanishing at infinity. If G is abelian, let $\widehat{G}$ be the dual group of G, and let $A(G)$ be all $\widehat{f}$ (Fourier transform of f), $f \in L^1(\widehat{G})$ (the group algebra of the dual group $\widehat{G}$); and let $B(G)$ be all $\widehat{\mu}$ (the Fourier-Stieltjes transform of μ), $\mu \in M(\widehat{G})$ (the measure algebra of $\widehat{G}$). Then $A(G)$ is a subalgebra of $C_0(G)$, and $B(G)$ is a subalgebra of $C^b(G)$. Furthermore, $A(G)$ (respectively, $B(G)$) with norm from $L^1(\widehat{G})$ (respectively, $M(\widehat{G})$) is a commutative Banach algebra called the Fourier (respectively, Fourier-Stieltjes) algebra of G.

In Chapter 2, we shall introduce and study some basic properties of Fourier and Fourier-Stieltjes algebras, $A(G)$ and $B(G)$, associated to a locally compact group G based on the fundamental paper of Eymard [**73**]. $B(G)$ will be identified as the Banach space dual of the group C^*-algebra $C^*(G)$ and a fair number of basic functorial properties will be presented. Similarly, for the Fourier algebra $A(G)$, the elements will be shown to be precisely the convolution products of L^2-functions on G.

In Chapter 3, we shall study some further topics of $A(G)$ and $B(G)$. Generalizing the classical description of idempotents in the measure algebra of a locally compact abelian group, Host [**129**] has identified the integer-valued functions in $B(G)$. Host's idempotent theorem, which has numerous applications, will be shown in this chapter. A natural question is whether either of the Banach algebras $A(G)$ and $B(G)$ determines G as a topological group. This question has been affirmatively answered by Walter [**280**]. If G_1 and G_2 are locally compact groups and $B(G_1)$ and $B(G_2)$ (respectively, $A(G_1)$ and $A(G_2)$) are isometrically isomorphic, then G_1 and G_2 are topologically isomorphic or anti-isomorphic.

Amenable Banach algebras were introduced by B. E. Johnson. He showed the fundamental result that a locally compact group is amenable if and only if the group algebra $L^1(G)$ is amenable. We present a proof of the "only if" part of Johnson's result in Chapter 4. In particular, if G is abelian, then $A(G)$, being isometrically isomorphic to the L^1-algebra of the dual group $\widehat{G}$, is amenable. However, when G is nonabelian, then $A(G)$ need not be weakly amenable, even when G is compact.

In Chapter 4, we will also consider the completely bounded cohomology theory of the Fourier algebra $A(G)$ and of the Fourier-Stieltjes algebra $B(G)$. We will show that $A(G)$, equipped with the operator space structure inherited from being embedded into $VN(G)^*$, is a completely contractive Banach algebra. Using this, we establish in this chapter the fundamental result, due to Ruan [**245**], that a locally compact group G is amenable precisely when $A(G)$ is operator amenable.

An important object associated to any (nonunital) commutative Banach algebra A is the multiplier algebra $M(A)$ of A; that is, the algebra of all bounded linear maps $T : A \to A$ satisfying the equation $T(ab) = aT(b)$ for all $a, b \in A$. When A is faithful, then the map $a \to T_a$, where $T_a(b) = ab$ for $b \in A$, is a continuous embedding of A into $M(A)$.

Let G be a locally compact group. Then $M(A(G))$ consists of all bounded continuous functions u on G such that $uA(G) \subseteq A(G)$, and since $A(G)$ is an ideal in $B(G)$, $B(G)$ embeds continuously into $M(A(G))$. If G is abelian, then as shown by Wendel [**288**], $M(L^1(G)) = M(G)$, and hence $M(A(G)) = B(G)$. It is not difficult to see that this holds true, more generally, when G is amenable. One of the profound achievements in abstract harmonic analysis has been that the converse holds; that is, $M(A(G)) = B(G)$ forces G to be amenable. This was shown by Nebbia [**219**] for discrete groups G and by Losert [**201**] for nondiscrete G. We will present these results in Chapter 5.

In Chapter 6, we study spectral synthesis and ideal theory for $A(G)$. A famous theorem of Malliavin [**207**] states that spectral synthesis fails for $A(G)$ whenever G is any nondiscrete abelian locally compact group. Using this and a deep theorem of Zel′manov [**293**] ensuring the existence of infinite abelian subgroups of infinite compact groups, we prove that for an arbitrary locally compact group G, under a mild additional hypothesis, spectral synthesis holds for $A(G)$ if and only if G is discrete.

One of the most interesting problems in the ideal theory of a commutative Banach algebra is to identify the closed ideals with bounded approximate identities. For Fourier algebras this problem is also treated in Chapter 6.

The Hahn-Banach extension theorem asserts that if E is a normed linear space and F is a closed linear subspace of E, then each continuous linear functional on F extends to a continuous linear functional on E. From this it follows that given $x \in E \backslash F$, there exists a continuous linear functional ϕ on E such that $\phi = 0$ on F and $\phi(x) \neq 0$ (the Hahn-Banach separation theorem). In Chapter 7, we address the analogous properties for positive definite functions on locally compact groups.

Let G be an arbitrary locally compact group, and let H be a closed subgroup of G. We show in Chapter 2 that the restriction map $u \to u|_H$ from $A(G)$ into $A(H)$ is surjective. The corresponding problem for Fourier-Stieltjes algebras is much more delicate. We say that G has the extension property if for every closed subgroup H, each $\varphi \in P(H)$ admits an extension $\phi \in P(G)$ (equivalently, $B(H) = B(G)|_H$). The largest class of locally compact groups sharing this extension property is formed by the groups with small conjugation invariant neighbourhoods of the identity, the SIN-groups. The converse implication is true for connected Lie groups and for compactly generated nilpotent groups. More precisely, a connected Lie group has the extension property only if it is a direct product of a vector group and a compact group. On the other hand, there exists a compactly generated 2-step solvable group which has the extension property, but fails to be a SIN-group.

Acknowledgments

The first author was very pleased to be a Pacific Institute of Mathematical Sciences Distinguished Visiting Professor at the University of Alberta at Edmonton between 2007 and 2008 when the first thoughts of this work started. Subsequently, the first author's visit to the University of Alberta was funded by NSERC grant of Anthony To-Ming Lau, and the second author's visit to the University of Paderborn was supported by the funding of Eberhard Kaniuth from the university. The authors are very grateful to Dr. Liangjin Yao for generously helping in the final preparation of this manuscript. We are very grateful to our friends Garth H. Dales, Brian Forrest, Zhiguo Hu, Mehdi Monfared, Ali Ülger, and Matthew Wiersma for very careful reading of the manuscript with many valuable suggestions. Without their kind help, this book may not have been completed on time. We are also very grateful to the American Mathematical Society for accepting our book. We would like to thank Ina Mette for kindly inviting us to submit our book at a Canadian Mathematical Society meeting in winter 2009 held at University of Windsor, and for her patience. We would also like to thank Marcia Almeida and Becky Rivard of the American Mathematical Society for their very kind help in the final preparation of the manuscript for publication.

We would like to thank for referees of the first and second reviews of the book and their many valuable comments.

Eberhard Kaniuth passed away recently in April 2017—an enormous loss to the mathematical community. I enjoyed very much our over 20 years of research collaboration and his warm friendship. I would also like to thank Eberhard for his many years of hard work in preparation of the book.

CHAPTER 1

Preliminaries

This introductory chapter provides the notational conventions used throughout the book and presents all those basic results which are fundamental for developing the theory of Fourier and Fourier-Stieltjes algebras and the various special topics treated in this book. We refrain from giving proofs because these can be found in many existing monographs. Without claiming completeness, we list a number of sources for the following subjects.

- Gelfand theory of commutative Banach algebras: [**151**], [**223**]
- Haar measure and group algebras: [**78**], [**125**]
- Harmonic analysis on locally compact groups: [**78**], [**125**], [**198**], [**241**], [**247**],
- amenability of locally compact groups: [**107**], [**231**], [**236**]
- representation theory and positive definite functions: [**60**], [**95**]
- weak containment of representations: [**60**], [**74**], [**158**].

1.1. Banach algebras and Gelfand theory of commutative Banach algebras

Let A be an algebra A over the complex number field. A is called *unital* if it has an identity. An *involution* on A is a map $a \to a^*$ from A to A which satisfies

$$(a+b)^* = a^* + b^*, \ (\alpha a)^* = \overline{\alpha} a^*, \ (ab)^* = b^* a^* \text{ and } (a^*)^* = a$$

for all $a, b \in A$ and $\alpha \in \mathbb{C}$. A is then called a *$*$-algebra*. An element $a \in A$ is called *self-adjoint* if $a^* = a$, and every $a \in A$ can be uniquely decomposed as $a = a_1 + ia_2$, where a_1 and a_2 are self-adjoint:

$$a_1 = \frac{1}{2}(a + a^*) \quad \text{and} \quad a_2 = \frac{1}{2i}(a - a^*).$$

If A is a $*$-algebra and I is a $*$-ideal in A, that is, $a \in I$ implies $a^* \in I$, then the quotient algebra A/I is a $*$-algebra with the involution $(a+I)^* = a^* + I$.

The algebra A is called a *normed (Banach) algebra* if it is endowed with a submultiplicative norm with respect to which A is a normed (Banach) space. A is called a *normed $*$-algebra* (*Banach $*$-algebra*) if it is a normed (Banach) algebra with an involution such that $\|a^*\| = \|a\|$ for all $a \in A$. Finally, a Banach $*$-algebra is called a *C^*-algebra* if the norm satisfies $\|a^*a\| = \|a\|^2$ for all $a \in A$.

EXAMPLE 1.1.1. (1) For an arbitrary set X, we denote by $B(X)$ the set of all bounded complex-valued functions on X. With pointwise operations, complex conjugation as involution and the supremum norm

$$\|f\|_\infty = \sup\{|f(x)| : x \in X\},$$

$B(X)$ is a unital commutative C^*-algebra.

(2) For a topological space X, we denote by $C(X)$ the space of all continuous functions on X and let $C^b(X) = C(X) \cap B(X)$. Moreover, $C_0(X)$ (resp. $C_c(X)$) denotes the space all continuous function on X which vanish at infinity (resp. have compact support). Then $C^b(X)$, equipped with the supremum norm, is a C^*-algebra, $C_0(X)$ is a closed subalgebra of $C^b(X)$ and $C_c(X)$ is a dense subalgebra of $C_0(X)$.

(3) Let $\mathcal{H}$ be a Hilbert space and let $\mathcal{B}(\mathcal{H})$ denote the set of all bounded linear operators on $\mathcal{H}$. With $T \to T^*$, where T^* is the adjoint of T, and the operator norm, $\mathcal{B}(\mathcal{H})$ is a unital C^*-algebra.

THEOREM 1.1.2. [Gelfand-Naimark theorem] *For every C^*-algebra A, there exists a Hilbert space $\mathcal{H}$ such that A is isometrically $*$-isomorphic to some C^*-subalgebra of $\mathcal{B}(\mathcal{H})$.*

THEOREM 1.1.3. [Gelfand-Mazur theorem] *Let A be a unital Banach algebra in which every nonzero element is invertible. Then A is isomorphic to $\mathbb{C}$.*

An *approximate identity* for a normed algebra A is a net $(e_\alpha)_\alpha$ in A such that $\|e_\alpha a - a\| \to 0$ and $\|a e_\alpha - a\| \to 0$ for every $a \in A$. The net $(e_\alpha)_\alpha$ is said to be *bounded by* $c > 0$ if $\|e_\alpha\| \leqslant c$ for every α. Left and right approximate identities are defined similarly.

Let A be a Banach algebra and X a left A-module. Then X is called a *left Banach A-module*, if X is a Banach space and the module operation $A \times X \to X, (a, x) \to a \cdot x$ is a bounded bilinear mapping. The following theorem, which is due to Cohen when $X = A$ and to Hewitt in the general case, is usually referred to as the Cohen-Hewitt factorization theorem.

THEOREM 1.1.4. *Let A be a Banach algebra having a left approximate identity bounded by $d > 0$, and let X be a left Banach A-module. Then, given an element x of the closed linear span of $A \cdot X$ and $\epsilon > 0$, there exist $a \in A$ and $y \in X$ with the following properties: $x = a \cdot y$, $\|a\| \leqslant d$, $y \in \overline{A \cdot x}$ and $\|x - y\| \leqslant \epsilon$.*

Let I be an ideal of a normed algebra A. Then the closure $\overline{I}$ is an ideal, and if I is closed then the algebra A/I, equipped with the quotient norm $\|a + I\| = \inf\{\|a + b\| : b \in I\}$ is a normed algebra. If A is complete, then so is A/I.

PROPOSITION 1.1.5. (See [**248**].) *Let I and J be closed ideals of a normed algebra A.*

(i) *If A has a bounded approximate identity, then the same is true for A/I.*
(ii) *If both I and A/I have bounded approximate identities, then A has a bounded approximate identity.*
(iii) *If I and J have bounded approximate identities, then the ideals $I \cap J$ and $\overline{I + J}$ have bounded approximate identities. Furthermore, if one of the two ideals is closed, then $I + J$ is closed.*

The following proposition is an application of Urysohn's lemma.

PROPOSITION 1.1.6. *Let X be a locally compact Hausdorff space and, for each subset E of X let*

$$I(E) = \{f \in C_0(X) : f(x) = 0 \text{ for all } x \in E\}.$$

Then the map $E \to I(E)$ is a bijection between the collections of all closed subsets E of X and all closed ideals of $C_0(X)$. Furthermore, $C_0(X)/I(E)$ is unital if and only if E is compact, and $I(E)$ is maximal if and only if E is a singleton.

Let A and B be Banach algebras, and let $A \otimes B$ denote their algebraic tensor product. Then the *projective tensor norm* γ on $A \otimes B$ is defined by

$$\gamma(x) = \inf \left\{ \sum_{j=1}^{n} \|a_j\| \|b_j\| : x = \sum_{j=1}^{n} a_j \otimes b_j \right\},$$

where the infimum is taken over all such representations of $x \in A \otimes B$. The completion of $A \otimes B$ with respect to γ is the projective tensor product of A and B, and is denoted $A \otimes_\gamma B$. Every $x \in A \otimes_\gamma B$ can be represented in the form $x = \sum_{j=1}^{\infty} a_j \otimes b_j$, where $\sum_{j=1}^{\infty} \|a_j\| \|b_j\| < \infty$, and the norm of x is given by

$$\gamma(x) = \inf \left\{ \sum_{j=1}^{\infty} \|a_j\| \|b_j\| : x = \sum_{j=1}^{\infty} a_j \otimes b_j \right\},$$

where again the infimum is taken over all such representations of x.

PROPOSITION 1.1.7. (See [**61**].) *Let A and B be Banach algebras. Then the multiplication on $A \otimes B$ extends uniquely to $A \otimes_\gamma B$, turning $A \otimes_\gamma B$ into a Banach algebra. If A and B are $*$-algebras, then $A \otimes_\gamma B$ is a $*$-algebra, and if A and B are commutative, then so is $A \otimes_\gamma B$. Moreover $A \otimes_\gamma B$ has a bounded approximate identity if and only if A and B both have bounded approximate identities.*

Let A be a commutative Banach algebra and let $\sigma(A)$ denote the set of all algebra homomorphisms from A onto $\mathbb{C}$. Every $\varphi \in \sigma(A)$ is bounded, actually $\|\varphi\| \leqslant 1$ and $\|\varphi\| = 1$ if A is unital. The space $\sigma(A)$ is equipped with the relative w^*-topology of A^* and is called the *spectrum* or *Gelfand space* of A. Thus this topology on $\sigma(A)$ is the weakest topology for which all the functions $\varphi \to \varphi(a)$, $a \in A$, are continuous. The Gelfand space of the projective tensor product of two commutative Banach algebras turns out to be canonically homeomorphic to the product of the two Gelfand spaces.

The algebra A is said to be *regular* if the algebra of Gelfand transforms is a regular algebra of functions on $\sigma(A)$ in the sense that, given a closed subset E of $\sigma(A)$ and some $\varphi \in \sigma(A) \setminus E$, there exists $a \in A$ such that $\widehat{a} = 0$ on E and $\widehat{a}(\varphi) \neq 0$. Moreover, A is called *Tauberian* if the set of all $a \in A$ such that $\widehat{a}$ has compact support, is dense in A.

PROPOSITION 1.1.8. *Let A and B be commutative Banach algebras.*

(i) *Given $\varphi \in \sigma(A)$ and $\psi \in \sigma(B)$, there exists a unique element of $\sigma(A \otimes_\gamma B)$, denoted $\varphi \otimes_\gamma \psi$, such that*

$$(\varphi \otimes_\gamma \psi)(a \otimes b) = \varphi(a)\psi(b)$$

for all $a \in A$ and $b \in B$.

(ii) *When $\sigma(A) \times \sigma(B)$ carries the product topology, then the mapping*

$$\sigma(A) \times \sigma(B) \to \sigma(A \otimes_\gamma B), \quad (\varphi, \psi) \to \varphi \otimes_\gamma \psi$$

is a surjective homeomorphism.

(iii) *$A \otimes_\gamma B$ is regular if and only if A and B are regular.*

For $a \in A$, the continuous function $\widehat{a} : \sigma(A) \to \mathbb{C}$, $\varphi \to \widehat{a}(\varphi) = \varphi(a)$, is called the *Gelfand transform* of a, and the map $\Gamma : A \to C(\sigma(A))$, sending a to $\widehat{a}$, is usually referred to as the *Gelfand homomorphism* or *Gelfand representation* of A. Let $\|\widehat{a}\|_\infty$ denote the sup norm of $\widehat{a}, a \in A$.

Let A be a normed algebra. For $a \in A$, the number

$$r(a) = \inf\{\|a^n\|^{1/n} : n \in \mathbb{N}\}$$

is the *spectral radius* of a. Then the spectral radius formula $r(a) = \lim_{n\to\infty} \|a^n\|^{1/n}$ holds. Moreover, if A is a commutative Banach algebra, then

$$r(a) = \sup\{|\gamma(a)| : \gamma \in \sigma(A)\}.$$

PROPOSITION 1.1.9. *Let A be a commutative Banach algebra.*

(i) *$\sigma(A)$ is a locally compact Hausdorff space.*
(ii) *If A is unital, then $\sigma(A)$ is compact.*
(iii) *If $\sigma(A)$ is noncompact, then the w^*-closure of $\sigma(A)$ in A^* is equal to $\sigma(A) \cup \{0\}$.*

PROPOSITION 1.1.10. *The Gelfand homomorphism has the following properties.*

(i) *For every $a \in A$, $\|\widehat{a}\|_\infty = \lim_{n\to\infty} \|a^n\|^{1/n} \leqslant \|a\|$.*
(ii) *Γ maps A into $C_0(\sigma(A))$.*
(iii) *$\sigma(A)$ strongly separates the points of $\Gamma(A)$ if A is semisimple.*
(iv) *Γ is isometric if and only if $\|a^2\| = \|a\|^2$ for all $a \in A$.*
(v) *If A is unital, then an element $a \in A$ is invertible if and only if $\widehat{a}$ never vanishes on $\sigma(A)$.*

LEMMA 1.1.11. *Let A be a commutative Banach algebra and I a proper ideal in A such that A/I has an identity.*

(i) *The closure $\overline{I}$ of I is a proper ideal.*
(ii) *I is contained in a maximal proper ideal, that is, maximal proper ideal I of A for which A/I is unital.*
(iii) *If I is a maximal proper ideal, then it is closed.*

PROPOSITION 1.1.12. *For a commutative Banach algebra A, the mapping*

$$\varphi \to \ker\varphi = \{a \in A : \varphi(a) = 0\}$$

is a bijection between $\sigma(A)$ and the set of all maximal modular ideals of A.

For a closed ideal I of A, the Gelfand spaces of I and of A/I embed naturally into $\sigma(A)$. More precisely, the following holds.

For any subset M of A, the *hull* $h(M)$ of M is defined by

$$h(M) = \{\varphi \in \sigma(A) : \varphi(M) = \{0\}\}.$$

Associated with a subset E of $\sigma(A)$ is the closed ideal

$$k(E) = \{a \in A : \widehat{a}(\varphi) = 0 \text{ for all } \varphi \in E\}$$

of A. The assignment $E \to \overline{E} = h(k(E))$ defines a closure operation on the collection of all subsets E of $\sigma(A)$. Therefore, there exists a unique topology on $\sigma(A)$ such that $\overline{E}$ is the closure of E. This topology is called the *hull-kernel topology* on $\sigma(A)$. In general, it is weaker than the Gelfand topology.

PROPOSITION 1.1.13. *Let I be a closed ideal of A, $q : A \to A/I$ the quotient homomorphism and $h(I) = \{\varphi \in \sigma(A) : \varphi(I) = \{0\}\}$.*

(i) *The map $\varphi \to \varphi \circ q$ is a homeomorphism from $\sigma(A/I)$ onto the closed subset $h(I)$ of $\sigma(A)$.*

(ii) *The map* $\varphi \to \varphi|_I$ *is a homeomorphism from the open subset* $\sigma(A)\backslash h(I)$ *of* $\sigma(A)$ *onto* $\sigma(I)$.

The commutative Banach algebra A is called *semisimple* if, for every $a \in A$, $\widehat{a} = 0$ implies $a = 0$.

PROPOSITION 1.1.14. *Let* ϕ *be a homomorphism from a commutative Banach algebra into a semisimple Banach algebra. Then* ϕ *is continuous.*

The preceding proposition implies that on a semisimple commutative Banach algebra any two Banach algebra norms are equivalent.

PROPOSITION 1.1.15. *Let* A *be a commutative Banach algebra. Then* A *is regular if and only if the hull-kernel topology and the Gelfand topology on* $\sigma(A)$ *coincide.*

PROPOSITION 1.1.16. *Let* A *be regular commutative Banach algebra and suppose that* I *is an ideal of* A *and* C *is a compact subset of* $\sigma(A)$ *such that* $C \cap h(I) = \varnothing$. *Then there exists an element* $a \in I$ *such that* $\widehat{a} = 1$ *on* C *and* $\widehat{a}$ *vanishes on some neighbourhood of* $h(I)$.

Let A be a commutative Banach $*$-algebra. Then A is called *symmetric* (or *self-adjoint*) if $\widehat{a^*} = \overline{\widehat{a}}$ for all $a \in A$. In other words, A is symmetric if the Gelfand homomorphism is a $*$-homomorphism, when $C_0(\sigma(A))$ is endowed with complex conjugation as involution.

PROPOSITION 1.1.17. *Suppose that* A *is a commutative Banach* $*$*-algebra.*

(i) *If* A *is symmetric, then* $\Gamma(A)$ *is dense in* $C_0(\sigma(A))$.
(ii) *If* A *is a commutative* C^**-algebra, then* Γ *is an isometric isomorphism onto* $C_0(\sigma(A))$.

LEMMA 1.1.18. *Let* A *and* B *be commutative Banach algebras and* $\phi : A \to B$ *an algebra homomorphism such that* $\gamma \circ \phi \in \sigma(A)$ *for every* $\gamma \in \sigma(B)$. *Then the adjoint map*

$$\phi^* : \sigma(B) \to \sigma(A), \quad \gamma \to \gamma \circ \phi$$

is continuous and a homeomorphism provided it is a bijection.

THEOREM 1.1.19. *Let* A *be a regular commutative Banach algebra,* C *a closed subset of* $\sigma(A)$ *and* $a \in A$ *such that* $|\widehat{a}(\gamma)| \geqslant \delta > 0$ *for all* $\gamma \in C$. *Then there exists* $b \in A$ *such that* $\widehat{b}(\gamma) = \frac{1}{\widehat{a}(\gamma)}$ *for all* $\gamma \in C$.

A linear operator $T : A \to A$ of a commutative Banach algebra A is called a *multiplier* if $T(ab) = aT(b)$ holds for all $a, b \in A$. When A is without order, that is, for any $a \in A$ the condition $aA = \{0\}$ implies $a = 0$, then every multiplier is bounded and the set $M(A)$ of all multipliers of A is a unital commutative subalgebra of the algebra $\mathcal{B}(A)$ of all bounded linear operators on A, the so-called *multiplier algebra* of A. For each $T \in M(A)$, there exists a uniquely determined continuous function $\widehat{T}$ on $\sigma(A)$ such that $\widehat{T(a)}(\gamma) = \widehat{T}(\gamma)\widehat{a}(\gamma)$ for all $\gamma \in \sigma(A)$. Readers are referred for the book [**41**, **151**, **223**] for more details.

1.2. Locally compact groups and examples

A *topological group* G is a set with the structure of both a group and a topological space such that the group product is a continuous map from $G \times G$ into G and the group inverse is continuous on G. The group product of x and y in G will be written multiplicatively as xy and the inverse of x is x^{-1} except in a few specific cases such as the group of integers or the real numbers. In general, the identity element of G is denoted by e. If $y \in G$ is fixed, then each of the maps $R_y : x \to xy, L_y : x \to y^{-1}x$ and $x \to x^{-1}$ is a homeomorphisms of G. When the topology on G is Hausdorff and locally compact, we call G a *locally compact group*.

In this section we introduce basic notation and collect a number of standard results on locally compact groups which will be used throughout the book without any further comment. Moreover, we present a number of examples which will be dealt with later.

Let G be any group. For subsets A and B of G, let

$$AB = \{xy : x \in A, y \in B\}$$

and $A^{-1} = \{x^{-1} : x \in A\}$. The set A is called *symmetric* if $A^{-1} = A$. Also, for $k \in \mathbb{N}$, let $A^k = \{x_1 \cdots x_k : x_j \in A, 1 \leqslant j \leqslant k\}$. Instead of $\{x\}A$ and $A\{y\}$, we write xA and Ay, respectively. We now list a number of elementary properties of topological groups.

PROPOSITION 1.2.1. *Let G be a topological group with identity e. Let A and B be subsets of G.*

(i) *If $A \subseteq G$ is closed (compact) and $B \subseteq G$ is compact, then AB is closed (compact).*
(ii) *Every neighbourhood U of the identity e of G contains an open symmetric neighbourhood V of e such that $V^2 \subseteq U$.*
(iii) *If $U \subseteq G$ is open, then AU and UA are open for every $A \subseteq G$.*
(iv) *Let $\mathcal{V}$ be a neighbourhood basis of e in G. Then the closure $\overline{A}$ of $A \subseteq G$ equals $\bigcap\{AV : V \in \mathcal{V}\}$.*
(v) *Let C be a compact subset of G and U an open set containing C. Then there exists a neighbourhood V of e such that $CV \cup VC \subseteq U$.*

PROPOSITION 1.2.2. *Let G be a topological group.*

(i) *If H is a subgroup of G, then so is the closure $\overline{H}$ of H.*
(ii) *Every open subgroup of G is closed.*
(iii) *Let U be a symmetric neighbourhood of e. Then $H = \bigcup_{n\in\mathbb{N}} U^n$ is an open and closed subgroup of G.*

A topological group G is said to be *compactly generated* if there exists a compact subset C of G such that

$$G = \{e\} \cup \textstyle\bigcup_{n\in\mathbb{N}}(C \cup C^{-1})^n.$$

COROLLARY 1.2.3. *Let G be a locally compact group and C a compact subset of G. Then there exists an open compactly generated subgroup of G containing C.*

For a subgroup H of a group G, G/H will always denote the set of all left cosets xH, $x \in G$, of H and $q : G \to G/H$ will denote the quotient map $x \to xH$. If G is a topological group, then G/H is endowed with the quotient topology. Let H be a closed subgroup of G and let $G/H = \{xH : x \in G\}$, the space of *left H-cosets*. If

G is locally compact, then G/H is a locally compact Hausdorff space and q is an open as well as a continuous mapping.

PROPOSITION 1.2.4. *Let H be a closed subgroup of the locally compact group G and C a compact subset of G/H. Then there exists a compact subset K of G such that $q(K) = C$.*

PROPOSITION 1.2.5. *Let H be a subgroup of the topological group G.*

(i) *The quotient map is continuous and open.*
(ii) *If H is normal, then G/H is a topological group.*
(iii) *If H is closed and normal and G is locally compact, then G/H is a locally compact group.*

We now formulate the topological group versions of the two isomorphism theorems from algebraic group theory. Let G be a locally compact group, N a closed normal subgroup and H a closed subgroup of G.

THEOREM 1.2.6. *Suppose that H is also normal in G and $N \subseteq H$. Then the algebraic isomorphism between G/H and $(G/N)/(H/N)$ is a homeomorphism.*

If H is some topological group with identity element e and $\phi : G \to H$ is a continuous homomorphism, then $N = \{x \in G : \psi(x) = e\}$ is a closed normal subgroup of G. There is a unique injective homomorphism $\widetilde{\phi} : G/N \to H$ such that $\widetilde{\phi} \circ q = \phi$, and $\widetilde{\phi}$ is continuous since q is open. Note that $\widetilde{\phi}$ need not be open when G is not σ-compact.

LEMMA 1.2.7. *Suppose that G and H are locally compact groups and $\phi : G \to H$ is a continuous surjective homomorphism. If G is σ-compact, then $\widetilde{\phi}$ is a topological isomorphism.*

The proof of the preceding lemma is a straightforward application of Baire's category theorem and it implies the following isomorphism theorem.

THEOREM 1.2.8. *Suppose that H is σ-compact and HN is closed in G. Then the map $HN/N \to H/(H \cap N)$, $hN \to h(H \cap N)$ is a topological isomorphism.*

For any topological group G, let G_0 denote the connected component of the identity of G. Then G_0 is a closed normal subgroup of G and the quotient group G/G_0 is totally disconnected. G is said to be *almost connected* if G/G_0 is compact. Moreover, if G is totally disconnected, then every neighbourhood of the identity contains an open compact subgroup. A locally compact group G is called a *Lie group* if G_0 is open and G_0 is an analytic group in the usual sense (see [**26**]). In every locally compact group G, there exists a (unique) largest solvable connected normal subgroup, the so-called *radical* $\mathrm{rad}(G)$ of G.

Let I be an index set and for every $\iota \in I$, let G_ι be a topological group. Then the Cartesian product $G = \prod_{\iota\in I} G_\iota$ become a topological group under the product topology. Note that G is locally compact if and only if all but finitely many of the G_ι are compact and the remaining ones are locally compact.

PROPOSITION 1.2.9. *An almost connected locally compact group G is a projective limit of Lie groups. More precisely, there exists a directed system of compact normal subgroups K_α of G such that $\bigcap_\alpha K_\alpha = \{e\}$ and the map*

$$G \to \prod_\alpha G/K_\alpha, \quad x \to (xK_\alpha)_\alpha$$

is a topological embedding of G into the direct product group $\prod_\alpha G/K_\alpha$.

We continue by recalling some further notions from group theory. Let G be a locally compact group. For subsets A and B of G, $[A, B]$ will always denote the closed subgroup of G generated by all commutators $[a, b] = aba^{-1}b^{-1}$, $a \in A$, $b \in B$. Then $[G, G]$ is the closed commutator subgroup of G, and the *commutator series* or derived series of G is defined by

$$G = G_0 \supseteq G_1 = [G, G] \supseteq \ldots \supseteq G_n = [G_{n-1}, G_{n-1}] \supseteq \ldots.$$

G is called *solvable* if $G_n = \{e\}$ for some $n \in \mathbb{N}$ and *n-step solvable* if $G_n = \{e\}$, but $G_{n-1} \neq \{e\}$. In particular, G is 2-step solvable or metabelian if $[G, G]$ is abelian and $\neq \{e\}$. Of course, solvability of G is equivalent to the existence of a sequence $N_0 = \{e\} \subseteq N_1 \subseteq \ldots \subseteq N_r = G$ of normal subgroups of G such that N_j/N_{j-1} is abelian for each $1 \leqslant j \leqslant r$. By

$$\{e\} = Z_0(G) \subseteq Z_1(G) \subseteq \ldots \subseteq Z_m(G) \subseteq \ldots$$

we denote the *ascending central series* of G. That is, $Z_1(G)$ (also denoted $Z(G)$) is the centre of G and, for $j \geqslant 1$, $Z_{j+1}(G)$ is specified by $Z_{j+1}(G)/Z_j(G) = Z(G/Z_j(G))$. Note that every $Z_m(G)$ is a closed normal subgroup of G. The group G is said to be *n-step nilpotent* if $Z_n(G) = G$, but $Z_{n-1}(G) \neq G$ and G is called *nilpotent* if it is n-step nilpotent for some $n \in \mathbb{N}$. Nilpotent groups can equally well be defined in terms of the *descending central series* $G = C_0 \supseteq C_1 \supseteq \ldots \supseteq C_m \supseteq \ldots$, which is inductively defined by $C_1 = [G, G]$ and $C_{m+1} = [G, C_m]$ for $m \geqslant 1$. Then G is m-step nilpotent if and only if $C_m = \{e\}$, but $C_{m-1} \neq \{e\}$.

We now present a number of examples, which serve the dual purpose of not only providing an idea of the nature of locally compact groups and their variety but also introducing some of the particular groups or classes of groups that will be used later in the text.

EXAMPLE 1.2.10. The set $\mathbb{R}$ of real numbers with addition and the usual topology has already been mentioned as a locally compact group. The integers $\mathbb{Z}$ form a closed subgroup of $\mathbb{R}$. Indeed, for every $a \in \mathbb{R}$, $\mathbb{Z}a = \{ka : k \in \mathbb{Z}\}$ is a closed subgroup of $\mathbb{R}$, and any proper closed subgroup of $\mathbb{R}$ is of this form.

Let $\mathbb{R}^* = \{a \in \mathbb{R} : a \neq 0\}$ and $\mathbb{R}^+ = \{a \in \mathbb{R} : a > 0\}$. Equipped with multiplication of real numbers, both $\mathbb{R}^*$ and $\mathbb{R}^+$ are locally compact groups with 1 as identity and $a^{-1} = 1/a$ as the inverse of a generic element a. Note that $\mathbb{R}^+$ is an open subgroup of $\mathbb{R}^*$.

If $\mathbb{C}$ is the field of complex numbers, then the *circle group*,

$$\mathbb{T} = \{z \in \mathbb{C} : |z| = 1\},$$

is a compact group under multiplication. If $a > 0$, then $\psi_a(t) = \exp(2\pi it/a)$, for $t \in \mathbb{R}$, defines a continuous homomorphism of $\mathbb{R}$ onto $\mathbb{T}$. Since $\mathbb{Z}a = \ker \psi_a$, $\widetilde{\psi}_a : t + \mathbb{Z}a \to \exp(2\pi it/a)$ identifies $\mathbb{R}/\mathbb{Z}a$ with $\mathbb{T}$.

EXAMPLE 1.2.11. Let $GL(n, \mathbb{R})$, $n \geqslant 2$, denote the multiplicative group of invertible real $n \times n$-matrices. Equipped with the topology inherited from $\mathbb{R}^{n^2}$, $GL(n, \mathbb{R})$ becomes a Lie group. The closed subgroup H consisting of all upper triangular matrices is n-step solvable.

EXAMPLE 1.2.12. We can equip $\mathbb{R} \times \mathbb{R}^+$ with a different multiplication. For $(b_1, a_1), (b_2, a_2) \in \mathbb{R} \times \mathbb{R}^+$ let $(b_1, a_1)(b_2, a_2) = (b_1 + a_1 b_2, a_1 a_2)$. Notice that

$(0, 1)(b, a) = (b, a)(0, 1) = (b, a)$ and

$$(b, a)(-a^{-1}b, a^{-1}) = (-a^{-1}b, a^{-1})(b, a) = (0, 1).$$

Thus, this multiplication endows $\mathbb{R} \times \mathbb{R}^+$ with a group structure, and the resulting group will be denoted G. The operations of multiplication and inversion are clearly continuous for the product topology. Thus G is a locally compact group.

For $(b, a) \in G$, define a transformation of $\mathbb{R}$ by $(b, a) \cdot x = ax + b$, for all $x \in \mathbb{R}$. This is an affine transformation of $\mathbb{R}$, and every orientation-preserving affine transformation of $\mathbb{R}$ arises this way. Moreover, this action is consistent with the group product in G. The group G is called the *affine group* or, often, the *$ax + b$ group*.

The previous example is a special case of an important technique for constructing new locally compact groups from given ones. This is the semidirect product construction, which we present in detail because many of our examples dealt with later in the book arise in this manner. Let N and H be locally compact groups. Let $\mathrm{Aut}(N)$ denote the group of automorphisms of N. An automorphism of N is a topological group isomorphism of N with itself. Suppose that there is a homomorphism $\alpha : h \to \alpha_h$ of H into $\mathrm{Aut}(N)$ such that $(n, h) \to \alpha_h(n)$ is continuous from $N \times H$ to N. We use these data to form a locally compact group denoted $N \rtimes_\alpha H$, or simply $N \rtimes H$ if the homomorphism α is understood. As a set and topological space, $N \rtimes H = N \times H$. The group product of (n_1, h_1) with (n_2, h_2) in $N \rtimes H$ is given by

$$(n_1, h_1)(n_2, h_2) = (n_1\alpha_{h_1}(n_2), h_1h_2).$$

One checks easily that this product is associative, that (e_N, e_H) serves as the identity, where e_N and e_H are the identities of N and H, respectively, that $(n, h)^{-1} = (\alpha_{h^{-1}}(n^{-1}), h^{-1})$, and that the group operations are continuous. In short, $N \rtimes H$ is a locally compact group, called the *semidirect product* of N and H.

If we define $\widetilde{N} = \{(n, e_H) : n \in N\}$ and $\widetilde{H} = \{(e_N, h) : h \in H\}$, then $\widetilde{N}$ and $\widetilde{H}$ are closed subgroups of $N \rtimes H$ that satisfy $\widetilde{N} \cap \widetilde{H} = \{e\}$ and $\widetilde{N}\widetilde{H} = N \rtimes H$, and $\widetilde{N}$ is normal in G. Moreover, for $h \in H$ and $n \in N$,

$$(e_H, h)(n, e_H)(e_H, h)^{-1} = (\alpha_h(n), e_H).$$

In general, if G is a locally compact group, we are sometimes able to find two closed subgroups N and H such that N is normal in G, $N \cap H = \{e\}$ and $NH = G$. In such case we can view G as the semidirect product of N and H as follows. For $h \in H$, define α_h on N by $\alpha_h(n) = hnh^{-1}$, for all $n \in N$. Since N is normal, α_h is an automorphism of N and $h \to \alpha_h$ is a homomorphism of H into $\mathrm{Aut}(N)$. Also $(n, h) \to \alpha_h(n)$ is continuous from $N \times H$ to N. Thus, we can form the semidirect product $N \rtimes_\alpha H$, and the map $(n, h) \to nh$ is a topological isomorphism between $N \rtimes_\alpha H$ and G.

We next present two important examples of semidirect product groups.

EXAMPLE 1.2.13. Let $n \in \mathbb{N}$ and let $SO(n)$ denote the group of orthogonal $n \times n$-matrices of determinant one. Equipped with the topology inherited from embedding $SO(n)$ into $\mathbb{R}^{n^2}$, $SO(n)$ is a compact group. Form the semidirect product $G_n = \mathbb{R}^n \rtimes SO(n)$, where $SO(n)$ acts on $\mathbb{R}^n$ by rotation. G_n is called the *Euclidian motion group* of $\mathbb{R}^n$.

EXAMPLE 1.2.14. This example is known as the *Heisenberg group* of dimension $2n+1$ ($n \in \mathbb{N}$).

For $x, y \in \mathbb{R}^n$, $x = (x_1, \dots, x_n)$ and $y = (y_1, \dots, y_n)$, and $z \in \mathbb{R}$, let $[x, y, z]$ denote the matrix

$$\begin{pmatrix} 1 & x_1 & \cdots & x_n & z \\ 0 & 1 & \dots & 0 & y_1 \\ \vdots & \vdots & & \vdots & \vdots \\ 0 & 0 & \dots & 1 & y_n \\ 0 & 0 & \dots & 0 & 1 \end{pmatrix}.$$

Let $\mathbb{H}_n = \{[x, y, z] : x, y \in \mathbb{R}^n, z \in \mathbb{R}\}$. If $x \cdot y = \sum_{i=1}^n x_i y_i$, the matrix product of $[x, y, z]$ and $[x', y', z']$ is given by

$$[x, y, z][x', y', z'] = [x + x', y + y', z + z' + x \cdot y'].$$

Clearly, $\mathbb{H}_n$ is a closed subgroup of $GL(n+2, \mathbb{R})$.

It is easy to check that $N = \{[0, y, z] : y \in \mathbb{R}^n, z \in \mathbb{R}\}$ is a closed normal subgroup of $\mathbb{H}_n$, isomorphic to $\mathbb{R}^{n+1}$, and that $A = \{[x, 0, 0] : x \in \mathbb{R}^n\}$ is a closed subgroup of $\mathbb{H}_n$, isomorphic to $\mathbb{R}^n$. Since $\mathbb{H}_n = NA$ and $N \cap A = \{[0, 0, 0]\}$, $\mathbb{H}_n$ is isomorphic to the semidirect product $\mathbb{R}^{n+1} \rtimes_\alpha \mathbb{R}^n$, where α is defined by $\alpha_x(y, z) = (y, z + x \cdot y)$, for $(y, z) \in \mathbb{R}^n \times \mathbb{R}$ and $x \in \mathbb{R}^n$.

Let $Z = \{[0, 0, z] : z \in \mathbb{R}\}$, a closed subgroup of $\mathbb{H}_n$ which is isomorphic to $\mathbb{R}$. It is obvious that $[0, 0, z]$ commutes with every element of $\mathbb{H}_n$. Conversely, if $[x, y, z]$ commutes with every element of $\mathbb{H}_n$, then a quick look at the group product shows that $x = y = 0$. Thus Z is exactly the centre of $\mathbb{H}_n$. The map $\psi : \mathbb{H}_n \to \mathbb{R}^{2n}$ defined by $\psi([x, y, z]) = (x, y)$, for $[x, y, z] \in \mathbb{H}_n$, is a continuous homomorphism onto $\mathbb{R}^{2n}$ with $\ker \psi = Z$. So the quotient group $\mathbb{H}_n / Z$ is isomorphic to $\mathbb{R}^{2n}$.

There is an important discrete subgroup of $\mathbb{H}_1$. Let

$$D = \{[k, l, m] : k, l, m \in \mathbb{Z}\}.$$

Clearly, D is discrete in the relative topology. It is called the *discrete or integer Heisenberg group*.

Within the class of totally disconnected locally compact groups the most prominent example is the additive group of the p-adic number field which we now briefly introduce. Let p be a fixed prime number. Then every nonzero rational number x has a unique decomposition $x = p^m y$, where $m \in \mathbb{Z}$ and y is a rational number the numerator and denominator of which are both not divisible by p. The p-adic valuation of x is defined to be $|x|_p = p^{-m}$. Setting $|0|_p = 0$, we then have

$$|x_1 x_2|_p = |x_1|_p |x_2|_p \quad \text{and} \quad |x_1 + x_2|_p \leqslant \max\{|x_1|_p, |x_2|_p\} \tag{1.1}$$

for $x_1, x_2 \in \mathbb{Q}$. It follows that $d(x_1, x_2) = |x_1 - x_2|_p$ defines a metric on $\mathbb{Q}$ with respect to which the algebraic operations in $\mathbb{Q}$ are continuous. Consequently, these operations extend to the completion Ω_p of $\mathbb{Q}$, turning Ω_p into a field, the *p-adic number field*. The elements of Ω_p can be described explicitly as follows (see [**78**, Proposition 2.8]).

PROPOSITION 1.2.15. *Let $n \in \mathbb{Z}$ and $c_j \in \{0, 1, \dots, p-1\}$ for $j \in \mathbb{Z}$ with $j \geqslant n$. Then the series $\sum_{j=n}^\infty c_j p^j$ converges in Ω_p. Conversely, every p-adic number is the sum of a unique such series.*

For $x \in \Omega_p$ and $r > 0$, let $B(x,r)$ denote the closed ball of radius r around x, i.e.

$$B(x,r) = \{y \in \Omega_p : |y-x|_p \leqslant r\}.$$

Then $B(x,r)$ is also open. In fact, since $|\cdot|_p$ only attains the values p^n, $n \in \mathbb{Z}$, and 0, for any $r > 0$ and $\delta > 0$ sufficiently small, $|y-x|_p \leqslant r$ is equivalent to $|y-x|_p < r+\delta$. This shows that Ω_p is a totally disconnected field. Moreover, it follows from (1.1) that $B(0,r)$ is an additive subgroup of Ω_p for any $r > 0$ and it is a subring for $r \leqslant 1$. $B(0,1)$ is called the *ring of p-adic integers* and denoted by Δ_p. Let $K = \{x \in \Omega_p : |x|_p = 1\}$. Then K is a compact multiplicative group. Let $n \in \mathbb{Z}$, $x \in \Omega_p$ and $\epsilon > 0$ be given and choose $k \in \mathbb{Z}$ such that $k < n$ and $p^k \leqslant \epsilon$. It is then easy to verify that the ball $B(x,p^n)$ is the union of p^{n-k} balls of radius p^k. This shows that the balls $B(x,p^n)$ are totally bounded and hence, being closed in the complete metric space Ω_p, they are compact. Thus Ω_p is a locally compact field and Δ_p is a compact open subring.

We close this section by introducing certain function spaces. For a complex-valued function f on G and $x \in G$, the left and right translates through x are defined by

$$L_x f(y) = f(x^{-1}y) \quad \text{and} \quad R_x f(y) = f(yx), \quad y \in G.$$

Then the maps $x \to L_x$ and $x \to R_x$ are group homomorphisms. A space E of functions on G is said to be left (right, two-sided) translation invariant if $L_x f \in E$ ($R_x f \in E$, $L_x f \in E$ and $R_x f \in E$) for all $f \in E$ and $x \in G$. Moreover, we set $\widetilde{f}(x) = \overline{f(x^{-1})}$ and $\check{f}(x) = f(x^{-1})$.

A function $f \in C^b(G)$ is called left (right) uniformly continuous, if $\|L_x f - f\|_\infty \to 0$ ($\|R_x f - f\|_\infty \to 0$) as $x \to e$, and f is said to be uniformly continuous if it is both left and right uniformly continuous. The sets

$LUC(G) = \{f \in C^b(G) : f \text{ is left uniformly continuous}\}$,
$RUC(G) = \{f \in C^b(G) : f \text{ is right uniformly continuous}\}$,
$UC(G) = \{f \in C^b(G) : f \text{ is uniformly continuous}\}$

are closed translation invariant subalgebras of $C^b(G)$. Moreover, $C_0(G) \subseteq UC(G)$. These subspaces of $C^b(G)$ are of relevance in the study of amenability of G (Section 1.8).

Note that functions in $LUC(G)$ (respectively $RUC(G)$) are called *right* (respectively *left*) *uniformly continuous functions* in [**125**].

THEOREM 1.2.16. [Kakutani-Kodaira] *Let G be a σ-compact locally compact group, $(U_n)_n$ a sequence of neighbourhoods of the identity and f a uniformly continuous function on G. Then there exists a compact normal subgroup N of G such that $N \subseteq \bigcap_{n=1}^\infty U_n$, G/N is second countable and f is constant on cosets of N.*

Let G be a locally compact group. A bounded continuous function f on G is called *almost periodic* if the set $\{L_x R_y f : x, y \in G\}$ has compact closure in $C^b(G)$. A function $f \in C^b(G)$ is already almost periodic if either of the sets $\{L_x f : x \in G\}$ or $\{R_x f : x \in G\}$ has compact closure. The set $AP(G)$ of almost periodic functions on G is a self-adjoint closed subalgebra of $C^b(G)$. The group G is called *maximally almost periodic* if the functions in $AP(G)$ separate the points of G. The algebra $AP(G)$ leads to a compactification of G, the Bohr compactification $b(G)$, as follows. Let $b(G)$ denote the spectrum of the commutative C^*-algebra $AP(G)$, and let

$\phi : G \to b(G)$ be the mapping defined by $\phi(x)(f) = f(x)$ for $f \in AP(G)$. Then ϕ is continuous and $\phi(G)$ is dense in $b(G)$ because $AP(G) \subseteq C(b(G))$. The kernel

$$N = \{x \in G : f(x) = f(e) \text{ for all } f \in AP(G)\}$$

of ϕ is a closed normal subgroup of G, and $\phi^{-1}(\phi(x)) = xN$ for every $x \in G$.

PROPOSITION 1.2.17. *The group operations on $\phi(G) \subseteq b(G)$, obtained by transferring the ones from G/N, can be extended to all of $b(G)$, turning $b(G)$ into a compact topological group. Note that when G is abelian, $b(G)$ is topologically isomorphic to the dual group of G_d.*

Reader are referred to [**78, 125**] for more details.

1.3. Haar measure and group algebra

If X is a locally compact Hausdorff space, a positive Borel measure μ on X is called *regular* if, for any Borel subset E of X,

$$\begin{aligned}\mu(E) &= \inf\{\mu(U) : E \subseteq U, U \text{ open}\} \\ &= \sup\{\mu(K) : K \subseteq E, K \text{ compact}\}.\end{aligned}$$

A complex Borel measure ν on X is regular if its total variation $|\nu|$ is regular.

A *Radon measure* on X is a positive Borel measure on X such that $\mu(K) < \infty$ for any compact set $K \subseteq X$,

$$\mu(E) = \inf\{\mu(U) : E \subseteq U, U \text{ open}\},$$

for any Borel subset E of X, and, for every open subset U of X,

$$\mu(U) = \sup\{\mu(K) : K \subseteq U, K\text{compact}\}.$$

If μ is a σ-finite Radon measure, then μ is regular.

Now let X be a locally compact group G. The existence of a translation invariant Radon measure on G is of fundamental importance. A Borel measure μ on G is called *left invariant* (respectively, *right invariant*) if $\mu(xE) = \mu(E)$ (respectively, $\mu(Ex) = \mu(E)$), for any $x \in G$ and Borel subset E of G. The following theorem states the existence and essential uniqueness of a nonzero left invariant Radon measure on G.

THEOREM 1.3.1. *Let G be a locally compact group. Then there exists a nonzero left invariant Radon measure μ_G on G. It satisfies $\mu_G(U) > 0$ for any nonempty open subset U of G. If ν is another nonzero left invariant Radon measure on G, then there is a constant $c > 0$ such that $\nu = c\,\mu_G$.*

Such a measure μ_G is called *left Haar measure* on G. It is understood that a choice has been made out of the family $\{c\mu_G : c > 0\}$. Usually this choice is not made explicit, but if there is a distinguished compact neighbourhood V of e, we may assume that $\mu(V) = 1$. For example, if G is compact itself, we may assume $\mu_G(G) = 1$. If G is an infinite group equipped with the discrete topology, we may assume that $\mu_G(\{e\}) = 1$. Then the left invariance of μ_G implies that μ_G is simply the counting measure (use the left invariance).

Of course there also exists a right invariant Radon measure on G with the same kind of uniqueness, right Haar measure. In fact, the homeomorphism $x \to x^{-1}$ interchanges left and right Haar measures $\nu(E) = \mu_G(E^{-1})$, for all Borel subsets E of G, defines a right Haar measure on G). We shall consistently use left Haar measures and always assume without mentioning it that a left Haar measure is chosen. Actually, we will rarely use the notation μ_G. If A is a measurable subset of G, then $|A|$ will denote $\mu_G(A)$.

If $f \in C_c(G)$, then f is integrable with respect to μ_G and we usually write $\int_G f(x)d\mu_G(x)$ simply as $\int_G f(x)dx$. Indeed, we use the same simplification for any kind of function f on G (non-negative measurable, Haar integrable, or even vector-valued versions of integrability) for which $\int_G f(x)dx$ makes sense. The left invariance of μ_G implies that $\int_G f(yx)dx = \int_G f(x)dx$, for any $y \in G$. Note that if $f \in C_c^+(G) = \{g \in C_c(G) : g \geqslant 0\}$ and $f \neq 0$, then $\int_G f(x)dx > 0$.

On the other hand, let us fix a $y \in G$ and define a new measure ν by $\nu(E) = \mu_G(Ey)$, for all Borel subsets E of G. Then ν is a left invariant Radon measure on G that is positive on nonempty open sets. Thus, there is a positive constant $\Delta_G(y)$ so that $\mu_G(Ey) = \nu(E) = \Delta_G(y)\mu_G(E)$, for every Borel subset E of G. This gives a change of variables formula to use in integrals:

$$\int_G f(x)dx = \Delta_G(y)\int_G f(xy)dx,$$

for any function f where the integral makes sense and for any $y \in G$.

Letting y vary, $y \to \Delta_G(y)$ is a continuous homomorphism of G into $\mathbb{R}^+$, the multiplicative group of positive real numbers. It is called the *modular function* of G. The modular function enables a change of variables by inversion:

$$\int_G f(x^{-1})dx = \int_G f(x)\Delta_G(x^{-1})dx.$$

If $\Delta_G \equiv 1$ on G, that is if every left Haar measure is also right invariant, then G is called *unimodular*. Of course, if G is abelian, then right and left translations are the same, and if G has the discrete topology, then counting measure is both left and right invariant. Also, if G is compact, then $\Delta_G(G)$ is a compact subgroup of $\mathbb{R}^+$, so it must be trivial. Thus, each of these classes of groups, abelian, discrete or compact, is contained in the class of all unimodular groups. Nevertheless, one frequently encounters nonunimodular groups, and the modular function and functions related to it play an important role. For unimodular groups, left Haar measure is also right invariant and we usually just refer to *Haar measure* rather than left Haar measure.

Recall that for $1 \leqslant p < \infty$, $L^p(G, \mu_G)$ is the Banach space of all (equivalence classes of) Borel measurable functions f on G for which the norm $\|f\|_p = \left(\int_G |f(x)|d\mu_G(x)\right)^{1/p}$ is finite, and that $L^\infty(G, \mu_G)$ is the space of all μ_G-essentially bounded functions, endowed with the corresponding supremum norm. The Lebesgue space $L^p(G, \mu_G)$ will simply be denoted $L^p(G)$ for $1 \leqslant p \leqslant \infty$. Note that, if $1 \leqslant p < \infty$, then $C_c(G)$ is dense in $L^p(G)$ and that $L^2(G)$ is a Hilbert space, the inner product being given by

$$\langle f, g\rangle = \int_G f(x)\overline{g(x)}dx.$$

For $1 \leqslant p < \infty$, left and right translations of L^p-functions are continuous: Given $f \in L^p(G)$ and $\epsilon > 0$, there exists a neighbourhood U of the identity e in G such that $\|L_y f - f\|_p < \epsilon$ and $\|R_y f - f\|_p < \epsilon$ for all $y \in U$.

PROPOSITION 1.3.2. *Let* $1 \leqslant p \leqslant \infty$ *and suppose that* $f \in L^1(G)$ *and* $g \in L^p(G)$. *Then the integral*

$$f * g(x) = \int_G f(y)g(y^{-1}x)dy = \int_G f(xy)g(y^{-1})dy$$

is absolutely convergent for almost all $x \in G$ *and we have* $f * g \in L^p(G)$ *and* $\|f * g\|_p \leqslant \|f\|_1 \|g\|_p$. *Furthermore, suppose that* $p = \infty$. *Then* $f * g$ *is continuous.*

The resulting measurable function, $f * g$, is called the *convolution* of f and g. When $p = 1$, this implies that $L^1(G)$, equipped with convolution as multiplication, is a Banach algebra. For $f \in L^1(G)$ and $x \in G$, let

$$f^*(x) = \Delta_G(x^{-1})\overline{f(x^{-1})}.$$

Then $||f^*||_1 = ||f||_1$ and $f \to f^*$ is an involution on $L^1(G)$.

REMARK 1.3.3. Suppose that G possesses a compact neighbourhood V of the identity such that $x^{-1}Vx = V$ for all $x \in G$. Then G is unimodular. Indeed, $|V| > 0$ and

$$|V| = \int_G 1_{x^{-1}Vx}(y)dy = \int_G L_x(R_x 1_V)(y)dy = \int_G R_x 1_V(y)dy = \Delta(x^{-1})|V|$$

for every $x \in G$.

PROPOSITION 1.3.4. *Let* G *be a locally compact group. Then* $L^1(G)$ *is a Banach* $*$*-algebra when equipped with the convolution and involution defined above and the norm:* $||f^*||_1 = ||f||_1$, *for each* $f \in L^1(G)$.

If G is nondiscrete, $L^1(G)$ does not have an identity. However, it always possesses a two-sided (bounded) approximate identity in $C_c(G)$, which can be constructed as follows. Let $\mathcal{U}$ be a neighbourhood basis at e in G, and for each $U \in \mathcal{U}$ choose $f_U \in C_c^+(G)$ such that $f(x^{-1}) = f(x)$ for all $x \in G$, $\int_G f(x)dx = 1$ and with compact support contained in U. Then, for any $g \in L^p(G)$, $1 \leqslant p < \infty$, $\|f_U * g - g\|_p \to 0$ and $\|g * f_U - g\|_p \to 0$ as $U \to \{e\}$.

Using the existence of an approximate identity, it is straightforward to show the following

PROPOSITION 1.3.5. *Let* I *be a closed linear subspace of* $L^1(G)$. *Then* I *is an ideal in* $L^1(G)$ *if and only if* I *is closed under left and right translations.*

If X is a locally compact Hausdorff space, then $M(X)$ denotes the space of regular complex Borel measures on X equipped with the total variation norm and the pairing $(g, \mu) = \int_X g(t)\, d\mu(t)$, $g \in C_0(X), \mu \in M(X)$, identifies $M(X)$ with $C_0(X)^*$, the Banach space dual of $C_0(X)$. If $X = G$, a locally compact group, then $M(G)$ can be equipped with a convolution product. For $\mu, \nu \in M(G)$, there is a unique $\mu * \nu \in M(G)$ such that

$$\int_G \varphi(x)\, d(\mu * \nu)(x) = \int_G \int_G \varphi(xy)\, d\mu(x)\, d\nu(y),$$

for all $\varphi \in C_0(G)$, and we have $\|\mu * \nu\| \leqslant \|\mu\| \cdot \|\nu\|$. If δ_x denotes the point mass at $x \in G$, then

$$\int_G \varphi(y) d(\mu * \delta_x)(y) = \int_G R_x\varphi(y) d\mu(y)$$

and, similarly,

$$\int_G \varphi(y)d(\delta_x * \mu)(y) = \int_G L_{x^{-1}}\varphi(y)d\mu(y).$$

Moreover, for $\mu \in M(G)$, define $\mu^* \in M(G)$ such that $\mu^*(E) = \overline{\mu(E^{-1})}$, for any Borel $E \subseteq G$. Then $\mu \to \mu^*$ is an involution on $M(G)$ and $M(G)$ is a Banach $*$-algebra with identity δ_e, called the *measure algebra* of G. For each $f \in L^1(G)$, there is a measure $\mu_f \in M(G)$ such that $d\mu_f(x) = f(x)\,dx$. This embeds $L^1(G)$ as a closed two-sided ideal in $M(G)$. Indeed, if $\nu \in M(G)$ and $f \in L^p(G)$, then the function

$$\nu * f : y \to \int_G f(x^{-1}y)d\nu(x)$$

belongs to $L^p(G)$ and $\|\nu * f\|_p \leqslant \|\nu\| \cdot \|f\|_p$. When $p = 1$, we have

$$\int_G \varphi(y)d(\nu * \mu_f)(y) = \int_G \varphi(y)(\nu * f)(y)dy,$$

for all $\varphi \in C_c(G)$. Similarly, $f * \nu$ is defined and $\mu_f * \nu = f * \nu$, for $f \in L^1(G)$ and $\nu \in M(G)$.

Let G be a locally compact group and H a closed subgroup of G. The group G acts continuously on G/H by $G \times G/H \to G/H$, $(x, yH) \to xyH$. A measure μ on G/H is called *G-invariant* if $\mu(xE) = \mu(E)$ for every Borel subset E of G/H and $x \in G$. The existence of a G-invariant measure on G/H is clarified by the following

PROPOSITION 1.3.6. *There exists a nonzero positive $G-$ invariant regular Borel measure on G/H if and only if $\Delta_G(h) = \Delta_H(h)$ for all $h \in H$. When this is the case, the $G-$invariant measure is unique up to multiplication by a positive constant.*

PROPOSITION 1.3.7. *Let H be a closed subgroup of G. For $f \in C_c(G)$, define $T_H f$ on G/H by*

$$T_H f(\dot{x}) = \int_H f(xh)\,dh, \;\; \dot{x} = xH.$$

Then $T_H f \in C_c(G/H)$ and the map $T_H : C_c(G) \to C_c(G/H)$ is surjective. Moreover, given $\varphi \in C_c(G/H)$ (resp., $\varphi \in C_c^+(G/H)$), there exists $f \in C_c(G)$ (resp., $f \in C_c^+(G)$) such that $T_H f = \varphi$ and $q(\operatorname{supp} f) = \operatorname{supp}\varphi$.

REMARK 1.3.8. Let μ be a (left) Haar measure on G and H an open subgroup of G. Then the restriction of μ to Borel subsets of H is a (left) Haar measure on H. If E is a Borel subset of G and there are countably many distinct left cosets x_jH of H, $j \in \mathbb{N}$, such that $E \subseteq \bigcup_{j=1}^{\infty} x_j E$, then $\mu(E) = \sum_{j=1}^{\infty} \mu(E \cap x_j H)$.

Let μ be such an invariant measure on G/H. Then Haar measures on G and on H can be normalized in such a way that *Weil's formula*

$$\int_G f(x)dx = \int_{G/H} \left(\int_H f(xh)dh \right) d\mu(\dot{x})$$

holds for all $f \in C_c(G)$. Actually, the map T_H extends to all of $L^1(G)$. More precisely, given $f \in L^1(G)$, the integral $\int_H f(xh)dh$ exists for μ-almost all $\dot{x} \in G/H$, the function $\dot{x} \to \int_H f(xh)dh$ is μ-integrable on G/H and the above Weil's formula holds for all $f \in L^1(G)$. Finally, $\|T_H f\|_1 \leqslant \|f\|_1$, and T_H maps $L^1(G)$ onto $L^1(G/H, \mu)$. Furthermore, T_H is a $*$-homomorphism.

EXAMPLE 1.3.9. (1) The Haar measure $\mu_{\mathbb{R}}$ of $\mathbb{R}$ with $\mu_{\mathbb{R}}([0,1[) = 1$ is Lebesgue measure and Haar measure on $\mathbb{T}$ is given by $\mu_{\mathbb{T}}(E) = \mu_{\mathbb{R}}(\psi_1^{-1}(E) \cap [0,1))$, for any Borel subset E of $\mathbb{T}$. That is, for any nonnegative measurable function f on $\mathbb{T}$,

$$\int_{\mathbb{T}} f(z)dz = \int_{[0,1)} f(\exp(2\pi it))dt,$$

where the integral on the right is Lebesgue integral.

(2) It is easiest to describe Haar measure on the multiplicative group $\mathbb{R}^* = \mathbb{R}\backslash\{0\}$ by showing the formula for invariant integration. If f is a nonnegative measurable function on $\mathbb{R}^*$, then $\int_{\mathbb{R}^*} f(a)\,\frac{da}{|a|}$ satisfies

$$\int_{\mathbb{R}^*} f(ba)\,\frac{da}{|a|} = \int_{\mathbb{R}^*} f(a)\,\frac{da}{|a|},$$

for all $b \in \mathbb{R}^*$. Therefore, $\int_0^\infty f(a)\,\frac{da}{a}$ is integration with respect to Haar measure on $\mathbb{R}^+$.

(3) Haar measure on the multiplicative group $\mathbb{C}\backslash\{0\}$ is given by

$$d\mu(x+iy) = \frac{dxdy}{x^2+y^2}, \qquad x, y \in \mathbb{R}.$$

We will often specify left Haar measure on a given locally compact group G by writing out the expression for integration with respect to the measure in a convenient parametrization of the elements of G.

Since $\Delta_G|_N = \Delta_N$, a G-invariant measure on the quotient group G/N exists and is nothing but a Haar measure. Given any of the three Haar measures on G, N and G/N, the other two can be normalized so that

$$\int_G f(x)dx = \int_{G/N} \left(\int_N f(xn)dn\right) d(xN)$$

holds for all $f \in L^1(G)$. Furthermore, the map $T_N : L^1(G) \to L^1(G/N)$ is a $*$-homomorphism.

EXAMPLE 1.3.10. Let $n \in \mathbb{N}$ and $G_1, G_2, \ldots, G_n$ be locally compact groups. Then the Cartesian product, $G_1 \times G_2 \times \ldots \times G_n$, also denoted $\Pi_{j=1}^n G_j$, is a group when given the coordinatewise operations. With the product topology, $\prod_{j=1}^n G_j$ is a locally compact group called the *product group*. The left Haar measure on $\prod_{j=1}^n G_j$ is the product of the left Haar measures on the groups G_j. Thus, if f is a nonnegative measurable function on $G = \prod_{j=1}^n G_j$,

$$\int_G f(x)dx = \int_{G_1}\int_{G_2}\cdots\int_{G_n} f(x_1, x_2, \ldots, x_n)dx_n \ldots dx_2dx_1.$$

If $G_j = H$, a fixed locally compact group, for $j = 1, \ldots, n$, then $\prod_{j=1}^n G_j$ is denoted H^n. Thus, we have $\mathbb{R}^n, \mathbb{Z}^n, \mathbb{T}^n$ and combinations $\mathbb{R}^k \times \mathbb{Z}^l \times \mathbb{T}^m$, for any nonnegative integers k, l and m.

EXAMPLE 1.3.11. We identify the left Haar measure of the $ax+b$-group G by providing a left invariant integration formula. For a nonnegative measurable function f on G,

$$\int_G f(z)dz = \int_{\mathbb{R}}\int_{\mathbb{R}^+} f(b,a)a^{-2}da\,db.$$

One can check that, for any $(b',a') \in G$,

$$\int_{\mathbb{R}}\int_{\mathbb{R}} f((b',a')(b,a))a^{-2}da\,db = \int_{\mathbb{R}}\int_{\mathbb{R}} f(b,a)a^{-2}da\,db.$$

On the other hand,

$$\begin{aligned}\int_{\mathbb{R}}\int_{\mathbb{R}^+} f((b,a)(b',a'))a^{-2}da\,db &= \int_{\mathbb{R}}\int_{\mathbb{R}^+} f(b+ab',aa')a^{-2}da\,db \\ &= a'\int_{\mathbb{R}}\int_{\mathbb{R}^+} f(b,a)a^{-2}da\,db.\end{aligned}$$

Thus, G is nonunimodular and $\Delta_G(b,a) = a^{-1}$.

To find the left Haar integral on a semi-direct product group $N \rtimes H$, one uses those on N and H together with a factor which records the amount by which α_h scales the left Haar integral of N. More precisely, fix $h \in H$ and define a measure μ_N^h on N by $\mu_N^h(E) = \mu_N(\alpha_h(E))$, for Borel subsets E of N. Since

$$\mu_N^h(nE) = \mu_N(\alpha_h(nE)) = \mu_N(\alpha_h(n)\alpha_h(E)) = \mu_N(\alpha_h(E)) = \mu_N^h(E),$$

μ_N^h is a left Haar measure on N. Thus, there exists $\delta(h) > 0$ so that $\mu_N^h(E) = \delta(h)\mu_N(E)$. Then $\delta : H \to \mathbb{R}^+$ is a continuous homomorphism and

$$\int_N f(x)dx = \delta(h)\int_N f(\alpha_h(x))dx,$$

for any nonnegative measurable function f on N. Now the left Haar integral of any nonnegative measurable function f on $N \rtimes H$ is given by

$$\int_{N\rtimes H} f(n,h)d(n,h) = \int_H\int_N f(n,h)\delta(h)^{-1}dndh.$$

The reader should check the left invariance. We will compute the modular function $\Delta_{N\rtimes H}$. Let $(m,k) \in N \rtimes H$. Then

$$\begin{aligned}\int_{N\rtimes H} f((n,h)(m,k))d(n,h) &= \int_H\int_N f(n\alpha_h(m),hk)\delta(h)^{-1}dndh \\ &= \int_H\int_N f(\alpha_h(\alpha_{h^{-1}}(n)m),hk)\delta(h)^{-1}dndh \\ &= \int_H\int_N f(\alpha_h(nm),hk)dndh \\ &= \int_H\int_N f(\alpha_{hk^{-1}}(n),h)\Delta_N(m^{-1})\Delta_H(k^{-1})dndh \\ &= \int_H\int_N f(n,h)\delta(kh^{-1})\Delta_N(m^{-1})\Delta_H(k^{-1})dndh \\ &= \delta(k)(\Delta_N(m)\Delta_H(k))^{-1}\int_{N\rtimes H} f(n,h)d(n,h).\end{aligned}$$

This shows that, for $(m,k) \in N \rtimes H$,

$$\Delta_{N\rtimes H}(m,k) = \Delta_N(m)\Delta_H(k)\delta(k)^{-1}.$$

Using the above formula for the modular function of a semidirect product group, the reader can easily check that the motion group G_n is unimodular (see [**125**] for details).

EXAMPLE 1.3.12. Easy calculations show that $\mathbb{H}_n$ is unimodular with Haar integral given by

$$\int_{\mathbb{H}_n} f([x,y,z])d[x,y,z] = \int_{\mathbb{R}^n}\int_{\mathbb{R}^n}\int_{\mathbb{R}} f([x,y,z])\,dz\,dy\,dx.$$

THEOREM 1.3.13. *If G is a locally compact group and $T \in \mathcal{B}(L^1(G))$ is such that $T(f*g) = f*T(g)$ for all $f,g \in L^1(G)$, then there exists $\mu \in M(G)$ such that $T(f) = f*\mu$ for all $f \in L^1(G)$. Moreover, $\|T\| = \|\mu\|$.*

Readers are referred to [**78, 125, 288, 289**] for more details.

1.4. Unitary representations and positive definite functions

Let $\mathcal{H}$ be a Hilbert space and $\mathcal{U}(\mathcal{H})$ the group of unitary operators on $\mathcal{H}$. If G is a locally compact group, a *continuous unitary representation* of G is a pair $(\pi,\mathcal{H}(\pi))$, where $\mathcal{H}(\pi)$ is a Hilbert space and π is a homomorphism of G into $\mathcal{U}(\mathcal{H}(\pi))$ such that, for any $\xi,\eta \in \mathcal{H}(\pi)$, the function $\varphi_{\xi,\eta} : x \to \langle\pi(x)\xi,\eta\rangle$ is continuous on G. Thus the map $\pi : G \to \mathcal{U}(\mathcal{H}(\pi))$ is required to be continuous when $\mathcal{U}(\mathcal{H}(\pi))$ is equipped with the weak operator topology. Because the weak and the strong operator topologies coincide on $\mathcal{U}(\mathcal{H}(\pi))$, this is equivalent to continuity of the map $x \to \pi(x)\xi$ from G into $\mathcal{H}(\pi)$ for every $\xi \in \mathcal{H}(\pi)$. Throughout this book, we shall simply use the word *representation* to mean a continuous unitary representation of G and write π with the associated Hilbert space $\mathcal{H}(\pi)$.

EXAMPLE 1.4.1. (1) Let $\chi : G \to \mathbb{T}$ be a continuous homomorphism and note that $\mathbb{T}$ can be identified with the unitary group of the 1-dimensional Hilbert space $\mathbb{C}$. In particular, the so-called trivial representation of G is given by $\chi(x) = 1$ for all $x \in G$, and every character of an abelian group is a representation.

(2) The *left regular representation* λ_G of G on the Hilbert space $L^2(G)$ is defined by $\lambda_G(x)f(y) = f(x^{-1}y)$ for $f \in L^2(G)$ and almost all $y \in G$.

(3) For $x \in G$ and $f \in L^2(G)$, define $\rho_G(x)f$ by $\rho_G(x)f(y) = \Delta_G(x)^{1/2}f(yx)$ for almost all $y \in G$. Since

$$\int_G |\rho_G(x)f(y)|^2 dy = \int_G \Delta(x)|f(yx)|^2 dy = \int_G |f(y)|^2 dy,$$

each $\rho_G(x)$ is a unitary operator on $L^2(G)$. The representation $\rho_G : x \to \rho_G(x)$ is called the *right regular representation* of G.

Two representations π and σ of G are said to be *(unitarily) equivalent* if there exists a unitary map $U : \mathcal{H}(\pi) \to \mathcal{H}(\sigma)$ such that $U\pi(x) = \sigma(x)U$ for all $x \in G$. We then write $\pi \cong \sigma$ and call U an *intertwining operator* for π and σ.

EXAMPLE 1.4.2. The left and right regular representations of G are equivalent. In fact, for $f \in L^2(G)$, define the function Uf by $Uf(y) = \Delta_G(y^{-1})^{1/2}f(y^{-1})$ for almost all $y \in G$. Then the map $f \to Uf$ is a unitary map from $L^2(G)$ onto itself, and it is easily verified that $(\lambda_G(x)U)(f) = (U\rho_G(x))(f)$ for all $x \in G$ and $f \in L^2(G)$.

A linear subspace V of $\mathcal{H}(\pi)$ is called *π-invariant* if $\pi(x)\xi \in V$ for all $x \in G$ whenever $\xi \in V$. The representation π is called *irreducible* if $\mathcal{H}(\pi)$ and $\{0\}$ are the only closed π-invariant subspaces of $\mathcal{H}(\pi)$.

Given a subset $\delta \subseteq \mathcal{H}(\pi)$, let

$$\delta' = \{T \in \mathcal{B}(\mathcal{H}(\pi)) \text{ such that } TS = ST \text{ for all } S \in \delta\}$$

PROPOSITION 1.4.3. (See [**125**].) *For a representation π of G, the following are equivalent.*

(i) *π is irreducible.*
(ii) *$\pi(G)' = \mathbb{C}I$, where I denotes the identity operator on $\mathcal{H}(\pi)$.*
(iii) *For $\xi, \eta \in \mathcal{H}(\pi)$, $\xi \neq 0$ and $\eta \neq 0$ imply that $\varphi_{\xi,\eta} \neq 0$.*

If G is abelian and π is any representation of G, then all the operators $\pi(x)$, $x \in G$, commute with each other. It follows from Proposition 1.4.3 that if π is irreducible, then $\pi(x)$ is a scalar multiple of the identity operator on $\mathcal{H}(\pi)$. Thus every 1-dimensional subspace is π-invariant and so $\dim \mathcal{H}(\pi) = 1$. Thus the irreducible representations of G are in one-to-one correspondence with the characters of G. The following theorem tells us that, for any locally compact group G, there are enough irreducible representations to separate the elements of G.

THEOREM 1.4.4. [Gelfand-Raikov theorem](See [**125**].) *Let G be a locally compact group, and let $x, y \in G$, $x \neq y$. Then there exists an irreducible representation π of G such that $\pi(x) \neq \pi(y)$.*

If V is π-invariant and $\eta \in V^\perp$, the orthogonal complement of V, then for all $\xi \in V$ and $x \in G$, $\langle \pi(x)\eta, \xi\rangle = \langle \eta, \pi(x^{-1})\xi\rangle = 0$, whence $\pi(x)\eta \in V^\perp$. Then, if V is closed in $H(\pi)$, π is the direct sum of the two subrepresentations $x \to \pi(x)|_V$ and $x \to \pi(x)|_{V^\perp}$. More generally, we have the following definition.

DEFINITION 1.4.5. Let I be a nonempty index set. For each $\iota \in I$, let $\mathcal{H}_\iota$ be a Hilbert space and π_ι a representation of G with $\mathcal{H}(\pi_\iota) = \mathcal{H}_\iota$. For $\xi = (\xi_\iota)_\iota \in \oplus_{\iota \in I}\mathcal{H}_\iota$, the Hilbert space direct sum of the $\mathcal{H}_\iota$, let

$$(\oplus_{\iota\in I}\pi_\iota)(x)\xi = (\pi_\iota(x)\xi_\iota)_{\iota\in I}, \quad x \in G.$$

This defines a representation $\oplus_{\iota\in I}\pi_\iota$ of G, the so-called *direct sum* of the representation π_ι. If all the π_ι are identical, say $\pi_\iota = \sigma$ and m denotes the cardinality of I, then $\oplus_{\iota\in I}\pi_\iota$ is denoted simply $m\sigma$ simply.

PROPOSITION 1.4.6. *Let π be a representation of G, and let $\{\mathcal{H}_\iota : \iota \in I\}$ be a family of closed π-invariant subspaces of $\mathcal{H}(\pi)$ such that $\mathcal{H}_\iota \perp \mathcal{H}_\lambda$ for $\iota \neq \lambda$ and the linear span of $\bigcup_{\iota\in I} \mathcal{H}_\iota$ is dense in $\mathcal{H}(\pi)$. Then π is equivalent to the direct sum $\oplus_{\iota\in I}\pi_\iota$, where π_ι is defined by $\pi_\iota(x) = \pi(x)|_{\mathcal{H}_\iota} : \mathcal{H}_\iota \to \mathcal{H}_\iota$.*

An element $\xi \in \mathcal{H}(\pi)$ is called a *cyclic vector* for the representation π if the set $\{\pi(x)\xi : x \in G\}$ is total in $\mathcal{H}(\pi)$. If there exists such a cyclic vector for π, then π is called a *cyclic representation*.

PROPOSITION 1.4.7. *Every representation π of G is equivalent to a direct sum of cyclic representations.*

DEFINITION 1.4.8. Let G and H be locally compact groups with representations π and ρ. Then the *outer tensor product* $\pi \times \rho$ of π and ρ is the representation of $G \times H$ acting on the Hilbert space tensor product $\mathcal{H}(\pi) \otimes \mathcal{H}(\rho)$ defined by

$$(\pi \times \rho)(x, y)(\xi \otimes \eta) = \pi(x)\xi \otimes \rho(y)\eta,$$

for $x \in G$, $y \in H$, $\xi \in \mathcal{H}(\pi)$ and $\eta \in \mathcal{H}(\rho)$. If π and ρ are both representations of the same group G, then the *(inner) tensor product* $\pi \otimes \rho$ is the representation of G on defined on $\mathcal{H}(\pi) \otimes \mathcal{H}(\rho)$ defined by

$$(\pi \otimes \rho)(x) = (\pi \times \rho)(x, x), \quad x \in G.$$

For a representation π of G, the conjugate or contragredient representation $\overline{\pi}$ of π is defined on the conjugate Hilbert space $\overline{\mathcal{H}(\pi)}$ by $\overline{\pi(x)} = \pi(x)$, that is

$$\langle \overline{\pi}(x)\xi, \eta \rangle = \overline{\langle \pi(x)\xi, \eta \rangle} = \langle \pi(x^{-1}\eta), \xi \rangle$$

for $x \in G$ and $\xi, \eta \in \mathcal{H}(\pi)$.

Every representation π of G determines a nondegenerate $*$-representation $\widetilde{\pi}$ of $L^1(G)$ in the following manner. For $f \in L^1(G)$, define the bounded linear operator $\widetilde{\pi}(f)$ on $\mathcal{H}(\pi)$ by

$$\langle \widetilde{\pi}(f)\xi, \eta \rangle = \int_G f(x)\langle \pi(x)\xi, \eta \rangle dx, \ \xi, \eta \in \mathcal{H}(\pi).$$

We then write $\widetilde{\pi}(f) = \int_G f(x)\pi(x)dx$, an operator-valued integral. It is clear that $\widetilde{\pi}$ is a representation of $L^1(G)$ and $\|\widetilde{\pi}(f)\| \leqslant \|f\|_1$ for $f \in L^1(G)$. Observe that

$$\pi(x)\widetilde{\pi}(f) = \widetilde{\pi}(L_x f) \quad \text{and} \quad \widetilde{\pi}(f)\pi(x^{-1}) = \Delta(x)\widetilde{\pi}(R_x f)$$

for $x \in G$ and $f \in L^1(G)$, where Δ is the modular function for G.

THEOREM 1.4.9. *For each representation π of G, let $\widetilde{\pi}$ denote the associated $*$-representation of $L^1(G)$. Then the assignment $\pi \to \widetilde{\pi}$ has the following properties.*

(i) *π is irreducible if and only if $\widetilde{\pi}$ is irreducible.*
(ii) *π is cyclic if and only if $\widetilde{\pi}$ is cyclic.*
(iii) *$\pi_1 \equiv \pi_2$ if and only if $\widetilde{\pi_1} \cong \widetilde{\pi_2}$.*
(iv) *For each non-degenerate representation σ of $L^1(G)$, there exists a unique representation π of G such that $\sigma = \widetilde{\pi}$.*

Apart from forming tensor products of representations and restricting representations to subgroups, there is another important procedure of constructing new representations from given ones. We briefly indicate the construction, referring the reader to [**158**] for details. Let H be a closed subgroup of G, and let π be a unitary representation of H.

Let $\mathcal{F}(G, \pi)$ denote the space of continuous mappings $\xi : G \to \mathcal{H}(\pi)$ such that $q(\operatorname{supp} \xi)$ is compact in G/H and

$$\xi(xh) = \Delta_H(h)^{1/2}\Delta_G(h)^{-1/2}\xi(x)$$

for all $x \in G$ and $h \in H$. An inner product can be defined on $\mathcal{F}(G, \pi)$ by

$$\langle \xi, \eta \rangle = \int_G \psi(x)\langle \xi(x), \eta(x) \rangle \, dx,$$

where $\psi \in C_c(G)$ is such that $T_H\psi = 1$ on the compact set $q(\operatorname{supp}\xi) \cup q(\operatorname{supp}\eta)$. To each $x \in G$, one can associate a unitary operator, denoted $\operatorname{ind}_H^G \pi(x)$, on the completion of $\mathcal{F}(G,\pi)$ defined, for $\xi \in \mathcal{F}(G,\pi)$, by

$$\left(\operatorname{ind}_H^G \pi(x)\xi\right)(y) = \xi(x^{-1}y), y \in G.$$

This representation is called the representation of G induced by π and denoted $\operatorname{ind}_H^G \pi$. We list a few properties of induced representations, which will be used later in the book. A very important result is the following *induction in stages* theorem.

THEOREM 1.4.10. *Let K and H be closed subgroups of the locally compact group G such that $K \subseteq H$, and let π be a unitary representation of K. Then the representations $\operatorname{ind}_K^G \pi$ and $\operatorname{ind}_H^G(\operatorname{ind}_K^H \pi)$ are unitarily equivalent.*

THEOREM 1.4.11. *Let H be a closed subgroup of G. Then, for arbitrary representations ρ of G and π of H,*

$$\rho \otimes \operatorname{ind}_H^G \pi \cong \operatorname{ind}_H^G(\rho|_H \otimes \pi).$$

It follows from the preceding theorem that if H is a closed subgroup of G, then $\operatorname{ind}_H^G \lambda_H$ is equivalent to λ_G.

Let N be an abelian closed normal subgroup of G. Then G acts on the dual group $\widehat{N}$ of N by $(x,\chi) \to x \cdot \chi$, where $x \cdot \chi(n) = \chi(x^{-1}nx)$ for $\chi \in \widehat{N}$, $n \in N$ and $x \in G$. Then $G_\chi = \{x \in G : x \cdot \chi = \chi\}$ is a closed subgroup of G, the stabilizer of χ, and $G(\chi) = \{x \cdot \chi : x \in G\}$ is called the *G-orbit of χ*.

We are going to describe a procedure for determining the irreducible representations of G (up to unitary equivalence) in terms of $\widehat{N}$, the action of G on $\widehat{N}$ and the irreducible representations of the stabilizer groups G_χ, $\chi \in \widehat{N}$. This procedure was developed by Mackey. It is often referred to as the Mackey machine and it is based on the so-called imprimitivity theorem, which is far beyond the scope of this book. In order for the procedure to work, certain technical conditions have to be placed on the manner in which N is embedded in G. These conditions are fulfilled at least when G is second countable and the G-orbits in $\widehat{N}$ are locally compact (equivalently, open in their closures).

THEOREM 1.4.12. *Suppose that G is second countable and that $G(\chi)$ is locally compact for every $\chi \in \widehat{N}$. Let π be an irreducible representation of G. Then there exist $\chi \in \widehat{N}$ and an irreducible representation σ of G_χ such that $\sigma(n) = \chi(n)I_{\mathcal{H}(\sigma)}$ for all $n \in N$ and π is unitarily equivalent to the induced representation $\operatorname{ind}_{G_\chi}^G \sigma$.*

In the preceding theorem, for a given π, the orbit in $\widehat{N}$ is uniquely determined, but the choice of χ in this orbit is arbitrary. If $\gamma = x\cdot\chi \in G(\chi)$, then $G_\gamma = xG_\chi x^{-1}$, and the assignment $\sigma \to x \cdot \sigma$, where $x \cdot \sigma(y) = \sigma(x^{-1}yx)$ for $y \in G_\gamma$, defines a bijection between the representations of G_χ and those of G_γ, and the induced representations $\operatorname{ind}_{G_\chi}^G \sigma$ and $\operatorname{ind}_{G_\gamma}^G (x \cdot \sigma)$ are equivalent.

THEOREM 1.4.13. *Suppose that G is second countable and that $G(\chi)$ is locally compact for every $\chi \in \widehat{N}$. Let $\chi \in \widehat{N}$ and let σ be an irreducible representation of G_χ such that $\sigma(n) = \chi(n)I_{\mathcal{H}(\sigma)}$ for every $n \in N$. Then $\operatorname{ind}_{G_\chi}^G \sigma$ is irreducible. If τ is another irreducible representation of G_χ such that $\tau(n) = \chi(n)I_{\mathcal{H}(\sigma)}$ for all $n \in N$ and $\operatorname{ind}_{G_\chi}^G \sigma$ and $\operatorname{ind}_{G_\chi}^G \tau$ are equivalent, then σ and τ are equivalent.*

Now suppose that G is a semidirect product $G = N \rtimes H$, where N is abelian, and for $\chi \in \widehat{N}$, let $H_\chi = G_\chi \cap N$. Then, if σ is an irreducible representations of H_χ, $(\chi \times \sigma)(n, h) = \chi(n)\sigma(h)$, $n \in N$, $h \in H_\chi$, defines an irreducible representation of G_χ with $\mathcal{H}(\chi \times \sigma) = \mathcal{H}(\sigma)$. Conversely, every irreducible representation ρ of G_χ with $\rho(n) = \chi(n)I_{\mathcal{H}(\rho)}$ for all $n \in N$ is of this form. Moreover, $\chi \times \sigma$ is equivalent to $\chi \times \tau$ if and only if σ and τ are equivalent.

Combing Theorems 1.4.12 and 1.4.13, one obtains the following description of the irreducible representations of G in terms of the characters χ of N and the irreducible representations of their stability groups H_χ.

THEOREM 1.4.14. *Let G be a semidirect product $G = N \rtimes H$, where N is abelian, and suppose that G is second countable and the H-orbits in $\widehat{N}$ are locally compact.*

(i) *If $\chi \in \widehat{N}$ and σ is an irreducible representation of the stability group H_χ, then the induced representation $\mathrm{ind}_{G_\chi}^G(\chi \times \sigma)$ of G is irreducible.*
(ii) *Every irreducible representation of G is equivalent to one of this form.*
(iii) *If also $\gamma \in \widehat{N}$ and τ is an irreducible representation of H_γ, then $\mathrm{ind}_{G_\chi}^G(\chi \times \sigma)$ and $\mathrm{ind}_{G_\gamma}^G(\gamma \times \tau)$ are equivalent if and only if there exists $x \in H$ such that $\gamma = x \cdot \chi$ and the two representations σ and $h \to \tau(x^{-1}hx)$ of H_χ are equivalent.*

Let H be a closed subgroup of G, π a unitary representation of H and $x \in G$. Then $x \cdot \pi$ denotes the representation of xHx^{-1} on $\mathcal{H}(\pi)$ defined by $x \cdot \pi(y) = \pi(x^{-1}yx)$. The two induced representations $\mathrm{ind}_{xHx^{-1}}^G(x \cdot \pi)$ and $\mathrm{ind}_H^G \pi$ are equivalent.

DEFINITION 1.4.15. A complex-valued function φ on a group G is *positive definite* if the inequality

$$\sum_{j=1}^{n}\sum_{k=1}^{n} \overline{\lambda_j}\lambda_k \varphi(x_j^{-1}x_k) \geqslant 0$$

holds for any finitely many $x_1, \ldots, x_n \in G$ and $\lambda_1, \ldots, \lambda_n \in \mathbb{C}$.

PROPOSITION 1.4.16. *Let φ be a positive definite function on G such that $\varphi(e) = 1$. Then*

(i) $|\varphi(x)| \leqslant 1$ *for all* $x \in G$.
(ii) $\varphi(x^{-1}) = \overline{\varphi(x)}$ *for all* $x \in G$.
(iii) $|\varphi(x) - \varphi(y)|^2 \leqslant 2[1 - \mathrm{Re}(\varphi(x^{-1}y))]$ *for all* $x, y \in G$. *In particular,* $\varphi(xy) = \varphi(x)$ *whenever* $\varphi(y) = 1$.
(iv) $|\varphi(y^{-1}z) - \varphi(y^{-1}x)\varphi(x^{-1}z)|^2 \leqslant [1 - |\varphi(x^{-1}y)|^2][1 - |\varphi(x^{-1}z)|^2]$ *for all* $x, y, z \in G$.
(v) *If* $y \in G$ *is such that* $|\varphi(y)| = 1$, *then* $|\varphi(xy)| = |\varphi(x)|$ *for all* $x \in G$.

PROPOSITION 1.4.17. (See [**78**].) *Let φ be a bounded continuous function on the locally compact group G. Then the following are equivalent.*

(i) φ *is positive definite.*
(ii) $\int_G (f^* * f)(x)\varphi(x)dx \geqslant 0$ *for all* $f \in L^1(G)$.
(iii) *There exist a unitary representation π of G and a vector $\xi \in \mathcal{H}(\pi)$ such that $\varphi(x) = \langle \pi(x)\xi, \xi\rangle$ for all $x \in G$.*

DEFINITION 1.4.18. Let $P(G)$ denote the set of all continuous positive definite functions on G and let $P_1(G) = \{\varphi \in P(G) : \varphi(e) = 1\}$ and $P_{\leqslant 1}(G) = \{\varphi \in P(G) : \varphi(e) \leqslant 1\}$.

Since $\|\varphi\|_\infty = \varphi(e)$ for any $\varphi \in P(G)$, both $P_1(G)$ and $P_{\leqslant 1}(G)$ are bounded convex sets. Let $\mathrm{ex}(P_1(G))$ and $\mathrm{ex}(P_{\leqslant 1}(G))$ denote the set of all extreme points of $P_1(G)$ and $P_{\leqslant 1}(G)$, respectively.

LEMMA 1.4.19. (See [**126**].) *Let φ and ψ be positive definite functions on G. Then $\overline{\varphi}, \check{\varphi}$ and $\varphi\psi$ are also positive definite.*

$P_{\leqslant 1}(G)$ is a w^*-closed convex subset of the unit ball of $L^\infty(G)$, hence w^*-compact. Assertion (i) of the following theorem is then a consequence of the Krein-Milman theorem.

THEOREM 1.4.20. *Let G be a locally compact group.*

(i) *$P_{\leqslant 1}(G)$ is the w^*-closure of the convex hull of* $\mathrm{ex}(P_1(G))$.
(ii) $\mathrm{ex}(P_{\leqslant 1}(G)) = \mathrm{ex}(P_1(G)) \cup \{0\}$.
(iii) *The convex hull of* $\mathrm{ex}(P_1(G))$ *is w^*-dense in $P_1(G)$.*

THEOREM 1.4.21. *The w^*-topology on $P_1(G) \subseteq L^\infty(G)$ coincides with the topology of uniform convergence on compact subsets of G.*

Let H be an open subgroup of G and $\varphi \in P(H)$. Then the trivial extension $\dot{\varphi}$ of φ to all of G, defined by $\dot{\varphi}(x) = \varphi(x)$ for $x \in H$ and $\varphi(x) = 0$ for $x \in G\backslash H$, is also positive definite.

COROLLARY 1.4.22. *Let φ be a positive definite function on G such that $\varphi(e) = 1$, and let $H = \{x \in G : |\varphi(x)| = 1\}$. Then H is a subgroup of G and*

$$\varphi(xy) = \varphi(yx) = \varphi(x)\varphi(y)$$

for all $x \in H$ and $y \in G$. In particular, $\varphi|_H$ is a character of H.

For $\xi, \eta \in \mathcal{H}(\pi)$, the coefficient function $\varphi_{\xi,\eta}$ on G is defined by $\varphi_{\xi,\eta}(x) = \langle \pi(x)\xi, \eta\rangle$. Then $\varphi_{\xi,\eta}$ is a bounded continuous function on G.

Note that if χ is a 1-dimensional representation of G, then $\mathcal{H}(\chi) = \mathbb{C}$ and hence $\mathcal{H}(\chi) \otimes \mathcal{H}(\pi) = \mathcal{H}(\pi)$ for any representation π of G. Thus $(\chi \otimes \pi)(x) = \chi(x)\pi(x)$ and $\langle (\chi \otimes \pi)(x)\xi, \eta\rangle = \chi(x)\langle \pi(x)\xi, \eta\rangle$ for all $x \in G$ and $\xi, \eta \in \mathcal{H}(\pi)$.

PROPOSITION 1.4.23. *Suppose that π and σ are cyclic representations of G with cyclic vectors ξ and η. If $\langle \pi(x)\xi, \xi\rangle = \langle \sigma(x)\eta, \eta\rangle$ for all $x \in G$, then there exists a unitary map U from $\mathcal{H}(\pi)$ onto $\mathcal{H}(\sigma)$ intertwining π and σ such that $U\xi = \eta$. In particular, π and σ are equivalent.*

We close this section by mentioning a result which will be used in Chapter 5. A complex-valued function φ on a group G is called *negative definite* if $\varphi(x^{-1}) = \overline{\varphi(x)}$ for all $x \in G$ and, for any finitely many $x_1, \ldots, x_n \in G$ and $c_1, \ldots, c_n \in \mathbb{C}$, $n \geqslant 2$, such that $\sum_{j=1}^n c_j = 0$,

$$\sum_{i,j=1}^{n} c_i\overline{c_j}\varphi(x_j^{-1}x_i) \leqslant 0.$$

The remarkable part of the following theorem is the 'only if' assertion.

THEOREM 1.4.24. [Schoenberg's theorem] *A complex-valued function φ on a group is negative definite if and only if $e^{-\alpha\varphi}$ is positive definite for every $\alpha > 0$.*

Readers are referred to [**60, 78, 95, 126**] for more details.

1.5. Abelian locally compact groups

We shall now specialize to abelian groups where a rich theory generalizing classical Fourier analysis emerges. As pointed out at the outset of this chapter, we shall only present the notation and basic results that are necessary for our later chapters.

Let G be a locally compact abelian group. A *character* of G is a continuous homomorphism of G into the circle group $\mathbb{T}$. Let $\widehat{G}$ denote the set of all such characters. Clearly, the pointwise product of two characters is again a character, and if $\chi \in \widehat{G}$, then $\chi^{-1} \in \widehat{G}$, where $\chi^{-1}(x) = \overline{\chi(x)}$ for $x \in G$. Thus $\widehat{G}$ is an abelian group. We endow $\widehat{G}$ with the topology of uniform convergence on compact subsets of G. Then a neighbourhood basis of $\chi_0 \in \widehat{G}$ is formed by the sets

$$V(\chi_0, C, \epsilon) = \{\chi \in \widehat{G} : |\chi(x) - \chi_0(x)| < \epsilon \text{ for all } x \in C\},$$

where $\epsilon > 0$ and C is a compact subset of G. This turns $\widehat{G}$ into a topological group, the *dual group* of G. The following proposition is a major step towards the Pontryagin duality theorem (Theorem 1.5.2 below) and the development of harmonic analysis on locally compact abelian groups.

PROPOSITION 1.5.1. *The dual group $\widehat{G}$ is a locally compact group and it separates the points of G, that is, if $x, y \in G$ are such that $\chi(x) = \chi(y)$ for all $\chi \in \widehat{G}$, then $x = y$.*

Elements of a locally compact abelian group may be considered as characters on the dual group. That is, for a locally compact abelian group G and $x \in G$, the map $\chi \to \chi(x)$ is a continuous homomorphism of $\widehat{G}$ into $\mathbb{T}$.

THEOREM 1.5.2. [Pontryagin duality theorem] *Let G be a locally compact abelian group. For each $x \in G$, let $\alpha_x(\chi) = \chi(x)$, for $\chi \in \widehat{G}$. Then the map $x \to \alpha_x$ is a topological group isomorphism of G with the dual group of $\widehat{G}$.*

The Pontryagin duality theorem allows us to derive a number of further important results. The first one is a duality between quotient groups of a locally compact abelian group G and subgroups of $\widehat{G}$. For a closed subgroup H of G, let $H^\perp = \{\chi \in \widehat{G} : \chi|_H \equiv 1\}$, the annihilator of H in $\widehat{G}$. This is a closed subgroup of $\widehat{G}$.

LEMMA 1.5.3. *Let $q : G \to G/H$ denote the quotient homomorphism. Then the map $\omega \to \omega \circ q$ is a topological isomorphism between $\widehat{G/H}$ and $H^\perp$.*

Identifying G with the dual group of $\widehat{G}$, we have $(H^\perp)^\perp = H$. Using this, it is easy to show the following.

PROPOSITION 1.5.4. (See [**125**].) *Let H be a closed subgroup of G and define a map $\psi : \widehat{G}/H^\perp \to \widehat{H}$ by $\psi(\chi H^\perp) = \chi|_H$. Then ψ is a topological group isomorphism. In particular, every character of H extends to a character of G.*

The second application of the duality theorem concerns the structure of compactly generated abelian groups. Let $\mathbb{N}_0$ denote the set of non-negative integers.

THEOREM 1.5.5. (See [**125**].) *Let G be a locally compact abelian group.*

(i) *G contains an open subgroup of the form $\mathbb{R}^k \times K$, where $k \in \mathbb{N}_0$ and K is a compact group.*

(ii) *If G is compactly generated, then $G = \mathbb{R}^k \times \mathbb{Z}^l \times K$, where $k, l \in \mathbb{N}_0$ and K is a compact group. In particular, if G is connected, then G is the direct product of a vector group and a compact connected group.*

Finally, Proposition 1.5.4 together with Theorem 1.5.2 yield

PROPOSITION 1.5.6. *Let G be a locally compact abelian group. Then*

(i) *G is discrete if and only if $\widehat{G}$ is compact.*

(ii) *G is compact if and only if $\widehat{G}$ is discrete.*

We now present a number of examples.

EXAMPLE 1.5.7. Let $G = \mathbb{R}$. For each $\gamma \in \mathbb{R}$ define $\chi_\gamma : \mathbb{R} \to \mathbb{T}$ by $\chi_\gamma(t) = e^{i\gamma t}$, for all $t \in \mathbb{R}$. Then $\gamma \to \chi_\gamma$ is a topological group isomorphism of $\mathbb{R}$ with $\widehat{\mathbb{R}}$. Note that it will often be convenient for us to parametrize $\widehat{\mathbb{R}}$ by scaling by 2π. That is, by considering $\chi_\gamma(t) = e^{2\pi i\gamma t}, t \in \mathbb{R}$, for each $\gamma \in \mathbb{R}$.

EXAMPLE 1.5.8. Let $G = \mathbb{T}$. For each $n \in \mathbb{Z}$ define $\psi_n(z) = z^n$, for all $z \in \mathbb{T}$. Then $n \to \psi_n$ identifies $\mathbb{Z}$ with $\widehat{\mathbb{T}}$. Let $q(t) = e^{2\pi it}$, for $t \in \mathbb{R}$. Then q is a continuous homomorphism of $\mathbb{R}$ onto $\mathbb{T}$ with kernel $\mathbb{Z}$. This identifies $\mathbb{T}$ with $\mathbb{R}/\mathbb{Z}$. For $\psi_n \in \widehat{\mathbb{T}} = \widehat{\mathbb{R}/\mathbb{Z}}$, note that $\psi_n \circ q(t) = e^{2\pi int}$, for $t \in \mathbb{R}$. This embeds $\widehat{\mathbb{R}/\mathbb{Z}}$ as a closed subset of $\widehat{\mathbb{R}}$ as in Proposition 1.5.4.

EXAMPLE 1.5.9. (i) Let $G = \mathbb{Z}$. For each $z \in \mathbb{T}$, let $\sigma_z(n) = z^n$, for all $n \in \mathbb{Z}$. Then $z \to \sigma_z$ identifies $\mathbb{T}$ with $\widehat{\mathbb{Z}}$.

(ii) Let $n \in \mathbb{N}, n > 1$ and $G = \mathbb{Z}/n\mathbb{Z}$. By Proposition 1.5.4,

$$\widehat{\mathbb{Z}/n\mathbb{Z}} = \{\sigma_z : z \in \mathbb{T}_n\},$$

where $\mathbb{T}_n = \{z \in \mathbb{T} : z^n = 1\}$, the group of nth roots of unity.

Many of the frequently occurring abelian locally compact groups are products of finitely many of the above examples, so the following identification is helpful to quickly determine dual groups for such products.

EXAMPLE 1.5.10. Let $G_1, \cdots, G_n$ be abelian locally compact groups and form $\prod_{i=1}^n G_i$ and $\prod_{i=1}^n \widehat{G_i}$, the product groups. For $(\chi_1, \ldots, \chi_n) \in \prod_{i=1}^n \widehat{G_i}$ and $(x_1, \ldots, x_n) \in \prod_{i=1}^n G_i$, let

$$(\chi_1, \ldots, \chi_n)(x_1, \ldots, x_n) = \chi_1(x_1) \ldots \chi_n(x_n).$$

Thus, we can naturally consider $(\chi_1, \ldots, \chi_n)$ as a character of $\prod_{i=1}^n G_i$. All characters of $\prod_{i=1}^n G_i$ arise this way. Therefore, $\widehat{\prod_{i=1}^n G_i} = \prod_{i=1}^n \widehat{G_i}$.

EXAMPLE 1.5.11. Let p be a prime number and consider the additive group Ω_p of the p-adic number field. Recall that every p-adic number x has a unique representation $x = \sum_{n=-\infty}^{\infty} c_n p^n$, where $c_n \in \{0, \ldots, p-1\}$ and $c_n \neq 0$ only for finitely many $n \in -\mathbb{N}$. Moreover $x \in \Delta_p$ if and only if $c_n = 0$ for all $n < 0$. We can define a character γ_1 of Ω_p by setting

$$\gamma_1\left(\sum_{n=-\infty}^{\infty} c_n p^n\right) = \exp\left(2\pi i \sum_{n=-\infty}^{\infty} c_n p^n\right).$$

Continuity of γ_1 follows from the fact that $\ker(\gamma_1) = \Delta_p$ and Δ_p is open in Ω_p. For an arbitrary $y \in \Omega_p$, define γ_y on Ω_p by $\gamma_y(x) = \gamma_1(yx)$, $x \in \Omega_p$. Then γ_y is a character of Ω_p with kernel equal to $\{x \in \Omega_p : |x|_p \leqslant |y|_p^{-1}\}$. It is less easy to show that every character of Ω_p is of this form. It is then clear that the map $y \to \gamma_y$ is a continuous group isomorphism from Ω_p onto $\widehat{\Omega_p}$. That this map is also open, can be seen as follows. The sets

$$U(j,k) = \{x \in \Omega_p : |\chi(x) - 1| < 1/j \text{ for all } x \text{ with } |x|_p \leqslant p^k\},$$

$j \in \mathbb{N}$, $k \in \mathbb{Z}$, form a neighbourhood basis of 1_{Ω_p} in $\widehat{\Omega_p}$. Moreover, $\chi_y \in U(j,k)$ if and only if $|y|_p \leqslant p^{-k}$. Thus the map $y \to \chi_y$ is a topological isomorphism between Ω_p and its dual group. For all this, see [**78**] or [**125**].

We now identify $\widehat{G}$ with the spectrum of the commutative Banach algebra $L^1(G)$. For $\chi \in \widehat{G}$, define $\gamma_\chi : L^1(G) \to \mathbb{C}$ by

$$\gamma_\chi(f) = \int f(x)\chi(-x)dx.$$

It is easily verified that $\gamma_\chi(f*g) = \gamma_\chi(f)\gamma_\chi(g)$ and $\gamma_\chi(f^*) = \overline{\gamma_\chi(f)}$ for $f, g \in L^1(G)$.

PROPOSITION 1.5.12. *The map $\chi \to \gamma_\chi$ is a homeomorphism from $\widehat{G}$ onto $\sigma(L^1(G))$, the latter space equipped with the Gelfand topology.*

DEFINITION 1.5.13. For $f \in L^1(G)$, the *Fourier transform* of f is the function $\widehat{f} : \widehat{G} \to \mathbb{C}$ defined by

$$\widehat{f}(\chi) = \gamma_\chi(f) = \int_G f(x)\chi(x)\,dx,$$

for all $\chi \in \widehat{G}$. The Fourier transformation extends to complex Radon measures on G. If $\mu \in M(G)$, then its *Fourier-Stieltjes transform* $\widehat{\mu}$ is the function on $\widehat{G}$ defined by

$$\widehat{\mu}(\chi) = \int_G \chi(x)d\mu(x), \ \chi \in \widehat{G}.$$

Many of the properties of the Fourier transform are summarized in the following statement. We view $C_0(\widehat{G})$ as a commutative C^*-algebra under pointwise operations and equipped with the norm $\|\cdot\|_\infty$.

THEOREM 1.5.14. *Let G be a locally compact abelian group.*

i) *The map $f \to \widehat{f}$ is an injective $*$-homomorphism of $L^1(G)$ into $C_0(\widehat{G})$ and the image of $L^1(G)$ under the Fourier transform is dense in $C_0(\widehat{G})$.*

(ii) *Given a compact subset C of $\widehat{G}$ and a neighbourhood U of C in $\widehat{G}$, there exists $f \in L^1(G)$ such that $\widehat{f} = 1$ on C and $\widehat{f} = 0$ on $\widehat{G}\backslash U$. Thus $L^1(G)$ is regular.*

(iii) *Given $f \in L^1(G)$ and $\epsilon > 0$, there exists $g \in L^1(G)$ such that $\widehat{g}$ has compact support and $\|f - g\|_1 \leqslant \epsilon$. Thus $L^1(G)$ is Tauberian.*

One can show that the Gelfand homomorphism $\Gamma : f \to \widehat{f}$ from $L^1(G)$ into $C_0(\widehat{G})$ is surjective only when G is finite.

THEOREM 1.5.15. [Plancherel theorem] *Let G be a locally compact abelian group. Then Haar measures on G and $\widehat{G}$ can be simultaneously chosen so that $\|\widehat{f}\|_2 = \|f\|_2$, for any $f \in L^1(G) \cap L^2(G)$. Moreover, $\{\widehat{f} : f \in L^1(G) \cap L^2(G)\}$ is dense in $L^2(\widehat{G})$ and so there is a unitary map $\mathcal{F} : L^2(G) \to L^2(\widehat{G})$ such that $\mathcal{F}(f) = \widehat{f}$, for all $f \in L^1(G) \cap L^2(G)$.*

Unless otherwise indicated, we will always assume that Haar measures are scaled so that the Plancherel equality holds. The unitary map $\mathcal{F}$ in Theorem 1.5.15 is called the *Plancherel transform*. The Plancherel formula and linearization imply

COROLLARY 1.5.16. *For $f, g \in L^2(G)$ the Parseval identity*

$$\int_G f(x)\overline{g(x)}dx = \int_{\widehat{G}} \widehat{f}(\chi)\overline{\widehat{g}(\chi)}d\chi$$

holds.

EXAMPLE 1.5.17. When $G = \mathbb{R}$, suppose $\widehat{\mathbb{R}}$ is parametrized by $\{\chi_\gamma : \gamma \in \mathbb{R}\}$, where $\chi_\gamma(t) = e^{2\pi i\gamma t}$, for $t \in \mathbb{R}$. Then we can write

$$\widehat{f}(\gamma) = \int_{\mathbb{R}} f(t)e^{2\pi i\gamma t}dt,$$

for $\gamma \in \mathbb{R}$, where we are writing γ in place of χ_γ. Then $\int_{\widehat{\mathbb{R}}} |\widehat{f}(\gamma)|^2 d\gamma = \int_{\mathbb{R}} |f(t)|^2 dt$, for all $f \in L^1(\mathbb{R}) \cap L^2(\mathbb{R})$. Therefore, both $\mathbb{R}$ and $\widehat{\mathbb{R}}$ can be equipped with unscaled Lebesgue measure and the Plancherel equality holds.

DEFINITION 1.5.18. For $\xi \in L^1(\widehat{G})$, define the *inverse Fourier transform* $\xi^\vee$ of ξ on G by $\xi^\vee(x) = \int_{\widehat{G}} \xi(\chi)\overline{\chi(x)}\,d\chi$, for $x \in G$. More generally, for $\mu \in M(\widehat{G})$, the *inverse Fourier-Stieltjes transform* $\mu^\vee$ of μ on G is defined by $\mu^\vee(x) = \int_{\widehat{G}} \overline{\chi(x)}d\mu(\chi)$.

The inverse Fourier transform terminology is justified by the following inversion result.

THEOREM 1.5.19. [Bochner's theorem] *Let G be an abelian locally compact group and φ a continuous function on $\widehat{G}$. Then φ is positive definite if and only if there exists a (unique) non-negative $\mu \in M(G)$ such that*

$$\varphi(\gamma) = \widehat{\mu}(\gamma) = \int_G \gamma(x)d\mu(x)$$

for all $\gamma \in \widehat{G}$. Thus the set $\{\widehat{\mu} : \mu \in M(G)\}$ coincides with the linear span of $P(\widehat{G})$.

An alternative characterization of the functions in the linear span of $P(\widehat{G})$ is given by the following theorem.

THEOREM 1.5.20. *For a continuous function φ on the dual group $\widehat{G}$, the following conditions are equivalent.*

(i) *φ is a linear combination of positive definite functions and $\|\varphi\| \leqslant C$.*
(ii) *For every trigonometric polynomial f on G of the form $f(x) = \sum_{j=1}^n c_j\gamma_j(x)$, $c_j \in \mathbb{C}, \gamma_j \in \widehat{G}$, we have*

$$\left|\sum_{j=1}^n c_j\varphi(\gamma_j)\right| \leqslant C \cdot \|f\|_\infty.$$

THEOREM 1.5.21. [Inversion theorem](See [**247**].) *Let G be a locally compact abelian group. Let $f \in L^1(G) \cap B(G)$, where $B(G) = \langle P(G)\rangle$. Then $\widehat{f} \in L^1(\widehat{G})$ and*

$$f(x) = \int_{\widehat{G}} \widehat{f}(\chi)\overline{\chi(x)}\,d\chi,$$

for all $x \in G$.

In view of the Pontryagin duality theorem, Bochner's theorem can be reformulated as follows. A function φ on G is a finite linear combination of continuous positive definite functions if and only if it equals the inverse Fourier-Stieltjes transform of some $\nu \in M(\widehat{G})$.

Readers are referred to [**78, 247**] for more details.

1.6. Representations and positive definite functionals

Let A be a normed $*$-algebra. A $*$-representation of A is a pair $(\pi, \mathcal{H}(\pi))$, where $\mathcal{H}(\pi)$ is a Hilbert space and π is a homomorphism of A into $\mathcal{B}(\mathcal{H}(\pi))$ such that $\pi(a^*) = \pi(a)^*$ for all $a \in A$. If A is complete, then such a π is automatically continuous. More precisely, $\|\pi(a)\| \leqslant \|a\|$ for all $a \in A$. If A is not assumed to be complete and π is continuous, then π extends uniquely to a $*$-representation of the completion of A and hence $\|\pi(a)\| \leqslant \|a\|$ holds. The representation π is called *nondegenerate* if the closed subspace

$$N = \{\xi \in \mathcal{H}(\pi) : \pi(a)\xi = 0 \text{ for all } a \in A\}$$

of $\mathcal{H}(\pi)$ is trivial. Equivalently, $\mathcal{H}(\pi)$ is the closed linear span of the set

$$\{\pi(a)\xi : a \in A, \xi \in \mathcal{H}(\pi)\}.$$

The definitions of equivalence, subrepresentation, irreducibility, direct sum, cyclicity etc. for $*$-representations are the obvious analogues to those for unitary representations of locally compact groups.

Every unitary representation π of G determines a nondegenerate $*$-representation $\tilde{\pi}$ of $L^1(G)$ in the following manner. For $f \in L^1(G)$, define the bounded linear operator $\tilde{\pi}(f)$ on $\mathcal{H}(\pi)$ by

$$\langle \tilde{\pi}(f)\xi, \eta\rangle = \int_G f(x)\langle \pi(x)\xi, \eta\rangle dx, \; \xi, \eta \in \mathcal{H}(\pi).$$

We then write $\tilde{\pi}(f) = \int_G f(x)\pi(x)dx$, an operator-valued integral. It is clear that $\tilde{\pi}$ is a $*-$representation of $L^1(G)$ and $\|\tilde{\pi}(f)\| \leqslant \|f\|_1$ for $f \in L^1(G)$. Observe that

$$\pi(x)\tilde{\pi}(f) = \tilde{\pi}(L_x f) \quad \text{and} \quad \tilde{\pi}(f)\pi(x^{-1}) = \Delta(x)\tilde{\pi}(R_x f)$$

for $x \in G$ and $f \in L^1(G)$.

If λ_G is the left regular representation of G, then an elementary calculation shows that, for $f \in L^1(G)$ and $\xi, \eta \in L^2(G)$,

$$\langle \lambda_G(f)\xi, \eta\rangle = \int_G f(x)\langle \lambda_G(x)\xi, \eta\rangle \, dx = \langle f * \xi, \eta\rangle.$$

Thus $\lambda_G(f)$ is just the operator of left convolution of f on $L^2(G)$.

A linear functional f on a Banach $*$-algebra A is said to be *positive* if $f(a^*a) \geqslant 0$ for every $a \in A$. Then, for $a, b \in A$,

$$f(b^*a) = \overline{f(a^*b)} \quad \text{and} \quad |f(b^*a)|^2 \leqslant f(a^*a)f(b^*b).$$

Moreover, if f is continuous and A has an approximate identity, then

$$f(a^*) = \overline{f(a)}, \; |f(b^*a)| \leqslant \|a\| f(b^*b) \text{ and } |f(a)|^2 \leqslant \|f\| f(a^*a)$$

for all $a, b \in A$.

Let π be a unitary representation of a locally compact group G on the Hilbert space $\mathcal{H}(\pi)$ and let $\xi, \eta \in \mathcal{H}(\pi)$. Then the *coefficient function* $x \to \varphi_{\xi,\eta}(x) = \langle \pi(x)\xi, \eta \rangle$ is clearly continuous and it is positive definite when $\eta = \xi$ since

$$\sum_{j=1}^{n} \sum_{k=1}^{n} \overline{\lambda_j} \lambda_k \varphi_{\xi,\xi}(x_j^{-1} x_k) = \left\| \sum_{k=1}^{n} \lambda_k \pi(x_k) \xi \right\|^2$$

for any $x_1, \ldots, x_n \in G$ and $\lambda_1, \ldots, \lambda_n \in \mathbb{C}$. As an example, taking $f \in L^2(G)$, the function

$$\varphi(x) = (f * \widetilde{f})(x) = \langle \lambda_G(x^{-1}) f, f \rangle,$$

$x \in G$, is a continuous positive definite function. Moreover, given a neighbourhood U of e in G, choosing a symmetric neighbourhood V of e such that $V^2 \subseteq U$ and taking $f = |V|^{-1/2} 1_V$, the resulting function φ satisfies $\varphi(e) = 1$ and $\operatorname{supp} \varphi \subseteq U$.

THEOREM 1.6.1. *Let ϕ be a positive linear functional on $L^1(G)$. Then there exists a unique $\varphi \in P(G)$ such that*

$$\phi(f) = \int_G f(x) \varphi(x) dx$$

for all $f \in L^1(G)$.

PROPOSITION 1.6.2. *If $\varphi \in P_1(G)$, then $\varphi \in \operatorname{ex}(P_1(G))$ if and only if the* GNS-*representation π_φ associated with φ is irreducible.*

We now indicate a fundamental method for constructing cyclic representations of G. This construction is usually referred to as the *GNS-construction*, named after Gelfand, Naimark and Segal.

A function of positive type on G is a function $\varphi \in L^\infty(G)$ which defines a positive linear functional on $L^1(G)$, that is, which satisfies

$$\int_G (f^* * f)(x) \varphi(x) dx \geqslant 0$$

for every $f \in L^1(G)$. An easy calculation shows that $\varphi \in L^\infty(G)$ is of positive type if and only if

$$\int_G \int_G f(x) \overline{f(y)} \varphi(y^{-1} x) \, dx dy \geqslant 0$$

for all $f \in L^1(G)$. Thus such a function φ defines a positive semidefinite sesquilinear form on $L^1(G)$ by

$$\langle f, g \rangle_\varphi = \int_G \int_G f(x) \overline{g(y)} \varphi(y^{-1} x) \, dx dy,$$

which satisfies $|\langle f, g \rangle_\varphi| \leqslant \|\varphi\|_\infty \|f\|_1 \|g\|_1$. Let $N_\varphi = \{f \in L^1(G) : \langle f, f \rangle_\varphi = 0\}$. It follows from the Cauchy-Schwarz inequality that $\langle f, f \rangle_\varphi = 0$ if and only if $\langle f, g \rangle_\varphi = 0$ for all $g \in L^1(G)$. Thus N_φ is a linear subspace of $L^1(G)$, and $\langle \cdot, \cdot \rangle_\varphi$ defines an inner product on the quotient space $L^1(G)/N_\varphi$. Let H_φ denote the Hilbert space completion of $L^1(G)/N_\varphi$. It is easily verified that $\langle L_x f, L_x g \rangle_\varphi = \langle f, g \rangle_\varphi$ for all $x \in G$ and $f, g \in L^1(G)$. In particular, $L_x(N_\varphi) \subseteq N_\varphi$, and hence for each $x \in G$ we can define a unitary operator $\pi_\varphi(x)$ on H_φ by

$$\pi_\varphi(x)(f + N_\varphi) = L_{x^{-1}} f + N_\varphi, \quad f \in L^1(G).$$

It is easily verified that π_φ is a unitary representation of G, the so-called *GNS-representation* associated with φ.

THEOREM 1.6.3. *Let φ be a function of positive type on G and let π_φ be the unitary representation constructed above. Then there exists a cyclic vector $\xi_\varphi \in H_\varphi$ such that*

$$\langle \pi_\varphi(x)\xi_\varphi, \xi_\varphi \rangle = \varphi(x)$$

for locally almost all $x \in G$.

COROLLARY 1.6.4. *If π is a cyclic representation of G with cyclic vector ξ and if $\varphi(x) = \langle \pi(x)\xi, \xi \rangle$ for all $x \in G$, then $\pi \cong \pi_\varphi$.*

REMARK 1.6.5. (1) Since the function $x \to \langle \pi_\varphi(x)\xi_\varphi, \xi_\varphi \rangle$ is continuous, Theorem 1.6.3 in particular shows that every function of positive type agrees locally almost everywhere with a continuous function.

(2) The $*$-representation $\widetilde{\pi_\varphi}$ of $L^1(G)$ associated with π_φ is given by

$$\widetilde{\pi_\varphi}(f)[g] = [f * g], g \in L^1(G),$$

where $[h] = h + N_\varphi$ for $h \in L^1(G)$.

Readers are referred to [**78**, **126**] for more details.

1.7. Weak containment of representations

Let A be a C^*-algebra. We start by introducing the notion of weak containment for representations of A. For such a representation π of A, the ideal

$$\ker \pi = \{a \in A : \pi(a) = 0\}$$

is called the *kernel of* π. Note that a representation of C^*-algebra is always assume to be a $*$-representation.

Let $\widehat{A}$ denote the set of equivalence classes of irreducible representations of A. If π and σ are equivalent representations, then $\ker(\pi) = \ker(\sigma)$. Therefore, one can abuse notation safely and write $\ker(\pi)$ for $\pi \in \widehat{A}$ and consider the canonical map $k : \widehat{A} \to \mathrm{Prim}(A), \pi \to \ker(\pi)$.

DEFINITION 1.7.1. The set $\widehat{A}$ equipped with the pull-back of the hull-kernel topology on $\mathrm{Prim}(A)$ is called the *dual space* of A. Thus a subset U of $\widehat{A}$ is open in $\widehat{A}$ if and only if $U = k^{-1}(V)$ for some open subset V of $\mathrm{Prim}(A)$. This topology on $\widehat{A}$ is called the *dual space topology* or *Fell topology.*

DEFINITION 1.7.2. Let S and T be two sets of representations of A. Then S is *weakly contained* in T ($S \prec T$) if $\bigcap_{\tau\in T} \ker \tau \subseteq \bigcap_{\sigma\in S} \ker \sigma$. If S is a singleton, say $\{\pi\}$, we simply write $\pi \prec T$. We say that S and T are *weakly equivalent* ($S \sim T$) if $S \prec T$ and $T \prec S$. For an arbitrary representation π of A, the *support* of π is defined to be the set of all $\sigma \in \widehat{A}$ such that $\sigma \prec \pi$.

If π is a representation of A on $\mathcal{H}(\pi)$, then for every $\xi \in \mathcal{H}(\pi)$, the function φ defined by $\varphi(a) = \langle \pi(a)\xi, \xi \rangle$, is a positive linear functional on A associated with π. We also say that a positive linear functional φ is associated with a set S of representations of A if φ is associated with some $\pi \in S$. It is important to characterize weak containment in terms of positive functionals.

THEOREM 1.7.3. *Let $S \subseteq \widehat{A}$ and $\pi \in \widehat{A}$. Then the following conditions are equivalent.*

(i) *π is weakly contained in S.*

(ii) *Every nonzero positive functional associated with* π *is the* w^**-limit of finite linear combinations of positive functionals associated with* S.
(iii) *Every nonzero positive functional* φ *associated with* π *is the* w^**-limit of finite sums* ψ *of positive functionals associated with* S *which satisfy* $\|\psi\| \leqslant \|\varphi\|$.

Suppose that π and all the representations in S are irreducible. Then Theorem 1.7.3 can be sharpened as follows.

THEOREM 1.7.4. *Let* $S \subseteq \widehat{A}$ *and* $\pi \in \widehat{A}$. *Then the following conditions are equivalent.*

(i) π *is weakly contained in* S.
(ii) *Some nonzero positive functional associated with* π *is the* w^**-limit of positive functionals associated with* S.
(iii) *Every nonzero positive functional* φ *associated with* π *is the* w^**-limit of positive functionals* ψ *associated with* S *such that* $\|\psi\| \leqslant \|\varphi\|$.

DEFINITION 1.7.5. Let A be a C^*-algebra. An ideal I of A is called a *primitive ideal* if $I = \ker(\pi)$ for some irreducible representation π of A, and the collection $\mathrm{Prim}(A)$ of all primitive ideals of A is called the *primitive ideal space* of A. $\mathrm{Prim}(A)$ is endowed with the *hull-kernel* or *Jacobson topology*, which can most easily be defined by describing the closure operation. For any subset $\mathcal{I}$ of $\mathrm{Prim}(A)$, let

$$\overline{\mathcal{I}} = \{P \in \mathrm{Prim}(A) : P \supseteq \bigcap_{I \in \mathcal{I}} I\}.$$

Then the assignment $\mathcal{I} \to \overline{\mathcal{I}}$ is a closure operation, and hence there is a unique topology on $\mathrm{Prim}(A)$ such that $\overline{\mathcal{I}}$ equals the closure of $\mathcal{I}$.

Let G be a locally compact group. Given any positive linear functional ϕ on $C^*(G)$, there is a unique $\varphi \in P(G)$ such that $\phi(f) = \int_G f(x)\varphi(x)dx$ for all $f \in L^1(G)$. Conversely, if $\varphi \in P(G)$, then the functional on $L^1(G)$ given by

$$f \to \int_G f(x)\varphi(x)dx$$

extends uniquely to a positive linear functional on $C^*(G)$. Since there is a bijection between (equivalence classes of) unitary representations of a locally compact group G and (equivalence classes of) nondegenerate $*$-representations of the group C^*-algebra $C^*(G)$, the notion of weak containment can simply be transferred from the C^*-algebra context to the group context.

Since the w^*-topology on $P_1(G)$ coincides with the topology of uniform convergence on compact sets, weak containment can be expressed as follows.

PROPOSITION 1.7.6. *Let* S *be a set of representations of* G *and* π *a representation of* G. *Then* $\pi \prec S$ *if and only if, for any positive definite function* φ *associated with* π, *there exists a net* $(\psi_\alpha)_\alpha$, *where each* ψ_α *is a linear combination of positive definite functions associated with* S, *such that* $\psi_\alpha \to \varphi$ *uniformly on compact subsets of* G.

Theorem 1.7.4 can now be reformulated as follows

COROLLARY 1.7.7. *For* $S \subseteq \widehat{G}$ *and* $\pi \in \widehat{G}$, *the following are equivalent.*

(i) π *is weakly contained in* S.

(ii) *At least one nonzero function of positive type associated with π is the uniform limit on compacta of a net of functions of positive type associated with S.*

(iii) *Every nonzero function of positive type associated with π is the uniform limit on compacta of a net of functions of positive type associated with S.*

DEFINITION 1.7.8. Let G be a locally compact group and let $\widehat{G}$ denote the set of unitary equivalence classes of irreducible representations of G. The *dual space topology* or *Fell topology* on $\widehat{G}$ is the topology which makes the bijection $\widehat{G} \to \widehat{C^*(G)}$ a homeomorphism. Endowed with this topology, $\widehat{G}$ is called the *dual space* of G. The set of all $\pi \in \widehat{G}$ such that $\pi \prec \lambda_G$ is called the *reduced dual* of G and denoted $\widehat{G}_r$. This is a closed subset of $\widehat{G}$. The *support* of a representation π of G is the set of all $\rho \in \widehat{G}$ which are weakly contained in π.

Weak containment of group representations is inherited by restricting representations to subgroups and by forming tensor products. More precisely, we have the following.

(1) If S and T are sets of unitary representations of G such that $S \prec T$ and H is a closed subgroup of G, then
$$S|_H = \{\sigma|_H : \sigma \in S\} \prec T|_H = \{\tau|_H : \tau \in T\}.$$

(2) If $S \prec T$ and $S' \prec T'$, then
$$S \otimes S' = \{\sigma \otimes \sigma' : \sigma \in S, \sigma' \in S'\} \prec T \otimes T' = \{\tau \otimes \tau'\}.$$

For $f \in L^1(G)$, let $\|f\|_* = \sup\{\|\pi(f)\| : \pi \in \widehat{G}\}$. Since we always have $\|\sigma(f)\| \leqslant \|f\|_1$ for any representation σ, $\|f\|_* \leqslant \|f\|_1$. Moreover,
$$\|f^* * f\|_* = \sup\{\|\pi(f)\|^2 : \pi \in \widehat{G}\} = \|f\|_*^2$$
as well as
$$\|f * g\|_* \leqslant \|f\|_* \|g\|_* \quad \text{and} \quad \|f^*\|_* = \|f\|_*$$
for $f, g \in L^1(G)$. Thus $\|\cdot\|_*$ defines a C^*-norm on $L^1(G)$. The completion of $(L^1(G), \|\cdot\|_*)$ is called the *group C^*-algebra* of G and denoted $C^*(G)$.

The norm $\|\cdot\|_*$ on $L^1(G)$ is designed so that every representation π of $L^1(G)$ extends uniquely to a representation, also denoted π, of $C^*(G)$. The concepts of irreducibility and equivalence of representations mean the same whether one is considering the representations as being of G or of $C^*(G)$. For a representation π of G, $\pi(C^*(G))$ is a norm closed $*$-subalgebra of $\mathcal{B}(\mathcal{H}(\pi))$. Indeed,
$$\pi(C^*(G)) = \overline{\pi(L^1(G))} = \overline{\text{span}(\pi(G))},$$
where the closure is in the operator norm.

Another norm on the Banach $*$-algebra $L^1(G)$ satisfying the C^*-condition is given by $f \to \|\lambda_G(f)\|$, where λ_G denotes the left regular representation of $L^1(G)$ on $L^2(G)$. The completion of $L^1(G)$ with respect to this norm is called the reduced group C^*-algebra.

Let N be a closed normal subgroup of G, $q : G \to G/N$ the quotient homomorphism and $T_N : C^*(G) \to C^*(G/N)$ the canonical homomorphism of C^*-algebras. Then the map $\pi \to \pi \circ q$ is a bijection between the unitary representations of G/N and the unitary representations σ of G such that $\sigma(N) = \{I_{\mathcal{H}_\sigma}\}$. Equivalently, $\sigma = \pi \circ q$ if and only if $\ker \sigma \supseteq \ker T_N$.

PROPOSITION 1.7.9. *The map* $\pi \to \pi \circ q$ *gives a homeomorphism from* $\widehat{G/N}$ *and the closed subset of all* $\sigma \in \widehat{G}$ *which annihilate* N.

Note that if $1_G \prec \lambda_G$, then $\pi \prec \lambda_G$ for every representation π of G and hence $\widehat{G}_r = \widehat{G}$. In fact, since weak containment is preserved under forming tensor products,

$$\begin{aligned}\pi &\cong \pi \otimes 1_G \prec \pi \otimes \lambda_G = \pi \otimes \operatorname{ind}_{\{e\}}^G 1_{\{e\}} \\ &= \operatorname{ind}_{\{e\}}^G (\pi|_{\{e\}} \otimes 1_{\{e\}}) \sim \operatorname{ind}_{\{e\}}^G = \lambda_G.\end{aligned}$$

We shall see in Section 1.8 that $\widehat{G}_r$ equals $\widehat{G}$ if and only if G is amenable.

Then, for any representation π of G/N, $\xi, \eta \in \mathcal{H}(\pi)$ and $f \in L^1(G)$,

$$\begin{aligned}\langle \pi \circ q(f)\xi, \eta\rangle &= \int_G f(x)\langle \pi \circ q(x)\xi, \eta\rangle dx \\ &= \int_{G/N} \left(\int_N f(xn)\langle \pi(xN)\xi, \eta\rangle dn \right) d(xN) \\ &= \int_{G/N} T_N f(xN)\langle \pi(xN)\xi, \eta\rangle d(xN) \\ &= \langle \pi(T_N(f)\xi, \eta\rangle.\end{aligned}$$

This implies that $\|T_N f\|_{C^*(G/N)} \leqslant \|f\|_{C^*(G)}$ and hence $T_N : L^1(G) \to L^1(G/N)$ extends to a $*$-homomorphism, also denoted T_N, from $C^*(G)$ onto $C^*(G/N)$.

Readers are referred to [**60, 74, 158**] for more details.

1.8. Amenable locally compact groups

Amenability of a locally compact group can also be characterized through the existence of left invariant means on certain, much smaller, subspaces of $L^\infty(G)$ which have been introduced in Section 1.2.

DEFINITION 1.8.1. Let E be a linear subspace of $L^\infty(G)$ containing the constant functions. A *mean* on E is an element of E^* satisfying $\langle m, 1\rangle = \|m\| = 1$. Suppose that E is left (right) translation invariant. Then a mean on E is called *left invariant* (*right invariant*) if $m(L_x f) = m(f)$ ($m(R_x f) = m(f)$) for all $f \in E$ and $x \in G$.

DEFINITION 1.8.2. A locally compact group is *amenable* if there exists a left invariant mean on $L^\infty(G)$.

PROPOSITION 1.8.3. *For a locally compact group* G*, the following are equivalent.*

(i) G *is amenable.*
(ii) *There exists a left invariant mean on* $C^b(G)$.
(iii) *There exists a left invariant mean on* $LUC(G)$.
(iv) *There exists a left invariant mean on* $RUC(G)$.
(v) *There exists a left invariant mean on* $UC(G)$.

REMARK 1.8.4. Let G_d denote a locally compact group G with the discrete topology. If G_d is amenable, then so is G.

PROPOSITION 1.8.5. *For a locally compact group* G*, the following are equivalent.*

(i) G *is amenable.*

(ii) *There exists a right invariant mean on* $L^\infty(G)$.
(iii) *There exists a two-sided invariant mean on* $L^\infty(G)$.

PROPOSITION 1.8.6. *Amenability of a locally compact group* G *is equivalent to the so-called Folner's condition: Given any compact subset* C *of* G *and* $\epsilon > 0$, *there exists a compact subset* U *of* G *such that* $C \subseteq U$ *and*

$$|(xU\backslash U) \cup (U\backslash xU)| \leqslant \epsilon|U|, \qquad \text{for all } x \in C.$$

THEOREM 1.8.7 (Day's fixed point theorem). *For a locally compact group* G, *the following are equivalent.*

(i) G *is amenable.*
(ii) *If* G *acts affinely on a non-empty compact, convex subset* K *of a separated locally convex vector space* E, *i.e., the map* $G \times K \to K$, $(g, x) \to g \cdot x$ *satisfies*

$$g \cdot (tx + (1-t)y) = t(g \cdot x) + (1-t)g \cdot y, \; x, y \in E, t \in [0,1],$$

and such that $(g, x) \to g \cdot x$ *is separately continuous, then there exists some* $x \in E$ *such that* $g \cdot x = x$ *for all* $g \in G$.

The Markov-Kakutani fixed point theorem asserts that if G is commutative, then G has the fixed point properties (ii). In particular, G is amenable.

THEOREM 1.8.8. *Let* G *be a locally compact group and* N *a closed normal subgroup. If* N *and* G/N *are amenable, then so is* G.

THEOREM 1.8.9. *Let* G *be an amenable locally compact group. Then every closed subgroup of* G *is amenable.*

PROPOSITION 1.8.10. *Let* G *and* H *be locally compact groups and* $\varphi : G \to H$ *a continuous homomorphism with dense range. If* G *is amenable, then* H *is amenable as well. In particular, every quotient group of an amenable group is amenable.*

PROPOSITION 1.8.11. *Let* $(G_\alpha)_\alpha$ *be an upwards directed family of closed subgroups of* G *such that* $\bigcup_\alpha G_\alpha$ *is dense in* G *and each* G_α *is amenable. Then* G *is amenable.*

DEFINITION 1.8.12. Let $(g_\alpha)_\alpha$ be a net in $L^1(G)$ such that $g_\alpha \geqslant 0$ and $\|g_\alpha\|_1 = 1$ for each α. Then this net is *strongly convergent to left invariance* if $\|L_x(g_\alpha) - g_\alpha\|_1 \to 0$ for each $x \in G$.

EXAMPLE 1.8.13. (1) Let G be a compact group and let μ denote normalized Haar measure on G. Then $\langle m, f\rangle = \int_G f(x)d\mu(x)$, $f \in L^\infty(G) \subseteq L^1(G)$, defines an invariant mean on $L^\infty(G)$.

(2) The free group $\mathbb{F}_2$ on two generators a and b is not amenable. Towards a contradiction, assume that there exists a left invariant mean on $\ell^\infty(\mathbb{F}_2)$. Write every element of $\mathbb{F}_2$ as a reduced word in $\{a, b, a^{-1}, b^{-1}\}$, and for $x \in \{a, b, a^{-1}, b^{-1}\}$, let E_x denote the set of all elements of $\mathbb{F}_2$ beginning with x. Then

$$1 = m(1_G) = m(\delta_e) + m(1_{E_a}) + m(1_{E_b}) + m(1_{E_{a^{-1}}}) + m(1_{E_{b^{-1}}}).$$

On the other hand, since $\mathbb{F}_2$ is the disjoint union of the sets E_a and $a^{-1}E_a$ as well as E_b and $b^{-1}E_b$,

$$1 = m(1_{E_a}) + m(1_{a^{-1}E_a}) = m(1_{E_b}) + m(1_{b^{-1}E_b}).$$

Obviously, left invariance of m now leads to a contradiction.

(3) By (1), the orthogonal groups $SO(N)$ are all amenable. However, for $N \geqslant 3$, $SO(N)$ contains a subgroup isomorphic to $\mathbb{F}_2$. Thus $SO(N)_d$ is not amenable. This shows that the converse of Remark 1.8.4 does not hold.

It follows from Theorem 1.8.9 that any locally compact group, which contains $\mathbb{F}_2$ as a closed subgroup, cannot be amenable. This implies that all the groups $SL(n,\mathbb{R}), SL(n,\mathbb{C}), GL(n,\mathbb{R})$ and $GL(n,\mathbb{C})$ are not amenable. If G is a locally compact group, then radical of G (denoted by rad(G)) is the largest solvable connected normal subgroup in G (see [**107**, A52]). More generally, the following theorem holds:

THEOREM 1.8.14. *Let G be an almost connected locally compact group. Then G is amenable if and only if G modulo its radical, the maximal connected solvable normal subgroup of G, is compact.*

PROPOSITION 1.8.15. *A locally compact group G is amenable if and only if there exists a net $(g_\alpha)_\alpha$ in $L^1(G)$ such that $g_\alpha \geqslant 0$, $\|g_\alpha\|_1 = 1$ for each α and $(g_\alpha)_\alpha$ is strongly convergent to left invariance.*

Since abelian groups are amenable, it follows from Theorem 1.8.8 that solvable locally compact groups are amenable. In particular, every nilpotent locally compact group is amenable. A group G is *locally finite* if every finite subset of G generates a finite subgroup. Every locally finite group is amenable.

Suppose that G contains an abelian closed normal subgroup N such that G/N is compact. Then G is amenable (Theorem 1.8.8). For example, the classical Euclidean motion groups $\mathbb{R}^n \rtimes SO(n)$ are amenable.

PROPOSITION 1.8.16. *Suppose that E is a subspace of $L^\infty(G)$ which contains the constant functions and is closed under complex conjugation. Let m be a linear functional on E such that $\langle m, 1\rangle = 1$. Then m is a mean on E if and only if m is positive, i.e. $\langle f, m\rangle \geqslant 0$ for every $f \in E$, $f \geqslant 0$.*

THEOREM 1.8.17. *For a locally compact group, the following are equivalent.*

(i) *G is amenable.*
(ii) *$1_G \prec \overline{\pi} \otimes \pi$ for every unitary representation π of G.*
(iii) *$1_G \prec \overline{\pi} \otimes \pi$ for every irreducible unitary representation π of G.*

A locally compact group G is said to satisfy *Reiter's condition* (P_1) if given a compact subset K of G and $\epsilon > 0$, there exists $u \in L^1(G)$ with $u \geqslant 0$, $\|u\|_1 = 1$ and $\|L_x u - u\|_1 \leqslant \epsilon$ for all $x \in K$. We shall need in the next chapter that amenable groups share Reiter's condition (P_1). Actually, the two properties are equivalent (see [**59, 158, 231**] for details).

THEOREM 1.8.18. *Let G be a locally compact group. The following are equivalent:*

(i) *G is amenable.*
(ii) *G satisfies Reiter's condition (P_1).*
(iii) *The trivial representation of G is weakly contained in λ_G.*

PROPOSITION 1.8.19. *Let G be an amenable locally compact group and H a closed subgroup of G. Then, for any unitary representation π of G,*

$$\pi \prec \mathrm{ind}_H^G(\pi|_H).$$

PROPOSITION 1.8.20. *Suppose that* $\|\lambda_G(f)\| = \|f\|_1$ *for all nonnegative* $f \in L^1(G)$. *Then the trivial representation of* G *is weakly contained in* λ_G.

REMARK 1.8.21. Let $C^*_{\lambda_G}(G)$ or $C^*_\lambda(G)$ denote the C^* subalgebra of $\mathcal{B}(L^2(G))$ generated by $\{\lambda_G(f) : f \in L^1(G)\}$, and $VN(G)$ denote the von Neumann algebra generated by $\{\lambda_G(f) : f \in L^1(G)\}$ in $\mathcal{B}(L^2(G))$. When G is abelian, then

$$C^*_\lambda(G) = C_0(\widehat{G}) \quad \text{and} \quad VN(G) = L^\infty(\widehat{G}) = L^1(\widehat{G})^*.$$

Readers are referred to [**107, 231, 236, 249**] for more details.

CHAPTER 2

Basic Theory of Fourier and Fourier-Stieltjes Algebras

In this chapter the Fourier and Fourier-Stieltjes algebras, $A(G)$ and $B(G)$, associated to a locally compact group G, are introduced and studied almost to the extent of Eymard's fundamental paper [**73**]. In particular, $B(G)$ is identified as the Banach space dual of the group C^*-algebra $C^*(G)$ and a fair number of basic functorial properties are presented. Similarly, for the Fourier algebra $A(G)$, the elements of which are shown to be precisely the convolution products of L^2-functions on G.

Given a commutative Banach algebra A, immediate problems arising are to determine the spectrum (or Gelfand space) of A and to check whether the range of the Gelfand transform is a regular function algebra. As we show in Section 2.3, the spectrum $\sigma(A(G))$ turns out to be homeomorphic to G and the Gelfand homomorphism is then nothing but the identity mapping. Moreover, $A(G)$ is regular.

We next identify, following Eymard's seminal paper [**73**], the Banach space dual of $A(G)$ as the von Neumann subalgebra $VN(G)$ of $\mathcal{B}(L^2(G))$ generated by the left regular representation of G. The fact that $A(G)$ is the predual of a von Neumann algebra will prove to be of great importance. For instance, it allows us to equip $A(G)$ with a natural operator space structure and employing the theory of operator spaces has led to significant progress, as will be shown in Chapters 4 and 6.

In Section 2.5 the very important notion of support of an operator in $VN(G)$ is introduced and several properties of these supports, which are extremely useful later on, are shown. An immediate consequence of one of the results about the support is that singletons in G are sets of synthesis for $A(G)$.

Let H be a closed subgroup of the locally compact group G. A challenging problem is whether functions in $A(H)$ and $B(H)$ extend to functions in $A(G)$ and $B(G)$, respectively. For the Fourier algebras there is a very satisfactory solution to the effect that every function in $A(H)$ extends to a function in $A(G)$ of the same norm (Section 2.6). For Fourier-Stieltjes algebras, however, the problem is considerably more difficult and its investigation will cover a major portion of Chapter 7.

If A is a nonunital Banach algebra, then often the existence of a bounded approximate identity in A proves useful. In Section 2.7 we present Leptin's theorem [**191**] saying that $A(G)$ has a bounded approximate identity precisely when the group G is amenable. The proof uses several different characterizations of amenability of a locally compact group.

The notion of Fourier algebra has been generalized by Arsac [**5**]. He associated to any unitary representation π of G a closed subspace $A_\pi(G)$ of $B(G)$ and studied these spaces extensively. When π is the left regular representation of G, then

$A_\pi(G)$ equals $A(G)$. We present in Section 2.8 those results from [**5**] which either will be needed in Chapter 3 or used in Section 2.9 to show that for certain examples of locally compact groups the Fourier-Stieltjes algebra $B(G)$ decomposes into the direct sum of $A(G)$ and $B(G/N)$ for a large normal subgroup N of G.

2.1. The Fourier-Stieltjes algebra $B(G)$

Let G be a locally compact group. In this section we introduce the Fourier-Stieltjes algebra $B(G)$ and its subspaces $B_{\mathcal{S}}(G)$, where $\mathcal{S}$ is a collection of (equivalence classes of) unitary representations of G, and prove a number of basic results on these spaces.

Let Σ denote the equivalence classes of continuous unitary representations of G. For $\mathcal{S} \subseteq \Sigma$ and $\mu \in M(G)$, let $\|\mu\|_{\mathcal{S}} = \sup\{\|\pi(\mu)\| : \pi \in \mathcal{S}\}$. Then the assignment $\mu \to \|\mu\|_{\mathcal{S}}$ is a semi-norm on $M(G)$, and for $\mu, \nu \in M(G)$, $f \in L^1(G)$ and $x, y \in G$, we have

(1) $\|\mu\|_{\mathcal{S}} \leqslant \|\mu\|$ and $\|\mu * \nu\|_{\mathcal{S}} \leqslant \|\mu\|_{\mathcal{S}}\|\nu\|_{\mathcal{S}}$
(2) $\|\mu^*\|_{\mathcal{S}} = \|\mu\|_{\mathcal{S}}$ and $\|\mu * \mu^*\|_{\mathcal{S}} = \|\mu^*\|_{\mathcal{S}}^2$
(3) $\|L_x f\|_{\mathcal{S}} = \|f\|_{\mathcal{S}}$ and $\|R_y f\|_{\mathcal{S}} = \Delta(y^{-1})\|f\|_{\mathcal{S}}$.

Let $N_{\mathcal{S}} = \{f \in L^1(G) : \pi(f) = 0 \text{ for all } \pi \in \mathcal{S}\}$. Then $N_{\mathcal{S}}$ is a closed $*$-ideal of $L^1(G)$, and if $\dot{f} = f + N_{\mathcal{S}} \in L^1(G)/N_{\mathcal{S}}$, then $\|\dot{f}\|_{\mathcal{S}} = \|f\|_{\mathcal{S}}$ defines a norm on $L^1(G)/N_{\mathcal{S}}$. It is clear that $L^1(G)/N_{\mathcal{S}}$ becomes a normed $*$-algebra, and the norm satisfies $\|\dot{f} * \dot{f}^*\|_{\mathcal{S}} = \|\dot{f}\|_{\mathcal{S}}^2$. Let $C^*_{\mathcal{S}}(G)$ denote the completion of $L^1(G)/N_{\mathcal{S}}$. The group G acts on $L^1(G)/N_{\mathcal{S}}$, and we have

$$\|L_x\dot{f}\|_{\mathcal{S}} = \|\dot{f}\|_{\mathcal{S}} = \Delta(y)\|R_y\dot{f}\|_{\mathcal{S}}$$

for all $x, y \in G$. These actions extend uniquely to the C^*-algebra $C^*_{\mathcal{S}}(G)$.

LEMMA 2.1.1. *The mapping $f \to \dot{f} = f + k(\mathcal{S})$ from $L^1(G)$ onto $(L^1(G) + k(\mathcal{S}))/k(\mathcal{S})$ extends to a homomorphism from $C^*(G)$ onto $C^*_{\mathcal{S}}(G)$ with kernel $k(\mathcal{S})$, where $k(\mathcal{S})$ denotes for all $g \in \mathcal{S}$ such that $\pi(g) = 0$.*

PROOF. Since $\|\dot{f}\|_{\mathcal{S}} = \|f\|_{\mathcal{S}} \leqslant \|f\|_{C^*}$ for $f \in L^1(G)$, the map $f \to \dot{f}$ extends uniquely to a $*$-homomorphism $\phi : C^*(G) \to C^*_{\mathcal{S}}(G)$, and $\|\phi\| \leqslant 1$. Since $\phi(C^*(G)) \supseteq \phi(L^1(G))$ and a homomorphism between C^*-algebras with dense range is surjective, $\phi(C^*(G)) = C^*_{\mathcal{S}}(G)$. It remains to show that $\ker\phi = k(\mathcal{S})$.

Let $g \in C^*(G)$ such that $\phi(g) = 0$. Then there exist $f_n \in L^1(G)$, $n \in \mathbb{N}$, such that

$$\sup_{\pi\in\mathcal{S}} \|\pi(g) - \pi(f_n)\| \leqslant \|g - f_n\|_{C^*} \to 0$$

and $\|\dot{f}_n\|_{\mathcal{S}} = \|f_n\|_{\mathcal{S}} \to 0$. This implies that $\sup_{\pi\in\mathcal{S}}\|\pi(g)\| = 0$, i.e. $g \in k(\mathcal{S})$. Conversely, let $g \in k(\mathcal{S})$ and let $f_n \in L^1(G)$ such that $\|g - f_n\|_{C^*} \to 0$. Then $\lim_{n\to\infty}\|f_n\|_{\mathcal{S}} = 0$, and hence $\phi(g) = \lim_{n\to\infty}\phi(f_n) = 0$. □

Let G be a locally compact abelian group with dual group $\widehat{G}$ and let $\mathcal{S} \subseteq \widehat{G}$. Then, for $f \in L^1(G)$, $f \in k(\mathcal{S})$ if and only if $\widehat{f}$ vanishes on $\overline{\mathcal{S}} \subseteq \widehat{G}$. Thus $N_{\mathcal{S}} = \{0\}$ if and only if $\mathcal{S}$ is dense in $\widehat{G}$, and in this case $\|f\|_{\mathcal{S}} = \|\widehat{f}\|_\infty$, and the Fourier transform is an isometric isomorphism between $C^*_{\mathcal{S}}(G) = C^*(G)$ and $C_0(\widehat{G})$. If $\mathcal{S}$ is not dense in $\widehat{G}$, then $L^1(G)/N_{\mathcal{S}}$ is isometrically isomorphic to the subalgebra of $C_0(\overline{\mathcal{S}})$ consisting of all $\widehat{f}|_{\mathcal{S}}$, $f \in L^1(G)$, equipped with the uniform norm, and $C^*_{\mathcal{S}}(G)$ identifies with $C_0(\overline{\mathcal{S}})$.

LEMMA 2.1.2. *For $\mathcal{S} \subseteq \Sigma$ and $u \in P(G)$, the following are equivalent.*

(i) $\pi_u \prec \mathcal{S}$.

(ii) *There exists a positive linear functional φ on $C^*_{\mathcal{S}}(G)$ such that, for each $f \in L^1(G)$,*

$$\varphi(f + k(\mathcal{S})) = \int_G f(x)u(x)dx.$$

PROOF. Let $u(x) = \langle \pi_u(x)\xi, \xi\rangle$, where ξ is a cyclic vector for π_u. Then $\pi_u \prec \mathcal{S}$ is equivalent to $\langle \pi_u(g)\xi, \xi\rangle = 0$ for all $g \in k(\mathcal{S})$. In fact, the sufficiency of this latter condition is immediate from the facts that $k(\mathcal{S})$ is a two-sided ideal of $C^*(G)$ and that ξ is a cyclic vector. Now, by Lemma 2.1.1, the positive linear functionals on $C^*_{\mathcal{S}}(G)$ are exactly the positive linear functionals on $C^*(G)$ which are zero on $k(\mathcal{S})$. It follows that (i) and (ii) are equivalent. □

From now on, $P_{\mathcal{S}}(G)$ will denote the set of all $u \in P(G)$ which satisfy any of the equivalent conditions in Lemma 2.1.2.

LEMMA 2.1.3. *For collections $\mathcal{S}$ and $\mathcal{T}$ of representations of G, the following conditions are equivalent.*

(i) $\mathcal{S} \prec \mathcal{T}$.

(ii) *For every $\mu \in M(G)$, $\|\mu\|_{\mathcal{S}} \leqslant \|\mu\|_{\mathcal{T}}$.*

(iii) *For every $f \in L^1(G)$, $\|f\|_{\mathcal{S}} \leqslant \|f\|_{\mathcal{T}}$.*

PROOF. First, assume (iii) and let $u \in P_{\mathcal{S}}(G)$. Then, for any $f \in L^1(G)$,

$$\left|\int_G f(x)u(x)dx\right| \leqslant \|f\|_{\mathcal{S}}u(e) \leqslant \|f\|_{\mathcal{T}}u(e).$$

Thus u defines a positive linear functional on $L^1G)/N_{\mathcal{T}}$, which extends to a positive linear functional on $C^*_{\mathcal{T}}(G) = C^*(G)/k(\mathcal{T})$. This implies that $\pi_u \prec \mathcal{T}$. Since this holds for every $u \in P_{\mathcal{S}}(G)$, (i) follows.

Now suppose that (i) holds. Since $k(\mathcal{T}) \subseteq k(\mathcal{S})$, we get

$$\|f\|_{\mathcal{S}} = \inf\{\|f+g\|_{C^*(G)} : g \in k(\mathcal{S})\} \leqslant \inf\{\|f+g\|_{C^*(G)} : g \in k(\mathcal{T})\} = \|f\|_{\mathcal{T}}$$

for every $f \in L^1(G)$. From this inequality we are going to deduce $\|\mu\|_{\mathcal{S}} \leqslant \|\mu\|_{\mathcal{T}}$ for $\mu \in M(G)$. Let $\mathcal{V}$ be a neighbourhood basis of the identity, and for each $V \in \mathcal{V}$, choose a nonnegative continuous function g_V with $\operatorname{supp} g_V \subseteq V$, $\|g_V\|_1 = 1$ and let $f_V = \mu * g_V$. Then the bounded net $(f_V)_V$ in $L^1(G)$ converges to μ in the weak topology $\sigma(M(G), C^b(G))$ of G. Let π be an arbitrary unitary representation and $\xi, \eta \in \mathcal{H}(\pi)$. Then

$$\begin{aligned}\langle \pi(\mu)\xi, \eta\rangle &= \int_G \langle \pi(x)\xi, \eta\rangle d\mu(x) = \lim_V \int_G \langle \pi(x)\xi, \eta\rangle f_V(x)dx \\ &= \lim_V \langle \pi(f_V)\xi, \eta\rangle,\end{aligned}$$

and hence the net of operators $\pi(f_V)$ converges to $\pi(\mu)$ in the weak operator topology of $\mathcal{B}(\mathcal{H}(\pi))$ of G. Since $\|\pi(f_V)\| \leqslant \|\pi(\mu)\| \cdot \|\pi(g_V)\| \leqslant \|\pi(\mu)\|$ and the ball of radius $\|\pi(\mu)\|$ in $\mathcal{B}(\mathcal{H}(\pi))$ is weakly closed and the weak topology agrees with the ultraweak topology, it follows that $\|\pi(\mu)\| = \sup_{V\in\mathcal{V}} \|\pi(f_V)\|$. Since $\mathcal{S} \prec \mathcal{T}$, for

$\pi \in \mathcal{S}$ we get

$$\begin{aligned} \|\pi(\mu)\| &= \sup_{V \in \mathcal{V}} \|\pi(f_V)\| \leqslant \sup_{V \in \mathcal{V}} \|\pi(f_V)\|_{\mathcal{S}} \\ &\leqslant \sup_{V \in \mathcal{V}} \|\pi(f_V)\|_{\mathcal{T}} = \sup_{\sigma \in \mathcal{T}} \left(\sup_{V \in \mathcal{V}} \|\sigma(f_V)\| \right) = \|\mu\|_{\mathcal{T}}. \end{aligned}$$

Since $\pi \in \mathcal{S}$ was arbitrary, we conclude that $\|\mu\|_{\mathcal{S}} \leqslant \|\mu\|_{\mathcal{T}}$. This completes the proof. □

LEMMA 2.1.4. *For a function u on G and a collection $\mathcal{S}$ of unitary representations of G the following assertions are equivalent.*

(i) *u is a finite linear combination of functions in $P_{\mathcal{S}}(G)$.*

(ii) *There exist a unitary representation π of G, which is weakly contained in the $\mathcal{S}$, and vectors $\xi, \eta \in \mathcal{H}(\pi)$ such that $u(x) = \langle \pi(x)\xi, \eta\rangle$ for all $x \in G$.*

(iii) *u is a bounded continuous function and*

$$\sup_{f \in L^1(G), \|f\|_{\mathcal{S}} \leqslant 1} \left| \int_G f(x)u(x)dx \right| < \infty.$$

PROOF. (i) $\Rightarrow$ (ii) Since every $u \in P_{\mathcal{S}}(G)$ can be represented as in (ii), it suffices to observe that the functions in (ii) form a linear space. So let π_1 and π_2 be representations that are both weakly contained in $\mathcal{S}$ and let $\xi_j, \eta_j \in \mathcal{H}(\pi_j)$, $j = 1, 2$. Then $\pi_1 \oplus \pi_2$ is weakly contained in $\mathcal{S}$ and

$$\langle (\pi_1 \oplus \pi_2)(\xi_1 \oplus \xi_2), \eta_1 \oplus \eta_2 \rangle = \sum_{j=1}^{2} \langle \pi_j(x)\xi_j, \eta_j \rangle.$$

(ii) $\Rightarrow$ (iii) Since $\pi \prec \mathcal{S}$, for each $f \in L^1(G)$,

$$\left| \int_G f(x)u(x)dx \right| = |\langle \pi(f)\xi, \eta\rangle| \leqslant \|\pi(f)\| \cdot \|\xi\| \cdot \|\eta\| \leqslant \|f\|_{\mathcal{S}} \|\xi\| \cdot \|\eta\|.$$

(iii) $\Rightarrow$ (i) Condition (iii) implies that u defines a bounded linear functional on $C^*_{\mathcal{S}}(G)$, which then can be written as a linear combination of positive functionals on $C^*_{\mathcal{S}}(G)$. By Lemma 2.1.2(iii), each of the latter functionals is given by a function in $P_{\mathcal{S}}(G)$. □

DEFINITION 2.1.5. Let $B_{\mathcal{S}}(G)$ denote the set of functions satisfying the equivalent conditions of Lemma 2.1.4. Of course $B_{\mathcal{S}}(G)$ is translation invariant. We simply write $B(G)$ for $B_{\Sigma(G)}(G)$. Thus $B(G)$ consists of all finite linear combinations of continuous positive definite functions and hence equals the collection of all coefficient functions $\langle \pi(\cdot)\xi, \eta\rangle$, where $\pi \in \Sigma(G)$ and $\xi, \eta \in \mathcal{H}(\pi)$.

It follows from Lemma 2.1.4 that $B(G)$ identifies with the Banach space dual of $C^*(G)$ through the pairing

$$\langle f, u\rangle = \int_G f(x)u(x)\, dx, \quad f \in L^1(G), u \in B(G). \tag{2.1}$$

The norm on $B(G)$ is then given by

$$\|u\| = \sup \left\{ \left| \int_G f(x)u(x)dx \right| : f \in L^1(G), \|f\|_{C^*} \leqslant 1 \right\}. \tag{2.2}$$

Note that if $u(\cdot) = \langle \pi(\cdot)\xi, \eta\rangle$, then $\langle g, u\rangle = \langle \pi(g)\xi, \eta\rangle$ for all $g \in C^*(G)$.

REMARK 2.1.6. (1) For any $\mathcal{S} \subseteq \Sigma(G)$, $C^*_{\mathcal{S}}(G)$ is isometrically isomorphic to a quotient of $C^*(G)$ and hence $B_{\mathcal{S}}(G) = C^*_{\mathcal{S}}(G)^*$ is a closed subspace of $B(G)$. Thus the norms of u, considered as an element of $C^*_{\mathcal{S}}(G)$ and of $C^*(G)$, respectively, are equal. In particular, if $u \in B_\lambda(G)$, then

$$\|u\| = \sup\left\{\left|\int_G f(x)u(x)dx\right| : f \in C_c(G), \|\lambda_G(f)\| \leqslant 1\right\}.$$

(2) If $u \in B_{\mathcal{S}}(G)$, then u admits Jordan decompositions

$$u = u^+ - u^-,\ u^+, u^- \in P(G),\ \|u\| = u^+(e) + u^-(e) \tag{2.3}$$

as an element of $B(G)$, and

$$u = u^+_{\mathcal{S}} - u^-_{\mathcal{S}},\ u^+_{\mathcal{S}}, u^-_{\mathcal{S}} \in P_{\mathcal{S}}(G),\ \|u\| = u^+_{\mathcal{S}}(e) + u^-_{\mathcal{S}}(e)$$

as an element of $B_{\mathcal{S}}(G)$. It follows from the uniqueness of the decomposition (2.3) that $u^+_{\mathcal{S}} = u^+$ and $u^-_{\mathcal{S}} = u^-$. Similarly, for $u \in B_{\mathcal{S}}(G) = C^*_{\mathcal{S}}(G)^*$, the absolute value $|u|$ does not depend on S. In fact, this follows from the uniqueness of the polar decomposition. This in particular shows that if $u \in B_{\mathcal{S}}(G)$, then $|u| \in P_{\mathcal{S}}(G)$ and if $u = \tilde{u}$, then also u^+ and u^- belong to $P_{\mathcal{S}}(G)$.

(3) Let ω denote the universal representation of G (see [**270**, page 122]). Then $C^*(G)^{**}$ is the von Neumann subalgebra of $\mathcal{B}(\mathcal{H}(\omega))$ generated by either the operators $\omega(f)$, $f \in L^1(G)$, or the operators $\omega(x)$, $x \in G$, since the sets $\omega(L^1(G))$ and $\omega(G)$ have the same commutant in $\mathcal{B}(\mathcal{H}(\omega))$. Let $\mu \in M(G)$ and let $\mathcal{V}$ be a neighbourhood basis of e. For each $V \in \mathcal{V}$, let g_V and $f_V = \mu * g_V$ be as in the proof of Lemma 2.1.3. Then $\omega(\mu)$ is the ultraweak limit of the net $(\omega(f_V))_V$ and hence $\omega(\mu) \in C^*(G)^{**}$. Since $B(G) = C^*(G)^*$ and

$$\langle \omega(f_V), u\rangle = \int_G f_V(x)u(x)dx,$$

for all V, passing to the limit, by duality of $B(G)$ and $C^*(G)^{**}$, we conclude that

$$\langle \omega(\mu), u\rangle = \int_G u(x)d\mu(x), \tag{2.4}$$

and in particular

$$\langle \omega(x), u\rangle = u(x) \tag{2.5}$$

for $u \in B(G)$ and $x \in G$.

(4) Let $u \in B(G)$ and let π be a representation of G and $\xi, \eta \in \mathcal{H}(\pi)$ such that $u(x) = \langle \pi(x)\xi, \eta\rangle$ for all $x \in G$. Then, for any $g \in C^*(G)$,

$$\langle \omega(g)\xi, \eta\rangle = \langle \pi(g)\xi, \eta\rangle = \langle \pi''(\omega(g))\xi, \eta\rangle.$$

Again, passing to the ultraweak limit, we get

$$\langle T, u\rangle = \langle \pi''(T)\xi, \eta\rangle \tag{2.6}$$

for all $T \in C^*(G)^{**}$.

LEMMA 2.1.7. *Let $u = \tilde{u} \in B(G)$. Then the functions u^+, u^- and $|u|$ are uniform limits on G of finite linear combinations of right translates of u.*

PROOF. Let $u = V|u|$ be the polar decomposition of u. Then, since $u^+ = \frac{1}{2}(|u| + u)$ and $u^- = \frac{1}{2}(|u| - u)$, it suffices to prove the assertion for $|u|$. Let A denote the $*$-subalgebra of $C^*(G)^{**}$ generated by the operators $\omega(x)$, $x \in G$. Since V^* is a partial isometry, V^* is contained in the unit ball of $C^*(G)^{**}$. Then, by

Kaplansky's density theorem, V^* is the limit in the strong operator topology of a net $(S_\alpha)_\alpha$ in A such that $\|S_\alpha\| \leqslant 1$ for all α. Each of the functions

$$x \to S_\alpha u(x) = \langle \omega(x), S_\alpha u \rangle = \langle \omega(x) S_\alpha, u \rangle$$

is a linear combination of right translates of u. Now, using that $u = \widetilde{u}$, formula (2.5) and the estimate $|\langle T^*, u\rangle|^2 \leqslant \|u\| \langle T^*T, |u|\rangle$ for every $T \in C^*(G)^{**}$, we obtain for $x \in G$

$$\begin{aligned} ||u|(x) - S_\alpha u(x)|^2 &= |\langle \omega(x), |u|\rangle - \langle \omega(x) S_\alpha, u\rangle|^2 \\ &= |\langle \omega(x), V^* u\rangle - \langle \omega(x) S_\alpha, u\rangle|^2 \\ &= |\langle \omega(x)(V^* - S_\alpha), u\rangle|^2 \\ &= |\langle (V^* - S_\alpha)^* \omega(x^{-1}), u\rangle|^2 \\ &\leqslant \|u\| \langle (V^* - S_\alpha)^* \omega(x^{-1}) \omega(x) (V^* - S_\alpha), |u|\rangle \\ &= \|u\| \langle (V^* - S_\alpha)^* (V^* - S_\alpha), |u|\rangle. \end{aligned}$$

Now $S_\alpha \to V^*$ in the strong operator topology, and therefore the bounded net $[(V^* - S_\alpha)^*(V^* - S_\alpha)]_\alpha$ in $C^*(G)^{**}$ converges in the weak operator topology and hence ultraweakly to zero. Since the weak operator topology on $C^*(G)^{**}$ coincides with the topology $\sigma(C^*(G)^{**}, B(G))$ on bounded sets, the above estimate implies that the net $(S_\alpha u)_\alpha$ converges uniformly on G to $|u|$. □

The next two lemmas provide additional expressions for the norms of elements in $B_{\mathcal{S}}(G)$.

LEMMA 2.1.8. *Let $\mathcal{S} \subseteq \Sigma(G)$ and $u \in B_{\mathcal{S}}(G)$. Then*

$$\|u\| = \sup \left\{ \left| \sum_{j=1}^n c_j u(x_j) \right| : x_i \in G, c_i \in \mathbb{C}, 1 \leqslant i \leqslant n, n \in \mathbb{N}, \left\| \sum_{j=1}^n c_j \delta_{x_j} \right\|_{\mathcal{S}} \leqslant 1 \right\}.$$

PROOF. Suppose first that $\mathcal{S} = \Sigma(G)$ and let A denote the C^*-subalgebra of $C^*(G)^{**}$ generated by $\omega(G)$. Then, by Kaplansky's density theorem, the unit ball of A is ultraweakly dense in the unit ball of $C^*(G)^{**}$. By (2.5) this implies

$$\begin{aligned} \|u\| &= \sup \{ |\langle T, u\rangle| : T \in C^*(G)^{**}, \|T\| \leqslant 1\} \\ &= \sup \{ |\langle S, u\rangle| : S \in A, \|S\| \leqslant 1\} \\ &= \sup \left\{ \left| \sum_{j=1}^n c_j u(x_j) \right| \right\}, \end{aligned}$$

where the supremum extends over all $x_1, \ldots, x_n \in G, c_1, \ldots, c_n \in \mathbb{C}$ such that $\left\| \sum_{j=1}^n c_j \delta_j \right\|_{\mathcal{S}} \leqslant 1$.

Now let $\mathcal{S}$ be arbitrary and consider G_d, the group G made discrete. Since $P_{\mathcal{S}}(G) \subseteq P_{\mathcal{S}}(G_d)$, $u \in B_{\mathcal{S}}(G_d)$ and Remark 2.1.6(1), applied to G_d, shows that

$$\sup \left\{ \left| \sum_{j=1}^n c_j u(x_j) \right| : \left\| \sum_{j=1}^n c_j \delta_j \right\|_{\mathcal{S}} \leqslant 1 \right\} = \sup \left\{ \left| \sum_{j=1}^n c_j u(x_j) \right| : \left\| \sum_{j=1}^n c_j \delta_j \right\|_{\Sigma(G)} \leqslant 1 \right\}.$$

The statement of the lemma now follows from the first part of the proof. □

LEMMA 2.1.9. *Let $u \in B(G)$ and $u(x) = \langle \pi(x)\xi, \eta\rangle$, $x \in G$. Then $\|u\| \leqslant \|\xi\| \cdot \|\eta\|$. Conversely, if $u \in B_{\mathcal{S}}(G)$ then there exist a representation π which is weakly contained in $\mathcal{S}$ and $\xi, \eta \in \mathcal{H}(\pi)$ such that*

$$u(x) = \langle \pi(x)\xi, \eta\rangle,\ x \in G, \quad \text{and} \quad \|u\| = \|\xi\| \cdot \|\eta\|.$$

More precisely, if $u = V|u|$, $V \in C^*(G)^{**}$, denotes the polar decomposition of u, it suffices to take π and η such that $|u|(x) = \langle \pi(x)\eta, \eta\rangle$, where η is a cyclic vector in $\mathcal{H}(\pi)$, and then put $\xi = \pi''(V)\eta$.

PROOF. The first statement follows from

$$\begin{aligned}\|u\| &= \sup\left\{\left|\int_G f(x)u(x)dx\right| : f \in L^1(G), \|f\|_{C^*} \leqslant 1\right\}\\ &= \sup\left\{|\langle \pi(f)\xi, \eta\rangle| : f \in L^1(G), \|f\|_{C^*} \leqslant 1\right\}\\ &\leqslant \|\xi\| \cdot \|\eta\|.\end{aligned}$$

Now choose π, η and ξ as announced. Then, for every $x \in G$,

$$\begin{aligned}u(x) &= \langle \omega(x), u\rangle = \langle \omega(x), V|u|\rangle = \langle \omega(x)V, |u|\rangle\\ &= \langle \pi''(\omega(x)V)\eta, \eta\rangle = \langle \pi(x)\pi''(V)\eta, \eta\rangle\\ &= \langle \pi(x)\xi, \eta\rangle.\end{aligned}$$

Since $|u| \in P_{\mathcal{S}}(G)$, π is weakly contained in $\mathcal{S}$ (compare Lemma 2.1.2). So it only remains to show that $\|u\| \geqslant \|\xi\| \cdot \|\eta\|$. Now, since V is a partial isometry, $\|V\| = 1$ and hence $\|\pi''(V)\| \leqslant 1$ and $\|\xi\| = \|\pi''(V)\eta\| \leqslant \|\eta\|$. Consequently,

$$\|u\| = \||u|\| = |u|(e) = \|\eta\|^2 \geqslant \|\xi\| \cdot \|\eta\|,$$

as required. □

REMARK 2.1.10. (1) For each $u \in B(G)$, we have $\|u\|_\infty \leqslant \|u\|$. In fact, since $\|f\|_{C^*} \leqslant \|f\|_1$ for every $f \in L^1(G)$, (2.2) implies

$$\begin{aligned}\|u\| &= \sup\left\{\left|\int_G f(x)u(x)dx\right| : f \in L^1(G), \|f\|_{C^*} \leqslant 1\right\}\\ &\geqslant \sup\left\{\left|\int_G f(x)u(x)dx\right| : f \in L^1(G), \|f\|_1 \leqslant 1\right\} = \|u\|_\infty.\end{aligned}$$

(2) Let $\mu \in M(G)$. Then, for any $\mathcal{S} \subseteq \sum(G)$,

$$\|\mu\|_{\mathcal{S}} = \sup\left\{\left|\int_G u(x)d\mu(x)\right| : u \in B_{\mathcal{S}}(G), \|u\| \leqslant 1\right\}.$$

To see this, let ϕ denote the continuous linear functional on $B_{\mathcal{S}}(G)$ defined by $\langle \phi, u\rangle = \int_G u(x)d\mu(x)$ and let $u(x) = \langle \pi(x)\xi, \eta\rangle$, where π is weakly contained in $\mathcal{S}$ and $\xi, \eta \in \mathcal{H}(\pi)$ satisfy $\|u\| = \|\xi\| \cdot \|\eta\|$ (Lemma 2.1.9). Then

$$|\langle \phi, u\rangle| = |\langle \pi(\mu)\xi, \eta\rangle| \leqslant \|\pi(\mu)\| \cdot \|\xi\| \cdot \|\eta\| \leqslant \|\mu\|_S\|u\|$$

and hence $\|\mu\|_S \geqslant \|\phi\|$. Conversely, consider any $\pi \in S$ and $\xi, \eta \in \mathcal{H}(\pi)$ with $\|\xi\| \leqslant 1$ and $\|\eta\| \leqslant 1$ and let $v(x) = \langle \pi(x)\xi, \eta\rangle$. Then $v \in B_{\mathcal{S}}(G)$ and

$$|\langle \pi(\mu)\xi, \eta\rangle| = \left|\int_G v(x)d\mu(x)\right| \leqslant \|\phi\| \cdot \|v\| \leqslant \|\phi\| \cdot \|\xi\| \cdot \|\ \eta\| \leqslant \|\phi\|.$$

Since $\|\mu\|_S$ is the supremum of all such values $|\langle \pi(\mu)\xi, \eta\rangle|$, it follows that $\|\mu\|_S \leqslant \|\phi\|$.

(3) Let $u \in B(G)$, (respectively, $u \in B_\lambda(G)$). Then the functions $\widetilde{u}$, $\overline{u}$ and $\check{u}$ all belong to $B(G)$ (respectively, $B_\lambda(G)$) and

$$\|u\| = \|\widetilde{u}\| = \|\overline{u}\| = \|\check{u}\|.$$

The first statement follows from the fact that it holds for $P(G)$ (respectively, $P_\lambda(G)$). For the equality of the norms, let $u(x) = \langle \pi(x)\xi, \eta\rangle$ with $\|u\| = \|\xi\| \cdot \|\eta\|$. Then

$$\widetilde{u}(x) = \langle \pi(x)\eta, \xi\rangle \quad \text{and} \quad \overline{u}(x) = \langle \eta, \pi(x)\xi\rangle$$

and therefore $\|\widetilde{u}\| \leqslant \|\eta\| \cdot \|\xi\| = \|u\|$ and $\|\overline{u}\| \leqslant \|u\|$. Since $(\widetilde{u})^\sim = u$ and $(\overline{u})^- = u$, the reverse norm inequalities follow. Finally, as $\check{u} = (\overline{u})^\sim$, all four norms have to be equal.

THEOREM 2.1.11. *Let G be a locally compact group. Then $B(G)$, equipped with pointwise multiplication and the norm*

$$\|u\| = \sup\left\{\left|\int_G f(x)u(x)dx\right| : f \in L^1(G), \|f\|_{C^*} \leqslant 1\right\},$$

is a unital commutative Banach algebra, called the Fourier-Stieltjes algebra of G, and $B(G)$ contains $B_\lambda(G)$ as a closed ideal.

PROOF. Let u and v be elements of $B(G)$. By Lemma 2.1.9 u and v admit representations

$$u(x) = \langle \pi(x)\xi, \eta\rangle \quad \text{and} \quad v(x) = \langle \pi'(x)\xi', \eta'\rangle,$$

where $\|u\| = \|\xi\| \cdot \|\eta\|$ and $\|v\| = \|\xi'\| \cdot \|\eta'\|$. Thus

$$u(x)v(x) = \langle (\pi \otimes \pi')(x)(\xi \otimes \xi'), \eta \otimes \eta'\rangle$$

and hence $uv \in B(G)$. Also,

$$\|uv\| \leqslant \|\xi \otimes \xi'\| \cdot \|\eta \otimes \eta'\| = \|\xi\| \cdot \|\eta\| \cdot \|\xi'\| \cdot \|\eta'\|.$$

So the norm $\|\cdot\|$ on $B(G)$ is submultiplicative.

Finally, $B_\lambda(G)$ is a closed linear subspace of $B(G)$. To see that $B_\lambda(G)$ is an ideal in $B(G)$, it suffices to verify that $P(G)P_\lambda(G) \subseteq P_\lambda(G)$ because $B(G)$ and $B_\lambda(G)$ are spanned by $P(G)$ and $P_\lambda(G)$, respectively. Now, $P(G)P_\lambda(G) \subseteq P_\lambda(G)$ follows from the fact that for any representation π of G, $\pi \otimes \lambda \sim \lambda$ (see Section 1.6). □

PROPOSITION 2.1.12. *Let $\mathcal{T}$ and $\mathcal{S}$ be collections of unitary representations of G such that $P_\mathcal{T}(G)P_\mathcal{S}(G) \subseteq P_\mathcal{S}(G)$. If $u \in B_\mathcal{T}(G)$ and $\mu \in M(G)$, then $u\mu \in B_\mathcal{S}(G)$ and*

$$\|u\mu\|_\mathcal{T} \leqslant \|u\| \cdot \|\mu\|_\mathcal{S}.$$

In particular, $\|u\mu\|_{\Sigma(G)} \leqslant \|u\| \cdot \|\mu\|_{\Sigma(G)}$ for any $u \in B(G)$, and if $u \in B_\lambda(G)$, then $\|u\mu\|_{\Sigma(G)} \leqslant \|u\| \cdot \|\lambda(\mu)\|$.

PROOF. The hypothesis that $P_\mathcal{T}(G)P_\mathcal{S}(G) \subseteq P_\mathcal{S}(G)$ implies $B_\mathcal{T}(G)B_\mathcal{S}(G) \subseteq B_\mathcal{S}(G)$. Then, by Remark 2.1.10(2),

$$\begin{aligned}\|u\mu\|_\mathcal{T} &= \sup\left\{\left|\int_G u(x)v(x)d\mu(x)\right| : v \in B_\mathcal{T}(G), \|v\| \leqslant 1\right\} \\ &\leqslant \sup\{\|uv\| \cdot \|\mu\|_S : v \in B_\mathcal{T}(G), \|v\| \leqslant 1\} \\ &\leqslant \|u\| \cdot \|\mu\|_\mathcal{S}.\end{aligned}$$

The remaining statements follow since $B_\lambda(G)$ is an ideal in $B(G)$. □

LEMMA 2.1.13. *Let $\mathcal{S}$ be a collection of equivalence classes of unitary representations of G and let $u \in B_{\mathcal{S}}(G)$.*

(i) *If $\mu \in M(G)$, then $\mu * u \in B_{\mathcal{S}}(G)$ and*

$$\|\mu * u\| \leqslant \|u\| \cdot \sup\{\|\sigma(\mu)\| : \sigma \in S\}.$$

(ii) *If $\Delta^{-1}\mu \in M(G)$, then $u * \mu \in B_{\mathcal{S}}(G)$ and*

$$\|u * \mu\| \leqslant \|u\| \cdot \sup\{\|\sigma(\Delta^{-1}\mu)\| : \sigma \in S\}.$$

PROOF. By Lemma 2.1.9, there exist a unitary representation π of G which is weakly contained in $\mathcal{S}$ and $\xi, \eta \in \mathcal{H}(\pi)$ such that $u(x) = \langle \pi(x)\xi, \eta \rangle$ for all $x \in G$ and $\|u\| = \|\xi\| \cdot \|\eta\|$. Then

$$\begin{aligned} \mu * u(x) &= \int_G u(y^{-1}x)d\mu(y) = \int_G \langle \pi(x)\xi, \pi(y)\eta \rangle d\mu(y) \\ &= \langle \pi(x)\xi, \pi(\mu)\eta \rangle. \end{aligned}$$

This shows that $\mu * u \in B_{\mathcal{S}}(G)$ and $\|\mu * u\| \leqslant \|\xi\| \cdot \|\eta\| \cdot \|\pi(\mu)\|$. Since π is subordinate to S, (i) follows.

The proof of (ii) is of course similar. In fact,

$$\begin{aligned} u * \mu(x) &= \int_G u(xy^{-1})\Delta(y^{-1})d\mu(y) \\ &= \int_G \langle \pi(xy^{-1})\xi, \eta \rangle \Delta^{-1}(y)d\mu(y) \\ &= \langle \pi(x)\pi(\Delta^{-1}\mu)^*\xi, \eta \rangle. \end{aligned}$$

This implies that $u * \mu \in B_{\mathcal{S}}(G)$ and

$$\|u * \mu\| \leqslant \|\xi\| \cdot \|\eta\| \cdot \|\pi(\Delta^{-1}\mu)\|,$$

so that (ii) follows. □

Taking $\mu = \delta_x$ in (i) and $\mu = \Delta\delta_{x^{-1}}$ in (ii) of Lemma 2.1.13, respectively, we get

COROLLARY 2.1.14. *Let $u \in B_{\mathcal{S}}(G)$ and $x \in G$. Then $L_x u \in B_{\mathcal{S}}(G)$ and $R_x u \in B_{\mathcal{S}}(G)$ and $\|L_x u\| = \|u\| = \|R_x u\|$.*

REMARK 2.1.15. Let G be a locally compact abelian group and $\widehat{G}$ its dual group. We claim that $B(\widehat{G})$ is isometrically isomorphic to the measure algebra $M(G)$ via the Fourier-Stieltjes transform $\mu \to \widehat{\mu}$, $\mu \in M(G)$. In fact, this can be seen as follows.

Given $u \in B(\widehat{G})$, by Bochner's theorem there exists $\mu \in M(G)$ such that

$$u(\chi) = \widehat{\mu}(\chi) = \int_G \chi(x)d\mu(x)$$

for all $\chi \in \widehat{G}$. Then, using the inversion formula,

$$\begin{aligned}
\|u\| &= \sup\left\{\left|\int_{\widehat{G}} u(\chi)f(\chi)d\chi\right| : f \in C_c(\widehat{G}), \|f\|_{C^*(\widehat{G})} \leqslant 1\right\} \\
&= \sup\left\{\left|\int_{\widehat{G}} u(\chi)\widehat{g}(\chi)d\chi\right| : g \in C_c(G), \|g\|_\infty \leqslant 1\right\} \\
&= \sup\left\{\left|\int_{G}\left(\int_{\widehat{G}} \widehat{g}(\chi)\chi(x)\right)d\mu(x)\right| : g \in C_c(G), \|g\|_\infty \leqslant 1\right\} \\
&= \sup\left\{\left|\int_{G} g(x)d\mu(x)\right| : g \in C_c(G), \|g\|_\infty = 1\right\} \\
&= \|\mu\|.
\end{aligned}$$

Recall that by the Pontryagin duality theorem, a locally compact abelian group G is topologically isomorphic to the dual group of $\widehat{G}$. Therefore, the above can be restated as follows. For any locally compact abelian group G, $B(G)$ is isometrically isomorphic to the measure algebra $M(\widehat{G})$, the isomorphism being performed by the inverse Fourier-Stieltjes transform $\mu \to \check{\mu}$, $\mu \in M(\widehat{G})$.

2.2. Functorial properties of $B(G)$

Let G and H be locally compact groups and $\phi : H \to G$ a continuous homomorphism. If π is a unitary representation of G, then $\pi \circ \phi$ is a unitary representation of H in the same Hilbert space. If u is a positive definite function associated with π, that is, $u(x) = \langle \pi(x)\xi, \xi\rangle$ for some $\xi \in \mathcal{H}(\pi)$ and all $x \in G$, then $u \circ \phi(y) = \langle \pi \circ \phi(y)\xi, \xi\rangle$ defines a positive definite function of H associated with $\pi \circ \phi$. Moreover, if $\mathcal{S}$ is a set of equivalence classes of representations of G, then we denote by $\mathcal{S} \circ \phi$ the set of equivalence classes of representations $\pi \circ \phi$, $\pi \in \mathcal{S}$. Note that if π and π' are equivalent representations of G then $\pi \circ \phi$ and $\pi' \circ \phi$ are equivalent.

THEOREM 2.2.1. *Let ϕ be a continuous homomorphism from H into G.*

(i) *The map $j : u \to u \circ \phi$ is a norm decreasing homomorphism from $B(G)$ into $B(H)$, and for any $\mathcal{S}$,*

$$j(P_{\mathcal{S}}(G)) \subseteq P_{\mathcal{S}\circ\phi}(H) \quad \text{and} \quad j(B_{\mathcal{S}}(G)) \subseteq B_{\mathcal{S}\circ\phi}(H).$$

(ii) *Suppose that $\phi(H)$ is dense in G. Then j is isometric and*

$$j(B(G)) = B(H) \cap j(C(G)) = B_{\Sigma(H)\circ\phi}(H) \cap j(C(G)). \tag{2.7}$$

Moreover, if $u = \widetilde{u} \in B(G)$, then

$$(u \circ \phi)^+ = u^+ \circ \phi \quad \text{and} \quad (u \circ \phi)^- = u^- \circ \phi.$$

(iii) *Suppose that ϕ is surjective and that given any compact subset K of G, there exists a compact subset C of H such that $\phi(C) = K$. Then*

$$j(B_{\mathcal{S}}(G)) = B_{\mathcal{S}\circ\phi}(H) \cap j(C(G)) \tag{2.8}$$

for any subset $\mathcal{S}$ of $\Sigma(G)$.

PROOF. (i) It is clear that if u is positive definite, then so is $u \circ \phi$. Now assume that $u \in P_{\mathcal{S}}(G)$. To show that $u \circ \phi \in P_{\mathcal{S}\circ\phi}(H)$, let a compact subset C of H and $\epsilon > 0$ be given. Since $K = \phi(C)$ is compact, there exist elements $u_1, \ldots, u_n$

of $P(G)$ associated with representations $\pi_1, \ldots, \pi_n$ in S, respectively, such that $|u(x) - \sum_{j=1}^n u_j(x)| \leqslant \epsilon$ for all $x \in K$. Thus

$$\left| u \circ \phi(y) - \sum_{j=1}^n u_j \circ \phi(y) \right| \leqslant \epsilon$$

for all $y \in C$. Since $u_j \circ \phi$ is associated to $\pi_j \circ \phi$, it follows that $u \circ \phi \in P_{S \circ \phi}(H)$. By linearity, we obtain that $j(B_{\mathcal{S}}(G)) \subseteq B_{\mathcal{S} \circ \phi}(H)$.

Taking $\mathcal{S} = \Sigma(G)$, we conclude that j is a homomorphism from $B(G)$ into $B(H)$. To show that j is norm decreasing, let $u \in B(G)$ such that $u(x) = \langle \pi(x)\xi, \eta \rangle$ with $\|u\| = \|\xi\| \cdot \|\eta\|$. Then $u \circ \phi(y) = \langle \pi \circ \phi(y)\xi, \eta \rangle$ and hence $\|u \circ \phi\| \leqslant \|\xi\| \cdot \|\eta\| = \|u\|$.

(ii) Suppose that $\phi(H)$ is dense in G. Let ω denote the universal representation of G and A the subalgebra of $C^*(G)^{**}$ consisting of all finite linear combinations of operators $\omega(x)$, $x \in \phi(H)$. Since the mapping $x \to \omega(x)$ from G into $\mathcal{B}(\mathcal{H}(\omega))$ is continuous with respect to the strong operator topology, A is strongly dense in $C^*(G)^{**}$. Then the Kaplansky density theorem assures that the unit ball of A is ultra-weakly dense in the unit ball of $C^*(G)^{**}$.

Now, let $u \in B(G) = B_\omega(G)$. Then $u \circ \phi \in B_{\omega \circ \phi}(H)$ by (i), and hence, by Lemma 2.1.8,

$$\begin{aligned}
\|u \circ \phi\| &= \sup \left\{ \left| \sum_{j=1}^n c_j (u \circ \phi)(y_j) \right| : y_j \in H, c_j \in \mathbb{C}, \left\| \sum_{j=1}^n c_j (\omega \circ \phi)(y_j) \right\| \leqslant 1 \right\} \\
&= \sup \left\{ \left| \sum_{j=1}^n c_j u(x_j) \right| : x_j \in \phi(H), c_j \in \mathbb{C}, \left\| \sum_{j=1}^n c_j \omega(x_j) \right\| \leqslant 1 \right\} \\
&= \sup\{|\langle T, u \rangle| : T \in A, \|T\| \leqslant 1\} \\
&= \sup\{|\langle T, u \rangle| : T \in C^*(G)^{**}, \|T\| \leqslant 1\} \\
&= \|u\|.
\end{aligned}$$

This shows that j is isometric.

Let $u = \widetilde{u} \in B(G)$. Then

$$u \circ \phi = (u \circ \phi)^\sim = u^+ \circ \phi - u^- \circ \phi.$$

On the other hand, since j is isometric,

$$\|u \circ \phi\| = \|u\| = u^+(e_G) + u^-(e_G) = u^+ \circ \phi(e_H) - u^- \circ \phi(e_H).$$

These two equations imply that $(u \circ \phi)^+ = u^+ \circ \phi$ and $(u \circ \phi)^- = u^- \circ \phi$. Since, by (i),

$$j(B(G)) \subseteq B_{\Sigma(H) \circ \phi}(H) \cap j(C(G)) \subseteq B(H) \cap j(C(G)),$$

to finish the proof of (ii), it remains to show that if $v = u \circ \phi$, where $u \in C(G)$ and $v \in B(H)$, then $u \in B(G)$. Obviously, we can assume that $v = \widetilde{v}$. By Lemma 2.1.7, v^+ and v^- are uniform limits on H of linear combinations of right translates of $v = u \circ \phi$. Therefore, given $m \in \mathbb{N}$, there exist $y_1, \ldots, y_n \in H$ and $c_1, \ldots, c_n \in \mathbb{C}$ such that, for every $y \in H$,

$$\left| v^+(y) - \sum_{j=1}^n c_j R_{y_j}(u \circ \phi)(y) \right| \leqslant 1/m.$$

Setting $v_m = \sum_{j=1}^n c_j R_{\phi(y_j)} u$ and observing that

$$\sum_{j=1}^n c_j R_{y_j}(u \circ \phi) = \sum_{j=1}^n c_j (R_{\phi(y_j)} u) \circ \phi = \left(\sum_{j=1}^n c_j R_{\phi(y_j)} u \right) \circ \phi = v_m \circ \phi,$$

we see that for any $m \in \mathbb{N}$, there exists v_m, a linear combination of translates of u, such that

$$|v^+(y) - v_m \circ \phi(y)| \leqslant 1/m, \quad y \in H. \tag{2.9}$$

By (2.9) the sequence of functions v_m, $m \in \mathbb{N}$, converges uniformly on $\phi(H)$ and hence, by density, uniformly on G, to a function w^+ on G. Being a linear combination of translates of u, v_m is a continuous function on G and therefore w^+ is continuous. It follows from (2.9) that w^+ satisfies $v^+ = w^+ \circ \phi$. Moreover, since v^+ is positive definite on H, w^+ is positive definite on $\phi(H)$ and hence positive definite on G by continuity. So we have shown that $v^+ = w^+ \circ \phi$ for some $w^+ \in P(G)$.

Similarly, it is shown that there exists $w^- \in P(G)$ such that $v^- = w^- \circ \phi$. Since

$$u \circ \phi = v = v^+ - v^- = w^+ \circ \phi - w^- \circ \phi = (w^+ - w^-) \circ \phi,$$

we get $u = w^+ - w^- \in B(G)$. This shows (2.7).

(iii) Suppose that ϕ is onto and satisfies the condition on compact sets in (iii). Let $v \in B_{\mathcal{S}\circ\phi}(H)$, so that v^+ and v^- belong to $P_{\mathcal{S}\circ\phi}(H)$. Let K be a compact subset of G and $\epsilon > 0$, and choose a compact subset C of H such that $\phi(C) = K$. Since v^+ and v^- can be uniformly approximated on C, up to ϵ, by sums of continuous positive definite functions associated to $\pi \circ \phi$, where $\pi \in \mathcal{S}$, w^+ and w^- (see (ii)) can be uniformly approximated on K, up to ϵ, by continuous positive definite functions associated with representation π, $\pi \in \mathcal{S}$. Since K and $\epsilon > 0$ are arbitrary, it follows that $w^+, w^- \in P_{\mathcal{S}}(G)$ and hence $u = w^+ - w^- \in B_{\mathcal{S}}(G)$. This shows that

$$B_{\mathcal{S}\circ\phi} \cap j(C(H)) \subseteq j(B_{\mathcal{S}}(H)),$$

and hence (2.8) follows. □

As an immediate consequence of the preceding theorem we obtain the following corollary. It extends, to arbitrary locally compact groups, a theorem which is due to Bochner and Schoenberg for $\mathbb{R}$ and to Eberlein for general locally compact abelian groups G and characterizes the Fourier-Stieltjes transforms of measures in $M(G)$.

COROLLARY 2.2.2. *Let G be a locally compact group and G_d the same group equipped with the discrete topology. A function u on G belongs to $B(G)$ if and only if u is continuous and $u \in B(G_d)$. In that case, the norms of u in $B(G)$ and in $B(G_d)$ are equal.*

COROLLARY 2.2.3. *The unit ball of $B(G)$ is closed in $C(G)$ with respect to the topology of pointwise convergence.*

PROOF. If $v \in B(G)$ is such that $\|v\| \leqslant 1$, then by Lemma 2.1.8

$$\left| \sum_{j=1}^n c_j v(x_j) \right| \leqslant \left\| \sum_{j=1}^n c_j \delta_{x_j} \right\|_{\Sigma(G)}$$

for any finitely many $x_1, \ldots, x_n \in G$ and $c_1, \ldots, c_n \in \mathbb{C}$.

Now, let $u \in C(G)$ be a pointwise limit of such functions v. Then the same inequality holds for u. It then follows for any $f \in l^1(G_d)$ that

$$\left| \sum_{x \in G} f(x)u(x) \right| \leqslant \|f\|_{\Sigma(G)}$$

and hence $u \in B(G_d)$. Since u is continuous, Corollary 2.2.2 shows that $u \in B(G)$. □

COROLLARY 2.2.4. *Let G be a locally compact group, N a closed normal subgroup of G and $q : G \to G/N$ the quotient homomorphism. Then the map $u \to u \circ q$ is an isometric isomorphism from $B(G/N)$ onto the subspace of $B(G)$ consisting of all functions in $B(G)$ which are constant on cosets of N. For each $\mathcal{S} \subseteq \Sigma(G/N)$, the image of $B_{\mathcal{S}}(G/N)$ under this map is $B_{\mathcal{S} \circ q}(G)$.*

PROOF. The homomorphism q satisfies the hypotheses of Theorem 2.2.1(iii). Moreover, a function u on G/N is continuous if and only if $u \circ q$ is continuous on G. The statement now follows from Theorem 2.2.1(iii). □

Alternatively, Corollary 2.2.4 can be obtained more directly by using duality arguments for Banach spaces as follows. Let T_N denote the canonical $*$-homomorphism from $C^*(G)$ onto $C^*(G/N)$. Then the dual map, T_N^*, from $B(G/N)$ into $B(G)$ is given by

$$\begin{aligned} \langle T_N^*(u), f \rangle &= \langle u, T_N f \rangle \\ &= \int_{G/N} u(xN) \left(\int_N f(xn)dn \right) d(xN) \\ &= \int_G u(q(x)) f(x) dx \\ &= \langle u \circ q, f \rangle, \end{aligned}$$

so that T_N^* equals j, where j denotes the map $u \to u \circ q$.

COROLLARY 2.2.5. *Let G be a locally compact group, bG its Bohr compactification and $\phi : G \to bG$ the canonical homomorphism. Let $AP(G)$ denote the space of almost periodic functions on G. Then the map $j : u \to u \circ \phi$ is an isometry from $B(bG)$ onto $B(G) \cap AP(G)$.*

PROOF. Since $\phi(G)$ is dense in bG, by Theorem 2.2.1(ii) j is an isometry and $j(B(bG)) = B(G) \cap j(C(bG))$. Now, just note that $j(C(bG)) = AP(G)$. □

LEMMA 2.2.6. *Let H be an open subgroup of the locally compact group G, and for a function f on H, let $\mathring{f}$ denote the trivial extension of f to all of G. Then the map $f \to \mathring{f}$ from $C_c(H)$ into $C_c(G)$ extends uniquely to an isometric $*$-homomorphism of $C^*_\lambda(H)$ into $C^*_\lambda(G)$. Its adjoint map, which is the restriction map $u \to u|_H$, is a norm decreasing algebra homomorphism from $B_\lambda(G)$ onto $B_\lambda(H)$ and it maps $P_{\lambda_G}(G)$ onto $P_{\lambda_H}(H)$.*

PROOF. The map $f \to \mathring{f}$ is clearly an injective $*$-homomorphism of $C_c(H)$ into $C_c(G)$, and

$$\begin{aligned}\|\lambda_H(f)\| &= \sup\{\|f * g\|_2 : g \in C_c(H), \|g\|_2 \leqslant 1\} \\ &= \sup\{\|\mathring{f} * \mathring{g}\|_2 : g \in C_c(H), \|g\|_2 \leqslant 1\} \\ &\leqslant \sup\{\|\mathring{f} * h\|_2 : h \in C_c(G), \|h\|_2 \leqslant 1\} \\ &= \|\lambda_G(\mathring{f})\|.\end{aligned}$$

Conversely, since $B_\lambda(G)|_H \subseteq B_\lambda(H)$ and $\|v|_H\|_{B(H)} \leqslant \|v\|_{B(G)}$ for any $v \in B(G)$,

$$\begin{aligned}\|\lambda_G(\mathring{f})\| &= \sup\left\{\left|\int_G \mathring{f}(x)v(x)dx\right| : v \in B_\lambda(G), \|v\|_{B(G)} \leqslant 1\right\} \\ &= \sup\left\{\left|\int_H f(x)v|_H(x)dx\right| : v \in B_\lambda(G), \|v\|_{B(G)} \leqslant 1\right\} \\ &\leqslant \sup\left\{\left|\int_H f(x)u(x)dx\right| : u \in B_\lambda(H), \|u\|_{B(H)} \leqslant 1\right\} \\ &= \|\lambda_H(f)\|.\end{aligned}$$

So $\|\lambda_H(f)\| = \|\lambda_G(\mathring{f})\|$ for $f \in C_c(H)$, and hence the map $f \to \mathring{f}$ extends uniquely to an isometric $*$-homomorphism ϕ from $C^*_\lambda(H)$ into $C^*_\lambda(G)$. Since

$$\int_G \mathring{f}(x)v(x)dx = \int_H f(x)v|_H(x)dx$$

for $f \in C_c(H)$ and $v \in B_\lambda(G)$, the adjoint map of ϕ is simply the map $v \to v|_H$. Finally, ϕ^* is surjective since ϕ is an isometry. □

2.3. The Fourier algebra $A(G)$, its spectrum and its dual space

Let G be a locally compact group. In this section we first introduce the Fourier algebra of G and then identify its Gelfand spectrum and its Banach space dual.

LEMMA 2.3.1. *Let $f, g \in L^2(G)$. Then $f * \widetilde{g} \in B_\lambda(G)$ [Section 2.1] and $\|f * \widetilde{g}\| \leqslant \|f\|_2\|g\|_2$.*

PROOF. For $x \in G$, we have

$$(f * \widetilde{g})(x) = \int_G f(xy)\overline{g(y)}dy = \langle \lambda_G(x^{-1})f, g\rangle.$$

The statement now follows from Lemma 2.1.9 and Remark 2.1.10. □

PROPOSITION 2.3.2. *Let G be a locally compact group, C a compact subset of G and U an open subset of G such that $C \subseteq U$. Then there exists a function u on G which is a finite linear combination of functions in $P(G) \cap C_c(G)$ and satisfies*

$$0 \leqslant u \leqslant 1, \ u|_C = 1 \text{ and } u|_{G\setminus U} = 0.$$

PROOF. Since C is compact there exists a compact symmetric neighbourhood V of the identity such that $CV^2 \subseteq U$. Let

$$u(x) = |V|^{-1}(1_{CV} * \check{1}_V)(x) = |V|^{-1} \cdot |xV \cap CV|,$$

$x \in G$. Then $0 \leqslant u(x) \leqslant 1$ for all $x \in G$. If $x \in C$ then $|xV \cap CV| = |xV| = |V|$ and hence $u(x) = 1$, whereas if $x \notin CV^2$, then $xV \cap CV = \varnothing$ and so $u(x) = 0$. In particular, supp $u \subseteq CV^2$, which is compact, and $u(x) = 0$ for $x \in G \backslash U$.

Finally, the identity

$$\begin{aligned} 4(f * g^*) &= (f+g)*(f+g)^* - (f-g)*(f-g)^* \\ &\quad + i(f+ig)*(f+ig)^* - i(f-ig)*(f-ig)^*, \end{aligned}$$

$f, g \in L^1(G)$, ensures that u is a finite linear combination of functions of the form $h * h^*$, where $h \in L^\infty(G)$ and h has compact support and hence $h * h^* \in P(G) \cap C_c(G)$. □

PROPOSITION 2.3.3. *For $1 \leqslant j \leqslant 10$, define a subset M_j of $B(G)$ as follows:*
$M_1 = \{f * \widetilde{g} : f, g \in C_c(G)\}$;
$M_2 = \{h * \widetilde{h} : h \in C_c(G)\}$;
$M_3 = \{f * \widetilde{g} : f, g \in L^\infty(G)$ *with compact support*$\}$;
$M_4 = \{h * \widetilde{h} : h \in L^\infty(G)$ *with compact support*$\}$;
$M_5 = B(G) \cap C_c(G)$;
$M_6 = P(G) \cap C_c(G)$;
$M_7 = \{u \in P(G) : \Delta^{-1/2}u \in L^1(G)\}$;
$M_8 = P(G) \cap L^2(G)$;
$M_9 = \{h * \widetilde{h} : h \in L^2(G)\}$;
$M_{10} = \{f * \widetilde{g} : f, g \in L^2(G)\}$.
Let E_j denote the linear span of M_j, $1 \leqslant j \leqslant 10$. Then

$$E_1 = E_2 \subseteq E_3 = E_4 \subseteq E_5 = E_6 \subseteq E_7 \subseteq E_8 \subseteq E_9 = E_{10} \subseteq B_\lambda(G),$$

and all these subspaces of $B_\lambda(G)$ have the same closure, denoted $A(G)$, in $B_\lambda(G)$. Moreover, $A(G)$ is an ideal in $B(G)$.

PROOF. The equalities $E_1 = E_2$, $E_3 = E_4$ and $E_9 = E_{10}$ all follow from the polar identity. The inclusions $E_2 \subseteq E_3$, $E_4 \subseteq E_5$ and $E_6 \subseteq E_7$ are evident.

To see that $E_5 \subseteq E_6$, let $v \in B(G) \cap C_c(G)$ and $v = v_1 - v_2 + i(v_3 - v_4)$, where $v_k \in P(G)$, $1 \leqslant k \leqslant 4$. By Proposition 2.3.2 there exists a function u of the form $u = \sum_{j=1}^n c_j u_j$, where $c_j \in \mathbb{C}$ and $u_j \in P(G) \cap C_c(G)$, such that $u = 1$ on supp v. Then

$$v = uv = \sum_{j=1}^n c_j u_j [v_1 - v_2 + i(v_3 - v_4)],$$

which shows that $v \in E_6$. The inclusion $E_7 \subseteq E_8$ is a consequence of [**100**, Proposition 12], and $E_8 \subseteq E_9$ follows from the fact that every $u \in P(G) \cap L^2(G)$ can be written in the form $u = f * \widetilde{f}$ with $f \in L^2(G)$ [**60**, Theorem 13.8.6].

Finally, all these subspaces will have the same closure once we have shown that E_1 is dense in E_{10}. To this end, let $f, g \in L^2(G)$ and $\epsilon > 0$ be given. Then there exist functions $h, k \in C_c(G)$ such that $\|f - h\|_2 \leqslant \epsilon$ and $\|g - k\|_2 \leqslant \epsilon$. Now, by Lemma 2.3.1,

$$\begin{aligned} \|f * \widetilde{g} - h * \widetilde{k}\|_{B(G)} &= \|(f-h) * \widetilde{g} + h * (\widetilde{g} - \widetilde{k})\|_{B(G)} \\ &\leqslant \|(f-h) * \widetilde{g}\|_{B(G)} + \|h * (\widetilde{g} - \widetilde{k})\|_{B(G)} \\ &\leqslant \|f - h\|_2 \|g\|_2 + \|h\|_2 \|g - k\|_2 \\ &\leqslant \epsilon(\|g\|_2 + \|f\|_2 + \epsilon). \end{aligned}$$

This shows that $M_{10} \subseteq \overline{M_1}$ and hence $E_{10} \subseteq \overline{E_1}$. □

DEFINITION 2.3.4. The algebra $A(G)$, defined in Proposition 2.3.3, is called the *Fourier algebra* of the locally compact group G.

COROLLARY 2.3.5. *Let G be a locally compact group.*

(i) *Every $u \in A(G)$ vanishes at infinity.*
(ii) *$A(G)$ is uniformly dense in $C_0(G)$.*

PROOF. (i) follows from $\|u\|_\infty \leqslant \|u\|$ and the density of the subspace $B(G) \cap C_c(G) = A(G) \cap C_c(G)$ in $A(G)$ (Proposition 2.3.3).

(ii) $A(G) \cap C_c(G)$ is a self-adjoint subalgebra of $C_0(G)$ which, by Proposition 2.3.2, strongly separates the points of G. Thus $A(G)$ is uniformly dense in $C_0(G)$ by the Stone-Weierstrass theorem. □

LEMMA 2.3.6. *Let $u \in A(G)$ and $x, y \in G$. Then the functions $L_x u$, $R_y u$, $\overline{u}$, $\check{u}$ and $\widetilde{u}$ all belong to $A(G)$.*

PROOF. By Remark 2.1.10(3) and Corollary 2.1.14, all the linear maps of $B(G)$ into itself in question are continuous. Since they clearly map $B(G) \cap C_c(G)$ into itself and $B(G) \cap C_c(G)$ is dense in $A(G)$, the statements of the lemma follow. □

Our next purpose in this section is to show that the spectrum of $A(G)$ can be canonically identified with G and that $A(G)$ is a regular algebra of functions on G. To that end, we need the following lemma.

LEMMA 2.3.7. *Let $a \in G$ and $f \in A(G)$ such that $f(a) = 0$. Then, given $\epsilon > 0$, there exists $h \in A(G) \cap C_c(G)$ vanishing in a neighbourhood of a such that $\|h - f\|_{A(G)} \leqslant \epsilon$.*

PROOF. Notice first that, since $A(G) \cap C_c(G)$ is dense in $A(G)$, without loss of generality we can assume that $f \neq 0$, f has compact support and $\epsilon \leqslant \|f\|_\infty$ and $\epsilon < 1$. Let

$$W = \{y \in G : \|f - R_y f\|_{A(G)} \leqslant \epsilon\}.$$

Then W is a compact neighbourhood of e in G. Choose an open neighbourhood V of e such that $V \subseteq W$ and $\sup\{|f(ay)| : y \in V\} \leqslant \epsilon$. By regularity of Haar measure, there exists a compact neighbourhood U of e such that $U \subseteq V$ and $|U| \geqslant |V|(1-\epsilon)$. Now, define functions u, g and h on G by setting $u = |U|^{-1} 1_U, g = 1_{aV} f$ and

$$h = (f - g) * \check{u} \in A(G).$$

Then h has compact support since W is compact and f has compact support. For any $x \in G$,

$$h(x) = |U|^{-1} \int_U f(xy)[1 - 1_{aV}(xy)]dy.$$

It follows that, if $x \in G$ satisfies $a^{-1}xU \subseteq V$, then $h(x) = 0$. Thus h vanishes in a neighbourhood of a. Moreover,

$$\|u\|_2 = |U|^{-1/2} \leqslant |V|^{-1/2} \left(\frac{1}{1-\epsilon}\right)^{1/2},$$

$$\|g\|_2 = \left(\int_{aV} |f(y)|^2 dy\right)^{1/2} \leqslant \epsilon |V|^{1/2},$$

and

$$\begin{aligned}\|f - f * \check{u}\|_{A(G)} &= \left\| f - |U|^{-1} \int_U (R_y f) dy \right\|_{A(G)} \\ &\leqslant \sup_{y \in U} \|f - R_y f\|_{A(G)} \leqslant \epsilon.\end{aligned}$$

Combining all these estimates, we obtain

$$\|f - h\|_{A(G)} \leqslant \|f - f * \check{u}\|_{A(G)} + \|g\|_2 \|\check{u}\|_2 \leqslant \epsilon + \epsilon \left(\frac{1}{1-\epsilon} \right)^{1/2}.$$

This finishes the proof. □

THEOREM 2.3.8. *Let G be a locally compact group. For each $x \in G$, let*

$$\varphi_x : A(G) \to \mathbb{C}, \ u \to u(x).$$

Then the map $x \to \varphi_x$ is a homeomorphism from G onto $\sigma(A(G))$. Moreover, $A(G)$ is regular.

PROOF. It is obvious that $\varphi_x \in \sigma(A(G))$ and that the map $x \to \varphi_x$ is injective. Now let $\varphi \in \sigma(A(G))$ be given and suppose that $\varphi \neq \varphi_x$ for all $x \in G$. Then, for each $x \in G$ there exists $f_x \in A(G)$ such that $\varphi(f_x) = 1$, but $\varphi_x(f_x) = 0$. By Lemma 2.3.6, every $g \in A(G)$ vanishing at x is the limit of a sequence $(g_n)_n$ in $A(G)$ with the property that each g_n vanishes in a neighbourhood of x. Therefore we can assume that f_x vanishes in a neighbourhood V_x of x.

Since $A(G) \cap C_c(G)$ is dense in $A(G)$, there exists $f_0 \in C_c(G) \cap A(G)$ such that $\varphi(f_0) = 1$. Choose $x_1, \ldots, x_n \in \operatorname{supp} f_0$ such that

$$\operatorname{supp} f_0 \subseteq \textstyle\bigcup_{j=1}^n V_{x_j}$$

and let

$$f = f_0 f_{x_1} \cdots f_{x_n} \in A(G).$$

Then $f(x) = 0$ for every $x \in G$, whereas

$$\varphi(f) = \varphi(f_0) \textstyle\prod_{j=1}^n \varphi(f_{x_j}) = 1.$$

This contradiction shows that $\varphi = \varphi_x$ for some $x \in G$.

Now, since the subalgebra $A(G)$ of $C_0(G)$ is uniformly dense in $C_0(G)$, the topology on G coincides with the weak topology defined by the set of functions $x \to f(x) = \varphi_x(f), f \in A(G)$. Thus the map $x \to \varphi_x$ from G to $\sigma(A(G))$ is a homeomorphism.

Finally, Proposition 2.3.2 implies that $A(G)$ is a regular algebra of functions on G. □

Of course, after identifying $\sigma(A(G))$ with G, the Gelfand homomorphism of $A(G)$ is nothing but the identity mapping. In particular, $A(G)$ is a semisimple commutative Banach algebra.

We remind the reader that if G is a locally compact abelian group with dual group $\widehat{G}$, then $A(G)^* = L^1(\widehat{G})^* = L^\infty(\widehat{G})$ and that $VN(G) = L^\infty(\widehat{G})$ (see Remark 1.8.21). The following theorem shows that $VN(G)$ is isometrically isomorphic to the dual space of $A(G)$ for general locally compact groups G.

THEOREM 2.3.9. *Let G be a locally compact group. For any $\varphi \in A(G)^*$ there exists a unique operator $T_\varphi \in VN(G)$ such that*

$$\langle T_\varphi(f), g\rangle_2 = \langle \varphi, \overline{g} * \check{f}\rangle = \langle \varphi, (f * \tilde{g})\check{}\rangle$$

for all $f, g \in L^2(G)$. The mapping $\varphi \to T_\varphi$ from $A(G)^*$ to $VN(G)$ is a surjective linear isometry and has the following additional properties.

(i) *If $u = \sum_{j=1}^\infty (g_j * \check{f}_j)$, $f_j, g_j \in L^2(G)$, with $\sum_{j=1}^\infty \|f_j\|_2\|g_j\|_2 < \infty$, then*

$$\langle \varphi, u\rangle = \sum_{j=1}^\infty \langle T_\varphi(f_j), \overline{g}_j\rangle_2.$$

(ii) *If $\mu \in M(G)$ and φ_μ is the element of $A(G)^*$ defined by*

$$\langle \varphi_\mu, u\rangle = \int_G u(x)d\mu(x), \quad u \in A(G),$$

then $T_{\varphi_\mu} = \lambda_G(\mu)$.

(iii) *$\varphi \to T_\varphi$ is a homeomorphism for the w^*-topology on $A(G)^*$ and the ultraweak topology on $VN(G)$.*

PROOF. If $\varphi \in A(G)^*$, then for any $f, g \in L^2(G)$,

$$|\langle \varphi, \overline{g} * \check{f}\rangle| \leqslant \|\varphi\| \cdot \|\overline{g} * \check{f}\|_{A(G)} \leqslant \|\varphi\| \cdot \|g\|_2\|f\|_2.$$

Thus, for each $f \in L^2(G)$, the assignment $g \to \langle \varphi, \overline{g} * \check{f}\rangle$ defines a conjugate linear functional on $L^2(G)$. Hence there exists a unique $f_\varphi \in L^2(G)$ such that $\langle \varphi, \overline{g} * \check{f}\rangle = \langle f_\varphi, g\rangle_2$ for all $g \in L^2(G)$. Define $T_\varphi : L^2(G) \to L^2(G)$ by $T_\varphi(f) = f_\varphi$. Then T_φ is linear and $\|T_\varphi(f)\|_2 \leqslant \|\varphi\| \cdot \|f\|_2$ for all $f \in L^2(G)$, so that $\|T_\varphi\| \leqslant \|\varphi\|$. For any $f, g \in L^2(G)$ and $h \in L^1(G)$, we have

$$\begin{aligned} \langle T_\varphi(\lambda_G(h)f), g\rangle_2 &= \langle T_\varphi(f * h), g\rangle_2 = \langle \varphi, \overline{g} * (f * h)\check{}\rangle \\ &= \langle \varphi, \overline{g} * \check{h} * \check{f}\rangle = \langle \varphi, \overline{g * h^*} * \check{f}\rangle \\ &= \langle T_\varphi(f), g * h^*\rangle_2 = \langle T_\varphi(f), \lambda_G(h)^* g\rangle_2 \\ &= \langle \lambda_G(h)T_\varphi(f), g\rangle_2. \end{aligned}$$

So T_φ commutes with the right regular representation operators and therefore $T_\varphi \in VN(G)$. If u is as in (i), the series $\sum_{j=1}^\infty (g_j * \check{f}_j)$ is absolutely convergent and hence

$$\langle \varphi, u\rangle = \sum_{j=1}^\infty \langle \varphi, g_j * \check{f}_j\rangle = \sum_{j=1}^\infty \langle T_\varphi(f_j), \overline{g}_j\rangle_2$$

and $|\langle \varphi, u\rangle| \leqslant \|T_\varphi\| \cdot \sum_{j=1}^\infty \|f_j\|_2\|g_j\|_2$. Since this holds for all such representations of $u \in A(G)$, we conclude that $|\langle \varphi, u\rangle| \leqslant \|T_\varphi\| \cdot \|u\|_{A(G)}$. It follows that $\varphi \to T_\varphi$ is an isometry, and the above equation yields statement (iii). If $\mu \in M(G)$, then for all $f, g \in L^2(G)$,

$$\begin{aligned} \langle T_{\varphi_\mu}(f), g\rangle_2 &= \langle \varphi_\mu, \overline{g} * \check{f}\rangle = \int_G \int_G \overline{g(xy)} f(y)dyd\mu(x) \\ &= \int_G \int_G \overline{g(y)} f(x^{-1}y)d\mu(x)dy \\ &= \int_G (\mu * f)(y)\overline{g(y)}dy = \langle \lambda_G(\mu)f, g\rangle_2, \end{aligned}$$

so that $T_{\varphi_\mu} = \lambda_G(\mu)$.

It remains to show that every $T \in VN(G)$ is of the form T_φ for some $\varphi \in A(G)^*$. Thus, let $T \in VN(G)$ and note first that given $f, g \in C_c(G)$, the function $T(u) = T(f * \widetilde{g}) = T(f) * \widetilde{g}$ is continuous. Therefore, for $u \in E_1$, we can put $\langle \varphi_T, u \rangle = T(\check{u})(e)$, and this definition does not depend on the representation of u. To show that φ_T is a bounded linear functional on E_1, recall that by Kaplansky's density theorem there exists a net $(h_\alpha)_\alpha$ in $C_c(G)$ such that $\|\lambda_G(h_\alpha)\| \leqslant \|T\|$ for all α and $\|Tg - \lambda_G(h_\alpha)g\| \to 0$ for every $g \in L^2(G)$. For u of the form $u = \sum_{j=1}^n (f_j * g_j^*)$, where $f_j, g_j \in C_c^2(G)$, it follows that

$$\langle \varphi_T, u \rangle = Tu(e) = \sum_{j=1}^n \langle Tf_j, g_j \rangle = \lim_\alpha \sum_{j=1}^n \langle \lambda_G(h_\alpha) f_j, g_j \rangle = \lim_\alpha \langle \varphi_{\lambda_G(h_\alpha)}, u \rangle$$

and hence

$$|\langle \varphi_T, u \rangle| = \lim_\alpha |\langle \varphi_{\lambda_G(h_\alpha)}, u \rangle| \leqslant \|u\| \cdot \sup_\alpha \|\lambda_G(h_\alpha)\| \leqslant \|u\| \cdot \|T\|.$$

Since E_1 is dense in $A(G)$, φ_T extends uniquely to a bounded linear functional on $A(G)$, also denoted φ_T, of norm $\leqslant \|T\|$. By definition of φ_T,

$$\begin{aligned} \langle T_{\varphi_T}(f), g \rangle_2 &= \langle f_{\varphi_T}, g \rangle_2 = \langle \varphi_T, \overline{g} * \check{f} \rangle \\ &= \langle \varphi_T, (f * \widetilde{g})\check{} \rangle = T(f * \widetilde{g})(e) \\ &= (T(f) * \widetilde{g})(e) = \langle T(f), g \rangle_2 \end{aligned}$$

and hence $T_{\varphi_T} = T$. □

Given $T \in VN(G)$, we let $\check{T}$ denote the operator in $VN(G)$ defined by $\langle \varphi_{\check{T}}, u \rangle = \langle \varphi_T, \check{u} \rangle$, $u \in A(G)$. Thus $T \to \check{T}$ is the transpose of the isometry $u \to \check{u}$ of $A(G)$. Then, for $\mu \in M(G)$,

$$\langle \varphi_{\widecheck{\lambda_G(\mu)}}, u \rangle = \int_G \check{u}(x) d\mu(x) = \langle \varphi_{\lambda_G(\check{\mu})}, u \rangle,$$

where $d\check{\mu}(x) = d\mu(x^{-1})$. Thus $\widecheck{\lambda_G(\mu)} = \lambda_G(\check{\mu})$, and passing to ultraweak limits, we deduce that the map $T \to \check{T}$ is an isometric and ultraweakly continuous involution on $VN(G)$.

DEFINITION 2.3.10. For $T \in VN(G)$ and $u \in A(G)$, let Tu denote the unique element of $A(G)$ such that

$$\langle S, Tu \rangle = \langle \check{T}S, u \rangle$$

for all $S \in VN(G)$.

It is easily verified that the assignment $(T, u) \to Tu$ turns $A(G)$ into a left $VN(G)$-module.

LEMMA 2.3.11. *Let G be a locally compact group and $T \in VN(G)$.*

(i) *The map $u \to Tu$ is a bounded linear operator on $A(G)$ with norm $\|T\|$.*
(ii) *For $u \in A(G)$ and $x \in G$, $T\check{u}(x) = \langle T, L_x u \rangle$.*
(iii) *If $u \in A(G) \cap L^2(G)$, then $Tu \in A(G) \cap L^2(G)$ and $Tu = T(u)$, the action of T on $L^2(G)$.*

PROOF. (i) For every $u \in A(G)$, we have

$$\|Tu\| = \sup_{\|S\| \leqslant 1} |\langle S, Tu \rangle| = \sup_{\|S\| \leqslant 1} |\langle \check{T}S, u \rangle| \leqslant \|\check{T}\| \cdot \|u\| = \|T\| \cdot \|u\|.$$

So $u \to Tu$ is a bounded linear operator on $A(G)$. By definition, its transpose is the map $S \to \check{T}S$ which has norm $\|\check{T}\| = \|T\|$. Since the two maps have the same norm, (i) follows.

(ii) Consider first elements u of $A(G)$ of the form $u = (f * \widetilde{g})\widehat{\ }$, where $f, g \in C_c(G)$. Then

$$T(\check{u}) = T(f * \widetilde{g}) = T(f) * \widetilde{g}$$

is a continuous function and, for each $x \in G$,

$$\begin{aligned} T(\check{u})(x) &= (T(f) * \widetilde{g})(x) = \langle T(f), L_x g\rangle \\ &= \varphi_T(\overline{L_x g} * \check{f}) = \varphi_T(L_x[(f * \widetilde{g})\check{\ }]) \\ &= \langle T, L_x u\rangle. \end{aligned}$$

Using that $S(\check{u})(e) = \langle S, u\rangle$, it follows that

$$\begin{aligned} \langle T, L_x u\rangle &= T(\check{u})(x) = (\rho(x^{-1})T)(\check{u})(e) \\ &= \langle \rho(x^{-1})T, u\rangle = \langle \check{T}\rho(x), \check{u}\rangle = \langle \rho(x), T\check{u}\rangle \\ &= T\check{u}(x). \end{aligned}$$

By linearity, this shows that $T(\check{u}) = T\check{u}$ for all $u \in E_1$, where E_1 is the space defined in Proposition 2.3.3. Since E_1 is dense in $A(G)$ and the maps $u \to \check{u}$ and $u \to L_x u$ are continuous, we conclude that $T\check{u}(x) = \langle T, L_x u\rangle$ for all $u \in A(G)$.

(iii) Let $\mathcal{A}$ be the $*$-subalgebra of $VN(G)$ consisting of all operators T of the form $T = \lambda_G(\mu)$, where μ is a finite linear combination of Dirac measures. Note that, for $u \in A(G) \cap L^2(G)$ and $x \in G$, by (ii)

$$\lambda_G(x)u(y) = \langle \lambda_G(x), L_y\check{u}\rangle = L_y\check{u}(x) = u(x^{-1}y) = L_x u(y)$$

for all $y \in G$. Let $\mu = \sum_{j=1}^n c_j \delta_{x_j}$, $T = \lambda_G(\mu)$ and $f \in C_c(G)$. Then

$$\begin{aligned} \left|\int_G Tu(x)\overline{f(x)}dx\right| &= \left|\int_G \left(\sum_{j=1}^n c_j u(x_j^{-1}x)\overline{f(x)}\right) dx\right| \\ &= \left|\int_G u(x)\left(\sum_{j=1}^n c_j \overline{f(x_j x)}\right) dx\right| \\ &= \left|\int_G (\mu^* * f)(x)\overline{u(x)}dx\right| \\ &= \left|\int_G T^*(f)(x)\overline{u(x)}dx\right| \\ &\leqslant \|T\| \cdot \|f\|_2 \|u\|_2 \end{aligned}$$

by Schwarz' inequality.

Now let $T \in VN(G)$ be arbitrary. By Kaplansky's density theorem, there exists a net $(T_\alpha)_\alpha$ in $\mathcal{A}$ such that $\|T_\alpha\| \leqslant \|T\|$ for all α and $T_\alpha \to T$ in the ultra weak topology. Then

$$\int_G T_\alpha u(x)\overline{f(x)}dx = \langle \lambda_G(f), T_\alpha u\rangle = \langle \check{T_\alpha}\lambda_G(f), u\rangle,$$

and passing to the limit,

$$\int_G Tu(x)\overline{f(x)}dx = \langle \lambda_G(f), Tu\rangle = \langle \check{T}\lambda_G(f), u\rangle.$$

The above estimate then yields

$$\left| \int_G Tu(x)\overline{f(x)}dx \right| \leqslant \|T\| \cdot \|u\|_2 \|f\|_2$$

for all $f \in C_c(G)$ and $u \in A(G) \cap L^2(G)$. This shows that $Tu \in L^2(G)$ and at the same time that the mapping $u \to Tu$ from $A(G) \cap L^2(G)$ into $L^2(G)$ is continuous with respect to the L^2-norms. On the other hand, we know that $Tu = T(u)$ for all $u \in E_1$. Since E_1 is dense in $A(G)$, continuity implies that $Tu = T(u)$ for all $u \in A(G) \cap L^2(G)$. This finishes the proof of (iii). □

2.4. Functorial properties and a description of $A(G)$

Let H be an open subgroup of G and let Haar measure on H be the one induced by Haar measure of G. As before, we denote by $\mathring{f}$ the trivial extension of a function f of H to all of G. To $T \in VN(G)$ and $f \in L^2(H)$ we associate the function $T|_H(f) = T(\mathring{f})|_H$ on H. Since

$$\int_H |T|_H(f)(x)|^2 dx \leqslant \int_G |T(\mathring{f})(x)|^2 dx \leqslant \|T\|^2 \|\mathring{f}\|_2^2 = \|T\|^2 \|f\|_2^2,$$

the map $T|_H : f \to T|_H(f)$ is an operator on $L^2(H)$ with $\|T|_H\| \leqslant \|T\|$. Moreover, $T|_H \in VN(H)$ since, for any $g \in C_c(H)$,

$$\begin{aligned} T|_H(f * g) &= T(f * g)^{\circ}|_H = T(\mathring{f} * \mathring{g})|_H = (T(\mathring{f}) * \mathring{g})|_H \\ &= T(\mathring{f})|_H * g = T_H(f) * g. \end{aligned}$$

Let H be a closed subgroup of G. Let $VN_H(G)$ denote the w^*-closure of the linear span of the set $\{\lambda_G(h) : h \in H\}$. Then $VN_H(G)$ is a von Neumann algebra.

PROPOSITION 2.4.1. *Let H be an open subgroup of the locally compact group G.*

(i) *The map $\phi : u \to \mathring{u}$ is an isometric isomorphism of $A(H)$ into $A(G)$ and $\phi(P(H) \cap A(H)) \subseteq P(G) \cap A(G)$. The map $T \to T|_H$ is the adjoint ϕ^* of ϕ and*

$$\phi^*(VN(G)) = VN(H) \quad \text{and} \quad \phi^*(C^*_\lambda(G)) = C^*_\lambda(H).$$

(ii) *The restriction map $r : u \to u|_H$ maps $A(G)$ onto $A(H)$. Its adjoint r^* is an isomorphism of $VN(H)$ onto the von Neumann subalgebra of $VN(G)$ generated by the operators $\lambda_G(x)$, $x \in H$. Moreover, $r^*(C^*_\lambda(H)) \subseteq C^*_\lambda(G)$.*

PROOF. (i) We know from Theorem 2.2.1(i) that $r : u \to u|_H$ is a norm decreasing map from $B(G)$ into $B(H)$. Since $r(B(G) \cap C_c(G)) \subseteq B(H) \cap C_c(H)$, continuity of r implies that $r(A(G)) \subseteq A(H)$. Let $\phi : T \to \phi(T)$ denote the adjoint of the map $r|_{A(G)} : A(G) \to A(H)$. Then ϕ is a norm decreasing linear map of $VN(H)$ into $VN(G)$ and it is also continuous for the ultraweak topologies. For $f \in C_c(H)$ and $u \in A(G)$ we have

$$\begin{aligned} \langle \phi(\lambda_H(f)), u \rangle &= \langle \lambda_H(f), u|_H \rangle = \int_H f(x)u(x)dx \\ &= \langle \lambda_G(\mathring{f}), u \rangle, \end{aligned}$$

and hence $\phi(\lambda_H(f)) = \lambda_G(\mathring{f})$. Since $\lambda_H(C_c(H))$ is uniformly dense in $C^*_\lambda(H)$, ϕ maps $C^*_\lambda(H)$ into $C^*_\lambda(G)$. Moreover, since $C^*_\lambda(H)$ is ultraweakly dense in $VN(H)$, it follows that $\phi(T^*) = \phi(T)^*$ and $\phi(ST) = \phi(S)\phi(T)$ for all $S, T \in VN(H)$. So ϕ is a von Neumann algebra homomorphism from $VN(H)$ into $VN(G)$. More precisely, the range of ϕ is contained in $VN_H(G)$ since $\phi(\lambda_H(h)) = \lambda_G(h)$ for all $h \in H$.

Let $f, g \in C_c(H)$ and put

$$u = f * \tilde{g} \in E_1(H) \subseteq A(H) \subseteq VN(H)^*.$$

Then, for $T \in VN(G)$,

$$\begin{aligned}
\langle \varphi_T, \check{\mathring{u}} \rangle &= \langle \varphi_T, \overline{g} * \check{f} \rangle = \langle \varphi_T, \overline{\mathring{g}} * \check{\mathring{f}} \rangle \\
&= \langle T(\mathring{f}), \mathring{g} \rangle_2 = \langle T(\mathring{f})|_H, g \rangle_2 \\
&= \langle T|_H(f), g \rangle_2 = \langle \varphi_{T|_H}, \overline{g} * \check{f} \rangle \\
&= \langle \varphi_{T|_H}, \check{u} \rangle.
\end{aligned}$$

This shows that $R^*(u) = u_G \in E_1(G)$ for each $u \in E_1(H)$. Since R^* is continuous and $E_1(H)$ is dense in $A(H)$, it follows that $R^*(A(H)) \subseteq A(G)$. Actually, $R^*(u) = u_G$ for every $u \in A(H)$. In fact, this equation holds for all $u \in E_1(H)$, $E_1(H)$ is dense in $A(H)$ and the map $u \to u_G$ is continuous for the topologies of pointwise convergence. In particular, this implies that $u_G \in A(G)$ for every $u \in A(H)$. On the other hand,

$$\|u_G\| = \|R^*(u)\| \leqslant \|u\| = \|u_G|_H\| \leqslant \|u_G\|,$$

so that $\|u_G\| = \|u\|$. Observe next that if $u \in P(H) \cap A(H)$, then $u_G \in P(G) \cap A(G)$ since u_G is hermitian and

$$\|u_G\| = \|u\| = u(e) = u_G(e).$$

This proves (i).

(ii) Because the restriction map $r : v \to v|_H$ is surjective, it follows from duality theory that r^* is a topological isomorphism for the ultraweak topologies between $VN(H)$ and its range in $VN(G)$. Thus $r^*(VN(H))$ is ultraweakly closed in $VN(G)$ and hence coincides with $VN_H(G)$ since it contains all the operators $\lambda_G(h)$, $h \in H$. This completes the proof of (ii). □

Let K be a compact normal subgroup of G and let $q_K : G \to G/K$ denote the quotient homomorphism and μ_K the normalized Haar measure of K. Then the map $j_K : f \to f \circ q_K$ is a Hilbert space isomorphism between $L^2(G/K)$ and $L^2_K(G)$, the subspace of $L^2(G)$ consisting of all functions which are constant almost everywhere on xK for almost every coset $xK \in G/K$. In other words, $g \in L^2_K(G)$ if and only if $g \in L^2(G)$ and $g = g * \mu_K$. It follows that $T(L^2_K(G)) \subseteq L^2_K(G)$ for every $T \in VN(G)$. Therefore, to each $T \in VN(G)$ we can associate an operator T_K on $L^2(G/K)$ defined by

$$T_K(f) = (j_K^* \circ T \circ j_K)(f), \quad f \in L^2(G/K).$$

It is easy to see that $T_K \in VN(G/K)$. In addition,

$$(ST)_K = S_K T_K \quad \text{and} \quad (T^*)_K = (T_K)^*$$

for any $S, T \in VN(G)$.

PROPOSITION 2.4.2. *Let K be a compact normal subgroup of the locally compact group G.*

(i) *The map $u \to u \circ q_K$ is an isometric isomorphism from $A(G/K)$ onto the subalgebra $A_K(G)$ of $A(G)$ consisting of all $v \in A(G)$ such that $v(xk) = v(x)$ for all $x \in G$ and $k \in K$.*

(ii) *The map $T \to T_K$ is the adjoint of the map $u \to u \circ q_K$ and it is an ultraweakly continuous homomorphism of $VN(G)$ onto $VN(G/K)$.*

PROOF. (i) By Corollary 2.2.4, $\phi : u \to u \circ q_K$ maps $B(G/K)$ isometrically onto $B_K(G)$, the algebra of all $u \in B(G)$ which are constant on cosets of K. Clearly,

$$\phi(B(G/K) \cap C_c(G/K)) = B_K(G) \cap C_c(G).$$

Since $A(G/K)$ is the closure of $B(G/K) \cap C_c(G/K)$, it suffices to show that $B_K(G) \cap C_c(G)$ is dense in $A_K(G)$. Thus let $v \in A_K(G)$ and let $(v_n)_n$ be a sequence in $B(G) \cap C_c(G)$ converging to v. Then $v_n * \mu_K \in B_K(G) \cap C_c(G)$ and

$$\|v_n * \mu_K - v\|_{B(G)} = \|(v_n - v) * \mu_K\|_{B(G)} \leqslant \|\mu_K\| \cdot \|v_n - v\|_{B(G)} \to 0,$$

whence $v \in \overline{B_K(G) \cap C_c(G)}$.

(ii) The map $T \to T_K$ is the adjoint of the map $u \to u \circ q_K$. In fact, for $T \in VN(G)$ and $u \in A(G/K) \cap L^2(G/K)$ we have

$$\begin{aligned} \langle T, u \circ q_K \rangle &= T((u \circ q_K)\check{\ }(e) = T(\check{u} \circ q_K)(e) \\ &= T(j_K(\check{u}))(e) = j_K(T_K(\check{u}))(e) \\ &= T_K(\check{u})(e) = \langle T_K, u \rangle. \end{aligned}$$

The second statement in (ii) then follows from duality theory of Banach spaces. □

THEOREM 2.4.3. *Let G be a locally compact group. Then $A(G)$ is precisely the set of all functions $f * \tilde{g}$, where $f, g \in L^2(G)$.*

PROOF. Let us first assume that G is second countable. Then the space $L^2(G)$ is separable and hence the von Neumann algebra $VN(G)$ is countably generated. It follows from [**60**, Proposition 14.5.1] that every normal positive linear functional on $VN(G)$ is of the form $T \to \langle Tf, f \rangle$, where $f \in L^2(G)$. This in turn implies that every ultraweakly continuous linear functional on $VN(G)$ is of the form

$$T \to \langle Tf, g \rangle = \varphi_T(\overline{g} * \check{f}), \ \ f, g \in L^2(G).$$

On the other hand, as we have seen in Theorem 2.3.9, the ultraweakly continuous linear functionals on $VN(G)$ are exactly given by $T \to \varphi_T(u)$, where $u \in A(G)$. This establishes the theorem for second countable G.

Now suppose that G is σ-compact, and let $u \in A(G)$. Since u is continuous and vanishes at infinity, u is uniformly continuous. Then, by a theorem of Kakutani and Kodaira (Theorem 1.2.16), there exists a compact normal subgroup K of G such that G/K is second countable and u is constant on cosets of K. Then u is of the form $u = v \circ q$, where $v \in A(G/K)$ and $q : G \to G/K$ denotes the quotient homomorphism. The first part of the proof shows that there exist $f, g \in L^2(G/K)$ such that $v = f * \tilde{g}$. However, this implies

$$u = v \circ q = (f \circ q) * (g \circ q)^{\sim},$$

where $f \circ q, g \circ q \in L^2(G)$.

Finally, let G be an arbitrary locally compact group. Since $u \in C_0(G)$, we find a sequence $(C_n)_n$ of compact subsets of G such that, for all n, $C_n \subseteq C_{n+1}$, C_n contains a neighbourhood of e in G and also the set of all $x \in G$ such that $|u(x)| \geqslant 1/n$. Let H_n denote the open subgroup generated by C_n and let $H = \bigcup_{n=1}^{\infty} H_n$. Then H is an open subgroup of G since $H_n \subseteq H_{n+1}$, and H is σ-compact because each H_n is σ-compact. Moreover, $u = 0$ on $G\backslash H$. Since $u|_H \in A(H)$, by the second part of the proof there exist $f, g \in L^2(H)$ such that $u|_H = f * \widetilde{g}$. Denoting by f_1 and g_1 the trivial extensions of f and g to all of G, it follows that $u = f_1 * \widetilde{g_1}$. This finishes the proof. □

Since $(f * \widetilde{g})(x) = \int_G f(xy)\overline{g(y)}dy = \langle \lambda_G(x^{-1})f, g\rangle$, the preceding theorem tells us that $A(G)$ coincides with the collection of coefficient functions of the left regular representation.

COROLLARY 2.4.4. *On $VN(G) \subseteq \mathcal{B}(L^2(G))$ the weak and the ultraweak operator topologies coincide.*

PROOF. Let $(f_n)_n$ and $(g_n)_n$ be sequences in $L^2(G)$ such that $\sum_{n=1}^{\infty} \|f_n\|_2\|g_n\|_2 < \infty$. Then, by Theorem 2.4.3, there exist $f, g \in L^2(G)$ such that

$$\sum_{n=1}^{\infty} \langle \lambda_G(x)f_n, g_n\rangle = \langle \lambda_G(x)f, g\rangle$$

for all $x \in G$, and hence

$$\sum_{n=1}^{\infty} \langle Tf_n, g_n\rangle = \langle Tf, g\rangle$$

for all $T \in VN(G)$. This implies the statement. □

REMARK 2.4.5. Let G be a locally compact abelian group. Then

$$A(\widehat{G}) = \widehat{L^1(G)} = \{\widehat{f} : f \in L^1(G)\}$$

and $\|\widehat{f}\|_{A(G)} = \|f\|_1$. Since $\|\widehat{\mu}\|_{B(G)} = \|\mu\|$ for all $\mu \in M(G)$ (Remark 2.1.15), we only have to verify that $A(\widehat{G}) = \widehat{L^1(G)}$. Clearly, $\widehat{L^1(G)} \subseteq A(\widehat{G})$ since the set of all $f \in L^1(G)$ such that $\widehat{f} \in C_c(\widehat{G})$ is dense in $L^1(G)$ and $B(\widehat{G}) \cap C_c(\widehat{G}) \subseteq A(\widehat{G})$. Conversely, $u \in A(\widehat{G})$ is of the form $u = \xi * \widetilde{\eta}$, where $\xi, \eta \in L^2(\widehat{G})$. By the Plancherel theorem, $\xi = \widehat{f}$ and $\eta = \widehat{g}$ for certain $f, g \in L^2(G)$. Then $f\breve{g} \in L^1(G)$ and $\widehat{f\breve{g}} = \widehat{f} * \widehat{\breve{g}} = \xi * \widetilde{\eta} = u$.

2.5. The support of operators in $VN(G)$

The main theme of this section is to associate to each $T \in VN(G)$ a closed subset of G, the so-called support of T. This notion, which turns out to be a major tool in the sequel, will be studied thoroughly. We start by defining an action of $B(G)$ on $VN(G)$. Recall that, for $u \in B(G)$ and $T \in VN(G)$, the assignment $v \to \langle T, uv\rangle$ defines a bounded linear functional on $A(G)$.

DEFINITION 2.5.1. Let $u{\cdot}T$ denote the operator in $VN(G)$ defined by $\langle u{\cdot}T, v\rangle = \langle T, uv\rangle$ for $v \in A(G)$.

It is clear that $\|u \cdot T\| \leqslant \|u\| \cdot \|T\|$ and that with this action $VN(G)$ becomes a left $B(G)$-module. Note that if $u \in B(G)$ and $v \in A(G)$, then by Lemma 2.3.11,

$(u \cdot T)v \in A(G)$ and, for every $x \in G$,

$$\begin{aligned}[(u \cdot T)v](x) &= \varphi_{u \cdot T}(L_x \check{v}) \\ &= \varphi_{u \cdot T}((R_x v)\check{}) = \varphi_T((R_x v)\check{}u) \\ &= \varphi_T([(R_x v)\check{u}]\check{}) \\ &= T[vR_{x^{-1}}(\check{u})](x).\end{aligned}$$

REMARK 2.5.2. (1) If $u \in A(G)$ and $T \in VN(G)$, the reader should be careful to not mix up the operator $u \cdot T$ in $VN(G)$ with the function $Tu \in A(G)$.

(2) Let $T = \lambda_G(\mu)$ for some $\mu \in M(G)$. Then $u \cdot \lambda_G(\mu) = \lambda_G(u\mu)$ for $u \in B(G)$, where $u\mu$ denotes the product of the function u and the measure μ in the usual sense. Indeed, for any $x \in G$ and $v \in A(G)$, by the above formula

$$\begin{aligned}[u \cdot \lambda_G(\mu)]v(x) &= \lambda_G(\mu)[vR_{x^{-1}}(\check{u})](x) \\ &= \mu * [vR_{x^{-1}}(\check{u})](x) = \int_G \check{u}(y^{-1}x)v(y^{-1}x)d\mu(y) \\ &= \int_G v(y^{-1}x)u(y)d\mu(y) = [u\mu * v](x) \\ &= [\lambda_G(u\mu)v](x).\end{aligned}$$

Let $T \in C^*_\lambda(G) \subseteq VN(G)$ and $T = T_\varphi$, $\varphi \in A(G)^*$. Then the proof of Theorem 2.3.9 shows that $\langle u, T\rangle = \langle \varphi, u\rangle$ for all $u \in A(G) \subseteq B_\lambda(G)$, where $\langle u, T\rangle$ refers to the duality $C^*_\lambda(G)^* = B_\lambda(G)$.

PROPOSITION 2.5.3. *Let $T \in VN(G)$ and $a \in G$. Then the following three conditions are equivalent.*

(i) *The operator $\lambda_G(a)$ is the w^*-limit in $VN(G)$ of operators of the form $v \cdot T$, where $v \in A(G)$.*
(ii) *For every neighbourhood V of a in G, there exists $v \in A(G)$ such that* $\operatorname{supp} v \subseteq V$ *and $\langle T, v\rangle \neq 0$.*
(iii) *If $u \in A(G)$ is such that $u \cdot T = 0$, then $u(a) = 0$.*

PROOF. (i) $\Rightarrow$ (ii) Let $\lambda_G(a)$ be the w^*-limit of a net $(v_\alpha \cdot T)_\alpha$, $v_\alpha \in A(G)$, and let V be a neighbourhood of a in G. Since $A(G)$ is regular (Lemma 2.3.7), there exists $w \in A(G)$ such that $\operatorname{supp} w \subseteq V$ and $w(a) \neq 0$. Then

$$\langle v_\alpha \cdot T, w\rangle = \langle T, v_\alpha w\rangle \to \langle \lambda_G(a), w\rangle = w(a) \neq 0,$$

and hence $\langle T, v_\alpha w\rangle \neq 0$ eventually.

(ii) $\Rightarrow$ (iii) Towards a contradiction, assume that there exists $u \in A(G)$ with $u \cdot T = 0$, but $u(a) \neq 0$. Then we can find $\delta > 0$ and a compact neighbourhood V of a such that $|u(x)| \geqslant \delta$ for all $x \in V$. Since $A(G)$ is regular, by Theorem 1.1.19 there exists $w \in A(G)$ such that $w(x) = \frac{1}{u(x)}$ for all $x \in V$. Now, by (ii), there exists $v \in A(G)$ with $\operatorname{supp} v \subseteq V$ and $\langle T, v\rangle \neq 0$. Then $v = vwu$ since $\operatorname{supp} v \subseteq V$. It follows that

$$\langle T, v\rangle = \langle u \cdot T, vw\rangle = 0.$$

This contradiction proves (iii).

(iii) $\Rightarrow$ (i) Let $I = \{u \in A(G) : u \cdot T = 0\}$. Then I is a closed ideal in $A(G)$ since $\langle (uv) \cdot T, w\rangle = \langle u \cdot T, wv\rangle$ for all $v, w \in A(G)$ and $u \in I$. On the other hand, since $\langle u \cdot T, v\rangle = \langle v \cdot T, u\rangle$ for $v \in A(G)$, I is the annihilator in $A(G)$ of the subspace $A(G) \cdot T$ of $VN(G) = A(G)^*$. Consequently, $T^\perp$ is the w^*-closure of $A(G) \cdot T$

in $VN(G)$. Now, if $a \in G$ satisfies (iii) then $\lambda_G(a) \in I^\perp$ and hence $\lambda_G(a)$ is the w^*-limit of operators of the form $v \cdot T$, $v \in A(G)$. □

DEFINITION 2.5.4. Let $T \in VN(G)$. Then the *support of* T, supp T, is the set of all elements $a \in G$ satisfying one and hence all three conditions in Proposition 2.5.3.

It is clear that supp $0 = \varnothing$. On the other hand, we have

LEMMA 2.5.5. *If $T \in VN(G)$, $T \neq 0$, then* $\operatorname{supp} T$ *is a nonempty closed subset of G.*

PROOF. Let $a \in \overline{\operatorname{supp} T}$ and let V be an open neighbourhood of a. Then $V \cap \operatorname{supp} T \neq \varnothing$ and hence by Proposition 2.5.3(ii) there exists $v \in A(G)$ with $\operatorname{supp} v \subseteq V$ and $\langle T, v\rangle \neq 0$. Thus $a \in \operatorname{supp} T$.

Now, assume that $T \neq 0$. Since $A(G) \cap C_c(G)$ is dense in $A(G)$, there exists $v \in A(G) \cap C_c(G)$ with $\langle T, v\rangle \neq 0$. Towards a contradiction, suppose that $\operatorname{supp} T = \varnothing$. Then for each $a \in G$ there exists $u_a \in A(G)$ such that $u_a \cdot T = 0$, but $u_a(a) \neq 0$. Since $\operatorname{supp} v$ is compact, we find $a_1, \ldots, a_n \in \operatorname{supp} v$ such that $u_{a_i} \cdot T = 0$ and $u(x) = \sum_{i=1}^n u_{a_i}(x)^2 > 0$ for all $x \in \operatorname{supp} v$. There exists $w \in A(G)$ such that $w(x) = \frac{1}{u(x)}$ for all $x \in \operatorname{supp} v$ (Theorem 1.1.19). Then $v = vwu$ and hence

$$\langle T, v\rangle = \sum_{i=1}^n \langle u_{a_i} \cdot T, u_{a_i} wv\rangle = 0.$$

This contradiction finishes the proof. □

We now collect a number of useful facts about the supports of operators in $VN(G)$.

PROPOSITION 2.5.6. *Let G be a locally compact group and $T \in VN(G)$.*

(i) *For $v \in B(G)$, we have*

$$\operatorname{supp}(v \cdot T) \subseteq \operatorname{supp} T \cap \operatorname{supp} v.$$

In particular, $v \cdot T = 0$ whenever v vanishes in some neighbourhood of $\operatorname{supp} T$.

(ii) $\operatorname{supp} T$ *is the smallest closed subset C of G with the following property: If $v \in A(G) \cap C_c(G)$ vanishes in a neighbourhood of C, then $\langle T, v\rangle = 0$.*

(iii) $\operatorname{supp} T$ *is the smallest closed subset C of G with the following property: Given any closed neighbourhood V of C such that $G\backslash V$ is relatively compact, the operator T is a w^*-limit in $VN(G)$ of finite linear combinations of operators $\lambda_G(x)$, where $x \in V$.*

(iv) *Let F be a closed subset of G. Suppose that $(T_\alpha)_\alpha$ is a net in $VN(G)$ converging to T in the w^*-topology and satisfying* $\operatorname{supp} T_\alpha \subseteq F$ *for all α. Then* $\operatorname{supp} T \subseteq F$.

(v) *Let $T, T_1, T_2 \in VN(G)$ and $\lambda \in \mathbb{C}$, $\lambda \neq 0$. Then*

(1) $\operatorname{supp}(\lambda T) = \operatorname{supp} T$;

(2) $\operatorname{supp} T^* = (\operatorname{supp} T)^{-1}$;

(3) $\operatorname{supp}(T_1 + T_2) \subseteq \operatorname{supp} T_1 \cup \operatorname{supp} T_2$, *and equality holds whenever* $\operatorname{supp} T_1 \cap \operatorname{supp} T_2 = \varnothing$.

(4) $\operatorname{supp}(T_1 T_2) \subseteq (\operatorname{supp} T_1)(\operatorname{supp} T_2)$ *provided that one of* $\operatorname{supp} T_1$ *and* $\operatorname{supp} T_2$ *is compact.*

PROOF. (i) The inclusion $\mathrm{supp}(v \cdot T) \subseteq \mathrm{supp}\, T$ follows immediately from the description of the support in Proposition 2.5.3(i). In order to show that $\mathrm{supp}(v \cdot T) \subseteq \mathrm{supp}\, v$, let $a \in G \setminus \mathrm{supp}\, v$ and choose a neighbourhood V of a such that $v|_V = 0$. Then $uv = 0$ for every $u \in A(G)$ with $\mathrm{supp}\, u \subseteq V$ and hence $\langle v \cdot T, u \rangle = \langle T, vu \rangle = 0$ for every such u. Proposition 2.5.3(ii) implies that $a \notin \mathrm{supp}\; v \cdot T$.

If v vanishes in a neighbourhood of $\mathrm{supp}\, T$, then $\mathrm{supp}\, v \cap \mathrm{supp}\, T = \varnothing$ by (i) and therefore $\mathrm{supp}(v \cdot T) = \varnothing$. By Lemma 2.5.5, $v \cdot T = 0$.

(ii) Let $v \in A_c(G)$ and suppose that v vanishes in a neighbourhood of $\mathrm{supp}\, T$. There exists $w \in A(G)$ such that $w|_{\mathrm{supp}\, v} = 1$ and $w = 0$ in a neighbourhood of $\mathrm{supp}\, T$. Then $w \cdot T = 0$ by (i). On the other hand, $wv = v$ and hence

$$\langle T, v \rangle = \langle T, wv \rangle = \langle w \cdot T, v \rangle = 0.$$

This shows that $\mathrm{supp}\, T$ has the indicated property.

Now, let C be any closed subset of G satisfying the condition in (ii) and let $a \in G \setminus C$. Then there exist neighbourhoods V of a and U of C such that V is compact and $V \cap U = \varnothing$. If $v \in A(G)$ is such that $\mathrm{supp}\, v \subseteq V$, then $v \in A_c(G)$ and v vanishes in a neighbourhood of C. It follows that $\langle T, v \rangle = 0$ and consequently $a \notin \mathrm{supp}\, T$ by Proposition 2.5.3(ii).

(iii) Let C be a closed subset of G and V a closed neighbourhood of C such that $G \setminus V$ is relatively compact, and suppose that T is a w^*-limit of finite linear combinations of operators $\lambda_G(x), x \in V$. If v belongs to the ideal $I(V) = \{u \in A(G) : u|_V = 0\}$, then $\langle \lambda_G(x), v \rangle = v(x) = 0$ for all $x \in V$ and by w^*-continuity this implies $\langle T, v \rangle = 0$. It follows now from (ii) that $\mathrm{supp}\, T \subseteq C$, as required.

(iv) Let $u \in A(G) \cap C_c(G)$ be such that u vanishes in a neighbourhood of F. Then, for all α, $\langle T_\alpha, u \rangle = 0$ since $\mathrm{supp}\, T_\alpha \subseteq F$. Since $T_\alpha \to T$ in the w^*-topology, $\langle T, u \rangle = 0$, and this implies $\mathrm{supp}\, T \subseteq F$ by (ii).

(v) (1) is evident and (2) follows from (iii) taking into account that the map $T \to T^*$ is weakly continuous and that $\lambda_G(x)^* = \lambda_G(x^{-1})$.

To show (3) we apply (ii). Thus let $v \in A_c(G)$ vanish in a neighbourhood of $\mathrm{supp}\, T_1 \cup \mathrm{supp}\, T_2$. Then, by (ii),

$$\langle T_1 + T_2, v \rangle = \langle T_1, v \rangle + \langle T_2, v \rangle = 0$$

and hence, by (ii) again,

$$\mathrm{supp}(T_1 + T_2) \subseteq \mathrm{supp}\, T_1 \cup \mathrm{supp}\, T_2.$$

Suppose that in addition $\mathrm{supp}\, T_1 \cap \mathrm{supp}\, T_2 = \varnothing$. Let $a \in \mathrm{supp}\, T_1$ and let $v \in A(G)$ such that $v \cdot (T_1 + T_2) = 0$. Then $v \cdot T_1 = -v \cdot T_2 = S$, say. By (i),

$$\mathrm{supp}\, S \subseteq \mathrm{supp}\, T_1 \cap \mathrm{supp}\, T_2 = \varnothing.$$

This implies $S = 0$ by Lemma 2.5.5. Thus $v \cdot T_1 = 0$ and therefore $v(a) = 0$. So

$$\mathrm{supp}\, T_1 \subseteq \mathrm{supp}(T_1 + T_2)$$

by Proposition 2.5.3(iii). Similarly, $\mathrm{supp}\, T_2 \subseteq \mathrm{supp}(T_1 + T_2)$.

For (4), suppose that $\mathrm{supp}\, T_2$ is compact. It is a straightforward consequence of (iii) that

$$\mathrm{supp}(\lambda_G(x) T_2) = (\mathrm{supp}(\lambda_G(x))(\mathrm{supp}\, T_2)$$

for any $x \in G$. Hence, by (1) and (3), we get

$$\mathrm{supp}(T_1 T_2) \subseteq (\mathrm{supp}\, T_1)(\mathrm{supp}\, T_2)$$

whenever T_1 is a finite linear combination of operators $\lambda_G(x)$. If T_1 is arbitrary, then given any closed neighbourhood V of $\operatorname{supp} T_1$ with relatively compact complement, by (iii) there exists a net $(S_\alpha)_\alpha$ in $VN(G)$ such that $T_1 = w^*\text{-}\lim_\alpha S_\alpha$ and each S_α is a finite linear combination of operators $\lambda_G(x)$, $x \in V$. Then

$$\operatorname{supp}(S_\alpha T_2) \subseteq V \cdot \operatorname{supp} T_2$$

for all α, and since $V \cdot \operatorname{supp} T_2$ is closed and $T_1 T_2 = w^*\text{-}\lim_\alpha (S_\alpha T_2)$, it follows from (iv) that

$$\operatorname{supp}(T_1 T_2) \subseteq V \cdot \operatorname{supp} T_2.$$

Now let $\mathcal{V}$ denote the collection of all closed neighbourhoods of $\operatorname{supp} T_1$ with relatively compact complement. Then $\bigcap\{V : V \in \mathcal{V}\} = \operatorname{supp} T_1$, and since $\operatorname{supp} T_2$ is compact, it is easily verified that

$$\textstyle\bigcap_{V \in \mathcal{V}}(V \cdot \operatorname{supp} T_2) = (\bigcap_{V \in \mathcal{V}} V) \cdot \operatorname{supp} T_2.$$

This implies $\operatorname{supp}(T_1 T_2) \subseteq (\operatorname{supp} T_1)(\operatorname{supp} T_2)$. □

LEMMA 2.5.7. *Let $T \in VN(G)$ and $u \in A(G) \cap C_c(G)$. Then*

$$\operatorname{supp}(Tu) \subseteq (\operatorname{supp} T)(\operatorname{supp} u).$$

PROOF. Let $a \notin (\operatorname{supp} T)(\operatorname{supp} u)$ and hence $a(\operatorname{supp} u)^{-1} \cap \operatorname{supp} T = \varnothing$. Since $\operatorname{supp} u$ is compact, there exist closed neighbourhoods V of a and U of $\operatorname{supp} T$ such that $V(\operatorname{supp} u)^{-1} \cap U = \varnothing$ and $G \backslash U$ is relatively compact. By Proposition 2.5.6(iii) there exists a net $(S_\alpha)_\alpha$ in $VN(G)$ such that $T = w^*\text{-}\lim_\alpha S_\alpha$ and each S_α is a finite linear combination of operators $\lambda_G(x)$, $x \in U$.

Fix α and let $S_\alpha = \sum_{j=1}^n c_j \lambda_G(x_j)$, $x_j \in U$. Then, for any $x \in G$,

$$\langle \varphi_{S_\alpha}, (R_x u)\check{}\,\rangle = \sum_{j=1}^n c_j \langle \lambda_G(x_j), (R_x u)\check{}\,\rangle = \sum_{j=1}^n c_j u(x_j^{-1}x).$$

Now, if $x \in V$ then $x_j^{-1}x \in U^{-1}V$ and hence $x_j^{-1}x \notin \operatorname{supp}\, u$. Thus $\langle \varphi_{S_\alpha}, (R_x u)\check{}\,\rangle = 0$ for all $x \in V$ and all α. Passing to the w^*-limit, it follows that

$$Tu(x) = \langle \varphi_T, (R_x u)\check{}\,\rangle = 0$$

for all $x \in V$. This shows that $a \notin \operatorname{supp}(Tu)$ and hence proves the statement of the lemma. □

The formula $T\check{u}(x) = \langle T, L_x u \rangle$ for $u \in A(G)$ and $x \in G$ in Lemma 2.3.11(ii) implies for any $\mu \in M(G)$

$$\begin{aligned} [\lambda(\mu)u](x) &= \langle \varphi_{\lambda(\mu)}, (R_x u)\check{}\,\rangle = \int_G (r_x u)\check{}(y) d\mu(y) \\ &= \int_G u(y^{-1}x) d\mu(y) = (\mu * u)(x) \end{aligned}$$

for all $x \in G$, so that $\lambda(\mu)u = \mu * u$.

LEMMA 2.5.8. *Let $T \in VN(G)$ and suppose that* $\operatorname{supp}(Tu) \subseteq \operatorname{supp} u$ *for all $u \in A(G) \cap C_c(G)$. Then $T = \lambda I$ for some $\lambda \in \mathbb{C}$.*

PROOF. The proof is divided into several steps.

We first show if U is a relatively compact open subset of G, then for any $u \in A(G) \cap C_c(G)$ such that u is constant on U, Tu is also constant on U. To verify this, fix any two points a and b in U and choose an open neighbourhood V

of the identity such that $Va \cup Vb \subseteq U$. If $x \in Vb$, then $x \in U$ and $xb^{-1}a \in U$. So the function $u - R_{b^{-1}a}u \in A(G) \cap C_c(G)$ vanishes on Vb and hence, by hypothesis, $T(u - R_{b^{-1}a}u)$ also vanishes on Vb. In particular,

$$Tu(b) = TR_{b^{-1}a}u(b) = R_{b^{-1}a}Tu(b) = Tu(a),$$

since T commutes with right translations.

Next we prove the existence of a constant λ, depending only on T, with the following property: For any relatively compact open subset U of G and any $u \in A(G) \cap C_c(G)$ which is identically 1 on U, the function Tu is identically λ on U. By what we have seen above, for every pair (U, u) there exists a constant $\lambda(U, u)$ such that $Tu(x) = \lambda(U, u)$ for all $x \in U$. Now, fix U and let $u_1, u_2 \in A(G) \cap C_c(G)$ such that $u_1|_U = u_2|_U = 1$. Then $u_1 - u_2$ vanishes on U and hence so does $Tu_1 - Tu_2$ by hypothesis. Thus $\lambda(U, u_1) = \lambda(U, u_2)$. Finally, let U_1 and U_2 be two relatively compact open subsets of G and let $u \in A(G) \cap C_c(G)$ be such that $u = 1$ on $U_1 \cup U_2$. Then

$$\lambda(U_j, u) = Tu(x) = \lambda(U_1 \cup U_2, u)$$

for $x \in U_j$, $j = 1, 2$, and hence $\lambda(U_1, u) = \lambda(U_2, u)$. This shows that $\lambda(U, u)$ does neither depend on U nor on u.

Let λ denote the constant associated to T by the preceding discussion. We proceed to show that $T(1_C) = \lambda 1_C$ for every compact subset of G with the property that the boundary of C has measure zero. Fix such a set C. Since Haar measure is regular, there exists an ascending sequence $(V_n)_n$ of open subsets of G such that $\overline{V_n} \subseteq C^\circ$ and

$$|C \backslash \overline{V_n}| = |C^\circ \backslash \overline{V_n}| \leqslant 1/n$$

for all n. For each n, we find $u_n \in A(G) \cap C_c(G)$ such that $0 \leqslant u_n \leqslant 1$, $u_n = 1$ on $\overline{V_n}$ and $u_n = 0$ on $G \backslash C$. Then the bounded continuous function Tu_n vanishes on $G \backslash C$ and takes the value λ on $\overline{V_n}$. Since $u_n \to 1_C$ in $L^2(G)$, it follows that

$$T(1_C) = \lim_{n \to \infty} Tu_n = \lambda 1_C.$$

Since the characteristic functions 1_C, where C is an arbitrary compact subset of G, form a total set in $L^2(G)$, to finish the proof it suffices to show that the conclusion of the preceding paragraph also holds if the hypothesis that the boundary of C be of measure zero is dropped.

Thus let C be an arbitrary compact subset of G. Then there exists a sequence $(U_n)_n$ of relatively compact open sets U_n in G such that, for each n, $C \subseteq U_n$, $|U_n| \leqslant |C| + \frac{1}{n}$ and the boundary of U_n is of measure zero. Then $1_C = \lim_{n \to \infty} 1_{U_n}$ in $L^2(G)$ and hence $T(1_C) = \lim_{n \to \infty} T(1_{U_n})$. Since we already know that $T(1_{U_n}) = \lambda 1_{U_n}$, we conclude that $T(1_C) = \lambda 1_C$, as was to be shown. $\square$

COROLLARY 2.5.9. *Let G be a locally compact group and $a \in G$. If* $\operatorname{supp} T = \{a\}$, *then $T = \alpha \lambda_G(a)$ for some $\alpha \in \mathbb{C}$.*

PROOF. Since, by Proposition 2.5.6(v)(4),

$$\operatorname{supp}(\lambda_G(a^{-1})T) \subseteq \operatorname{supp} \lambda_G(a^{-1}) \cdot \operatorname{supp} T = \{e\},$$

we can assume that $a = e$. Then $\operatorname{supp}(Tu) \subseteq \operatorname{supp} u$ for all $u \in A(G)$ by Lemma 2.5.7 since $\operatorname{supp} T = \{e\}$. Lemma 2.5.8 now shows that $T = \alpha \cdot I = \alpha \lambda_G(e)$, as required. $\square$

The preceding corollary allows a quick application to the ideal theory of Fourier algebras. Recall that an ideal I of a commutative Banach algebra A is called primary if the hull of I in $\sigma(A)$ is a singleton.

COROLLARY 2.5.10. *Every closed primary ideal of $A(G)$ is maximal.*

PROOF. Let I be a closed ideal of $A(G)$ such that $h(I) = \{x\}$ for some $x \in G = A(G)$. The annihilator $I^\perp$ of I in $VN(G) = A(G)^*$ is weakly closed in $VN(G)$ and invariant under the transformations $T \to u \cdot T$, $u \in A(G)$, because I is an ideal. For $T \in VN(G)$, let $J_T = \{u \in A(G) : \langle T, u \rangle = 0\}$. Then, for $T \in I^\perp$, $J_T \supseteq I$ and hence $h(J_T) \subseteq \{x\}$. Thus, if $T \in I^\perp$ and $T \neq 0$, then $h(J_T) = \{x\}$ and so $\operatorname{supp} T = \{x\}$. Corollary 2.5.9 implies that T is a multiple of $\lambda_G(x)$. This shows that $I^\perp$ is 1-dimensional and therefore I is maximal and $I = k(x)$. □

COROLLARY 2.5.11. *Let $x \in G$ and $u \in A(G)$ such that $u(x) = 0$. Then there exists a sequence $(u_n)_n$ in $A(G)$ such that $\|u_n - u\|_{A(G)} \to 0$ and u_n vanishes on some neighbourhood of x.*

PROOF. By Corollary 2.5.10, $k(x)$ is the only closed ideal in $A(G)$ with hull $\{x\}$. The statement follows now from Section 1.1 since $A(G)$ is regular and semisimple. □

COROLLARY 2.5.12. *Let $u \in A(G)$ be such that $u(e) = 0$ and let $\epsilon > 0$. Then there exists $w \in P^1(G) \cap C_c(G)$ such that $\|uw\| \leqslant \epsilon$.*

PROOF. By Corollary 2.5.11 there exists $v \in A(G)$ such that $\|v - u\| \leqslant \epsilon$ and $v = 0$ in a neighbourhood U of e in G. Choose a compact symmetric neighbourhood V of e such that $V^2 \subseteq U$ and set

$$w = \|1_V\|_2^{-2}(1_V * \widetilde{1}_V) \in P(G) \cap C_c(G).$$

Then $w(e) = 1$, $\operatorname{supp} w \subseteq V^2$ and

$$\|wu - wv\| \leqslant \|w\| \cdot \|u - v\| = \|u - v\| \leqslant \epsilon.$$

Since $\operatorname{supp} w \cap \operatorname{supp} v = \emptyset$, $wv = 0$ and hence $\|wu\| \leqslant \epsilon$. □

2.6. The restriction map from $A(G)$ onto $A(H)$

Let H be a closed subgroup of the locally compact group G. In the study of Fourier algebras, a natural question arising is whether functions in $A(H)$ extend to functions in $A(G)$. In this section we show that this question admits an affirmative answer. The corresponding problem for Fourier-Stieltjes algebras is much more involved and will be investigated in Chapter 7.

In the following, let H be a closed subgroup of G and let $G\backslash H$ denote the space of all right cosets of H in G, endowed with the quotient topology defined through the map $p : G \to G\backslash H, x \to Hx$.

LEMMA 2.6.1. *Retain the above notation and let C be a compact, second countable subset of G. Then there exists a Borel subset M of C such that $p(M) = p(C)$ and p is one-to-one on M.*

PROOF. From general topology, it is known that there exist a perfect subset T of $[0, 1]$ and a continuous map φ from T onto the metric space C. For each $n \in \mathbb{N}$, define T_n to be the set of all $t \in T$ with the following property: for every $s \in T$ with $s \leqslant t - 1/n$, we have $p(\varphi(s)) \neq p(\varphi(t))$. Using that T is closed in $[0, 1]$, it is easily verified that T_n is relatively open in T. Then $M = \bigcap_{n=1}^{\infty} \varphi(T_n)$ is a Borel subset of C. To show that $p(M) = p(C)$, fix $x \in C$. Since $\varphi(T) = C$ and T is closed in $[0,1]$, the real number

$$t = \inf\{s \in T : p(\varphi(s)) = p(x)\} \in T$$

satisfies $p(\varphi(t)) = p(x)$. It is also clear that $t \in T_n$ for all n and hence $\varphi(t) \in M$.

It remains to show that φ is one-to-one on M. To see this, assume that $y \in C$ is such that $y \neq x$ and $p(y) = p(x)$ and set $S = \{s \in T : \varphi(s) = y\}$. Then, since $t \in T_n$ for all n, $S \subseteq [t,1]$, and since S is closed in T and $y \neq x$, it even follows that $S \subseteq [t+\delta, 1]$ for some $\delta > 0$. For every $n > 1/\delta$ we then have $S \subseteq T \backslash T_n$ and consequently

$$y \in \varphi(S) \subseteq \varphi(T) \backslash \varphi(T_n) = C \backslash \varphi(T_n) \subseteq C \backslash M.$$

This contradiction completes the proof. □

PROPOSITION 2.6.2. *Let G be a second countable locally compact group and let H be a closed subgroup of G. Then there exists a Borel set S in G with the following properties.*

(i) *S intersects each right coset of H in exactly one point.*
(ii) *For each compact subset C of G, $HC \cap S$ has a compact closure.*
(iii) *$H \cap S = \{e\}$.*

Moreover, there is a closed neighbourhood V of e in G such that $HV = V$ and $V \cap S$ is relatively compact.

PROOF. We first show the existence of a set S satisfying properties (i), (ii) and (iii). Choose a compact symmetric neighbourhood V of e in G. Then $L = \bigcup_{n=1}^{\infty} V^n$ is an open subgroup of G and L has at most countably many right cosets in G. Since every compact subset of G is contained in the union of finitely many cosets of L, we can find a sequence $C_1 \subseteq C_2 \subseteq \cdots$ of compact subsets of G such that every compact subset of G is contained in some C_j. By Lemma 2.6.1, for each j there exists a Borel subset $S_j \subseteq C_j$ such that $p(S_j) = p(C_j)$ and p is one-to-one on S_j. Observe next that the sets S_j can be chosen so that $S_j \subseteq S_{j+1}$ for every $j \in \mathbb{N}$. Indeed, suppose that we already have arranged for $S_1 \subseteq S_2 \subseteq \cdots \subseteq S_j$. Then choose any Borel set $M_{j+1} \subseteq C_{j+1}$ such that p is one-to-one on M_{j+1} and $p(M_{j+1}) = p(C_{j+1})$ and set

$$S_{j+1} = (M_{j+1} \backslash p^{-1}(p(S_j))) \bigcup S_j.$$

Then S_{j+1} is a Borel set. To see this, since S_j and M_{j+1} are Borel sets, it suffices to verify that $p^{-1}(p(S_j))$ is a Borel set. Since S_j is a Borel set in the complete metric space C_j and p is continuous and one-to-one on S_j, $p(S_j)$ is a Borel set [**170**], and hence $p^{-1}(p(S_j))$ is a Borel set as well.

Now, set $S = \bigcup_{j=1}^{\infty} S_j$. Then S is a Borel set satisfying (i) and (ii). Indeed,

$$p(S) = \bigcup_{j=1}^{\infty} p(S_j) = \bigcup_{j=1}^{\infty} p(C_j) = G \backslash H,$$

and since p is one-to-one on each S_j and $S_j \subseteq S_{j+1}$ for each j, p is one-to-one on S. To verify (ii), let $(s_n)_n$ be a sequence in $HC \cap S$ and choose $j \in \mathbb{N}$ such that $C \subseteq C_j$. Then, since $p(S_j) = p(C_j)$, for each n there exist $x_n \in H$ and $t_n \in S_j$ such that $s_n = x_n t_n$. This implies that $s_n = t_n$ and, since $S_j \subseteq C_j$, the sequence $(t_n)_n$ has a convergent subsequence. This shows that $HC \cap S$ is relatively compact.

Finally, translating S if necessary, we can arrange for $S \cap H = \{e\}$.

For the remaining statement, choose an open neighbourhood U of e in G such that $\overline{U}$ is compact and set $V = H\overline{U}$. Then V is a closed neighbourhood of e in G such that $HV = H(H\overline{U}) = H\overline{U} = V$. We also have $\overline{V \cap S} = \overline{(H\overline{U}) \cap S}$, which is compact by (ii). □

The following proposition, the proof of which is fairly long and technical, is a major step towards Theorem 2.6.4 below.

PROPOSITION 2.6.3. *Let G, H, S and V be as in Proposition 2.6.2 and for $x \in G$, let $\beta(x)$ be the unique element in H such that $x = \beta(x)s$ for some $s \in S$. For any complex-valued function f on H define f_V on G by*

$$f_V(x) = f(\beta(x))1_V(x), \; x \in G.$$

Then the following hold.

(i) *If f is a measurable function on H, then f_V is measurable on G.*
(ii) *If f has compact support, then so does f_V.*
(iii) *There exists a constant $c > 0$ such that $f \to c f_V$ is a linear isometry of $L^2(H)$ into $L^2(G)$.*
(iv) *The mapping $f \to f_V$ is a linear isometry of $L^\infty(H)$ into $L^\infty(G)$.*
(v) *If f and g are in $L^2(H)$, then for all $h \in H$, we have*

$$c^2(f_V * \widetilde{g_V})(h) = (f * \check{g})(h),$$

the convolution on the left and on the right being over G and H, respectively.

PROOF. In the sequel, m_G and m_H will denote left Haar measures on G and H, respectively.

(i) Assume first that f is real-valued. Then f_V is real-valued, and for every $r \in \mathbb{R}$, we have

$$\begin{aligned} \{x \in G : f(\beta(x)) > r\} &= \{x \in G : \beta(x) \in \{h \in H : f(h) > r\}\} \\ &= \{h \in H : f(h) > r\} \cdot S. \end{aligned}$$

Since f is measurable, $T = \{h \in H : f(h) > r\}$ is measurable in H and hence TS is a measurable subset of G. Thus $f \circ \beta$ is measurable, and so is 1_V since V is closed in G. So f_V is measurable. If f is an arbitrary complex-valued measurable function, then standard arguments on Ref and Imf show that f_V is measurable.

(ii) If $f_V(x) \neq 0$, then $x \in V$ and $f(\beta(x)) \neq 0$. This implies

$$\operatorname{supp} f_V \subseteq \overline{V \cap \beta^{-1}(\operatorname{supp} f)} = \overline{V \cap (\operatorname{supp} f)S}.$$

If $y \in V \cap (\operatorname{supp} f)S$, then $\beta(y) \in \operatorname{supp} f$ and $\gamma(y) = \beta(y)^{-1}y \in HV = V$, and therefore $y \in (\operatorname{supp} f)(S \cap V)$. This shows that

$$\operatorname{supp} f_V \subseteq \overline{V \cap (\operatorname{supp} f)S} \subseteq (\operatorname{supp} f) \cdot \overline{S \cap V},$$

which is compact since both $\operatorname{supp} f$ and $\overline{S \cap V}$ are compact.

(iii) It will be convenient to prove the following fact first. Let f be any real-valued measurable function on H, let $\delta > 0$ and suppose that

$$m_H(\{h \in H : f(h) > \delta\}) > 0.$$

Then $m_G(\{x \in G : f_V(x) > \delta\}) > 0$. To see this, let C be a compact neighbourhood of e in G such that $C \subseteq V$ and set $D = \{h \in H : f(h) > \delta\}$. Since $m_H(D) > 0$, we have $m_G(D(CH \cap S)) > 0$. Because $C \subseteq V$ and $D \subseteq V$,

$$D(CH \cap S) \subseteq D(V \cap S) \subseteq DS \cap V,$$

so that $m_G(V \cap DS) > 0$ (clearly, $DS \cap V$ is a measurable set). Now note that if $y \in DS \cap V$, then $\beta(y) \in D$ and $x \in V$. Therefore,

$$DS \cap V \subseteq \{x \in G : f_V(x) > \delta\},$$

and hence the latter set has positive measure.

We next define a positive and additive functional on $C_c^+(H)$ as follows. If $f \in C_c^+(H)$, then f_V is bounded and measurable by (i) and f_V has compact support by (ii). Thus we may define

$$I(f) = \int_G f_V(x)dx.$$

If $f \neq 0$, then for some $\delta > 0$ the set $\{h \in H : f(h) > \delta\}$ is nonempty and open in H. The previous paragraph then shows that $m_G(\{y \in G : f_V(y) > \delta\}) > 0$, whence $I(f) > 0$. Furthermore, since $HV = V$, for $h \in H$ we have

$$\begin{aligned} I(L_hf) &= \int_G 1_V(x)(L_hf)(\beta(x))dx = \int_G 1_V(x)f(h^{-1}\beta(x))dx \\ &= \int_G 1_V(h^{-1}x)f(\beta(h^{-1}x))dx = \int_G 1_V(x)f(\beta(x))dx \\ &= I(f). \end{aligned}$$

Hence I is left invariant on $C_c^+(H)$. By the uniqueness theorem for the left Haar integral there exists a constant $c > 0$ such that, for all $f \in C_c^+(H)$,

$$\int_H f(h)dh = c\int_G f_V(x)dx.$$

From this equation, (iii) follows by routine arguments from integration theory.

(iv) Since G and H are σ-compact, in both groups there is no distinction between locally null sets and null sets. Let $f \in L^\infty(H)$. Then, for each $\delta < \|f\|_\infty$, the set $\{h \in H : |f(h)| > \delta\}$ is not a null set. Thus by what we have shown in the first paragraph of the proof of (iii), $\{x \in G : |f_V(x)| > \delta\}$ is not a null set. Since this holds for all $\delta < \|f\|_\infty$, it follows that $\|f\|_\infty \leqslant \|f_V\|_\infty$.

For the reverse inequality, let $\delta < \|f_V\|_\infty$. There is a measurable subset M of G such that $m_G(M) > 0$ and $|f_V(x)| > \delta$ for all $x \in M$. Then $M \subseteq V$ and $|f(x)| > \delta$ for all $x \in \beta(M)$. Since $M \subseteq \beta(M) \subseteq S$, we have $m_G(\beta(M)S) \geqslant m_G(M) > 0$, which in turn implies $m_H(\beta(M)) > 0$. Therefore, the set $\{h \in H : f(h) > \delta\}$ contains $\beta(M)$ which has positive measure, and this implies $\|f\|_\infty > \delta$. As before, since this holds for all $\delta < \|f_V\|_\infty$, we conclude that $\|f\|_\infty \geqslant \|f_V\|_\infty$.

(v) Let $f, g \in L^2(H)$. Then $f_V, g_V \in L^2(G)$ by (iii). Therefore, the convolution products $(f * \check{g})(h)$ and $(f_V * \widecheck{g_V})(x)$ exist for all $h \in H$ and $x \in G$, respectively.

For $h \in H$, it follows from (iii) that

$$\begin{aligned}
(f_V * \widetilde{g_V})(h) &= \int_G f_V(hx) g_V(x) dx \\
&= \int_G 1_V(x) f(h\beta(x)) g(\beta(x)) dx \\
&= \int_G 1_V(x) (L_{h^{-1}} f)(\beta(x)) g(\beta(x)) dx \\
&= \int_G 1_V(x) (L_{h^{-1}} f) \cdot g)(\beta(x)) dx \\
&= \int_G ((L_{h^{-1}} f) g)_V(y)\, dy \\
&= \frac{1}{c^2} \int_H (L_{h^{-1}} f) \cdot g)(y) dy \\
&= \frac{1}{c^2} (f * \check{g})(h).
\end{aligned}$$

This completes the proof of the proposition. □

With Proposition 2.6.3 at hand, we are now ready for the main result of this section.

THEOREM 2.6.4. *Let G be a locally compact group and H a closed subgroup of G. For every $u \in A(H)$ there exists $v \in A(G)$ such that $v|_H = u$ and $\|v\|_{A(G)} = \|u\|_{A(H)}$. If u is positive definite, then v can be chosen to be positive definite.*

PROOF. (a) To begin with, we consider the case when G is second countable. Let $u \in A(H)$ and let $f, g \in L^2(H)$ be such that $u = f * \check{g}$ and $\|u\|_{A(H)} = \|f\|_2 \|g\|_2$. Define $v : G \to \mathbb{C}$ by $v(x) = c^2(f_V * \widetilde{g_V})(x)$, $x \in G$. Then $v \in A(G)$ and, by Proposition 2.6.3(v), $v(h) = (f * \check{g})(h) = u(h)$ for all $h \in H$. Moreover, by assertion (iii) of Proposition 2.6.3,

$$\|v\|_{A(G)} \leqslant c^2 \|f_V\|_2 \|g_V\|_2 = \|f\|_2 \|g\|_2 = \|u\|_{A(H)}.$$

Since $\|u\|_{A(H)} = \|v|_H\|_{A(H)} \leqslant \|v\|_{A(G)}$, it follows that $\|v\|_{A(G)} = \|u\|_{A(H)}$.

(b) Next, suppose that G is σ-compact. Since G is a normal topological space, u extends to some uniformly continuous function h on G. The theorem of Kakutani and Kodaira 1.2.16 then assures that there exists a compact normal subgroup K of G such that G/K is second countable and h is constant on cosets of G/K. Then u is constant on cosets of $H \cap K$ in H. Indeed, if $x, y \in H$ are such that $y^{-1}x \in K$, then

$$u(x) = h(x) = h(y(y^{-1}x)) = h(y) = u(y).$$

Since K is compact, HK is a closed subgroup of G. Moreover, since H is σ-compact, the map $x(H \cap K) \to xK$ is a topological isomorphism from $H/H \cap K$ onto HK/K. Therefore we can define a function w on HK/K by $w(xK) = u(x)$, $x \in H$. Then w belongs to $A(HK/K)$. Since G/K is second countable, (a) applies to the closed subgroup HK/K and w and yields the existence of some $w' \in A(G/K)$ with $w'|_{HK/K} = w$ and $\|w'\|_{A(G/K)} = \|w\|_{A(HK/K)}$. Now define $v \in A(G)$ by $v(x) = w'(xK)$. For $h \in H$, we then have

$$v(h) = w'(hK) = w(hK) = u(h).$$

It is also clear that

$$\|v\|_{A(G)} = \|w'\|_{A(G/K)} = \|w\|_{A(HK/K)} = \|u\|_{A(H)}.$$

(c) Now let G be an arbitrary locally compact group and let $u = f * \check{g} \in A(H)$. We prove the existence of a σ-compact open subgroup K of G such that f and g both vanish almost everywhere on $H \backslash K$. Since $C_c(H)$ is dense in $L^2(H)$ and $C_c(H) = C_c(G)|_H$, there exist sequences $(f_n)_n$ and $(g_n)_n$ in $C_c(G)$ such that

$$\|f_n|_H - f\|_2 \to 0 \quad \text{and} \quad \|g_n|_H - g\|_2 \to 0.$$

Now simply take for K the open subgroup of G generated by the σ-compact set

$$\bigcup_{n=1}^{\infty}(\text{supp } f_n \cup \text{supp } g_n).$$

Then K has the required properties. It follows that $u = 0$ on $H \backslash K$ and

$$u|_{H \cap K} = f|_{H \cap K} * \check{g}|_{H \cap K}.$$

Now, since K is σ-compact, by (b) $u|_{H \cap K}$ admits an extension $w \in A(K)$ with $\|w\|_{A(K)} = \|u|_{H \cap K}\|_{A(H \cap K)}$.

Finally, define v to be the trivial extension of w to all of G. Then $v|_H = u$ since both functions agree on $H \cap K$ and vanish on $H \backslash K$. Furthermore,

$$\|v\|_{A(G)} = \|w\|_{A(K)} = \|u|_{H \cap K}\|_{A(H \cap K)} = \|u\|_{A(H)},$$

and we are done.

Following the construction performed in (a), (b) and (c), it is easily seen that v is positive definite whenever u is positive definite, that is, if $u = f * \check{f}$ for some $f \in L^2(H)$ with $\|u\|_{A(H)} = \|f\|_2 \|\check{f}\|_2$. □

COROLLARY 2.6.5. *Let G be a locally compact group and H a closed subgroup of G. Let $I(H) = \{u \in A(G) : u|_H = 0\}$. Then the restriction map $u \to u|_H$ from $A(G)$ onto $A(H)$ induces an isometric isomorphism $u + I(H) \to u|_H$ from the quotient algebra $A(G)/I(H)$ onto $A(H)$.*

PROOF. The map $u + I(H) \to u|_H$ is an algebra isomorphism of $A(G)/I(H)$ into $A(H)$. Proposition 2.4.1(ii) shows that this map is onto and an isometry since

$$\|u|_H\|_{A(H)} = \inf\{\|v\|_{A(G)} : v \in A(G), v - u \in I(H)\} = \|u + I(H)\|$$

for all $u \in A(G)$. □

Let H be a closed subgroup of G and let

$$r : A(G) \to A(H), \; u \to r(u) = u|_H$$

be the restriction map. We conclude this section with briefly studying the adjoint map

$$r^* : A(H)^* = VN(H) \to A(G)^* = VN(G)$$

given by $\langle r^*(S), u \rangle = \langle S, r(u) \rangle$ for $u \in A(G)$ and $S \in VN(H)$. Since r is surjective by Proposition 2.4.1(ii), r^* is injective. Recall that λ_G and λ_H denote the regular representation of G and H, respectively.

PROPOSITION 2.6.6. *Let H be a closed subgroup of the locally compact group G. The map r^* is a w^*-w^*-continuous isomorphism from $VN(H)$ onto $VN_H(G)$ satisfying $r^*(\lambda_H(x)) = \lambda_G(x)$ for all $x \in H$.*

Proof. It is clear that r^* is w^*-w^*-continuous. To see that r^* is a homomorphism, we first observe that if $x \in H$, then $r^*(\lambda_H(x)) = \lambda_G(x)$. So r^* preserves products on the set $D = \{\lambda_H(x) : x \in H\}$. Since multiplication in a von Neumann algebra is separately continuous in the w^*-topology, it follows that $r^*(ST) = r^*(S)r^*(T)$ for all $S, T \in VN(H)$.

Observe next that r^* also preserves involution. Indeed, if $T \in D$ then clearly $r^*(T^*) = (r^*(T))^*$. If $T \in VN(H)$ is arbitrary, let $(T_\alpha)_\alpha$ be a net in the linear span of D such that $T_\alpha \to T$ in the w^*-topology. Then $T_\alpha^* \to T^*$ in the w^*-topology and hence

$$r^*(T^*) = \lim_\alpha r^*(T_\alpha^*) = \lim_\alpha (r^*(T_\alpha))^* = (r^*(T))^*.$$

To see that r^* is surjective, it suffices to show that $X = r^*(VN(H))$ is w^*-closed in $VN(G)$. Since r^* is a $*$-homomorphism of the C^*-algebra $VN(H)$ into the C^*-algebra $VN(G)$, X must be norm-closed. By the open mapping theorem, the unit ball X_1 of X is contained in $r^*(VN(H)_\delta)$ for some $\delta > 0$, where

$$VN(H)_\delta = \{S \in VN(H) : \|S\| \leqslant \delta\}.$$

We claim that X_1 is w^*-closed. For that, let $(T_\alpha)_\alpha$ be a net in X_1 converging to some $T \in VN(G)$ in the w^*-topology. Then, for each α, there exists $S_\alpha \in VN(H)_\delta$ such that $r^*(S_\alpha) = T_\alpha$. After passing to a subnet if necessary, we can assume that $S_\alpha \to S$ in the w^*-topology for some $S \in VN(H)_\delta$. Since r^* is w^*-w^*-continuous, it follows that $T_\alpha = r^*(S_\alpha) \to r^*(S)$ and hence that $T = r^*(S) \in X$. As $VN(G)_1$ is w^*-closed, $T \in X_1$. Thus X_1 is w^*-closed, and consequently X must be w^*-closed by the Krein-Šmulian theorem. This finishes the proof. □

2.7. Existence of bounded approximate identities

Given any nonunital Banach algebra A, it is important to know whether A at least admits a bounded approximate identity. The theme of this section is to solve this problem for the Fourier algebra of a locally compact group G. It turns out that the existence of a bounded approximate identity in $A(G)$ is equivalent to the amenability of G (Theorem 2.7.2).

Of course, the reader will be aware of that there are several different properties that are equivalent to amenability of a locally compact group (Section 1.8). We are going to present a very much focused approach to Theorem 2.7.2 mainly using that amenability is equivalent to that the trivial representation is weakly contained in the left regular representation.

The following proposition provides the main step towards showing that amenability of G implies the existence of a bounded approximate identity.

Proposition 2.7.1. *Let G be a locally compact group and let $(u_\alpha)_\alpha$ be a net in $P(G)$ that converges to some $u \in P(G)$ uniformly on compact subsets of G. Then*

$$\lim_\alpha \|(u_\alpha - u)v\|_{A(G)} = 0$$

for every $v \in A(G)$.

Proof. Note first that since $u_\alpha(e) \to u(e)$, we can assume that the net $(u_\alpha(e))_\alpha$ is bounded. Because $P(G) \cap C_c(G)$ spans a dense subspace of $A(G)$ and

$$\sup_\alpha \|(u_\alpha - u)w\|_{A(G)} \leqslant (u(e) + \sup_\alpha u_\alpha(e))\|w\|_{A(G)}$$

for every $w \in A(G)$, it is sufficient to show that

$$\lim_\alpha \|(u_\alpha - u)v\|_{A(G)} = 0$$

for each $v \in P(G) \cap C_c(G)$. Fix such v and put $w = uv$ and $w_\alpha = u_\alpha v$. Then $w_\alpha, w \in P(G)$ and the supports of w_α and w are all contained in the compact set $K = \operatorname{supp} v$. Since $u_\alpha \to u$ uniformly on K, we have $\|w_\alpha - w\|_\infty \to 0$. Since $w \in P(G) \cap C_c(G)$, $\lambda_G(w)$ is a positive bounded operator on $L^2(G)$ and there exists $g \in L^2(G)$ such that

$$w = g * \widetilde{g} \quad \text{and} \quad g * f = \lambda_G(w)^{1/2} f$$

for all $f \in C_c(G)$. Similarly, there exist functions $g_\alpha \in L^2(G)$ such that

$$w_\alpha = g_\alpha * \widetilde{g_\alpha} \quad \text{and} \quad g_\alpha * f = \lambda_G(w_\alpha)^{1/2} f$$

for all $f \in C_c(G)$.

We are going to show that $\|g_\alpha - g\|_2 \to 0$. For $f \in C_c(G)$, we have $\|R_y f\|_2 = \Delta(y)^{1/2}\|f\|_2$ and therefore, using vector-valued integration,

$$\begin{aligned}
\|(w_\alpha - w) * f\|_2 &= \left\| x \to \int_G [w_\alpha(xy) - w(xy)] f(y^{-1})\, dy \right\|_2 \\
&= \left\| x \to \int_G [w_\alpha(y) - w(y)] \Delta(y^{-1}) L_{y^{-1}} f(x)\, dy \right\|_2 \\
&\leqslant \int_G |w_\alpha(y) - w(y)| \cdot \|f\|_2 \, dy \\
&= \int_K |w_\alpha(y) - w(y)| \cdot \|f\|_2 \, dy \\
&\leqslant \|w_\alpha - w\|_\infty \|f\|_2 |K|.
\end{aligned}$$

This implies $\|\lambda_G(w_\alpha) - \lambda_G(w)\| \to 0$ and hence also that $c = \sup_\alpha \|\lambda_G(w_\alpha)\| < \infty$. Now, employing the continuous functional calculus for C^*-algebras and approximating the function $\sqrt{t}$ through polynomials uniformly on the interval $[0, c]$, we conclude that also

$$\|\lambda_G(w_\alpha)^{1/2} - \lambda_G(w)^{1/2}\| \to 0.$$

This in turn yields

$$\lim_\alpha \|(w_\alpha - w) * g\|_2 = \lim_\alpha \|(\lambda_G(w_\alpha)^{1/2} - \lambda_G(w)^{1/2}) g\|_2 = 0$$

for all $g \in C_c(G)$, and hence also

$$\lim_\alpha \langle g_\alpha - g, f * h \rangle = \lim_\alpha \langle (g_\alpha - g) * h^*, f \rangle = 0$$

for all $f, h \in C_c(G)$. Since $C_c(G) * C_c(G)$ is dense in $L^2(G)$ and $\sup_\alpha \|g_\alpha\|_2^2 = \sup_\alpha w_\alpha(e) < \infty$, it follows that

$$\lim_\alpha \langle g_\alpha, f \rangle = \langle g, f \rangle$$

for all $f \in L^2(G)$. Since also $\|g_\alpha\|_2^2 = w_\alpha(e) \to w(e) = \|g\|_2^2$, we get

$$\lim_\alpha \|g_\alpha - g\|_2^2 = 2\,\|g\|_2^2 - 2 \lim_\alpha \langle g_\alpha, g \rangle = 0.$$

This finally implies that

$$\begin{aligned}\|u_\alpha v - uv\|_{A(G)} &= \|w_\alpha - w\|_{A(G)} \\ &= \|g_\alpha * \tilde{g}_\alpha - g * \tilde{g}\|_{A(G)} \\ &= \|g_\alpha * (\tilde{g}_\alpha - \tilde{g}) + (g_\alpha - g) * \tilde{g}\|_{A(G)} \\ &\leqslant \|g_\alpha - g\|_2(\|g_\alpha\|_2 + \|g\|_2),\end{aligned}$$

which converges to 0, as was to be shown. □

We are now ready for the main result of this section.

THEOREM 2.7.2. *For a locally compact group G, the following three conditions are equivalent.*

(i) *G is amenable.*
(ii) *$A(G)$ has an approximate identity bounded by 1 and consisting of compactly supported positive definite functions.*
(iii) *$A(G)$ has a bounded approximate identity.*

PROOF. (i) ⇒ (ii) Since G is amenable, by Theorem 1.8.18 the trivial representation of G is weakly contained in the left regular representation. Thus, given any $K \in \mathcal{K}(G)$, the collection of all compact subsets of G, and $\epsilon > 0$, there exists $u_{K,\epsilon} \in P(G)$ associated with λ_G such that $|u_{K,\epsilon}(x) - 1| \leqslant \epsilon$ for all $x \in K$. Clearly, since $C_c(G)$ is dense in $L^2(G)$, we can in addition assume that $u_{K,\epsilon}$ has compact support. Now, order the set of all pairs $\alpha = (K, \epsilon)$, $K \in \mathcal{K}, \epsilon > 0$, by $\alpha \geqslant \alpha' = (K', \epsilon')$ if $K \supseteq K'$ and $\epsilon \leqslant \epsilon'$. Then the net $(u_\alpha)_\alpha$ satisfies the hypotheses of Proposition 2.7.1 with $u = 1_G$ and therefore forms an approximate identity for $A(G)$. Finally, since $u_\alpha(e) \to 1$, replacing u_α by $u_\alpha(e)^{-1}u_\alpha$, we can assume that $1 = u_\alpha(e) = \|u_\alpha\|_{A(G)}$. This shows that (ii) holds.

It remains to prove (iii) ⇒ (i). Suppose that $(u_\alpha)_\alpha$ is an approximate identity for $A(G)$ with $\|u_\alpha\|_{A(G)} \leqslant c < \infty$ for all α. We are going to show that $\|\lambda_G(f)\| = \|f\|_1$ for every $f \in C_c^+(G)$. By Proposition 1.8.20 and Theorem 1.8.18 , this property implies that G is amenable.

Let K be any compact subset of G, choose a compact neighbourhood V of e in G, and put

$$u = |V|^{-1}(1_V * 1_{V^{-1}K}) \in A(G).$$

Then, for $x \in K$,

$$u(x) = |V|^{-1} \int_G 1_V(y) 1_{V^{-1}K}(y^{-1}x)dy = 1.$$

Since $\|u_\alpha u - u\|_{A(G)} \to 0$, it follows that $u_\alpha u \to u$ uniformly on K. Thus, given $\epsilon > 0$, there exists an index α such that $\mathrm{Re}(u_\alpha(x)) \geqslant 1 - \epsilon$ for all $x \in K$. We now apply the preceding with $K = \operatorname{supp} f$, where $f \in C_c^+(G)$. Then

$$\mathrm{Re}\,\langle u_\alpha, f\rangle = \int_G f(x)\mathrm{Re}(u_\alpha(x))dx \geqslant (1-\epsilon)\|f\|_1.$$

On the other hand, we have

$$|\langle u_\alpha, f\rangle| = |\langle \lambda_G(f), u_\alpha\rangle| \leqslant c\,\|\lambda_G(f)\|.$$

Since $\epsilon > 0$ is arbitrary, we conclude that $\|f\|_1 \leqslant c\,\|\lambda_G(f)\|$ for every $f \in C_c^+(G)$. Replacing f with the n-fold convolution product f^n in the preceding calculations, it follows that

$$\|f\|_1^n = \|f^n\|_1 \leqslant c\,\|\lambda_G(f^n)\| \leqslant c\,\|\lambda_G(f)\|^n$$

and therefore

$$\|f\|_1 \leqslant \|\lambda_G(f)\| \cdot \lim_{n\to\infty} c^{1/n} = \|\lambda_G(f)\| \leqslant \|f\|_1.$$

This finishes the proof of (iii) $\Rightarrow$ (i). □

Theorem 2.7.2 raises the question of whether the Fourier algebra $A(G)$ of a nonamenable locally compact group G possesses an approximate identity which is bounded in some norm weaker than the $A(G)$-norm or at least an unbounded approximate identity. This appears to be a very difficult problem, which we are going to touch in Chapter 5. However, it should be mentioned that even the weakest possible variant, namely whether $u \in \overline{uA(G)}$ holds for every $u \in A(G)$, appears to be a widely open problem.

We continue with several interesting applications of Theorem 2.7.2.

COROLLARY 2.7.3. *Let G be an amenable locally compact group and C a compact subset of G. Then, given $\epsilon > 0$, there exists $u \in A(G) \cap C_c(G)$ such that $u = 1$ on C and $\|u\| \leqslant 1 + \epsilon$.*

PROOF. Choose $0 < \delta < 1$ such that $(1 + 2\delta)(1 - \delta)^{-1} \leqslant 1 + \epsilon$. Since $A(G)$ is regular, there exists $w \in A(G)$ such that $w = 1$ on C. Because $A(G)$ has an approximate identity bounded by 1, by Cohen's factorization theorem [**126**] we can decompose w as $w = vw_1$, where $\|v\| \leqslant 1$ and $\|w_1 - w\| \leqslant \delta$. Let $u_1 = v - v(w - w_1) \in A(G)$. Then $u_1 = 1$ on C and $\|u_1\| \leqslant 1 + \delta$. Finally, choose $u_2 \in A(G) \cap C_c(G)$ such that $\|u_2 - u_1\| \leqslant \delta$ and define $u \in A(G)$ by the norm-convergent sum

$$u = u_2 \cdot \sum_{n=0}^{\infty} (u_1 - u_2)^n.$$

Then u has compact support, $u = 1$ on C and

$$\|u\| \leqslant \frac{\|u_2\|}{1 - \|u_1 - u_2\|} \leqslant \frac{1 + 2\delta}{1 - \delta} \leqslant 1 + \epsilon,$$

as required. □

COROLLARY 2.7.4. *Let H be a proper closed subgroup of a locally compact G such that the ideal $I(H)$ has a bounded approximate identity. Then H is amenable.*

PROOF. By Theorem 2.7.2, it suffices to show that $A(H)$ has a bounded approximate identity. Choose any $x \in G\backslash H$. The ideal $I(xH)$ also has a bounded approximate identity, say $(u_\alpha)_\alpha$. For each α, let $v_\alpha = u_\alpha|_H \in A(H)$. Then $\|v_\alpha\|_{A(H)} \leqslant \|u_\alpha\|_{A(G)}$, and since $A(H) \cap C_c(H)$ is dense in $A(H)$, it is enough to verify that $\|v_\alpha v - v\|_{A(H)} \to 0$ for every $v \in A(H) \cap C_c(H)$. Fix such a v and let $C = \operatorname{supp} v$ and choose a neighbourhood V of C such that $V \cap xH = \emptyset$. Since $A(G)$ is regular, there exists $u_0 \in A(G)$ such that $u_0|_C = 1$ and $\operatorname{supp} u_0 \subseteq V$. Since the restriction map $A(G) \to A(H)$ is surjective, there exists $u \in A(G)$ with $u|_H = v$. Then $w = u_0 u$ belongs to $I(xH)$ and $w|_H = v$. Finally,

$$\lim_\alpha \|v_\alpha v - v\|_{A(H)} \leqslant \lim_\alpha \|w_\alpha w - w\|_{A(G)} = 0,$$

which completes the proof. □

THEOREM 2.7.5. *For a locally compact group G, $G \neq \{e\}$, the following are equivalent.*

(i) *G is amenable.*
(ii) *The ideal $I = I(\{e\})$ has a bounded approximate identity.*
(iii) *There exists a proper closed subgroup H of G such that $I(H)$ has a bounded approximate identity.*
(iv) *There exists an amenable proper closed subgroup H of G such that $I(H)$ has a bounded approximate identity.*

PROOF. (i) $\Rightarrow$ (ii) Let $(u_\alpha)_\alpha$ be an approximate identity for $A(G)$ with $\|u_\alpha\| \leqslant 1$ for all α and let $(W_\beta)_\beta$ be a neighbourhood basis of e. For each β, there exists $w_\beta \in P(G)$ such that supp $w_\beta \subseteq W_\beta$ and $w_\beta(e) = 1$. For each pair (α, β), let

$$v_{\alpha,\beta} = u_\alpha - u_\alpha(e) w_\beta \in A(G).$$

Then $v_{\alpha,\beta} \in I(\{e\})$ and

$$\|v_{\alpha,\beta}\| \leqslant \|u_\alpha\| + |u_\alpha(e)| \cdot \|w_\beta\| \leqslant 2.$$

Now let $u \in I(\{e\})$ and $\epsilon > 0$ be given. Since the singleton $\{e\}$ is set of synthesis (see Chapter 6), there exists $v \in j(\{e\})$ such that $\|v - u\| \leqslant \epsilon$. For β large enough, $W_\beta \cap \text{supp}\, v = \varnothing$ and hence $w_\beta v = 0$. It follows that

$$\begin{aligned} \|v_{\alpha,\beta} u - u\| &\leqslant \|v_{\alpha,\beta}(u - v)\| + \|v_{\alpha,\beta}\, v - v\| + \|v - u\| \\ &\leqslant 3\epsilon + \|v_{\alpha,\beta} v - v\| = 3\epsilon + \|u_\alpha v - v\| \end{aligned}$$

since $w_\beta v = 0$. Since $(u_\alpha)_\alpha$ is an approximate identity for $A(G)$ and $\epsilon > 0$ is arbitrary, it follows that the net $(v_{\alpha,\beta})_{\alpha,\beta}$ is an approximate identity for $I(\{e\})$.

(ii) $\Rightarrow$ (iii) being trivial, assume that (iii) holds. Then the subgroup H must be amenable by Corollary 2.7.4, so (iv) holds.

(iv) $\Rightarrow$ (i) Since H is amenable, $A(H)$ has a bounded approximate identity. Then $A(G)/I(H)$, being isometrically isomorphic to $A(H)$ (Corollary 2.7.4), also has a bounded approximate identity. Since both $I(H)$ and $A(G)/I(H)$ have bounded approximate identities, the same is true for $A(G)$ (Proposition 1.1.5). Therefore, G is amenable (Theorem 2.7.2). □

Corollary 2.7.4 and Theorem 2.7.5 suggest the question of which impact on the group G the existence of just some closed ideal of $A(G)$ with bounded approximate identity might have. The remaining part of this section is devoted to clarify this question.

LEMMA 2.7.6. *Let I be a closed ideal of $A(G)$ with bounded approximate identity. Then the interior $h(I)^\circ$ of the hull $h(I)$ is closed in G and the boundary $\partial(h(I))$ has measure zero. Moreover, $1_{G \setminus h(I)^\circ} \in B_\lambda(G) = C^*_\lambda(G)^*$.*

PROOF. Let $E = h(I)$ and let $(u_\alpha)_\alpha$ be a bounded approximate identity for I. After passing to a subnet if necessary, we can assume that $w^* - \lim_\alpha u_\alpha = u$ for some $u \in C^*_\lambda(G)^* = B_\lambda(G)$. We claim that $u = 1_{G \setminus E^\circ}$. To see this, let $v \in I$, $f \in L^1(G)$ and $\epsilon > 0$. Then

$$\langle uv, f \rangle = \langle u, vf \rangle = \lim_\alpha \langle u_\alpha, vf \rangle = \lim_\alpha \langle u_\alpha v, f \rangle.$$

Now choose α such that $|\langle uv, f\rangle - \langle u_\alpha v, f\rangle| \leqslant \epsilon$ and $\|u_\alpha v - v\| \leqslant \epsilon$. Then

$$\begin{aligned} |\langle uv, f\rangle - \langle v, f\rangle| &\leqslant |\langle uv, f\rangle - \langle u_\alpha v, f\rangle| + |\langle u_\alpha v, f\rangle - \langle v, f\rangle| \\ &\leqslant \epsilon + \|f\|_1 \|u_\alpha v - v\| \leqslant \epsilon(1 + \|f\|_1). \end{aligned}$$

Since $\epsilon > 0$ is arbitrary, it follows that $\langle uv, f\rangle = \langle v, f\rangle$ for every $f \in L^1(G)$, and hence $uv = v$ for every $v \in I$. This implies that $u(x) = 0$ on E° and $u(x) = 1$ on $G\backslash E$, and therefore $u = 1$ on $G\backslash E^\circ$. Thus $u = 1_{G\backslash E^\circ}$. In particular, E° is closed in G.

It remains to observe that $|\partial(E)| = 0$. For a contradiction, assume that $|\partial(E)| > 0$. Choose a Borel subset V of $\partial(E)$ such that $0 < |V| < \infty$, and let $f = 1_V$. Then, since $u = 1$ on $\partial(E)$ and $(u_\alpha)_\alpha \subseteq I$,

$$0 < |V| = \langle u, f\rangle = \lim_\alpha \langle u_\alpha, f\rangle = 0.$$

This contradiction finishes the proof. □

LEMMA 2.7.7. *Let I be a non-zero closed ideal in $A(G)$ with bounded approximate identity. Then either G is amenable or $h(I)$ has positive measure.*

PROOF. Assume that $|h(I)| = 0$. To conclude that G is amenable, it suffices to show that $\|f\|_1 = \|\lambda_G(f)\|$ for every $f \in C_c(G)$ with $f \geqslant 0$, where λ_G denotes the left regular representation of G (Section 1.8). Let M be a norm bound for an approximate identity of I.

Let $f \in C_c^+(G)$ and let C be any compact subset of G such that $\operatorname{supp} f \subseteq C$, and let $\epsilon > 0$. Since $|h(I)| = 0$, there exists an open neighbourhood V of $h(I)$ such that $|V| \cdot \|f\|_\infty \cdot M \leqslant \epsilon$. Since the bounded approximate identity of I converges to 1 uniformly on compact subsets, which are disjoint from $h(I)$, there exists $u \in I$ such that

$$\inf\{\operatorname{Re}(u(x)) : x \in C\backslash V\} \geqslant 1 - \epsilon.$$

Then $|\langle u, f\rangle| \leqslant \|u\| \cdot \|\lambda_G(f)\| \leqslant M \cdot \|\lambda_G(f)\|$. On the other hand,

$$\begin{aligned} \operatorname{Re}(\langle u, f\rangle) &= \int_C \operatorname{Re}(u(x)) f(x)dx \\ &\geqslant (1-\epsilon)\|f\|_1 + \int_V \operatorname{Re}(u(x)) f(x)dx \\ &\geqslant (1-\epsilon)\|f\|_1 - |V| \cdot \|f\|_\infty \cdot M \geqslant (1-\epsilon)\|f\|_1 - \epsilon. \end{aligned}$$

Since $\epsilon > 0$ is arbitrary, it follows that

$$M \cdot \|\lambda_G(f)\| \geqslant |\langle u, f\rangle| \geqslant \|f\|_1$$

for every $f \in C_c^+(G)$. Thus, for $g \in C_c^+(G)$ and any $n \in \mathbb{N}$,

$$\|g\|_1^n = \|g^n\|_1 \leqslant M\|\lambda_G(g^n)\| \leqslant M \cdot \|\lambda_G(g)\|^n,$$

and hence $\|g\|_1 \leqslant \|\lambda_G(g)\|$, as required. □

COROLLARY 2.7.8. *Suppose that G is connected and I is a nonzero closed ideal of $A(G)$ with bounded approximate identity. Then G is amenable.*

PROOF. Assuming that G is not amenable, $|h(I)| > 0$ by Lemma 2.7.7. On the other hand, by Lemma 2.7.6, $|\partial(h(I))| = 0$ and $h(I)^\circ$ is closed in G. Since G is connected, either $h(I)^\circ = \varnothing$ or $h(I) = G$. However, since $I \neq \{0\}$, $h(I) \neq G$. Thus $0 < |h(I)| = |h(I)^\circ \cup \partial(h(I))| = |\partial(h(I))| = 0$, a contradiction. □

THEOREM 2.7.9. *Suppose that $A(G)$ has a nonzero closed ideal which possesses a bounded approximate identity. Then G has an amenable open subgroup.*

PROOF. We first observe that G can be assumed to be almost connected. In fact, since $I \neq \{0\}$, there exists $x \in G \backslash h(I)$ and, by translating if necessary, we can assume that $x \in G_0$, the connected component of the identity. Fix an open subgroup H of G such that H/G_0 is compact. Then $1_H \cdot I$ can be viewed as a closed ideal of $A(H)$, which is nonzero and has a bounded approximate identity. If G_0 has been shown to be amenable, then H is amenable as well because H/G_0 is compact. Thus we can assume that G is almost connected.

Recall that the restriction map $r : A(G) \to A(G_0), u \to u|_{G_0}$ is surjective and norm decreasing. Therefore, $J = \overline{r(I)}$ is a closed ideal of $A(G_0)$ and J has a bounded approximate identity. Finally, $J \neq \{0\}$ since $x \in G_0 \backslash h(I)$. Corollary 2.7.8 now implies that G_0 is amenable. □

In Section 6.5, employing operator space theory, we shall explicitly describe all the closed ideals of $A(G)$ when G is amenable in terms of the closed coset ring of G.

2.8. The subspaces $A_\pi(G)$ of $B(G)$

In this section we associate to any unitary representation π of the locally compact group G a closed linear subspace $A_\pi(G)$ of $B(G)$ and present several results about these spaces, which will be used to determine $B(G)$ for some specific groups G (Section 2.9) and also in Chapters 3 and 4.

DEFINITION 2.8.1. Let π be a unitary representation of G. Let

$$A_\pi(G) = \overline{\operatorname{span}\{\varphi_{\xi,\eta} : \xi, \eta \in \mathcal{H}(\pi)\}}^{\|\cdot\|} \subseteq B(G),$$

where $\varphi_{\xi,\eta}(x) = \langle \pi(x)\xi, \eta \rangle$, $x \in G$. The space $A_\pi(G)$ is often called the *Fourier space* associated with the representation π. Moreover, let

$$VN_\pi(G) = \{\pi(x) : x \in G\}'' = \overline{\operatorname{span}\{\pi(x) : x \in G\}}^{w^*} \subseteq \mathcal{B}(\mathcal{H}(\pi)).$$

Note that when $\pi = \lambda_G$, then the set $A(G)$ of all coordinate functions of λ_G is already a closed linear subspace of $B(G)$ and hence $A(G) = A_{\lambda_G}(G)$.

If σ is a subrepresentation of π, then $A_\sigma(G)$ is a subspace of $A_\pi(G)$ by the very definition. For a more precise statement, see Lemma 2.8.3 below. We first identify the dual space of $A_\pi(G)$.

LEMMA 2.8.2. *For any representation π of G, the dual space $A_\pi(G)^*$ of $A_\pi(G)$ is isometrically isomorphic to $VN_\pi(G)$.*

PROOF. Let $\mathcal{B}(\mathcal{H}(\pi))_*$ denote the norm closure in $\mathcal{B}(\mathcal{H}(\pi))^*$ of the linear span of all linear functionals on $\mathcal{B}(\mathcal{H}(\pi))$ of the form

$$\varphi_{\xi,\eta}(T) = \langle T\xi, \eta \rangle, \quad T \in \mathcal{B}(\mathcal{H}(\pi)),\ \xi, \eta \in \mathcal{H}(\pi).$$

Then, as is well known, $\mathcal{B}(\mathcal{H}(\pi))_*$ is the unique predual of $\mathcal{B}(\mathcal{H}(\pi))$. Moreover, let $\mathcal{E}_\pi = \{\varphi|_{VN_\pi(G)} : \varphi \in \mathcal{B}(\mathcal{H}(\pi))_*\}$; then $\mathcal{E}_\pi^* = VN_\pi(G)$. For each $\varphi \in \mathcal{E}_\pi$, $\varphi \circ \pi \in A_\pi(G)$ and

$$\|\varphi|_{VN_\pi(G)}\| = \sup\left\{\left|\sum_{j=1}^n \lambda_j \varphi(\pi(x_j))\right| : \left\|\sum_{j=1}^n \lambda_j \pi(x_j)\right\| \leqslant 1\right\} = \|\varphi \circ \pi\|$$

by Lemma 2.1.8. □

Let V be a closed subspace of $B(G) = C^*(G)^*$, it follows readily from [**270**, Theorem 2.7, p. 123] that V is invariant, that is, $f \cdot V \cup V \cdot f \subseteq V$ for every $f \in C^*(G)$, if and only if V is left and right translation invariant. In this case there exists a unique central projection p in $C^*(G)^{**}$ such that $V = B(G) \cdot p$ [**270**, Theorem 2.7(iii), p. 123]. Since $A_\pi(G)$ is two-sided translation invariant, it is an invariant subspace of $B(G)$. We shall see next that conversely every closed invariant subspace of $B(G)$ is of this form.

LEMMA 2.8.3. *Let π be a unitary representation of G.*

(i) *If σ is a subrepresentation of π, then there exists a unique central projection P in*

$$\{T \in VN_\pi(G) : T \cdot A_\pi(G) \cup A_\pi(G) \cdot T \subseteq A_\pi(G)\} \subseteq \mathcal{B}(\mathcal{H}(\pi))$$

such that $A_\sigma(G) = P \cdot A_\pi(G)$.

(ii) *If V is a closed translation invariant subspace of $A_\pi(G)$, then $V = A_\sigma(G)$ for some subrepresentation σ of π.*

PROOF. (i) follows from [**270**, Theorem 2.7(iii), p. 123].

(ii) Given V, there exists a central projection P in $VN_\pi(G)$ such that $V = P \cdot A_\pi(G)$. Let $\mathcal{L} = P(\mathcal{H}(\pi))$ and define σ by $\sigma(x) = \pi(x)|_{\mathcal{L}}$, $x \in G$. Then σ is a subrepresentation of π on L and $A_\sigma(G) = V$. □

The following theorem gives an explicit description of the functions in $A_\pi(G)$.

THEOREM 2.8.4. *Let π be a representation of the locally compact group G.*

(i) *Let $(\xi_n)_n$ and $(\eta_n)_n$ be sequences in $\mathcal{H}(\pi)$ such that $\sum_{n=1}^{\infty} \|\xi_n\| \cdot \|\eta_n\|_n < \infty$. Then*

$$u(x) = \sum_{n=1}^{\infty} \langle \pi(x)\xi_n, \eta_n \rangle, \; x \in G,$$

defines an element of $A_\pi(G)$ and $\|u\| \leqslant \sum_{n=1}^{\infty} \|\xi_n\| \cdot \|\eta_n\|_n$.

(ii) *For each $u \in A_\pi(G)$ there exist sequences $(\xi_n)_n$ and $(\eta_n)_n$ in $\mathcal{H}(\pi)$ such that*

$$u(\cdot) = \sum_{n=1}^{\infty} \langle \pi(\cdot)\xi_n, \eta_n \rangle \quad \text{and} \quad \|u\| = \sum_{n=1}^{\infty} \|\xi_n\| \cdot \|\eta_n\|_n.$$

PROOF. (i) is clear since the series defining $u(x)$ is absolutely convergent and $\|\langle \pi(\cdot)\xi, \eta \rangle\| \leqslant \|\xi\| \cdot \|\eta\|$ for each $\xi, \eta \in \mathcal{H}(\pi)$. (ii) As in Section 2.1, we now use the polar decomposition of elements in the predual of a von Neumann algebra. For every $u \in A_\pi(G) = VN_\pi(G)_*$, there exist a partial isometry $V \in VN_\pi(G)$ and an element $|u| \in A_\pi(G)$ such that $u = V \cdot |u|$, $\|u\| = \||u|\|$ and $|u|$ defines a positive normal linear functional on $VN_\pi(G)$. Since the linear functional $|u|$ is positive, there exists a sequence $(\eta_n)_n$ in $\mathcal{H}(\pi)$ such that $|u|(x) = \sum_{n=1}^{\infty} \langle \pi(x)\eta_n, \eta_n \rangle$ for all $x \in G$ [**59**, Theorem 1, p.54]. It follows that $\||u|\| = |u|(e) = \sum_{n=1}^{\infty} \|\eta_n\|^2$. Now,

$$u(x) = (V \cdot |u|)(x) = \sum_{n=1}^{\infty} \langle \pi(x)V\eta_n, \eta_n \rangle,$$

and hence

$$\||u|\| = \|u\| \leqslant \sum_{n=1}^{\infty} \|V\eta_n\| \cdot \|\eta_n\| \leqslant \sum_{n=1}^{\infty} \|\eta_n\| \cdot \|\eta_n\| = \||u|\|.$$

Thus the stated decomposition follows by setting $\xi_n = V\eta_n$. □

LEMMA 2.8.5. *Let H and G be locally compact groups, $\phi : H \to G$ a continuous homomorphism and $j(f) = f \circ \phi$ for every function f on G. Let π be any representation of G.*

(i) $A_{\pi\circ\phi}(H) = j(A_\pi(G))$.

(ii) *For every $v \in A_{\pi\circ\phi}(H)$ there exists $u \in A_\pi(G)$ such that $v = j(u)$ and $\|v\|_{B(H)} = \|u\|_{B(G)}$.*

PROOF. (i) By Theorem 2.8.4, $A_\pi(G)$ consists precisely of those functions u in $B(G)$ for which there exist sequences $(\xi_n)_n$ and $(\eta_n)_n$ in $\mathcal{H}(\pi)$ such that

$$\sum_{n=1}^{\infty} \|\xi_n\| \cdot \|\eta_n\| < \infty \quad \text{and} \quad u(x) = \sum_{n=1}^{\infty} \langle \pi(x)\xi_n, \eta_n\rangle$$

for all $x \in G$. Consequently, $j(A_\pi(G))$ coincides with the set of all functions of the form $v(y) = \sum_{n=1}^\infty \langle \pi(\phi(y))\xi_n, \eta_n\rangle$, $y \in H$. This shows that $j(A_\pi(G)) = A_{\pi\circ\phi}(H)$.

(ii) Given $v \in A_{\pi\circ\phi}(H)$, by Theorem 2.8.4 there exist sequences $(\xi_n)_n$ and $(\eta_n)_n$ in $\mathcal{H}(\pi)$ such that $v(y) = \sum_{n=1}^\infty \langle \pi(\phi(y))\xi_n, \eta_n\rangle$ for all $y \in H$ and $\|v\| = \sum_{n=1}^\infty \|\xi_n\| \cdot \|\eta_n\|$. Thus $v = j(u)$ for some $u \in A_\pi(G)$ with $\|u\| \leqslant \sum_{n=1}^\infty \|\xi_n\| \cdot \|\eta_n\| = \|j(u)\|$. On the other hand, by Theorem 2.2.1(ii) we always have $\|j(u)\| \leqslant \|u\|$. □

REMARK 2.8.6. (1) Let H be the group G equipped with a stronger locally compact topology than the given one (e.g., the discrete topology), and let $\phi : H \to G$ be the identity map. Then the preceding lemma tells us that $A_\pi(G) = A_{\pi\circ\phi}(H)$.

(2) Let H be a closed subgroup of G and $\phi : H \to G$ the embedding. Then, by Lemma 2.8.5, $A_{\pi\circ\phi}(H) = A_\pi(G)|_H$. In particular, $A_\pi(G)|_H$ is closed in $B(H)$ for every representation π of G. Taking for π the universal representation of G, we conclude that $B(G)|_H$ is closed in $B(G)$.

We now turn to various properties of the assignment $\pi \to A_\pi(G)$.

LEMMA 2.8.7. *Let π and ρ be representations of G. Then $A_\pi(G) \cap A_\rho(G) = \{0\}$ if and only π and ρ are disjoint.*

PROOF. Suppose that π and ρ are not disjoint. Then there exists a subrepresentation σ of π with $\mathcal{H}(\sigma) \neq \{0\}$, which is equivalent to a subrepresentation of ρ, implemented by a unitary map $U : \mathcal{H}(\sigma) \to \phi(\mathcal{H}(\sigma)) \subseteq \mathcal{H}(\rho)$. Then

$$\langle \sigma(x)\xi, \eta\rangle = \langle U(\sigma(x)\xi), U\eta\rangle = \langle \rho(x)(U\xi), U\eta\rangle$$

for all $\xi, \eta \in \mathcal{H}(\sigma)$ and $x \in G$. This contradicts $A_\pi(G) \cap A_\rho(G) = \{0\}$.

For the converse, assume that π and ρ are disjoint and let $\sigma = \pi \oplus \rho$. Let P_π and P_ρ be the central projections in $VN_\sigma(G)$ provided by Lemma 2.8.3. Then P_π and P_ρ are the orthogonal projections from $\mathcal{H}(\sigma)$ onto $\mathcal{H}(\pi)$ and $\mathcal{H}(\rho)$, respectively. Since π and ρ are disjoint, it follows that $P_\pi P_\rho = 0$, and this implies that

$$\begin{aligned} A_\pi(G) \cap A_\rho(G) &= P_\pi A_\sigma(G) \cap P_\rho A_\sigma(G) = P_\pi A_\sigma(G) \cap (I - P_\pi)P_\rho A_\sigma(G) \\ &\subseteq P_\pi A_\sigma(G) \cap (I - P_\pi)A_\sigma(G) = \{0\}, \end{aligned}$$

as was to be shown. □

PROPOSITION 2.8.8. *Let $(\pi_\iota)_{\iota\in I}$ be a family of unitary representations of G and $\pi = \oplus_{\iota\in I}\pi_\iota$ their direct sum. Then $A_\pi(G)$ consists precisely of those functions in $B(G)$ which can be written as $u = \sum_{\iota\in I} u_\iota$, where $u_\iota \in A_{\pi_\iota}(G)$ and $\sum_{\iota\in I} \|u_\iota\| < \infty$.*

PROOF. Since each $A_{\pi_\iota}(G)$ is a subspace of $A_\pi(G)$, every family $(u_\iota)_{\iota\in I}$ satisfying $u_\iota \in A_{\pi_\iota}(G)$ and $\sum_{\iota\in I}\|u_\iota\| < \infty$, defines an element of $A_\pi(G)$ by setting $u(x) = \sum_{\iota\in I} u_\iota(x)$ for $x \in G$.

Conversely, let $u \in A_\pi(G)$ be given. Then, by Theorem 2.8.4, there exist sequences $(\xi_n)_n$ and $(\eta_n)_n$ in $\mathcal{H}(\pi)$ such that $\sum_{n=1}^\infty \|\xi_n\|\cdot\|\eta_n\| < \infty$ and

$$u(x) = \sum_{n=1}^\infty \langle \pi(x)\xi_n, \eta_n\rangle, \quad x \in G.$$

Since $\mathcal{H}(\pi) = \oplus_{\iota\in I}\mathcal{H}(\pi_\iota)$, ξ_n and η_n can be decomposed as $\xi_n = \sum_{\iota\in I}\xi_{n\iota}and\eta_n = \sum_{\iota\in I}\eta_{n\iota}$, where $\xi_{n\iota}, \eta_{n\iota} \in \mathcal{H}(\pi_\iota)$ and

$$\|\xi_n\|^2 = \sum_{\iota\in I}\|\xi_{n\iota}\|^2 \quad \text{and} \quad \|\eta_n\|^2 = \sum_{\iota\in I}\|\eta_{n\iota}\|^2.$$

This implies, for each $x \in G$ and $n \in \mathbb{N}$,

$$\langle \pi(x)\xi_n, \eta_n\rangle = \sum_{\iota\in I}\langle \pi_\iota(x)\xi_{n\iota}, \eta_{n\iota}\rangle.$$

Now the family of complex numbers $(\langle \pi_\iota(x)\xi_{n\iota}, \eta_{n\iota}\rangle)_{\iota\in I, n\in\mathbb{N}}$ is absolutely summable. Indeed, since $|\langle \pi_\iota(x)\xi_{n\iota}, \eta_{n\iota}\rangle)| \leqslant \|\xi_{n\iota}\|\cdot\|\eta_{n\iota}\|$ and

$$\sum_{\iota\in I}\|\xi_{n\iota}\|\cdot\|\eta_{n\iota}\| \leqslant \left(\sum_{\iota\in I}\|\xi_{n\iota}\|^2\right)^{1/2}\left(\sum_{\iota\in I}\|\eta_{n\iota}\|^2\right)^{1/2},$$

it follows that

$$\sum_{n=1}^\infty\left(\sum_{\iota\in I}\|\xi_{n\iota}\|\cdot\|\eta_{n\iota}\|\right) \leqslant \sum_{n=1}^\infty \|\xi_n\|\cdot\|\eta_n\| < \infty.$$

Thus we can write, for $x \in G$,

$$u(x) = \sum_{\iota\in I}\left(\sum_{n=1}^\infty\langle \pi_\iota(x)\xi_{n\iota}, \eta_{n\iota}\rangle\right).$$

Since $\sum_{n=1}^\infty \|\xi_{n\iota}\|\cdot\|\eta_{n\iota}\| < \infty$, we can now define $u_\iota \in A_{\pi_\iota}(G)$ by

$$u_\iota(x) = \sum_{n=1}^\infty \langle \pi_\iota(x)\xi_{n\iota}, \eta_{n\iota}\rangle.$$

Then $\|u_\iota\| \leqslant \sum_{n=1}^\infty \|\xi_{n\iota}\|\cdot\|\eta_{n\iota}\|$ and therefore

$$\begin{aligned}\sum_{\iota\in I}\|u_\iota\| &\leqslant \sum_{n=1}^\infty\left(\sum_{\iota\in I}\|\xi_{n\iota}\|\cdot\|\eta_{n\iota}\|\right)\\ &\leqslant \sum_{n=1}^\infty\left(\sum_{\iota\in I}\|\xi_{n\iota}\|^2\right)^{1/2}\left(\sum_{\iota\in I}\|\eta_{n\iota}\|^2\right)^{1/2}\\ &= \sum_{n=1}^\infty \|\xi_n\|\cdot\|\eta_n\|.\end{aligned}$$

This finishes the proof. □

PROPOSITION 2.8.9. *Let π and ρ be* disjoint representations *of G. Then*

(i) $A_{\pi\oplus\rho}(G) = A_\pi(G)\oplus_1 A_\rho(G)$.

(ii) $V_{\pi\oplus\rho}(G) = V_\pi(G)\oplus_\infty V_\rho(G)$.

PROOF. (i) We know from Lemma 2.8.7 that $A_\pi(G) \cap A_\rho(G) = \{0\}$. On the other hand, $A_{\pi\oplus\rho}(G) = A_\pi(G) + A_\rho(G)$ by Proposition 2.8.8. It thus remains to show that if $u \in A_{\pi\oplus\rho}(G)$, $u = u_1 + u_2$, where $u_1 \in A_\pi(G)$ and $u_2 \in A_\rho(G)$, then $\|u\| \geqslant \|u_1\| + \|u_2\|$.

To that end, let $\sigma = \pi \oplus \rho$ and $\xi, \eta \in \mathcal{H}(\sigma)$ such that $u(x) = \langle \sigma(x)\xi, \eta\rangle$ for all $x \in G$ and $\|u\| = \|\xi\| \cdot \|\eta\|$. Then, with P_π and P_ρ as in the proof of Lemma 2.8.7,

$$u_1(x) = \langle \sigma(x)P_\pi\xi, P_\pi\eta\rangle \quad \text{and} \quad u_2(x) = \langle \sigma(x)P_\rho\xi, P_\rho\eta\rangle$$

for all $x \in G$. Since $P_\pi + P_\rho = I_{\mathcal{H}(\sigma)}$,

$$\begin{aligned}\|u_1\| + \|u_2\| &\leqslant \|P_\pi\xi\| \cdot \|P_\pi\eta\| + \|P_\rho\xi\| \cdot \|P_\rho\eta\| \\ &\leqslant \left(\|P_\pi\xi\|^2 + \|P_\pi\eta\|^2\right)^{1/2} \left(\|P_\rho\xi\|^2 + \|P_\rho\eta\|^2\right)^{1/2} \\ &= \|\xi\| \cdot \|\eta\| = \|u\|.\end{aligned}$$

(ii) follows from (i) and Lemma 2.8.2. since

$$A_{\pi\oplus\rho}(G)^* = (A_\pi(G) \oplus_1 A_\rho(G))^* = V_\pi(G) \oplus_\infty V_\rho(G) = V_{\pi\oplus\rho}(G)$$

by disjointness of π and ρ. □

COROLLARY 2.8.10. *Let V be the closed linear span of all coefficient functions of those unitary representations of G which are disjoint from the left regular representation λ of G. Then $V = A_\sigma(G)$ for some representation σ of G and $B(G) = A_{\lambda_G}(G) \oplus_1 A_\sigma(G)$.*

PROOF. Clearly, V is translation invariant. Therefore, by Lemma 2.8.3, $V = A_\sigma(G)$ for some representation σ of G. The second statement follows from Proposition 2.8.9. □

PROPOSITION 2.8.11. *Let $(\pi_\iota)_{\iota\in I}$ be a family of pairwise disjoint representations of G and let $\sigma = \oplus_{\iota\in I}\pi_\iota$ and $u \in A_\sigma(G)$. Then there exists a unique decomposition $u = \sum_{\iota\in I} u_\iota$, where $u_\iota \in A_{\pi_\iota}(G)$. Moreover, $\|u\| = \sum_{\iota\in I} \|u_\iota\|$.*

PROOF. To show the uniqueness, fix $\lambda \in I$ and set $\rho = \sum_{\iota\neq\lambda} \pi_\iota$. Then π_λ and ρ are disjoint and hence $A_\sigma(G) = A_{\pi_\lambda}(G) \oplus A_\rho(G)$ by Proposition 2.8.9. Let $u = \sum_{\iota\in I} u_\iota$ be any decomposition of u as in Proposition 2.8.8, and let $v = \sum_{\iota\in I,\iota\neq\lambda} u_\iota$. Then $u = u_\lambda + v$, $u_\lambda \in A_{\pi_\lambda}(G)$, $v \in A_\rho(G)$, and this decomposition is unique. Thus u_λ is uniquely determined for every $\lambda \in I$.

Since $\|u\| \leqslant \sum_{\iota\in I} \|u_\iota\|$, it suffices to show that $\|u\| \geqslant \sum_{\iota\in J} \|u_\iota\|$ for every finite subset J of I. Let $\pi = \oplus_{\iota\in J}\pi_\iota$ and $\rho \oplus_{\iota\in I\setminus J} \pi_\iota$, and let $v = \sum_{\iota\in J} u_\iota \in A_\pi(G)$ and $w = \sum_{\iota\in I\setminus J} u_\iota \in A_\rho(G)$. Since π and ρ are disjoint, $\|u\| = \|v\| + \|w\|$ by Proposition 2.8.9. Moreover, as J is finite and the π, $\iota \in I$, are pairwise disjoint, repeated application of Proposition 2.8.9 yields $\|v\| = \sum_{\iota\in J} \|u_\iota\|$ and hence $\|u\| \geqslant \sum_{\iota\in J} \|u_\iota\|$, as required. □

We conclude this section by answering the natural question of when, for two representations π and ρ of G, the spaces $A_\pi(G)$ and $A_\rho(G)$ coincide. To that end, we have to introduce the following notion. Two representations π and ρ of G are called *quasi-equivalent* if there is no subrepresenation of π which is equivalent to a subrepresentation of ρ and no subrepresentation of ρ which is equivalent to a subrepresentation of π. Obviously, equivalence implies quasi-equivalence, and the two notions coincide for irreducible representations.

PROPOSITION 2.8.12. *Let π and ρ be representations of G. Then $A_\pi(G) = A_\rho(G)$ if and only if π and ρ are quasi-equivalent.*

PROOF. By [**60**, Proposition 5.3.1], π and ρ are quasi-equivalent if and only if there exists an isomorphism ϕ from $VN_\pi(G)$ onto $VN_\rho(G)$ such that $\rho(f) = \phi(\pi(f))$ for every $f \in L^1(G)$.

Suppose first that $A_\pi(G) = A_\rho(G)$. Then $A_\pi(G)^* = A_\rho(G)^*$, and hence there exists a Banach space isomorphism ϕ from $VN_\pi(G)$ onto $VN_\rho(G)$. We have to show that ϕ is an algebra isomorphism. For $T \in VN_\pi(G)$, the operator $\phi(T)$ satisfies $\langle \phi(T), u\rangle = \langle T, u\rangle$ for every $u \in A_\pi(G) = A_\rho(G)$. Thus

$$\langle \phi(\pi(f)), u\rangle = \langle \pi(f), u\rangle = \int_G f(x)u(x)dx = \langle \rho(f), u\rangle$$

for every $f \in L^1(G)$, and hence $\phi(\pi(f)) = \rho(f)$. For $S = \pi(f)$ and $T = \pi(g)$, $f, g \in L^1(G)$, it follows that

$$\phi(ST) = \phi(\pi(f)\pi(g)) = \phi(\pi(f * g)) = \rho(f * g) = \phi(S)\phi(T).$$

Now, $\pi(L^1(G))$ is ultraweakly dense in $VN_\pi(G)$ and the map $(S, T) \to ST$ is separately continuous for the ultraweak topologies. Moreover, the map ϕ remains to be continuous when $VN_\pi(G)$ and $VN_\rho(G)$ are endowed with their ultraweak topologies, because these topologies equal the weak topologies $\sigma(VN_\pi(G), A_\pi(G))$ and $\sigma(VN_\rho(G), A_\rho(G))$, respectively. From these facts we conclude that $\phi(ST) = \phi(S)\phi(T)$ for all $S, T \in VN_\pi(G)$. Thus ϕ is an algebra isomorphism, and consequently π and ρ are quasi-equivalent by the criterion mentioned above.

Conversely, assume that π and ρ are quasi-equivalent, and let ϕ be an algebra isomorphism from $VN_\pi(G)$ onto $VN_\rho(G)$ satisfying $\phi(\pi(f)) = \rho(f)$ for every $f \in L^1(G)$. Since the von Neumann algebras $VN_\pi(G)$ and $VN_\rho(G)$ are isomorphic, so are their preduals. Thus there exists an isometry $\phi_* : A_\pi(G) \to A_\rho(G)$ which is defined by the following property: $\langle u, T\rangle = \langle \phi_*(u), \phi(T)\rangle$ for $u \in A_\pi(G)$ and $T \in VN_\pi(G)$. For every $f \in L^1(G)$, it follows that

$$\begin{aligned}\int_G u(x)f(x)dx &= \langle u, \pi(f)\rangle = \langle \phi_*(u), \phi(\pi(f))\rangle = \langle \phi_*(u), \rho(f)\rangle \\ &= \int_G \phi_*(u)(x)f(x)dx.\end{aligned}$$

Since u and $\phi_*(u)$ are continuous, this equation implies $u = \phi_*(u)$. This shows that $A_\pi(G) = A_\rho(G)$. □

2.9. Some examples

When G is a noncompact locally compact abelian group then, according to common understanding, the spectrum of $B(G) = M(\widehat{G})$ is an intractable object. However, there are locally compact groups G, actually certain semidirect products of abelian groups with compact groups, for which $B(G)$ turns out to be an extension of $A(G)$ by another Fourier algebra and consequently $\sigma(B(G))$ can be determined. The situation is as follows.

Let G be a semidirect product $G = N \rtimes K$, where

(1) K is a compact group and N is a locally compact abelian group and both are second countable, and

(2) the dual space $\widehat{G}$ of G is countable and decomposes as

$$\widehat{K} \circ q \cup \{\pi_k : k \in \mathbb{N}\},$$

where $q : G \to K$ is the quotient map and each π_k is a subrepresentation of the left regular representation of G.

PROPOSITION 2.9.1. *Let $G = N \rtimes K$ as above. Then*

$$B(G) = A(K) \circ q + A(G).$$

PROOF. Let $u \in B(G)$, so that $u = \langle \pi(\cdot)\xi, \eta\rangle$ for some unitary representation π of G and $\xi, \eta \in \mathcal{H}(\pi)$. By condition (2) above, π is completely decomposable. Thus

$$\pi = (\oplus_{\sigma\in\widehat{K}} m_\sigma(\sigma \circ q)) \oplus (\oplus_{k=1}^{\infty} n_k \pi_k),$$

where m_σ and n_k denote the multiplicity of $\sigma \circ q$ and π_k as subrepresentations of π. For $\sigma \in \widehat{K}$ and $k \in \mathbb{N}$, let P_σ and P_k denote the orthogonal projections associated with $m_\sigma(\sigma \circ q)$ and $n_k\pi_k$, respectively. By Proposition 2.8.11, there is a unique decomposition

$$u(x) = \sum_{\sigma\in\widehat{K}} \langle m_\sigma(\sigma \circ q)(x) P_\sigma \xi, P_\sigma \eta\rangle + \sum_{k=1}^{\infty} \langle n_k \pi_k(x) P_k\xi, P_k\eta\rangle.$$

By Proposition 2.8.11, this is an ℓ^1-direct sum, that is,

$$\|u\|_{B(G)} = \sum_{\sigma\in\widehat{K}} \|v_\sigma\|_{B(G)} + \sum_{k=1}^{\infty} \|w_k\|_{B(G)},$$

where $v_\sigma = \langle m_\sigma(\sigma \circ q)(x)P_\sigma\xi, P_\sigma\eta\rangle$ and $w_k = \langle n_k\pi_k(x)P_k\xi, P_k\eta\rangle$. Now each v_σ lies in $B(K) \circ q$ and each w_k belongs to $A(G)$. So

$$u_1 = \sum_{\sigma\in\widehat{K}} v_\sigma \in A(K) \circ q \quad \text{and} \quad u_2 = \sum_{k=1}^{\infty} w_k \in A(G),$$

and $u = u_1 + u_2$. □

For locally compact groups G as above the spectrum of $B(G)$ and its topology can be described as follows.

PROPOSITION 2.9.2. *Let $G = N \rtimes K$ as above with $B(G) = A(K) \circ q + A(G)$ and N noncompact. Then G and K embed topologically into $\sigma(B(G))$ by $x \to \varphi_x$, where $\varphi_x(v) = v(x)$ for $v \in B(G)$, and $a \to \psi_a$, where $\psi_a(v \circ q + u) = v(a)$ for $v \in A(K)$ and $u \in A(G)$, respectively. Moreover, $\sigma(B(G)) = G \cup K$, G is open in $\sigma(B(G))$ and K is closed in $\sigma(B(G))$, and a net $(\varphi_{x_\alpha})_\alpha$, $x_\alpha = y_\alpha a_\alpha$, $y_\alpha \in N$, $a_\alpha \in K$, converges to ψ_a for some $a \in K$ if and only if $a_\alpha \to a$ in K and $y_\alpha \to \infty$ in N.*

PROOF. Since $A(G)$ is an ideal in $B(G)$ and $A(K) \circ q = B(G)/A(G)$, it is clear from Theorem 2.3.7 and from general Gelfand theory that $x \to \varphi_x$ and $a \to \psi_a$ are both topological embeddings of G and of K into $\sigma(B(G))$, respectively, that $\sigma(B(G)) = G \cup K$, G is open in $\sigma(B(G))$ and K is closed in $\sigma(B(G))$.

Now let $(x_\alpha)_\alpha$ be a net in G such that $\varphi_{x_\alpha} \to \psi_a$ for some $a \in K$, and let $x_\alpha = y_\alpha a_\alpha$, $y_\alpha \in N$, $a_\alpha \in K$. For each $u \in A(G)$ and $v \in A(K)$ we then have

$$\begin{aligned} v(a) &= \psi_a(v \circ q) = \psi_a(v \circ q + u) \\ &= \lim_\alpha \varphi_{x_\alpha}(v \circ q + u) \\ &= \lim_\alpha (v(a_\alpha) + u(y_\alpha a_\alpha)). \end{aligned}$$

Taking $u = 0$, this gives $v(a_\alpha) \to v(a)$ for every $v \in A(K)$ and hence $u(y_\alpha a_\alpha) \to 0$ for each $u \in A(G)$ and regularity of $A(K)$ implies that $a_\alpha \to a$.

Towards a contradiction, assume that there exist a compact subset C of N and a subnet $(y_{\alpha_\beta})_\beta$ of $(y_\alpha)_\alpha$ such that $y_{\alpha_\beta} \in C$ for all β. Then, passing to a further subnet if necessary, we can assume that $y_{\alpha_\beta} \to y$ for some $y \in C$. Then $y_{\alpha_\beta} a_{\alpha_\beta} \to ya$ and hence

$$u(ya) = \lim_\beta u(y_{\alpha_\beta} a_{\alpha_\beta}) = 0$$

for all $u \in A(G)$. This contradiction shows that $y_\alpha \to \infty$ in N.

Conversely, if $y_\alpha \to \infty$ in N and $a_\alpha \to a$ in K, then

$$\varphi_{x_\alpha}(v \circ q + u) = v(a_\alpha) + u(x_\alpha) \to v(a) = \psi_a(v \circ q + u)$$

for all $v \in A(K)$ and $u \in A(G)$, so that $\varphi_{x_\alpha} \to \psi_a$ in $\sigma(B(G))$. □

We close this section by presenting a class of examples of locally compact groups which satisfy the conditions (1) and (2) set out ahead of Proposition 2.9.1.

EXAMPLE 2.9.3. Let p be a prime number, Ω_p the locally compact field of p-adic numbers and Δ_p the compact open subring of p-adic integers. Let $SL(n, \Delta_p)$ be the multiplicative group of $n \times n$ matrices with entries in Δ_p and determinant of valuation 1. This compact group acts on the vector space Ω_p^n by matrix multiplication. Form the semidirect product $G = \Omega_p^n \rtimes SL(n, \Delta_p)$. When $n = 1$, this group was presented by Fell as a noncompact group which nevertheless has a countable dual space, and it is therefore usually referred to as the *Fell group*.

Since $SL(n, \Delta_p)$ is compact, the semidirect product G satisfies the hypotheses that allow to determine $\widehat{G}$ by applying the Mackey machine outlined in Section 1.4. Since Ω_p^n is self-dual, we only have to determine the orbits in Ω_p^n under the matrix action of $SL(n, \Delta_p)$.

For $x = (x_1, \ldots, x_n) \in \Omega_p^n$, let $S(x)$ denote the l^∞-sphere through x, that is,

$$S(x) = \{y = (y_1, \ldots, y_n) \in \Omega_p^n : \max_{1 \leqslant j \leqslant n} |y_j| = \max_{1 \leqslant j \leqslant n} |x_j|\}.$$

We claim that the nontrivial orbits in Ω_p^n are such spheres. Note first that $SL(n, \Delta_p)$ preserves spheres. Indeed, if $A \in SL(n, \Delta_p)$ and $A(x_1, \ldots, x_n) = (y_1, \ldots, y_n)$, then

$$\max_{1 \leqslant j \leqslant n} |y_j| = \max_{1 \leqslant j \leqslant n} \left| \sum_{i=1}^n a_{ji} x_i \right| \leqslant \max_{1 \leqslant j \leqslant n} |x_j|$$

because $|a_{ji}| \leqslant 1$ for all j, i and the valuation is nonarchimedean, and then $\max_{1 \leqslant j \leqslant n} |x_j| = \max_{1 \leqslant j \leqslant n} |y_j|$ since $SL(n, \Delta_p)$ is a group. Furthermore, $SL(n, \Delta_p)$ acts transitively on the spheres. To see this, let $(x_1, \ldots, x_n)$ be a nonzero element of Ω_p^n and choose $w \in \Omega_p$ with $|w| = \max_{1 \leqslant j \leqslant n} |x_j|$. Then there exists $A \in SL(n, \Delta_p)$ such that

$$A(w, 0, \ldots, 0) = (x_1, \ldots, x_n).$$

In fact, set $a_{j1} = w^{-1}x_j$ for $1 \leqslant j \leqslant n$. Then $a_{j1} \in \Delta_p$ since $|w^{-1}x_j| \leqslant 1$, and since $\max_{1\leqslant j\leqslant n} |w^{-1}x_j| = 1$, it is possible to find a_{ji}, $1 \leqslant j \leqslant n$, $2 \leqslant i \leqslant n$, so that $A = (a_{ji})_{1\leqslant j\leqslant n, 1\leqslant i\leqslant n} \in SL(n, \Delta_p)$. The important fact now is that all these spheres are open in Ω_p^n and that they cover $\Omega_p^n \backslash \{(0, \dots, 0)\}$.

For each $k \in \mathbb{Z}$, choose $w_k \in \Omega_p$ with $|w_k| = p^{-k}$ and let $\chi_k \in \widehat{\Omega_p^n}$ be the character corresponding to $(w_k, 0, \dots, 0)$. Let S_k denote the stability group of χ_k in $SL(n, \Delta_p)$ and let $G_k = \Omega_p^n \rtimes S_k$. Then

$$\widehat{G} = \widehat{SL(n, \Delta_p)} \circ q \cup \left(\bigcup_{k\in\mathbb{Z}} \{\pi_{k,\tau} : \tau \in \widehat{S_k}\} \right),$$

where $q : G \to SL(n, \Delta_p)$ is the quotient map and

$$\pi_{k,\tau} = \mathrm{ind}_{G_k}^{G} (\chi_k \otimes \tau), \quad k \in \mathbb{Z}, \tau \in \widehat{S_k}.$$

Then each $\pi_{k,\tau}$ is a subrepresentation of the left regular representation of G since the restriction of $\pi_{k,\tau}$ to Ω_p^n is supported on the open orbit through χ_k. Thus the group G satisfies the hypotheses of Proposition 2.9.1.

2.10. Notes and references

The Fourier and Fourier-Stieltjes algebras of a locally compact group have been introduced by Eymard as generalizations of the L^1- and measure algebras, $L^1(G)$ and $M(G)$, respectively, of a locally compact abelian group G and they have been extensively studied in his seminal paper [**73**]. Actually, [**73**] has not only initiated, but also enormously influenced what has since become one of the most popular research areas in abstract harmonic analysis. All the material presented in Sections 2.1 to 2.5 is taken from [**73**], and our treatment follows very closely the excellent exposition in [**73**].

Given a commutative Banach algebra A, the first relevant problem is to determine the Gelfand spectrum of $\sigma(A)$ of A. It is a classical result that, for a locally compact abelian group G with dual group $\widehat{G}$, $\sigma(L^1(\widehat{G})) = G$. This is generalized by Proposition 2.3.2, which states that for an arbitrary locally compact group G, $\sigma(A(G))$ can be canonically identified with G and which may be considered as one of the most fundamental results of the subject area. For example, it forms the basis for Walter's [**280**] isomorphism theorems presented in Section 3.4 as well as for the study of ideal theory in $A(G)$ (Chapter 6). Equally important is the fact that $A(G)$ is a regular function algebra (Proposition 2.3.2).

In contrast, the Gelfand spectrum of the Fourier-Stieltjes algebra $B(G)$ is much less understood. It was extensively investigated by J.L. Taylor (see [**274**] and the references therein) for abelian groups and by Walter [**280**], [**281**] for general locally compact groups. It was shown, for instance, that $\sigma(B(G))$ is a semigroup with multiplication inherited from $W^*(G)$, the dual of $B(G)$, and that then G identifies with all unitary elements in $\sigma(B(G))$. Part of this is treated in detail in Section 3.2.

The identification of the Banach space dual of $A(G)$ with the von Neumann algebra $VN(G)$, generated by the left regular representation of G on the Hilbert space $L^2(G)$, plays a central role in the study of $A(G)$. In particular, the various results on the support of operators in $VN(G)$, established in this section, proved to be invaluable tools subsequently. The fact that $A(G)^* = VN(G)$ is also of great importance for the operator space structure of $A(G)$ as developed by Ruan [**245**]. It has also motivated the investigation of uniformly continuous and weakly almost periodic functionals on $A(G)$ [**102**], [**174**], which are analogues of weakly almost periodic functions on $\widehat{G}$ when G is abelian [**66**].

Let H be a closed subgroup of the locally compact group G. Then the assignment $r : u \to u|_H$ maps $A(G)$ into $A(H)$. It was proved by McMullen [**212**], and independently by Herz [**123**], that r is actually surjective. More precisely, any $v \in A(H)$ admits an extension $u \in A(G)$ with the same norm, $\|u\|_{A(G)} = \|v\|_{A(H)}$. The proof is fairly technical in that it involves the reduction to second countable groups and the existence of appropriate Borel cross-sections. Our presentation in Section 2.7 follows [**212**]. As a consequence one obtains that the adjoint map $r^* : VN(H) \to VN(G)$ is a w^*-w^*-continuous isomorphism from $VN(H)$ onto $VN_H(G)$, the w^*-closure in $VN(G)$ of the linear span of the set of all left regular representation operators $\lambda_G(x), x \in H$. This latter fact as well as surjectivity of r are frequently used later in the book, especially in Section 3.3 and Chapter 6.

There are various properties equivalent to amenability of a locally compact group G, which have been employed to give different proofs of Leptin's theorem (Theorem 2.7.2), such as

(1) $\|f\|_1 = \|\lambda_G(f)\|$ for every $f \in L^1(G)$, $f \geqslant 0$;

(2) given any compact subset K of G and $\delta > 1$, there exists a compact subset U of G such that $|KU| < \delta|U|$;

(3) the constant one function can be uniformly on compact subsets of G approximated by functions of the form $f * f^*$, $f \in C_c(G)$.

Leptin, who actually was the first to characterize a Banach algebraic property of $A(G)$ in terms of the group G, used (1) and (2). An alternative proof of Theorem 2.7.2 was given by Derighetti [**50**], based on (3) and on his own result that on the unit sphere of $B(G)$ the compact-open topology coincides with the weak topology $\sigma(B(G), L^1(G))$. The proof presented in Section 2.7 builds on (3) and on Proposition 2.7.1, which can be found in [**27**] and is attributed there to an unpublished thesis of Nielson. The important question for which locally compact groups G, the Fourier algebra $A(G)$ possesses an approximate identity which is bounded in some norm weaker than the $A(G)$-norm, has been studied by several authors (see Chapter 5). The remaining results of Section 2.7 (Theorem 2.7.5 and 2.7.9), dealing with the impact on G of the existence of some ideal in $A(G)$ with bounded approximate identity, are due to Forrest [**79, 80**].

The subspaces $A_\pi(G)$ of $B(G)$, which we treated in Section 2.8, were introduced and studied in [**5**]. Arsac proved many results beyond those we have, mainly following the exposition in [**5**], presented here. For instance, he intensively investigated the assignment $\pi \to A_\pi(G)$ under various aspects, such as forming tensor products and direct integrals of representations and inducing representations. He also clarified the natural question of when $A_\pi(G)$ is a subalgebra of $B(G)$.

Let $1 < p, q < \infty$ such that $\frac{1}{p} + \frac{1}{q} = 1$. For $f \in L^p(G)$ and $g \in L^q(G)$, the convolution product

$$(f * \check{g})(x) = \int_G f(xy)g(y)dy, \quad x \in G,$$

defines a function in $C_0(G)$ such that $\|f * \check{g}\|_\infty \leqslant \|f\|_p\|g\|_q$. Since $(f * \check{g})(x) = \int_G g(x^{-1}y)f(y)dy$, u can be viewed as a coefficient function of the left regular representation of G on $L^q(G)$. Define $A_p(G)$ to be the set of all functions $u \in C_0(G)$

for which there exist sequences $(f_n)_n$ in $L^p(G)$ and $(g_n)_n$ in $L^q(G)$ such that

$$\sum_{n=1}^{\infty} \|f_n\|_p \|g_n\|_q < \infty \quad \text{and} \quad u(x) = \sum_{n=1}^{\infty} (f_n * \widetilde{g_n})(x)$$

for all $x \in G$. For $u \in A_p(G)$, let

$$\|u\|_{A_p(G)} = \inf \left\{ \sum_{n=1}^{\infty} \|f_n\|_p \|g_n\|_q \right\},$$

where $(f_n)_n$ and $(g_n)_n$ are sequences with the above two properties. Then, of course, $\|u\|_\infty \leqslant \|u\|_{A_p(G)}$.

THEOREM 2.10.1. *$(A_p(G), \|\cdot\|_{A_p(G)})$ is a Banach algebra with respect to pointwise operations, $A_p(G) \cap C_c(G)$ is dense in $A_p(G)$ and $A_p(G)$ is uniformly dense in $C_0(G)$.*

THEOREM 2.10.2. *The spectrum of $A_p(G)$ can be canonically identified with G. More precisely, the map $x \to \gamma_x$, where $\gamma_x(u) = u(x)$ for $u \in A_p(G)$, is a homeomorphism from G onto $\sigma(A_p(G))$. Moreover, $A_p(G)$ is regular and Tauberian.*

The algebras $A_p(G)$ are usually referred to as the *Figà-Talamanca-Herz algebras*. Note that $A_2(G) = A(G)$. The preceding two theorems are due to Herz and have been shown in [**122**] and [**123**], respectively.

Several other results presented in this chapter for $A(G)$, essentially extend to the algebras $A_p(G)$, $1 < p < \infty$. We mention two significant ones. As shown by Herz [**123**], surjectivity of the restriction map $u \to u|_H$ remains true for $A_p(G)$. However, when $p \neq 2$, given $\epsilon > 0$, the existence of an extension $u \in A_p(G)$ of $v \in A_p(H)$ could only be shown to satisfy the norm inequality $\|u\|_{A_p(G)} \leqslant \|v\|_{A_p(H)} + \epsilon$. In addition, as also shown in [**123**], if H is normal and $v \in A_p(H)$ has compact support, then given an open subset U of G such that $\operatorname{supp} v \subseteq U \cap H$, there exists such an extension u of v satisfying $\operatorname{supp} u \subseteq \overline{U}$. If G is amenable, then $A_p(G)$ has an approximate identity of norm bound 1 for every p. Conversely, if $A_p(G)$ has a bounded approximate identity for some p, then G is amenable.

The notions of the Fourier and the Fourier-Stieltjes algebra of a locally compact group have been generalized in various different directions, which we now briefly indicate, confining ourselves to Fourier algebras.

Firstly, let K be a compact subgroup of the locally compact group G. Forrest [**83**] has introduced the Fourier algebra $A(G/K)$ of the left coset space G/K. This algebra can simultaneously be viewed as an algebra of functions on G/K and as the subalgebra of $A(G)$ consisting of all those functions in $A(G)$ which are constant on left cosets of K. In many respects, $A(G/K)$ behaves as nicely as does $A(G)$. For instance, as shown in [**83**], $A(G/K)$ is regular and semisimple, $\sigma(A(G/K)) = G/K$ and $A(G/K)$ has a bounded approximate identity if and only if G is amenable. The algebras $A(G/K)$ are precisely the norm closed left translation invariant $*$-subalgebras of $A(G)$ [**271**].

Secondly, let H be a locally compact hypergroup with left Haar measure. The Fourier space $A(H)$ was defined in analogy to the description of functions in $A(G)$ in terms of L^2-functions (Proposition 2.3.3). In general, $A(H)$ need not be closed under pointwise multiplication. However, it is an algebra for many important classes of hypergroups, such as double coset hypergroups. See [**217**] and the references therein.

Let G be a topological group, let $P(G)$ denote the collection of all continuous positive definite functions on G, and let $B(G)$ denote the linear span of $P(G)$.

By a σ-continuous representation of G into a W*-algebra M, we shall mean a pair (ω, M) such that ω is a homomorphism of G into

$$Mu := \{x \in M : x^*x = xx^* = 1\},$$

the group of unitaries in M, where 1 is the identity of M, and σ is the weak*-topology $\sigma(M, M_*)$ defined by the unique predual of M.

Let $\Omega(G)$ denote the collection of all σ-continuous representations $\alpha = (\omega, M)$ of G, such that $\overline{\langle\omega(G)\rangle}^{\alpha} = M$. Then $B(G)$ is precisely the collection of all complex-valued functions ϕ on G such that $\phi = \hat{f}_\alpha$ for some $f \in M_*$ and some $\alpha = (\omega, M) = (\omega_\alpha, M_\alpha)$ in $\Omega(G)$, where $\hat{f}_\alpha(a) = \langle\omega(a), f\rangle$ for all $a \in G$. For each $\phi \in B(G)$, define

$$\|\phi\| := \|\phi\|_{B(G)} = \inf\big\{\|f_\alpha\| : f_\alpha \in M_*,\ \phi = \hat{f}_\alpha, \text{ and } \alpha = (\omega, M) \in \Omega(G)\big\}.$$

Also let $M_\Omega := \sum \bigoplus M_{\omega_\alpha}$, the direct summand of the W*- algebras $M_\alpha := M_{\omega_\alpha}, \alpha \in \Omega(G)$. Define a σ-continuous homomorphism of G into M_Ω by $\omega_\Omega(a)(\alpha) := \omega_\alpha(a)$ for each $\alpha = (\omega_\alpha, M_\alpha)$ in $\Omega(G)$. Write

$$W^*(G) = \overline{\langle\omega_\Omega(G)\rangle}^{\sigma}.$$

THEOREM 2.10.3. (a) *$B(G)$ is a subalgebra of $WAP(G)$, the space of continuous weakly almost periodic functions on G, containing the constant functions. Furthermore, $\|\cdot\|$ is a norm on $B(G)$ and $(B(G), \|\cdot\|)$ is a commutative Banach algebra. More specifically, the map $\rho : W^*(G)_* \to B(G)$ defined by $\rho(f) := \hat{f}$, $f \in W^*(G)_*$, is a linear isometry from $W^*(G)_*$ onto $B(G)$. Furthermore, $\rho(f)$ is positive definite if and only if f is positive.*

(b) *If (ω, M) is any σ-continuous representation of G, then there is a W*-homomorphism h_ω from $W^*(G)$ into M such that the diagram*

$$\begin{array}{ccc} G & \xrightarrow{\omega_\Omega} & W^*(G) \\ & \searrow^{\omega} & \downarrow{h_\omega} \\ & & M \end{array}$$

is commutative. Also if $f \in M_*$, then $\hat{f}(x) = \langle h_\omega(x), f\rangle$ for all $x \in W^*(G)$.

(c) *If $\phi \in B(G)$ and $a \in G$, then the functions $l_a\phi$, $r_a\phi$, ϕ^*, $\bar{\phi}$ are all in $B(G)$ and*

$$\|l_a\phi\| = \|\phi\|, \quad \|r_a\phi\| = \|\phi\|, \quad \|\phi^*\| = \|\phi\|, \quad \|\bar{\phi}\| = \|\phi\|,$$

where $\phi^*(y) = \phi(y^*)$ for each $y \in M$.

Let A be a C^*-algebra and (π, H_π) be a non-degenerate representation of A. Let π^e be the unique extension of π to $M(A)$, the multiplier algebra of A is the locally convex topology on $M(A)$ defined by the semi-norms: $\{P_a : a \in A\}$, where

$$P_a(m) = \|a \cdot m\| + \|m \cdot a\|, \quad m \in M(A).$$

DEFINITION 2.10.4. Let G be a topological group. We call a *host algebra* of G a pair (A, η), where A is a C^*-algebra and $\eta : G \mapsto U(M(A))$ is a continuous homomorphism from G into the unitary group of $M(A)$, such that the mapping $\eta^* : \mathrm{Rep(A)} \to \mathrm{Rep(G)}, \eta^*(\pi) := \pi^{\mathrm{e}} \circ \eta$ is injective, where Rep(A) denotes the collection of $*$- representations of A, and Rep(G) the group of the continuous unitary representations of G.

Given any C*- algebra A, let $\widetilde{A}$ denote the universal enveloping von Neumann algebra A^{**} of A.

We say that (A, η) is a *full host algebra* of a topological group G, if it is a host algebra of G and if η^* is also surjective.

THEOREM 2.10.5. *Let G be a topological group with a full host algebra A. Then the von Neumann algebras $\widetilde{A}$ and $W^*(G)$ are isomorphic. In particular, $\widetilde{A}_*$ is isometrically isomorphic to $B(G)$. Furthermore, if A_1 is any C^*-algebra such that A_1^* is isometrically isomorphic to the Banach space $B(G)$, then $\widetilde{A_1}$ and $\widetilde{A}$ are either isomorphic or anti-isomorphic.*

REMARK 2.10.6. Theorem 2.10.3 was also proved in [**173**], and Theorem 2.10.5 in [**181**]. See also [**182**].

CHAPTER 3

Miscellaneous Further Topics

After having introduced in Chapter 2 the Fourier and the Fourier-Stieltjes algebras, $A(G)$ and $B(G)$, of a locally compact group G and studied some elementary, but basic functorial properties as well as the von Neumann algebra $VN(G)$, the Banach space dual of $A(G)$, we start in this chapter the more intrinsic and somewhat deeper investigation of both algebras. The core of the chapter is formed by results which are by now classical, whereas the more recent achievements are postponed to subsequent chapters.

Generalizing the classical description of idempotents in the measure algebra of a locally compact abelian group, Host [**129**] has identified the integer-valued functions in $B(G)$. Host's so-called idempotent theorem, which has found numerous applications, is shown in Section 3.1. A natural question is whether either of the Banach algebras $A(G)$ and $B(G)$ determines G as a topological group. This question has been affirmatively answered by Walter [**280**]. If G_1 and G_2 are locally compact groups and $B(G_1)$ and $B(G_2)$ (respectively, $A(G_1)$ and $A(G_2)$) are isometrically isomorphic, then G_1 and G_2 are topologically isomorphic or anti-isomorphic. These results are displayed in Section 3.2, and building on these, Pham [**234**] has determined the structure of certain homomorphisms between Fourier and Fourier-Stieltjes algebras of general locally compact groups. This is the content of Section 3.2.

Takesaki and Tatsuuma [**271**] have established Galois type correspondences between closed subgroups of G and right translation invariant von Neumann subalgebras of $L^\infty(G)$ and between closed subgroups of G and von Neumann subalgebras of $VN(G)$, which are invariant in some sense. These correspondences are given in Section 3.4. In a similar spirit, in Section 3.5 translation invariant subalgebras of $A(G)$ and of $B(G)$ are studied and related to closed subgroups of G. Let G_1 and G_2 be locally compact groups. It is an interesting problem of when $A(G_1 \times G_2)$ is algebraically isomorphic to the projective tensor product of $A(G_1)$ and $A(G_2)$. This isomorphism problem was treated by Losert [**200**], and his solution is presented in Section 3.6. Finally, in Section 3.7 various topologies on $B(G)$ are studied, the main emphasis being on comparing the relative topologies on the unit sphere of $B(G)$. It is also shown that G has to be compact whenever the w^*-topology and the norm topology coincide on the unit sphere in $B(G)$.

3.1. Host's idempotent theorem

One of the profound discoveries of Cohen [**33**] was the description of all idempotent measures on a locally compact abelian group. Since, for such G, $B(G) = M(\widehat{G})$, the natural, much more general problem was the characterization of all idempotents in the Fourier-Stieltjes algebra $B(G)$ of an arbitrary locally compact group G. This was achieved by Host [**129**]. In this section we are going to present his description of all integer-valued functions in $B(G)$, thus in particular of all idempotents in $B(G)$.

We start with some preliminary facts on operators on Hilbert spaces.

LEMMA 3.1.1. *Let $\mathcal{H}$ be a Hilbert space. Then*

(i) *If P is an orthogonal projection onto a closed subspace E of $\mathcal{H}$, then $P \geqslant 0$, $\|P\| = 1$ unless $E = \{0\}$ and $E = \{\xi \in \mathcal{H} : \|P\xi\| = \|\xi\|\}$.*

(ii) *On the set of orthogonal projections in $\mathcal{H}$, the weak operator topology and the strong operator topology agree.*

(iii) *If $(T_\alpha)_\alpha$ is a net in $\mathcal{B}(\mathcal{H})$ such that $T_\alpha \to T \in \mathcal{B}(\mathcal{H})$ in the weak operator topology and $T_\alpha^* T_\alpha = T^*T$ for all α, then T_α converges to T in the strong operator topology.*

(iv) *If $(T_\alpha)_\alpha$ is a monotone increasing net of self-adjoint operators in $\mathcal{H}$ such that $T_\alpha \leqslant I$ for all α, then $(T_\alpha)_\alpha$ converges in the strong operator topology to some self-adjoint operator T, and T is the least upper bound of $(T_\alpha)_\alpha$. In particular, if $S \in \mathcal{B}(\mathcal{H})$ is such that $0 \leqslant S \leqslant I$, then the sequence $(S^n)_n$ converges in the strong operator topology to some projection $P \in \mathcal{B}(\mathcal{H})$.*

PROOF. For (i) see [**147**, p. 109].

(ii) We have to show that if $(P_\alpha)_\alpha$ is a net of orthogonal projections in $\mathcal{H}$ and $P_\alpha \to P$ in the weak operator topology, then $\|P_\alpha\xi - P\xi\| \to 0$ for each $\xi \in \mathcal{H}$. Now,

$$\begin{aligned}\|P_\alpha\xi - P\xi\|^2 &= \langle P_\alpha\xi, P_\alpha\xi - P\xi\rangle - \langle P\xi, P_\alpha\xi - P\xi\rangle \\ &= \langle \xi, P_\alpha\xi - P_\alpha(P\xi)\rangle - \langle P\xi, P_\alpha\xi - P\xi\rangle \\ &\leqslant |\langle \xi, (P_\alpha - P)\xi\rangle| + |\langle \xi, (P - P_\alpha)P\xi\rangle| - \langle P\xi, (P_\alpha - P)\xi\rangle,\end{aligned}$$

which converges to 0.

(iii) If $\xi \in \mathcal{H}$, then

$$\begin{aligned}\|T_\alpha\xi - T\xi\|^2 &= \langle T_\alpha\xi, T_\alpha\xi - T\xi\rangle - \langle T\xi, T_\alpha\xi - T\xi\rangle \\ &= \langle T_\alpha^*T_\alpha\xi, \xi\rangle - \langle T_\alpha\xi, T\xi\rangle - \langle T\xi, T_\alpha\xi - T\xi\rangle \\ &= \langle T^*T\xi, \xi\rangle - \langle T_\alpha\xi, T\xi\rangle - \langle T\xi, T_\alpha\xi - T\xi\rangle,\end{aligned}$$

which converges to $\langle T^*T\xi, \xi\rangle - \|T\xi\|^2 = 0$.

For (iv) see [**147**, p. 307]. □

For a representation π of G, let $\overline{\pi(G)}$ denote the closure of $\pi(G)$ in $\mathcal{B}(\mathcal{H}(\pi))$ in the weak operator topology. Then $\overline{\pi(G)} \subseteq \mathcal{B}(\mathcal{H}(\pi))_1$ and $\overline{\pi(G)}$ with the weak operator topology is a compact semitopological semigroup, i.e., for each $S \in \overline{\pi(G)}$, the mappings $T \to TS$ and $T \to ST$ from $\overline{\pi(G)}$ into $\overline{\pi(G)}$ are continuous. Furthermore, $\pi : G \to \overline{\pi(G)}$ is a continuous mapping of G onto a dense subset of $\overline{\pi(G)}$.

LEMMA 3.1.2. *Let π be a unitary representation of G and let $\xi, \eta \in \mathcal{H}(\pi)$. Suppose that η is a cyclic vector and that the function u on G defined by $u(x) = \langle \pi(x)\xi, \eta\rangle$, $x \in G$, is integer-valued. Then*

(i) *For each $S \in \overline{\pi(G)}$, $\langle S\xi, \eta\rangle$ is also an integer.*

(ii) *The set $\{S\xi : S \in \overline{\pi(G)}\}$ is a discrete subset of $\mathcal{H}(\pi)$.*

PROOF. (i) follows readily by passing to limits.

(ii) Let $S, T \in \overline{\pi(G)}$ such that $\|S\xi - T\xi\| < 1/\|\eta\|$. For each $x \in G$, $\pi(x^{-1})S$ and $\pi(x^{-1})T$ belong to $\overline{\pi(G)}$, and hence $\langle S\xi, \pi(x)\eta\rangle$ and $\langle T\xi, \pi(x)\eta\rangle$ are both integers by (i). However

$$|\langle S\xi, \pi(x)\eta\rangle - \langle T\xi, \pi(x)\eta\rangle| \leqslant \|S\xi - T\xi\| \cdot \|\pi(x)\eta\| < 1.$$

Hence $\langle S\xi - T\xi, \pi(x)\eta\rangle = 0$. Since $x \in G$ is arbitrary and η is cyclic, it follows that $S\xi = T\xi$. So (ii) holds. □

We denote by $\overline{\pi(G)}_+$ the set of all positive operators in $\overline{\pi(G)}$. Then $\overline{\pi(G)}_+$ is weakly closed in $\overline{\pi(G)}$.

LEMMA 3.1.3. *Let π be a unitary representation of G and let $\xi, \eta \in \mathcal{H}(\pi)$ be cyclic vectors. Assume that the function $u(x) = \langle \pi(x)\xi, \eta\rangle$, $x \in G$, is integer-valued. Then*

(i) *$\overline{\pi(G)}_+$ is a commutative semigroup consisting entirely of projections. In particular, each $S \in \overline{\pi(G)}_+$ is a partial isometry.*

(ii) *$E = \{P\xi : P \in \overline{\pi(G)}_+\}$ is a finite subset of $\mathcal{H}(\pi)$.*

PROOF. (i) Let $S \in \overline{\pi(G)}_+$. By Lemma 3.1.1(iv), the sequence $(S^n)_n$ converges to some $P \in \overline{\pi(G)}_+$ in the strong operator topology and P is a projection. We are going to show that $P = S$. Fix $x \in G$. Then $S^n\pi(x) \in \overline{\pi(G)}$ for all n and $P\pi(x) \in \overline{\pi(G)}$. Since the sequence $(S^n\pi(x)\xi)_n$ converges to $P\pi(x)\xi$, by Lemma 3.1.2(ii) we can find $n \in \mathbb{N}$ such that $S^n\pi(x)\xi = P\pi(x)\xi$. We show now that the operators $S^n - P$ and $S - P$ have the same kernel. For that, let $\omega \in \mathcal{H}(\pi)$. If $(S - P)\omega = 0$, then clearly $(S^n - P)\omega = (S^n - P^n)\omega = 0$ since S and P commute. Conversely, let $(S^n - P)\omega = 0$. Then since $SP = P$, $(S^m - P)\omega = S^{m-n}(S^n - P)\omega = 0$ for all $m \geqslant n$. Let m be the smallest even integer such that $(S^m - P)\omega = 0$. Then

$$\langle (S^{m/2} - P)\omega, S^{m/2}\omega\rangle = \langle (S^m - P)\omega, \omega\rangle = 0$$

and therefore, since $PS = P = P^2$,

$$\begin{aligned} \|S^{m/2}(\omega)\|^2 &= \langle S^{m/2}(\omega), S^{m/2}(\omega)\rangle = \langle P(\omega), S^{m/2}(\omega)\rangle \\ &= \langle P(\omega), PS^{m/2}(\omega)\rangle = \langle PS^{m/2}(\omega), PS^{m/2}(\omega)\rangle \\ &= \|PS^{m/2}(\omega)\|^2. \end{aligned}$$

So $S^{m/2}(\omega)$ is in the range of P by Lemma 3.1.1(i) and hence

$$(S^{m/2} - P)(\omega) = S^{m/2}(\omega) - P(S^{m/2}(\omega)) = 0.$$

This contradicts the minimality of m unless $m = 2$. In this case, $(S - P)(\omega) = 0$. It follows that $S\pi(x)\xi = P\pi(x)\xi$. Since this holds for all $x \in G$ and since ξ is cyclic, we conclude that $S = P$. This shows that $\overline{\pi(G)}_+$ consists entirely of idempotents. For each $S \in \overline{\pi(G)}_+$, S^*S and SS^* both belong to $\overline{\pi(G)}_+$ and are projections. Hence S is a partial isometry.

To see that $\overline{\pi(G)}_+$ is a commutative semigroup, let $S, T \in \overline{\pi(G)}_+$. Then, by what we have seen above, ST and TS are partial isometries in $\overline{\pi(G)}$. Now, for $\xi \in \mathcal{H}$, in order that $\|ST\xi\| = \|\xi\|$, it is necessary and sufficient that $T\xi = \xi = S\xi$ and so $ST\xi = \xi$. Indeed, if $\|ST\xi\| = \|\xi\|$, then $\|T\xi\| \geqslant \|ST\xi\| = \|\xi\|$ and so $\|T\xi\| = \|\xi\|$, whence $T\xi = \xi$. Consequently, $\|\xi\| = \|ST\xi\| = \|S\xi\|$ and therefore $S\xi = \xi$. Similarly, $\|TS\xi\| = \|\xi\|$ if and only if $S\xi = \xi = T\xi$, and so $TS\xi = \xi$. This shows that $ST = TS$ and $ST \in \overline{\pi(G)}_+$.

(ii) Since $\overline{\pi(G)}_+$ consists of projections, the weak operator topology and the strong operator topology coincide on $\overline{\pi(G)}_+$ by Lemma 3.1.1(iii). Hence the set $E = \{P\xi : P \in \overline{\pi(G)}_+\}$ is a norm-compact subset of $\mathcal{H}(\pi)$. Then, by Lemma 3.1.2(ii), E must be finite. □

The following is the generalization, announced at the outset of this section, of Cohen's idempotent theorem for locally compact abelian groups to general locally compact groups.

THEOREM 3.1.4. *Let G be a locally compact group. The integer-valued functions in $B(G)$ are precisely the finite linear combinations with integer coefficients of translates of characteristic functions of open subgroups of G. In particular, the idempotents in $B(G)$ are exactly the characteristic functions of finite unions of translates of open subgroups of G.*

To start the proof of the theorem, let u be an integer-valued function in $B(G)$. We know that there exist a unitary representation π and cyclic vectors $\xi, \eta \in \mathcal{H}(\pi)$ such that $u(x) = \langle \pi(x)\xi, \eta\rangle$ for all $x \in G$. Let

$$A = \{E\xi : E \in \overline{\pi(G)}_+\}.$$

By Lemma 3.1.3, A is a finite subset of $\mathcal{H}(\pi)$. For each $\omega \in A$, let

$$S_\omega = \left\{E \in \overline{\pi(G)}_+ : E\xi = \omega\right\}.$$

Then S_ω is a closed subsemigroup of $\overline{\pi(G)}_+$. Indeed, S_ω is clearly closed and if $E, F \in S_\omega$, then

$$(FE)\xi = F\omega = F^2\xi = F\xi = \omega,$$

so that $FE \in \overline{\pi(G)}_+$.

For $\omega \in A$, let P_ω denote the unique minimal element of S_ω, and for every $B \subseteq A$, let

$$Q_B = \left(\prod_{\omega\in B} P_\omega\right)\left(\prod_{\omega\in A\setminus B}(I - P_\omega)\right)$$

and $\xi_B = Q_B(\xi)$. It follows from the commutativity of $\overline{\pi(G)}_+$ that Q_B is a projection which is a finite linear combination with integer coefficients of elements of $\overline{\pi(G)}_+$. Then $\sum_{B\subseteq A} Q_B = I$. This is clear if A is a singleton. So suppose that A

contains at least two elements and fix $\omega_0 \in A$ and let $A_0 = A\backslash\{\omega_0\}$. Then

$$\begin{aligned}
\sum_{B\subseteq A} Q_B &= \sum_{B\subseteq A_0} Q_B + \sum_{C\subseteq A_0} Q_{C\cup\{\omega_0\}} \\
&= \sum_{B\subseteq A_0}\left(\prod_{\omega\in B} P_\omega\right)\left(\prod_{\omega\in\{\omega_0\}\cup(A_0\backslash B)} (I-P_\omega)\right) \\
&\quad + \sum_{C\subseteq A_0}\left(\prod_{\omega\in C\cup\{\omega_0\}} P_\omega\right)\left(\prod_{\omega\in A_0\backslash C} (I-P_\omega)\right) \\
&= (I-P_{\omega_0})\sum_{B\subseteq A_0}\left(\prod_{\omega\in B} P_\omega\right)\left(\prod_{\omega\in A_0\backslash B} (I-P\omega)\right) \\
&\quad + P_{\omega_0}\sum_{C\subseteq A_0}\left(\prod_{\omega\in C} P_\omega\right)\left(\prod_{\omega\in A_0\backslash C} (I-P_\omega)\right) \\
&= \sum_{C\subseteq A_0}\left(\prod_{\omega\in C} P_\omega\right)\left(\prod_{\omega\in A_0\backslash C} (I-P_\omega)\right) \\
&= \sum_{C\subseteq A_0} Q_C.
\end{aligned}$$

Thus the statement $\sum_{B\subseteq A} Q_B = I$ follows by induction on the number of elements in A. In particular, $\sum_{B\subseteq A} \xi_B = \xi$. Therefore, if for $B \subseteq A$, we define u_B by

$$u_B(x) = \langle \pi(x)\xi_B, \eta\rangle, \quad x \in G,$$

then $u = \sum_{B\subseteq A} u_B$. Each u_B is a finite linear combination with integer coefficients of functions $x \to \langle \pi(x)Q\xi, \eta\rangle$, where $Q \in \overline{\pi(G)}_+$. All these latter functions in $B(G)$ are integer-valued by Lemma 3.1.2(ii) and hence so is u_B. Thus, in order to show that u is of the form indicated in Theorem 3.1.4, it suffices to verify this for each u_B, $B \subseteq A$. To that end, fix B and let $v = u_B$, $P = P_B$ and $\psi = \xi_B$. Then $P\psi = P_B^2(\xi) = P_B(\xi) = \psi$.

The crucial statement in the following lemma is (iii), and this constitutes the final essential step towards the proof of the theorem.

LEMMA 3.1.5. *Retain the above notation.*

(i) *For any $Q \in \overline{\pi(G)}_+$, either $Q\psi = 0$ or $Q \geqslant P$.*
(ii) *If $S \in \overline{\pi(G)}$ and $S\psi \neq 0$, then $S^*S \geqslant P$.*
(iii) *The subset $D = \{\varphi \in \overline{\pi(G)}(\psi) : \langle\varphi, \psi\rangle \neq 0\}$ of $\mathcal{H}(\pi)$ is finite.*

PROOF. (i) Note that $Q \geqslant P_{Q(\xi)}$ since $Q \in S_{Q(\xi)}$ and $P_{Q(\xi)}$ is the smallest element of $S_{Q(\xi)}$. If $Q(\xi) \in B$, then $Q \geqslant P_{Q(\xi)} \geqslant P_B = P$, whereas if $Q(\xi) \notin B$, then

$$Q(\psi) = Q(P_B(\xi)) = P_B(Q(\xi)) = 0$$

since $P_B \leqslant I - P_{Q(\xi)}$ and $(I - P_{Q(\xi)})Q(\xi) = 0$.

(ii) If $S\psi \neq 0$, then $\langle S^*S\psi, \psi\rangle = \|S\psi\|^2 \neq 0$ and S^*S is a projection in $\overline{\pi(G)}_+$. Then (i) shows that $S^*S \geqslant P$.

(iii) We know from Lemma 3.1.2(ii) that $\overline{\pi(G)(\psi)}$ is a discrete subset of $\mathcal{H}(\pi)$. Therefore it suffices to show that D is compact. To that end, let $(\psi_\alpha)_\alpha$ be a net in D and, for each α, choose $S_\alpha \in \overline{\pi(G)}$ such that $\psi_\alpha = S_\alpha \psi$. Since $\overline{\pi(G)}$ is compact in the weak operator topology, there exists a subnet $(S_{\alpha_\beta})_\beta$ which converges to some $S \in \overline{\pi(G)}$ in the weak operator topology. Since, by Lemma 3.1.2(i), $\langle S_{\alpha_\beta}\psi, \eta\rangle = \langle \psi_{\alpha_\beta}, \eta\rangle$ is a nonzero integer for every β, $\langle S\psi, \eta\rangle$ is also a nonzero integer. In particular, $S_{\alpha_\beta}\psi \neq 0$ and hence $S^*_{\alpha_\beta} S_{\alpha_\beta} \geqslant P$ by (ii), and then also $S^*S \geqslant P$.

Now SP is a weak operator limit of the net $(S_{\alpha_\beta}P)_\beta$ and all these operators have absolute value P. It follows that $S_{\alpha_\beta}P \to SP$ in the strong operator topology as well. In particular, $\|S_{\alpha_\beta}P\psi - SP\psi\| \to 0$. But $P\psi = \psi$ and $S\psi \in D$, and this shows that the subnet $(S_{\alpha_\beta}\psi)_\beta$ converges in D. So D is compact. □

To finish the proof of Theorem 3.1.4, let $G_0 = \{x \in G : \pi(x)\omega = \omega\}$. Then G_0 is a subgroup of G and G_0 is open in G since $\overline{\pi(G)\omega}$ is a discrete subset of $\mathcal{H}(\pi)$ and the map $x \to \pi(x)\omega$ from G into H is continuous. Since $\pi(x)\omega = \pi(y)\omega$ if and only if $xG_0 = yG_0$, the function $x \to v(x) = \langle \pi(x)\omega, \eta\rangle$ is constant on cosets of G_0. But, by Lemma 3.1.5(iii), the set $\{\pi(x)\omega : v(x) \neq 0\}$ is finite. So v is zero except on a finite number of cosets of G_0. In particular, v is a linear combination with integer coefficients of translates of the characteristic function of the open subgroup G_0 of G.

3.2. Isometric isomorphisms between Fourier-Stieltjes algebras

In this section it is shown that for a locally compact group G, each of the commutative Banach algebras $B(G)$ and $A(G)$ completely determines the group G. More precisely, two locally compact groups G_1 and G_2 are topologically isomorphic if $B(G_1)$ and $B(G_2)$ (respectively, $A(G_1)$ and $A(G_2)$) are isometrically isomorphic (Theorem 3.2.5 and Theorem 3.2.6). Recall that for a locally compact abelian group H, $B(H)$ and $A(H)$ is isometrically isomorphic to $M(\widehat{H})$ and $L^1(\widehat{H})$, respectively, and that $\widehat{\widehat{H}} = H$ by the Pontryagin duality theorem. Therefore, the above theorems, which are due to Walter [**280**], are far reaching generalizations of the corresponding results of Wendel [**288**] and of Johnson [**140**] for abelian groups. As is to be expected, establishing Theorems 3.2.5 and 3.2.6 requires considerably more effort than in the abelian case.

Let G be a locally compact group and $W^*(G) = B(G)^*$. Then $W^*(G)$ is the enveloping W^*-algebra of $C^*(G)$. Given $T \in W^*(G)$ and $u \in B(G)$, we define a bounded complex-valued function $T_l(u)$ on G by

$$T_l(u)(x) = \langle T, l_x u\rangle,$$

where $(l_x u)(y) = u(xy)$ for $x, y \in G$. Let $\omega(x)$ denote the element of $W^*(G)$ such that $\langle \omega(x), u\rangle = u(x)$ for all $u \in B(G)$. Moreover, let $L_T u$ and $R_T u$ in $B(G)$ be defined by

$$\langle L_T u, S\rangle = \langle u, TS\rangle \text{ and } \langle R_T u, S\rangle = \langle u, ST\rangle,$$

$S \in W^*(G)$. Then, as is readily checked, $l_x u = L_{\omega(x)} u$ and hence

$$\begin{aligned} T_l u(x) &= \langle T, l_x u\rangle = \langle T, L_{\omega(x)} u\rangle \\ &= \langle u, \omega(x)T\rangle = \langle R_T u, \omega(x)\rangle \\ &= (R_T u)(x) \end{aligned}$$

for all $x \in G$, $u \in B(G)$ and $T \in W^*(G)$. Consequently, $T_l u \in B(G)$ and $\|T_l u\| \leqslant \|T\| \cdot \|u\|$. Thus, for any $S, T \in W^*(G)$, we can define an element $S \circ T$ of $W^*(G)$ through

$$\langle S \circ T, u \rangle = \langle S, T_l(u) \rangle$$

for $u \in B(G)$.

LEMMA 3.2.1. *Let G be a locally compact group and $S, T \in W^*(G)$.*

(i) *$S \circ T = ST$, the product of S and T in $W^*(G)$.*
(ii) *$(ST)_l(u) = S_l(T_l(u))$ for all $u \in B(G)$.*

PROOF. (i) Fix $T \in W^*(G)$. The equation $S \circ T = ST$ evidently holds when $S = \omega(x)$ for some $x \in G$. Hence it also holds for all $S \in E$, the linear span of $\omega(G)$. Now, if S is an arbitrary element of $W^*(G)$, then there is a net $(S_\alpha)_\alpha$ in E converging to S in the w^*-topology. Thus, for each $u \in B(G)$,

$$\begin{aligned} \langle S \circ T, u \rangle &= \langle S, T_l(u) \rangle = \lim_\alpha \langle S_\alpha, T_l(u) \rangle \\ &= \lim_\alpha \langle S_\alpha \circ T, u \rangle = \lim_\alpha \langle S_\alpha T, u \rangle \\ &= \langle ST, u \rangle \end{aligned}$$

and so $S \circ T = ST$.

(ii) follows from (i) by using that $T_l(l_x u) = l_x(T_l u)$ for all $x \in G$. □

Clearly, $\omega(x) \in \sigma(B(G))$ for every $x \in G$. Furthermore, the linear span of the set $\{\omega(x) : x \in G\}$ is w^*-dense in $W^*(G)$ since otherwise, by the Hahn-Banach theorem, there exist $T \in W^*(G)$ and $u \in B(G)$ such that $\langle T, u \rangle \neq 0$ and $u(x) = \langle \omega(x), u \rangle = 0$ for all $x \in G$, which is impossible.

COROLLARY 3.2.2. *For any closed linear subspace E of $B(G) = C^*(G)^*$, the following three conditions are equivalent.*

(i) *$T \cdot E \subseteq E$ and $E \cdot T \subseteq E$ for all $T \in W^*(G)$.*
(ii) *$T \cdot E \subseteq E$ and $E \cdot T \subseteq E$ for all $T \in C^*(G)$.*
(iii) *E is translation invariant.*

PROOF. It is easily checked that

$$\omega(x) \cdot u = l_x u \text{ and } u \cdot \omega(x) = r_x u$$

for all $x \in G$ and $u \in B(G)$. Thus (iii) is equivalent to

$$\omega(x) \cdot V \subseteq V \text{ and } V \cdot \omega(x) \subseteq V$$

for all $x \in G$. Since $C^*(G)$ and the linear span of $\omega(G)$ are both w^*-dense in $W^*(G)$, it follows right away from Lemma 3.2.1 that (iii) is also equivalent to (i) and (ii). □

We are now in a position to identify the spectrum of $B(G)$ and the spectrum of its ideal $A(G)$ within $W^*(G) = B(G)^*$.

LEMMA 3.2.3. *Let G be a locally compact group.*

(i) *An element T of $W^*(G)$ belongs to $\sigma(B(G))$ if and only if T_l is a nonzero algebra homomorphism of $B(G)$.*
(ii) *If $T, S \in \sigma(B(G))$, then $TS \in \sigma(B(G))$ and $T^* \in \sigma(B(G))$.*
(iii) *If $T \in \sigma(B(G))$, but $T \notin \omega(G)$, then $\langle T, u \rangle = 0$ for all $u \in A(G)$.*

(iv) *If* $T \in \sigma(B(G))$ *and* T *is invertible in* $W^*(G)$, *then* $T = \omega(x)$ *for some* $x \in G$. *In particular,*

$$\omega(G) = \{T \in \sigma(B(G)) : T \text{ is unitary}\}.$$

PROOF. (i) Suppose that $T \in \sigma(B(G))$. Then

$$\begin{aligned} T_l(uv)(x) &= \langle T, l_x(uv)\rangle = \langle T, (l_x u)(l_x v)\rangle \\ &= \langle T, l_x u\rangle\langle T, l_x v\rangle \\ &= T_l(u)(x)T_l(v)(x) \end{aligned}$$

for all $u, v \in B(G)$ and $x \in G$. Moreover, if $u \in B(G)$ is such that $\langle T, u\rangle \neq 0$, then $T_l(u) \neq 0$. So T_l is a nonzero algebra homomorphism.

Conversely, if T is a nonzero algebra homomorphism, then $T \neq 0$ and T is multiplicative since $\langle T, u\rangle = T_l(u)(e)$ for $u \in B(G)$.

(ii) Since $T \in \sigma(B(G))$ and S_l is a homomorphism by (i), by Lemma 3.2.1(i) we have for all $u, v \in B(G)$

$$\begin{aligned} \langle TS, uv\rangle &= \langle T, S_l(uv)\rangle = \langle T, S_l(u)S_l(v)\rangle \\ &= \langle T, S_l(u)\rangle\langle T, S_l(v)\rangle \\ &= \langle TS, u\rangle\langle TS, v\rangle. \end{aligned}$$

Hence TS is multiplicative, and TS is nonzero since

$$\langle TS, 1_G\rangle = \langle T, S_l(1_G)\rangle = \langle T, 1_G\rangle = 1.$$

(iii) Assume that $\langle T, u_0\rangle \neq 0$ for some $u_0 \in A(G)$. Then $T|_{A(G)} \in \sigma(A(G))$ and hence there exists $x \in G$ such that $\langle T, u\rangle = u(x)$ for all $u \in A(G)$. Since $A(G)$ is an ideal in $B(G)$, it follows that

$$\langle T, v\rangle\langle T, u_0\rangle = \langle T, vu_0\rangle = (vu_0)(x) = v(x)\langle T, u_0\rangle$$

for all $v \in B(G)$ and hence $T = \omega(x)$, which is a contradiction.

(iv) Suppose that T is invertible in $W^*(G)$ and, towards a contradiction, assume that $T \notin \omega(G)$. Then $\langle T, u\rangle = 0$ for all $u \in A(G)$ by (iii). Since $A(G)$ is translation invariant, it follows that

$$T \in A(G)^\circ = (E - Z)W^*(G)$$

for some central projection Z in $W^*(G)$ which satisfies $B(G)Z = A(G)$. Thus $(E - Z)T = T$, which implies that

$$E - Z = (E - Z)TT^{-1} = TT^{-1} = E$$

and hence $Z = 0$. Thus $A(G) = \{0\}$, which is impossible. This contradiction shows that $T = \omega(x)$ for some $x \in G$. □

COROLLARY 3.2.4. *Let* G *be a locally compact group and endow* $\sigma(B(G))$ *with the* w^**-topology and the product inherited from* $W^*(G)$ *(Lemma 3.2.3(ii)). Then* $\sigma(B(G))$ *is a compact semitopological semigroup with separately continuous multiplication and it is invariant under involution. Furthermore, if* G_1 *and* G_2 *are locally compact groups and* $\sigma(B(G_1))$ *and* $\sigma(B(G_2))$ *are topologically isomorphic semigroups, then* G_1 *and* G_2 *are topologically isomorphic.*

PROOF. Since $B(G)$ is unital, $\sigma(B(G))$ is compact, and it is closed under multiplication and involution by Lemma 3.2.3(ii). If $\phi : \sigma(B(G_1)) \to \sigma(B(G_2))$ is an isomorphism, then $T \in \sigma(B(G_1))$ is invertible in $W^*(G_1)$ if and only if $\phi(T)$

is invertible in $W^*(G_2)$. By Lemma 3.2.3(iv), it follows that $\phi(\omega(G_1)) = \omega(G_2)$. Finally, if ϕ is also a homeomorphism then G_1 and G_2 are topologically isomorphic since $\omega : G_i \to \omega(G_i) \subseteq \sigma(B(G_i))$ is a topological isomorphism, $i = 1, 2$. □

The following theorem is the main result we were aiming at. It describes explicitly the structure of isometric isomorphisms between Fourier-Stieltjes algebras.

THEOREM 3.2.5. *Let G_1 and G_2 be locally compact groups. If ϕ is an isometric algebra isomorphism from $B(G_2)$ onto $B(G_1)$, then there exist a topological isomorphism or anti-isomorphism α from G_2 onto G_1 and an element $b \in G_2$ such that*

$$\phi(u)(x) = u(b\,\alpha(x))$$

for all $u \in B(G_2)$ and $x \in G_1$. In particular, G_1 and G_2 are topologically isomorphic.

PROOF. Let e_1 and e_2 be the identities of G_1 and G_2, respectively. Then $\omega(e_1)$ and $\omega(e_2)$ are the identities of $W^*(G_1)$ and $W^*(G_2)$, respectively. Let $S = \phi^*(\omega(e_1))$ and $T = S^*$. Since S is a unitary, so is T. By Lemma 3.2.3(iv), $T = \omega(b)$ for some $b \in G_2$. Then

$$T_l u(y) = \langle T, l_y u\rangle = \langle \omega(b), l_y u\rangle = l_y u(b) = u(yb)$$

for all $y \in G_2$ and $u \in B(G_2)$, that is, T_l is the right translation operator associated to b on $B(G_2)$. Consequently, $\phi \circ T_l$ is also an isometric isomorphism from $B(G_2)$ onto $B(G_1)$. Let $\Lambda = (\phi \circ T_l)^*$, then Λ is a linear isometry from $W^*(G_1)$ onto $W^*(G_2)$. Moreover,

$$\begin{aligned}
\langle \Lambda(\omega(e_1)), u\rangle &= \langle T_l^* \circ \phi^*(\omega(e_1)), u\rangle \\
&= \langle \phi^*(\omega(e_1)), T_l u\rangle = \langle S, T_l u\rangle = \langle S \circ T, u\rangle \\
&= \langle \omega(e_2), u\rangle
\end{aligned}$$

by Lemma 3.2.3 and since S is a unitary. So $\Lambda(\omega(e_1)) = \omega(e_2)$.

By [**145**, Theorem 7 and its proof], Λ is a Jordan isomorphism from $W^*(G_1)$ onto $W^*(G_2)$. Let Λ' denote the restriction of Λ to $\sigma(B(G_1))$. Then Λ' is a homeomorphism from $\sigma(B(G_1))$ onto $\sigma(B(G_2))$. Also $\Lambda'(T^*) = \Lambda'(T)^*$ for all $T \in \sigma(B(G_1))$ by [**145**, Lemma 8]. Moreover, if $T, S \in \sigma(B(G_1))$, then either $\Lambda'(TS) = \Lambda'(T)\Lambda'(S)$ or $\Lambda'(TS) = \Lambda'(S)\Lambda'(T)$. Indeed, if $TS = ST$, this follows from [**145**, Theorem 5]. Otherwise, using [**145**, Lemma 6], we have

$$\Lambda'(T)\Lambda'(S) + \Lambda'(S)\Lambda'(T) = \Lambda'(TS) + \Lambda'(ST),$$

and assuming that $\Lambda'(TS) \neq \Lambda'(T)\Lambda'(S)$ as well as $\Lambda'(TS) \neq \Lambda'(S)\Lambda'(T)$, it follows that

$$\Lambda'(TS),\ \Lambda'(ST),\ \Lambda'(T)\Lambda'(S) \text{ and } \Lambda'(S)\Lambda'(T)$$

are pairwise distinct elements of $\sigma(B(G_2))$. However, elements in $\sigma(B(G_2))$ are linearly independent. So the above equation is impossible. It now follows that Λ' maps the unitary elements of $\sigma(B(G_1))$ onto the unitary elements of $\sigma(B(G_2))$. Thus, by Lemma 3.2.3, Λ' maps $\omega(G_1)$ onto $\omega(G_2)$.

By [**145**, Theorem 10], we can find central projections $Z_1 \in W^*(G_1)$ and $Z_2 \in W^*(G_2)$ such that

$$\Lambda|_{W^*(G_1)Z_1} : W^*(G_1)Z_1 \to W^*(G_2)Z_2$$

is a $*$-isomorphism and

$$\Lambda|_{W^*(G_1)(I-Z_1)} : W^*(G_1)(I - Z_1) \to W^*(G_2)(I - Z_2)$$

is a $*$-anti-isomorphism.

For $x \in G_1$, let

$$H_x = \{y \in G_1 : (xy - yx)Z_1 = 0\}$$

and

$$K_x = \{y \in G_1 : (xy - yx)(I - Z_1) = 0\}.$$

Then H_x is a subgroup of G_1. In fact, if $y_1, y_2 \in H_x$, then

$$x(y_1y_2)Z_1 = (y_1xZ_1)y_2 = y_1(xy_2Z_1) = y_1(y_2xZ_1) = (y_1y_2)xZ_1,$$

and hence $y_1y_2 \in H_x$. Also, $e_1 \in H_x$ and $y \in H_x$ implies $y^{-1} \in H_x$. Similarly, it is shown that K_x is a subgroup of G_1.

Notice next that $H_x \cup K_x = G_1$. This can be seen as follows. If $y \in G_1$ then, as we have seen above, either $\Lambda(yx) = \Lambda(y)\Lambda(x)$ or $\Lambda(yx) = \Lambda(x)\Lambda(y)$. In the first case, recalling that $\Lambda|_{W^*(G_1)(I-Z_1)}$ is an anti-isomorphism, we get

$$\Lambda((yx - xy)(I - Z_1)) = \Lambda(yx(I - Z_1)) - \Lambda(x(I - Z_1))\Lambda(y(I - Z_1)) = 0,$$

and hence $(yx - xy)(I - Z_1) = 0$, whence $y \in K_x$. In the second case, since $\Lambda|_{W^*(G_1)Z_1}$ is an isomorphism,

$$\Lambda((yx - xy)Z_1) = \Lambda(xZ_1)\Lambda(yZ_1) - \Lambda(xyZ_1) = 0,$$

which implies that $y \in H_x$. Now, $H_x \cup K_x = G_1$ in turn implies that either $H_x = G_1$ or $K_x = G_1$. Indeed, if there exist $y_1 \in H_x \cap (G_1 \backslash K_x)$ and $y_2 \in K_x \cap (G_1 \backslash H_x)$, then $y_1y_2 \notin H_x \cup K_x$, which is impossible. Now, let

$$H = \{x \in G_1 : H_x = G_1\} \text{ and } K = \{x \in G_1 : K_x = G_1\}.$$

Then H and K are subgroups of G_1. Indeed, if $x_1, x_2 \in H$, then for any $x \in G_1$,

$$\begin{aligned}[x(x_1x_2) - (x_1x_2)x]Z_1 &= (xx_1)Z_1x_2 - x_1(xx_2)Z_1 \\ &= (x_1x)Z_1x_2 - (x_1x)x_2Z_1 = 0\end{aligned}$$

and hence $x_1x_2 \in H$. Moreover, $e_1 \in H$ and $x \in H$ implies $x^{-1} \in H$ since, for any $y \in G_1$, $x^{-1}yZ_1 = yx^{-1}Z_1$ if and only if $yZ_1 = xyx^{-1}Z_1$, and this in turn is equivalent to

$$yxZ_1 = yZ_1x = xyx^{-1}Z_1x = xyZ_1.$$

So H is a subgroup of G_1, and similarly it is shown that K is a subgroup. Given $x \in G_1$, as shown above either $H_x = G_1$ or $K_x = G_1$. Thus $H \cup K = G_1$ and as before it follows that either $H = G_1$ or $K = G_1$.

Suppose that $H = G_1$. We claim that then $\Lambda(xy) = \Lambda(y)\Lambda(x)$ for all $x, y \in G_1$. To see this, we can assume that $\Lambda(xy) = \Lambda(x)\Lambda(y)$. Then, as shown above, $(xy - yx)(I - Z_1) = 0$. On the other hand, $(xy - yx)Z_1 = 0$ since $x \in G_1 = H_y$. So $xy = yx$ and hence $\Lambda(xy) = \Lambda(y)\Lambda(x)$. Similarly, if $K = G_1$ then $\Lambda(xy) = \Lambda(x)\Lambda(y)$ for all $x, y \in G_1$.

Thus we have shown that Λ' is an isomorphism or an anti-isomorphism from G_1 onto G_2. Finally, note that if Λ' is an anti-isomorphism, then $x \to \Lambda'(x^{-1})$ is an isomorphism.

Finally, let $b \in G_2$ be as chosen at the beginning of the proof and let Λ be the linear isometry from $W^*(G_1)$ onto $W^*(G_2)$ defined above. Now define $\alpha : G_1 \to G_2$

by setting, for $x \in G_1$, $\alpha(x) = y$ where y is the unique element of G_2 such that $\Lambda(\omega(x)) = \omega(y)$. It then follows that

$$\phi(u)(x) = u(b\alpha(x))$$

for all $u \in B(G_2)$ and $x \in G_1$. □

Using arguments similar to those in the proof of Theorem 3.2.5 one can show that Theorem 3.2.5 remains valid if $B(G_i)$ is replaced by $B_\lambda(G_i)$, $i = 1, 2$. Also, the next theorem can be proved by very much the same arguments as those employed in the proof of Theorem 3.2.5, replacing $W^*(G_i)$ by $VN(G_i)$, $i = 1, 2$, and noticing that $\sigma(A(G)) = G$ for any locally compact group G.

THEOREM 3.2.6. *Let G_1 and G_2 be locally compact groups. If $\phi : A(G_1) \to A(G_2)$ is an isometric isomorphism, then there exist a topological isomorphism or anti-isomorphism $\alpha : G_2 \to G_1$ and $a \in G_1$ such that $\phi u(y) = u(a\,\alpha(y))$ for all $u \in A(G_1)$ and $y \in G_2$. In particular, G_1 and G_2 are topologically isomorphic.*

3.3. Homomorphisms between Fourier and Fourier-Stieltjes algebras

In Theorem 3.2.6 it was shown that if G and H are locally compact groups and $\phi : A(G) \to A(H)$ is an isometric isomorphism, then ϕ is implemented by a topological isomorphism or anti-isomorphism $\varphi : H \to G$ and an element $a \in G$. An analogous result holds for the Fourier-Stieltjes algebras (Theorem 3.2.5). In this section we determine the structure of continuous homomorphisms from $A(G)$ into $A(H)$ and from $B(G)$ into $B(H)$ which are either positive (Theorem 3.3.5) or contractive (Theorem 3.3.7). However, both of these results to some extent rest on Theorem 3.2.6.

Let G and H be locally compact groups, and let $\phi : A(G) \to B(H)$ be a homomorphism. Let

$$U = \{h \in H : \phi(u)(h) \neq 0 \text{ for some } u \in A(G)\}.$$

Then U is open in H, and for each $h \in U$, there exists a unique element $\alpha(h)$ of G such that $\phi(u)(h) = u(\alpha(h))$ for every $u \in A(G)$. Of course, the map $\alpha : U \to G$ is continuous. Defining $\phi_\alpha : A(G) \to \ell^\infty(H)$ by

$$\phi_\alpha(u)(h) = \begin{cases} u(\alpha(h)) & : \text{ if } h \in U \\ 0 & : \text{ if } H \in H \backslash U, \end{cases}$$

we have $\phi = \phi_\alpha$. Conversely, given any map $\alpha : U \to G$, the preceding formula defines a homomorphism from $A(G)$ into $\ell^\infty(H)$.

By the preceding remarks, this is equivalent to determining the maps α for which ϕ_α maps $A(G)$ into $B(H)$. We remind the reader that a map φ from a group U into a group G is called *anti-homomorphism* if $\varphi(xy) = \varphi(y)\varphi(x)$ for all $x, y \in U$. The easier direction of the desired description of homomorphisms is provided by the following proposition.

PROPOSITION 3.3.1. *Let G and H be locally compact groups and U an open subgroup of H. Suppose that φ is a continuous homomorphism or anti-homomorphism from U into G and that $h \in H$ and $g \in G$. Then the map defined by, for $u \in A(G)$,*

$$\phi(u)(x) = \begin{cases} u(g\varphi(hx)) & : \text{ if } x \in h^{-1}U \\ 0 & : \text{ if } x \in H \backslash h^{-1}U, \end{cases}$$

is a contractive homomorphism from $A(G)$ into $B(H)$. Moreover, if $g = e_G$ and $h = e_H$, then ϕ is positive.

PROOF. Assume first that φ is a continuous homomorphism. It is clear that then ϕ is a homomorphism from $A(G)$ into $B(H)$. To see that ϕ is contractive, we can assume that $g = e_G$. Note that $L_{\varphi(h)}(\phi(u))(y) = 0$ for $y \notin \varphi(U)$, whereas $L_{\varphi(h)}(\phi(u))(\varphi(x)) = \phi(u)(\varphi(h^{-1}x)) = u(\varphi(x))$ for $x \in U$. It follows that

$$\begin{aligned}\|\phi(u)\|_{B(H)} &= \|L_{\varphi(h)}\|_{B(H)} = \|L_{\varphi(h)}|_U\|_{B(U)} \\ &= \|u \circ \varphi\|_{B(U)} \leqslant \|u\|_{B(G)}.\end{aligned}$$

Finally, if $g = e_G$ and $h = e_H$, then

$$\sum_{i,j=1}^{n} \lambda_i \overline{\lambda_j} \phi(u)(x_j^{-1} x_i) = \sum_{i,j=1}^{n} \lambda_i \overline{\lambda_j} u(\varphi(x_j)^{-1} \varphi(x_i))$$

for $x_1, \ldots, x_n \in U$ and $\lambda_1, \ldots, \lambda_n \in \mathbb{C}$, and therefore $\phi(P(U)) \subseteq P(G)$.

If φ is an anti-homomorphism, the statement of the proposition follows from the homomorphism case by first applying the inverse map $H \to H, h \to h^{-1}$. $\square$

It will turn out that the problem of characterizing homomorphisms from $A(G)$ into $B(H)$ can to some extent reduced to discrete groups.

PROPOSITION 3.3.2. *Let G and H be discrete groups and $\phi : A(G) \to B(H)$ a homomorphism. Let $\varphi : U \to G$ be a map from a subset U of H into G such that $\phi = \phi_\varphi$ as in Proposition 3.3.1. Let F be a subgroup of G containing $\varphi(U)$. Then the map ψ defined by*

$$\psi(u)(y) = \begin{cases} u(\varphi(y)) & : \text{ if } y \in U \\ 0 & : \text{ if } y \in H \backslash U \end{cases}$$

is a homomorphism from $A(F)$ into $B(H)$ with $\|\psi\| \leqslant \|\phi\|$.

PROOF. For the main statement of the proposition, we have to show that if $u \in A(F)$ is such that $\|u\|_{A(F)} = 1$, then $\|\psi(u)\|_{B(H)} \leqslant \|\phi\|$. Since $A(F) = \ell^2(F) * \ell^2(F)\check{}$, by Lemma 2.1.6 we can assume that $u = f * \check{g}$, where

$$f = \sum_{j=1}^{m} \alpha_j \delta_{a_j}, \; g = \sum_{j=1}^{m} \beta_j \delta_{b_j} \quad \text{and} \quad \sum_{j=1}^{m} |\alpha_j|^2 = \sum_{j=1}^{m} |\beta_j|^2 = 1,$$

$a_j, b_j \in F, \alpha_j, \beta_j \in \mathbb{C}, m \in \mathbb{N}$. On the other hand, by Lemma 2.1.8,

$$\|\psi(u)\|_{A(H)} = \sup \left\{ \left| \sum_{i=1}^{n} c_i \psi(u)(h_i) \right| \right\},$$

where the supremum is taken over all $h_1, \ldots, h_n \in H$ and $c_1, \ldots, c_n \in \mathbb{C}$, $n \in \mathbb{N}$, such that $\| \sum_{i=1}^{n} c_i \omega_H(h_i) \| \leqslant 1$. Now

$$\begin{aligned}\left| \sum_{i=1}^{n} c_i \psi(u)(h_i) \right| &= \left| \sum_{h_i \in U} c_i (f * \check{g})(\varphi(h_i)) \right| \\ &= \left| \sum_{h_i \in U} c_i \left(\sum_{j,k=1}^{m} \alpha_j \beta_k \delta_{a_j b_k^{-1}}(\varphi(h_i)) \right) \right|.\end{aligned}$$

Now choose a compact neighbourhood V of the identity in G such that

$$VV^{-1} \bigcap \{a_j^{-1} a_k, b_j^{-1} b_k, a_j^{-1} \varphi(x_i) b_k : 1 \leqslant j, k \leqslant m, x_i \in U\} = \{e_G\}.$$

Consider the functions ξ and η in $\ell^2(G)$ defined by

$$\xi = \sum_{j=1}^{m} \alpha_j 1_{a_j V} \quad \text{and} \quad \eta = \sum_{j=1}^{m} \beta_j 1_{b_j V}.$$

Then $\|\xi\|_2 = \|\eta\|_2 = |V|^{1/2}$, and hence the element $u = \xi * \check{\eta}$ of $A(G)$ has norm $\leqslant |V|$. Consequently, $\|\phi(u)\| \leqslant \|\phi\| \cdot |V|$. It follows from Lemma 2.1.8 and Corollary 2.2.2 that

$$\begin{aligned} \|\phi\| \cdot |V| &\geqslant \left| \sum_{i=1}^{n} c_i \phi(u)(x_i) \right| = \left| \sum_{x_i \in U} c_i u(\varphi(x_i)) \right| \\ &= \left| \sum_{x_i \in U} c_i \left(\sum_{j,k=1}^{m} \alpha_j \beta_k \, |a_j V \cap \varphi(x_i) b_k V| \right) \right| \\ &= |V| \cdot \left| \sum_{x_i \in U} c_i \left(\sum_{j,k=1}^{m} \alpha_j \beta_k \delta_{a_j b_k^{-1}}(\varphi(x_i)) \right) \right|. \end{aligned}$$

This proves the required inequality. □

In the proof of Lemma 3.3.4 below we shall use the following fact where, for simplicity of notation, we set $\operatorname{Re} a = \frac{1}{2}(a + a^*)$ for an element a of a C^*-algebra A.

LEMMA 3.3.3. *Let a, b, c, d be elements of a C^*-algebra A satisfying*

$$\operatorname{Re}[\alpha^2 a + \beta^2 b + \alpha\beta c + \alpha\overline{\beta} d] \geqslant 0$$

for all $\alpha, \beta \in \mathbb{C}$. Then $a = b = c = d = 0$.

PROOF. If f is any positive linear functional on A, then

$$\operatorname{Re}[\alpha^2 f(a) + \beta^2 f(b) + \alpha\beta f(c) + \alpha\overline{\beta} f(d)] \geqslant 0$$

for all $\alpha, \beta \in \mathbb{C}$. This implies that $f(a) = f(b) = f(c) = f(d) = 0$ for all f and hence $a = b = c = d = 0$. □

LEMMA 3.3.4. *The set U is an open subgroup of H and, for each $x, y \in U$, the ordered pair $\{\varphi(xy), \varphi(yx)\}$ is a permutation of the pair $\{\varphi(x)\varphi(y), \varphi(y)\varphi(x)\}$. In particular, the set $\varphi^{-1}(e_G)$ is a closed normal subgroup of U and $\varphi(xy) = \varphi(y)$ for every $x \in \varphi^{-1}(e_G)$ and $y \in U$.*

PROOF. The adjoint map $\phi^* : W^*(H) \to VN(G)$ is a positive linear operator with $\phi^*(I)) = I$ and satisfies

$$\phi^*(\omega_H(x)) = \begin{cases} \lambda_G(\varphi(x)) & : \text{ for } x \in U \\ 0 & : \text{ for } x \in H \backslash U. \end{cases} \tag{3.1}$$

Let $x, y \in U$ and $\alpha, \beta \in \mathbb{C}$ be arbitrary and consider the element

$$T = \alpha\, \omega_H(x) + \beta\, \omega_H(y) + \overline{\alpha}\, \omega_H(x^{-1}) + \overline{\beta}\, \omega_H(y^{-1})$$

of $W^*(H)$. A straightforward, but lengthy calculation shows that

$$\begin{aligned} \phi^*(T^2) = {} & 2 \operatorname{Re}\{(|\alpha|^2 + |\beta|^2) I + \alpha^2 \phi^*(\omega_H(x^2)) + \beta^2 \phi^*(\omega_H(y^2)) \\ & + \alpha\beta \phi^*[\omega_H(xy) + \omega_H(yx)] + \alpha\overline{\beta} \phi^*[\omega_H(xy^{-1}) + \omega_H(y^{-1}x)]\}, \end{aligned}$$

and similarly

$$\begin{aligned}\phi^*(T)^2 &= 2\,\mathrm{Re}\{(|\alpha|^2+|\beta|^2)I+\alpha^2\lambda_G(\varphi(x)^2)+\beta^2\lambda_G(\varphi(y)^2)\\ &\quad+\alpha\beta\,[\lambda_G(\varphi(x)\varphi(y))+\lambda_G(\varphi(y)\varphi(x))]\\ &\quad+\alpha\overline{\beta}\,[\lambda_G(\varphi(x)\varphi(y)^{-1})+\lambda_G(\varphi(y)^{-1}\varphi(x))]\}.\end{aligned}$$

Since $T = T^*$, Kadison's generalized Schwarz inequality [**146**, Theorem 2] gives $\phi^*(T^2) \geqslant \phi^*(T)^2$. Setting now

$$a = \phi^*(\omega_H(x^2)) - \lambda_G(\varphi(x)^2), \quad b = \phi^*(\omega_H(y)^2) - \lambda_G(\varphi(y)^2),$$
$$c = \phi^*[\omega_H(xy) + \omega_H(yx)] - [\lambda_G(\varphi(x)\varphi(y)) + \lambda_G(\varphi(y)\varphi(x))]$$

and

$$d = \phi^*[\omega_H(xy^{-1}) + \omega_H(y^{-1}x)] - [\lambda_G(\varphi(x)\varphi(y)^{-1}) + \lambda_G(\varphi(y)^{-1}\varphi(x))],$$

it follows that $\mathrm{Re}[\alpha^2 a + \beta^2 b + \alpha\beta c + \alpha\overline{\beta}d] \geqslant 0$ for all $\alpha, \beta \in \mathbb{C}$ and hence $c = 0$ by Lemma 3.3.3. Thus

$$\phi^*[\omega_H(xy)] + \phi^*[\omega_H(yx)] = \lambda_G(\varphi(x)\varphi(y)) + \lambda_G(\varphi(y)\varphi(x))$$

for all $x, y \in U$. Since the set of operators $\lambda_G(t)$, $t \in G$, is linearly independent, the preceding equation together with (3.1) implies that xy and yx belong to U. Since $U^{-1} \subseteq U$, U is a subgroup of H. Moreover, taking into account (3.1) again, it follows that the ordered pair

$$\{\phi^*(\omega_H(xy)), \phi^*(\omega_H(yx))\} = \{\lambda_G(\varphi(xy)), \lambda_G(\varphi(yx))\}$$

is a permutation of the pair $\{\lambda_G(\varphi(x)\varphi(y)), \lambda_G(\varphi(y)\varphi(x))\}$. Since λ_G is injective, for all $x, y \in U$, we therefore either have $\varphi(x)\varphi(y) = \varphi(xy)$ or $\varphi(x)\varphi(y) = \varphi(yx)$. This of course implies that $\varphi^{-1}(e_G)$ is a normal subgroup of U and that $\varphi(xy) = \varphi(y)$ for every $x \in \varphi^{-1}(e_G)$ and $y \in U$. □

A homomorphism $\phi : A(G) \to B(H)$ is positive if $\phi(u)$ is positive definite whenever u is positive definite. We now consider a nonzero positive homomorphism $\phi : A(G) \to B(H)$. Let U and φ be defined as in Proposition 3.3.1. For every positive definite $u \in A(G)$, $\phi(u)(e_H) = \|\phi(u)\|$ since $\phi(u)$ is positive definite. As $\phi(u) \neq 0$ for some u, we conclude that $e_H \in U$ and $\varphi(e_H) = e_G$. Also, for each $x \in U$ and every positive definite $u \in A(G)$,

$$\phi(u)(x^{-1}) = \overline{\phi(u)(x)} = \overline{u(\varphi(x))} = u(\varphi(x)^{-1}),$$

which forces $x^{-1} \in U$ and $u(\varphi(x^{-1})) = u(\varphi(x)^{-1})$, whence $\varphi(x^{-1}) = \varphi(x)^{-1}$.

THEOREM 3.3.5. *Let G and H be locally compact groups, and let $\phi : A(G) \to B(H)$ be a positive homomorphism. Then there exists a continuous group homomorphism or anti-homomorphism φ from an open subgroup U of H into G such that, for all $u \in A(G)$,*

$$\phi(u)(x) = \begin{cases} u(\varphi(x)) & : \ \text{if } x \in U \\ 0 & : \ \text{if } x \in H\backslash U. \end{cases}$$

PROOF. For any topological group X, let X_d denote the same group X with the discrete topology. Since $\varphi^{-1}(e_G)$ is a normal subgroup of U by Lemma 3.3.4, we can form the quotient group $H_0 = U_d/\varphi^{-1}(e_G)$ of U_d. Moreover, let $G_0 = \varphi(U) \subseteq G_d$, and let $\varphi_0 : H_0 \to G_0$ be the map induced from φ. Then φ_0 is injective and inherits from φ the property that the ordered pair $\{\varphi_0(xy), \varphi_0(yx)\}$ is a permutation of the pair $\{\varphi_0(x)\varphi_0(y), \varphi_0(y)\varphi_0(x)\}$ for every $x, y \in U_0$.

Now let $\rho = \phi_{\varphi_0} : A(G_0) \to B(H_0)$ be defined by

$$\rho(u)(x) = u(\varphi_0(x))$$

for $x \in G_0$. Then ρ is a homomorphism by Proposition 3.3.1 with $\|\rho\| \leqslant \|\phi\|$. Since φ_0 is bijective, ρ maps the set of all finitely supported functions on G_0 onto the set of all finitely supported functions on H_0. Consequently, ρ maps $A(G_0)$ injectively onto a dense subalgebra of $A(H_0)$. Now consider ρ as map into $A(H_0)$. Since $\rho(A(G_0))$ is dense in $A(H_0)$, the adjoint ρ^* maps $VN(H_0)$ injectively onto a w^*-dense subspace of $VN(G_0)$.

Since $\rho^*(\lambda_{H_0}(x)) = \lambda_{G_0}(\varphi_0(x))$ for all $x \in H_0$, we get

$$\rho^*(ST) + \rho^*(TS) = \rho^*(S)\rho^*(T) + \rho^*(T)\rho^*(S)$$

for all $S, T \in \lambda_{H_0}(H_0)$. Extending by linearity and w^*-continuity, it follows that $\rho^* : VN(H_0) \to VN(G_0)$ is a Jordan $*$-homomorphism. By a result in [**145**], ρ^* is isometric. It follows that ρ^* is an isometry from $VN(H_0)$ onto $VN(G_0)$, and therefore ρ is an isometric isomorphism from $A(G_0)$ onto $A(H_0)$. Then, by Theorem 3.2.6, $\varphi_0 : H_0 \to G_0$ must be a group isomorphism or anti-isomorphism. This of course implies that $\varphi : U \to G$ is either a group homomorphism or a group anti-homomorphism. This finishes the proof. □

COROLLARY 3.3.6. *Suppose that $\phi : A(G) \to A(H)$ is a positive isomorphism. Then there exists a topological isomorphism or anti-isomorphism $\varphi : G \to H$ such that $\phi(u) = u \circ \varphi$ for all $u \in A(G)$.*

THEOREM 3.3.7. *Let G and H be locally compact groups, and let $\phi : A(G) \to B(H)$ be a contractive homomorphism. Then there exist an open subgroup U of H, a continuous homomorphism or anti-homomorphism φ from U into G and elements $g \in G$ and $h \in H$ such that, for every $u \in A(G)$,*

$$\phi(u)(x) = \begin{cases} u(g\varphi(hx)) & : \ \text{if } x \in h^{-1}U \\ 0 & : \ \text{if } x \in H\backslash h^{-1}U. \end{cases}$$

PROOF. As shown at the outset of this section, there exist an open subset V of H and a continuous map $\psi : V \to G$ such that

$$\phi(u)(x) = \begin{cases} u(\psi(x)) & : \ \text{if } x \in V \\ 0 & : \ \text{if } x \in H\backslash V. \end{cases}$$

Fix $h \in V^{-1}$ and put $g = \psi(h^{-1}) \in G$ and $U = hV$. Define $\varphi : U \to G$ by

$$\varphi(x) = g^{-1}\psi(h^{-1}x), \ x \in U.$$

Then ϕ_φ is a contractive homomorphism from $A(G)$ into $B(H)$. Indeed, ϕ is contractive and ϕ_φ is a composition of ϕ with translations by elements of G and H. Moreover, $e_H \in U$ and $\varphi(e_H) = g^{-1}\psi(h^{-1}) = e_G$. Therefore, if $u \in A(G)$ is positive definite, then

$$\phi_\varphi(u)(e_H) = u(e_G) = \|u\|_{A(G)} \geqslant \|\phi_\varphi(u)\|_{B(H)}.$$

Thus $\phi_\varphi(u)$ is also positive definite. The result now follows from Theorem 3.3.5. □

COROLLARY 3.3.8. *Suppose that $\phi : A(G) \to B(H)$ is a contractive homomorphism. If ϕ is also injective, then ϕ is an isometry.*

PROOF. By Theorem 3.3.7, after applying translations and the inverse map $x \to x^{-1}$ of H (if necessary), we can assume that $\phi = \phi_\varphi$, where φ is a continuous homomorphism from an open subgroup U of H into G.

Since φ is injective, $\varphi(U)$ must be dense in G because otherwise there exists $0 \neq u \in A(G)$ such that $u = 0$ on $\varphi(U)$ and hence $\phi_\alpha(u) = 0$. Let $\mathcal{M}$ denote the von Neumann subalgebra of $W^*(H)$ generated by the operators $\omega_H(y)$, $y \in U$. Since $\phi^*(\omega_H(y)) = \lambda_G(y)$ for $y \in U$ and $\varphi(U)$ is dense in G, it follows that $\phi^* : \mathcal{M} \to VN(G)$ is a surjective $*$-homomorphism. Because $\|\phi^*\| = 1$, ϕ^* maps the unit ball of $W^*(H)$ onto the unit ball of $VN(G)$. Thus, for any $u \in A(G)$,

$$\begin{aligned} \|u\|_{A(G)} &= \sup\{|\langle u, \phi^*(S)\rangle| : S \in W^*(H), \|S\| \leqslant 1\} \\ &= \sup\{|\langle \phi(u), S\rangle| : S \in W^*(H), \|S\| \leqslant 1\} \\ &= \|\phi(u)\|_{B(G)} \end{aligned}$$

since $W^*(H) = B(H)^*$. Hence ϕ is isometric. □

The following corollary generalizes Theorem 3.2.6. The proof, however, uses Theorem 3.2.6.

COROLLARY 3.3.9. *Suppose that $\phi : A(G) \to A(H)$ is a contractive homomorphism such that $\phi(A(G))$ separates the points of H, and assume that H has at least two elements. Then, for each $v \in A(H)$, there exists $u \in A(G)$ such that $\phi(u) = v$ and $\|u\| = \|v\|$.*

PROOF. Again, by Theorem 3.3.7, we can assume that $\phi = \phi_\varphi$, where φ is a continuous homomorphism from some open subgroup U of H into G. Observe next that φ is a proper map. Indeed, suppose that there exists a compact subset C of G such that $\varphi^{-1}(C)$ is not compact. Then, for $u \in A(G)$ with $u|_C = 1$ and any $x \in \varphi^{-1}(C)$ we would have $\phi(u)(x) = u(\varphi(x)) = 1$, which is impossible since u vanishes at infinity. Now φ, being proper and continuous, is a closed map.

Since $\phi(A(G))$ separates the points of H, φ must be injective and $H \backslash U$ can at most have two elements. As H has more than two elements, this implies that $U = H$. Thus we have seen that φ is a topological isomorphism from H onto the closed subgroup $\varphi(H)$ of G. Given $v \in A(H)$, we have $v \circ \varphi^{-1} \in A(\varphi(H))$ and $\|v \circ \varphi^{-1}\| = \|v\|$, and by Theorem 3.2.6 there exists $u \in A(G)$ such that $u|_{\varphi(H)} = v \circ \varphi^{-1}$ and $\|u\|_{A(G)} = \|v \circ \varphi^{-1}\|$. □

The proof of the preceding corollary shows that ϕ is the composition of the restriction map $A(G) \to A(\varphi(H))$ and the isometric isomorphism $A(\varphi(H)) \to A(H)$. The following corollary extends Theorem 3.2.6 and [**136**, Corollary 3.12].

COROLLARY 3.3.10. *Suppose that $\phi : A(G) \to A(H)$ is a contractive isomorphism. Then there exist a topological group isomorphism or anti-isomorphism $\varphi : H \to G$ and an element $g \in G$ such that, for all $u \in A(G)$ and $x \in H$,*

$$\phi(u)(x) = u(g\varphi(x)).$$

THEOREM 3.3.11. *Let G and H be locally compact groups, and let $\phi : B(G) \to B(H)$ be an isomorphism which is positive on $A(G)$. Then there exists a topological group isomorphism or anti-isomorphism φ from H onto G such that $\phi(u) = u \circ \varphi$ for all $u \in B(G)$. In particular, ϕ and ϕ^{-1} are both positive.*

PROOF. Recall first that, since Fourier-Stieltjes algebras are semisimple, ϕ is bounded. Since the restriction $\phi|_{A(G)} : A(G) \to B(H)$ is positive, by Theorem 3.3.5 there exist an open subgroup U of H and a continuous group homomorphism or anti-homomorphism $\varphi : U \to G$ such that $\phi(u) = \phi_\varphi(u)$ for all $u \in A(G)$. Applying the inversion $x \to x^{-1}$ if necessary, we can assume that φ is a homomorphism. Since ϕ is injective, it follows that $\varphi(U)$ is dense in G.

Now consider the adjoint $\phi^* : W^*(H) \to W^*(G)$, which is a bounded linear isomorphism. Then we have $\phi^*(\omega_H)(H) \subseteq \sigma(B(G))$, the spectrum of $B(G)$. For each $x \in U$, $\phi^*(\omega_H(x))$ is a multiplicative linear functional on $B(G)$ which maps $u \in A(G)$ to $u(\varphi(x))$. By Lemma 3.2.5, this implies that $\phi^*(\omega_H(x)) = \omega_G(y)$ for some $y \in G$. It follows that $y = \varphi(x)$. In addition, the map $\omega_G : G \to \omega_G(G)$ is a homeomorphism when $\omega_G(G) \subseteq \sigma(B(G))$ is equipped with the relative w^*-topology, and the analogous statement holds for H. Since ϕ^* is a homomorphism from $W^*(H)$ onto $W^*(G)$, where both $W^*(H)$ and $W^*(G)$ carry the w^*-topology, we conclude that φ is a homeomorphism from U onto $\varphi(U)$. Thus $\varphi(U)$ is a locally compact subgroup of G. On the other hand, $\varphi(U)$ is dense in G, and therefore $\varphi(U) = G$.

It remains to show that $U = H$. Let $\mathcal{M}$ denote the w^*-closed subspace of $W^*(H)$ generated by the set $\omega_H(H)$. Then ϕ^* maps $\mathcal{M}$ onto $W^*(G)$, and therefore $\mathcal{M} = W^*(H)$. This in turn implies that, for every $u \in B(H)$, $u|_U = 0$ forces $u = 0$. Regularity of $A(H)$ implies that $U = H$, as required. □

We conclude this section with a corollary which extends [**136**, Corollary 5.5(ii)] and Theorem 3.2.5.

COROLLARY 3.3.12. *Let G and H be locally compact groups and let $\phi : B(G) \to B(H)$ be an isomorphism such that $\phi|_{A(G)}$ is contractive. Then there exist a topological group isomorphism or anti-isomorphism ψ from H onto G and an element $a \in G$ such that $\phi(u)(x) = u(a\psi(x))$ for all $x \in H$. In particular, ϕ is isometric on $B(G)$ and $\phi(A(G)) = A(H)$.*

PROOF. Since ϕ is an isomorphism and $\phi|_{A(G)}$ is contractive, by Theorem 3.3.7 there exists a continuous homomorphism or anti-homomorphism φ from H onto G and elements $g \in G$ and $h \in H$ such that $\phi(u)(x) = u(g\varphi(hx))$ for all $u \in A(G)$ and $x \in H$. Then let $a = g\varphi(h)$ and define ψ by $\psi(x) = \varphi(x)$ or $\psi(x) = \varphi(h)^{-1}\varphi(x)\varphi(h)$ depending on whether φ is a homomorphism or anti-homomorphism. □

3.4. Invariant subalgebras of $VN(G)$ and subgroups of G

For a locally compact abelian group G, the Pontryagin duality theorem provides a beautiful and complete one-to-one correspondence between closed subgroups H of G and closed subgroups $H^\perp$ of the dual group $\widehat{G}$ such that $\widehat{H} = \widehat{G}/H^\perp$ and $\widehat{G/H} = H^\perp$. In the nonabelian situation one cannot expect such a duality within the category of groups. However, Takesaki and Tatsuuma [**271**] have established correspondences between closed subgroups of G and certain von Neumann subalgebras of $L^\infty(G)$ and $VN(G)$.

Let G be an arbitrary locally compact group. The first theme of this section is to provide a Galois type correspondence between closed subgroups of G and right translation invariant von Neumann subalgebras of $L^\infty(G)$. We start with two preparatory lemmas.

LEMMA 3.4.1. *Let E be a right translation invariant w^*-closed linear subspace of $L^\infty(G)$. Then $E \cap C(G)$ is w^*-dense in E.*

PROOF. Fix $f \in E$. We first show that if $u \in L^1(G)$ is such that $\|u\|_1 = 1$, $u \geqslant 0$ and $\widetilde{u} = u$, then $f * u \in E$. The function u can be regarded as a mean on $C^b(G)$. Hence there exists a net $(u_\alpha)_\alpha$ of means, where each u_α is a finite linear combination $u_\alpha = \sum_{j=1}^{n_\alpha} c_{\alpha,j}\delta_{x_{\alpha,j}}$ of point evaluations, such that $u_\alpha \to u$ in the w^*-topology of $C^b(G)^*$. Now, for $g \in L^1(G)$,

$$\begin{aligned}\langle f * \widetilde{u}, g\rangle &= \langle(\Delta^{-1}\widetilde{g}) * f, u\rangle = \lim_\alpha \langle(\Delta^{-1}\widetilde{g}) * f, u_\alpha\rangle \\ &= \lim_\alpha \sum_{j=1}^{n_\alpha} c_{\alpha,j}(\Delta^{-1}\widetilde{g}) * f)(x_{\alpha,j}) \\ &= \lim_\alpha \sum_{j=1}^{n_\alpha} c_{\alpha,j}\int_G f(yx_{\alpha,j})g(y)dy \\ &= \lim_\alpha \left\langle \sum_{j=1}^{n_\alpha} c_{\alpha,j}R_{x_{\alpha,j}}f, g\right\rangle.\end{aligned}$$

Thus $f * \widetilde{u}$ is the w^*-limit in $L^\infty(G)$ of the net

$$\left(\sum_{j=1}^{n_\alpha} c_{\alpha,j}R_{x_{\alpha,j}}f\right)_\alpha \subseteq E.$$

Since E is w^*-closed, $f * \widetilde{u} \in E$.

Now $L^1(G)$ admits an approximate identity consisting of functions u as above. Since $\langle f * \widetilde{u}, g\rangle = \langle f, g * u\rangle$ for all $g \in L^1(G)$, it follows that $f * \widetilde{u} \to f$ in the w^*-topology. This completes the proof. □

LEMMA 3.4.2. *Let $\mathcal{A}$ be a right translation invariant W^*-subalgebra of $L^\infty(G)$. If $\mathcal{A} \neq \{0\}$, then $\mathcal{A}$ contains the constant functions.*

PROOF. Let h be the identity of $\mathcal{A}$. Then, for all $f \in \mathcal{A}$ and $x \in G$,

$$(R_x h)f = R_x(hR_{x^{-1}}(f)) = R_x(R_{x^{-1}}(f)) = f = hf$$

and hence $R_x h = h$. Choose any nonzero $f \in \mathcal{A} \cap C(G)$ (Lemma 3.4.1) and let $V = \{y \in G : f(y) \neq 0\}$. Then $h = 1$ locally almost everywhere on V, and since $R_x h = h$ and every compact subset of G is contained in a finite union of sets $R_x V$, $x \in G$, it follows that $h = 1$ locally almost everywhere on G. □

The following theorem is the precise formulation of the Galois type correspondence announced above.

THEOREM 3.4.3. *There exists a one-to-one correspondence between closed subgroups H of G and right translation invariant von Neumann subalgebras $\mathcal{A}$ of $L^\infty(G)$ which is determined by*

$$\mathcal{A} = \{f \in L^\infty(G) : L_x f = f \text{ for all } x \in H\}$$

and

$$H = \{x \in G : L_x f = f \text{ for all } f \in \mathcal{A}\}.$$

PROOF. Let H be a closed subgroup of G. Set

$$\mathcal{A}(H) = \{f \in L^\infty(G) : L_x f = f \text{ for all } x \in H\}.$$

Clearly, $\mathcal{A}(H)$ is a right invariant von Neumann subalgebra of $L^\infty(G)$. Conversely, for a right invariant von Neumann subalgebra $\mathcal{A}$ of $L^\infty(G)$, put

$$H(\mathcal{A}) = \{x \in G : L_x f = f \text{ for all } f \in \mathcal{A}\}.$$

Then $H(\mathcal{A})$ is a subgroup of G, and since the map $x \to L_x f$, $f \in L^\infty(G)$, from G into $L^\infty(G)$ is continuous for the w^*-topology on $L^\infty(G)$ and $\mathcal{A}$ is w^*-closed, $H(\mathcal{A})$ is closed in G. We claim that

$$H(\mathcal{A}(H)) = H \quad \text{and} \quad \mathcal{A}(H(\mathcal{A})) = \mathcal{A}.$$

The inclusions $H \subseteq H(\mathcal{A}(H))$ and $\mathcal{A} \subseteq \mathcal{A}(H(\mathcal{A}))$ are both clear from the definition of $\mathcal{A}(H)$ and $H(\mathcal{A})$. Let x be an arbitrary element of $G \backslash H$. Choose $g \in C_c(G)$ with $g|_H = 0$, $0 \leqslant g \leqslant 1$ and $g(x^{-1}) = 1$ and put

$$f(y) = \int_H g(t^{-1}y)dt, \ y \in G.$$

Then $L_s f = f$ for all $s \in H$, $f(e) = 0$ and $f(x^{-1}) \neq 0$. Thus $f \in \mathcal{A}(H)$, but $L_x f \neq f$, and hence $x \notin H(\mathcal{A}(H))$. This shows that $H(\mathcal{A}(H)) = H$.

It remains to prove that $\mathcal{A}(H(\mathcal{A})) \subseteq \mathcal{A}$. By Lemma 3.4.1 it suffices to show that $\mathcal{A}(H(\mathcal{A})) \cap C(G) \subseteq \mathcal{A}$. For sake of brevity, put $\mathcal{B} = \mathcal{A}(H(\mathcal{A}))$. Let $G \backslash H$ denote the homogeneous space of right cosets Hx, $x \in G$. Then $G \backslash H$ is a locally compact Hausdorff space. To any $f \in \mathcal{B} \cap C(G)$ we associate the continuous function $\widetilde{f}$ on $G \backslash H$ defined by $\widetilde{f}(Hx) = f(x)$.

Fix a compact subset K of $G \backslash H$ and let $\mathcal{A}_K$ denote the algebra of functions $\widetilde{f}|_K$, where $f \in \mathcal{A} \cap C(G)$. Then $\mathcal{A}_K$ contains the constant functions (Lemma 3.4.2) and is closed under complex conjugation. Moreover, $\mathcal{A}_K$ separates the points of K. Indeed, if Hx_1 and Hx_2 are two distinct right cosets, then $x_1 x_2^{-1} \notin H$ and hence there exists $f \in \mathcal{A} \cap C(G)$ such that $L_{x_1 x_2^{-1}} f \neq f$. This implies that $f(x_1 y) \neq f(x_2 y)$ for some $y \in G$ and hence the function $R_y f \in \mathcal{A} \cap C(G)$ separates x_1 and x_2. The Stone-Weierstrass theorem now yields that $\mathcal{A}_K$ is uniformly dense in $C(K)$.

Observe next that on bounded spheres in $L^\infty(G)$ the w^*-topology coincides with the topology defined by the seminorms

$$f \to \left| \int_G f(x)\psi(x)dx \right|, \quad \psi \in C_c(G), \|\psi\|_1 \leqslant 1.$$

Therefore, since $\mathcal{A}$ is w^*-closed in $L^\infty(G)$, to show that $\mathcal{B} \cap C(G) \subseteq \mathcal{A}$, it suffices to prove that given $f \in \mathcal{B} \cap C(G)$, $\epsilon > 0$ and $\psi_1, \ldots, \psi_n \in C_c(G)$ with $\|\psi_j\|_1 \leqslant 1$, $1 \leqslant j \leqslant n$, there exists $g \in \mathcal{A}$ such that

$$\left| \int_G [f(x) - g(x)]\psi_j(x)dx \right| \leqslant \epsilon$$

for $j = 1, \ldots, n$. To that end, let $C = \cup_{j=1}^n \operatorname{supp} \psi_j$ and $K = HC \backslash H$. It follows from the density of $\mathcal{A}_K$ in $C(K)$ that there exists $h \in \mathcal{A} \cap C(G)$ such that

$$\sup_{x \in C} |f(x) - h(x)| \leqslant \epsilon.$$

Let $\varphi : \mathbb{C} \to \mathbb{C}$ be a continuous function such that $\varphi(z) = z$ for all z with $|z| \leqslant \|f\|_\infty + \epsilon$ and $|\varphi(z)| \leqslant \|f\|_\infty + \epsilon$ if $|z| \geqslant \|f\|_\infty + \epsilon$. Then $g = \varphi \circ h \in \mathcal{A}$ by [**256**, Proposition 1.18.1] and $\sup_{x\in C} |f(x) - g(x)| \leqslant \epsilon$ and therefore

$$\left|\int_G [f(x) - g(x)]\psi_j(x)dx\right| \leqslant \epsilon,\ 1 \leqslant j \leqslant n.$$

This completes the proof. □

Let $\gamma : L^\infty(G) \to \mathcal{B}(L^2(G))$ be the embedding of $L^\infty(G)$ into $\mathcal{B}(L^2(G))$ defined by $\gamma(f)g = fg$, $g \in L^2(G)$. Then γ is a w^*-w^*-continuous isomorphism onto $\gamma(L^\infty(G))$. In particular, $\gamma(L^\infty(G))$ is a von Neumann subalgebra of $\mathcal{B}(L^2(G))$ isomorphic to $L^\infty(G)$.

The following proposition is the main step towards establishing a one-to-one correspondence between right invariant von Neumann subalgebras of $\gamma(L^\infty(G))$ and von Neumann subalgebras of $VN(G)$ which are invariant in a sense to be defined later. The proof of part (ii) builds on the notion of induced representations and a theorem of Blattner and Mackey which provides an isomorphism between certain spaces of intertwining operators. Concerning the Blattner-Mackey theorem, we refer the reader to [**15**].

PROPOSITION 3.4.4. *Let $\mathcal{B}$ be a von Neumann subalgebra of $\gamma(L^\infty(G))$ which is right translation invariant in the sense that $R_x f \in \gamma^{-1}(\mathcal{B})$ for all $f \in \gamma^{-1}(\mathcal{B})$ and $x \in G$. Then there exists a closed subgroup H of G with the following properties.*

(i) $\gamma^{-1}(\mathcal{B}) = \{f \in L^\infty(G) : L_x f = f \text{ for all } x \in H\}$.

(ii) $VN(G) \cap \mathcal{B}' = VN_H(G)$, *where $\mathcal{B}'$ denotes the commutant of $\mathcal{B}$ in $\mathcal{B}(L^2(G))$.*

PROOF. (i) has been shown in Theorem 3.4.3 with H being defined by

$$H = \{x \in G : L_x f = f \text{ for all } f \in \gamma^{-1}(\mathcal{B})\}.$$

We claim that $H = \{x \in G : \lambda_G(x) \in \mathcal{B}'\}$. This can be seen as follows. If $x \in H$, $h \in L^2(G)$ and $\gamma(f) \in \mathcal{B}$, then

$$\lambda_G(x)\gamma(f)h = (L_x f)(L_x h) = f(L_x h) = \gamma(f)(\lambda_G(x)h).$$

Conversely, if $\lambda_G(x) \in \mathcal{B}'$ and $\gamma(f) \in \mathcal{B}$, then for each $h \in L^2(G)$,

$$\gamma(L_x f)(L_x h) = L_x(fh) = \lambda_G(x)(\gamma(f)h) = \gamma(f)(\lambda_G(x)h) = \gamma(f)(L_x h).$$

Since $L_x(L^2(G)) = L^2(G)$, this implies that $\gamma(L_x f) = \gamma(f)$, and hence $L_x f = f$, so that $x \in H$.

We shall now prove that $VN(G) \cap \mathcal{B}' = VN_H(G)$. For any locally compact group K, let ρ_K denote the right regular representation of K and recall that the commuting algebra of $\rho_K(K)$ in $\mathcal{B}(L^2(K))$ is equal to $VN(K)$. The unitary representation of G obtained by inducing ρ_K up to G, $\mathrm{ind}_K^G \rho_K$, can be realized in the Hilbert space $L^2(G)$, and $\mathrm{ind}_K^G \rho_K$ is then equivalent to ρ_G. The equivalence is furnished by the unitary operator $U : L^2(G) \to L^2(G)$ given by

$$U\xi(y) = \Delta_G(y^{-1})^{1/2}\xi(y^{-1}),\ \ \xi \in L^2(G), y \in G.$$

For $\varphi \in C_0(G/H)$, let $P(\varphi) \in \mathcal{B}((L^2(G))$ be the multiplication operator defined by $P(\varphi)\xi(y) = \varphi(yH)\xi(y)$. Let $\mathcal{P}$ denote the algebra of all operators $P(\varphi)$, $\varphi \in$

$C_0(G/H)$. To show that $VN(G) \cap \mathcal{B}' = VN_H(G)$, we now apply the Blattner-Mackey theorem for induced representations [**15**], which in the current situation asserts the existence of an isomorphism

$$\rho_H(H)' \to \operatorname{ind}_H^G \rho_H(G)' \cap \mathcal{P}'.$$

Since $\rho_G(x) = U(\operatorname{ind}_H^G \rho_H)(x)U^{-1}$ for all $x \in G$, we thereby obtain an isomorphism

$$\phi : VN(H) \to (U \operatorname{ind}_H^G \rho_H(G)U^{-1})' \cap (U\mathcal{P}U^{-1})'.$$

By construction of the above isomorphism, $\phi(\lambda_H(x)) = \lambda_G(x)$ for all $x \in H$. Note next that for $\xi \in L^2(G)$, $\varphi \in C_0(G/H)$ and $y \in G$,

$$(UP(\varphi)U^{-1}\xi)(y) = \Delta_G(y)^{-1/2}\varphi(y^{-1}H)(U^{-1}\xi)(y^{-1}) = \varphi(y^{-1}H)\xi(y).$$

It follows that the set of all operators $UP(\varphi)U^{-1}$, $\varphi \in C_0(G/H)$, equals the set of all operators $\gamma(f)$, where f is continuous and constant on right cosets of H and, viewed as a function on $G\backslash H$, vanishes at infinity. Let $\mathcal{C}$ denote the algebra of all these operators $\gamma(f)$. Then $\mathcal{C}$ is dense in $\mathcal{B}$ and this implies that

$$(U\mathcal{P}U^{-1})' = \mathcal{C}' = \mathcal{B}'.$$

Since $\phi(\lambda_H(x)) = \lambda_G(x)$ for all $x \in H$, we conclude that $\phi(VN(H)) = VN_H(G)$ and hence $VN_H(G) = VN(G) \cap \mathcal{B}'$. □

Now recall that for $T \in VN(G)$ and $u \in A(G)$, $Tu \in A(G)$ is defined by $\langle Tu, S\rangle = \langle u, \check{T}S\rangle$ for $S \in VN(G)$, and that Tu satisfies $\operatorname{supp}(Tu) \subseteq (\operatorname{supp} T)(\operatorname{supp} u)$ whenever u has compact support (Lemma 2.5.7).

LEMMA 3.4.5. *If $u \in A(G) \cap L^2(G)$, then $R_T u \in L^2(G)$ for every $T \in VN(G)$.*

PROOF. Let $\xi, \eta \in L^2(G)$ be such that $u(x) = \langle \lambda_G(x)\xi, \eta\rangle$ for all $x \in G$. Then, for any $S \in VN(G)$, $\langle S, u\rangle = \langle S(\xi), \eta\rangle$. Now, let $T \in VN(G)$ and let $(T_\alpha)_\alpha$ be a net in $VN(G)$ converging to T in the strong operator topology. Then

$$\begin{aligned}
\|T_\alpha u - Tu\| &= \sup\{|\langle T_\alpha u - Tu, S\rangle| : S \in VN(G), \|S\| \leqslant 1\}\\
&= \sup\{|\langle u, S(T_\alpha - T)\rangle| : S \in VN(G), \|S\| \leqslant 1\}\\
&= \sup\{|\langle S(T_\alpha - T)\xi, \eta\rangle| : S \in VN(G), \|S\| \leqslant 1\}\\
&= \sup\{|\langle (T_\alpha - T)\xi, S^*\eta\rangle| : S \in VN(G), \|S\| \leqslant 1\}\\
&\leqslant \|(T_\alpha - T)\xi\| \cdot \|\eta\|,
\end{aligned}$$

and hence $\|T_\alpha u - Tu\| \to 0$.

Now $\lambda_G(x)u = R_x u$ for all $x \in G$. It follows that $Su \in L^2(G)$ whenever S is a finite linear combination of operators $\lambda_G(x), x \in G$. Since these operators are dense in $VN(G)$ in the strong operator topology, the net $(T_\alpha)_\alpha$ above can be chosen so that $T_\alpha u \in L^2(G)$ for all α. This implies that $Tu \in L^2(G)$ for each $T \in VN(G)$. □

THEOREM 3.4.6. *Let H be a closed subgroup of G and $T \in VN(G)$. Then $T \in VN_H(G)$ if and only if* $\operatorname{supp} T \subseteq H$.

PROOF. If $T \in VN_H(G)$, then T is a w^*-limit of finite linear combinations of operators $\lambda_G(x)$, $x \in H$. Since $\operatorname{supp} \lambda_G(x) = \{x\}$, it follows from Proposition 2.5.6(iv) that $\operatorname{supp} T \subseteq H$.

Conversely, suppose that $\operatorname{supp} T \subseteq H$ and let V be any open subset of G with $V = HV$. For every $u \in A(G) \cap C_c(G)$ with $\operatorname{supp} u \subseteq V$, we then have

$$\operatorname{supp}(Tu) \subseteq \operatorname{supp} T \cdot \operatorname{supp} u \subseteq HV = V.$$

Let $P_V : L^2(G) \to L^2(G)$ denote the projection defined by $P_V h = 1_V h$. Since the set of all $u \in A(G) \cap C_c(G)$ with $\operatorname{supp} u \subseteq V$ forms a dense linear subspace of $P_V(L^2(G))$, it follows that $T(P_V(L^2(G))) \subseteq P_V(L^2(G))$, that is, $TP_V = P_V T P_V$. Similarly, since $\operatorname{supp} T^* = (\operatorname{supp} T)^{-1} \subseteq H$, we get that $T^* P_V = P_V T^* P_V$ as well. Consequently, $P_V T = P_V T P_V$. Therefore T and P_V commute whenever V is an open subset of G with $V = HV$. Such projections P_V generate a von Neumann subalgebra $\mathcal{B}_H$ of $\gamma(L^\infty(G))$ which is right translation invariant in the sense of Proposition 3.4.4 since $R_x 1_V = 1_{Vx^{-1}}$ and $Vx^{-1} = HVx^{-1}$. By Proposition 3.4.4 there exists a closed subgroup K of G such that

$$\gamma^{-1}(\mathcal{B}_H) = \{f \in L^\infty(G) : L_x f = f \text{ for all } x \in K\}$$

and $VN(G) \cap \mathcal{B}_H' = VN_K(G)$. Now, if $x \in G$ is such that $1_V = L_x 1_V = 1_{xV}$ for all open subsets V of G with $V = HV$, then $x \in H$. It follows that $K \subseteq H$ and hence $VN_K(G) \subseteq VN_H(G)$. Consequently, $T \in VN(G) \cap \mathcal{B}_H' \subseteq VN_H(G)$, as was to be shown. □

COROLLARY 3.4.7. *Let $\mathcal{M}$ be an invariant W^*-subalgebra of $VN(G)$ and let $H = \{x \in G : \lambda_G(x) \in \mathcal{M}\}$. Then $\mathcal{M} = VN_H(G)$.*

PROOF. Clearly, H is a closed subgroup of G and $VN_H(G) \subseteq \mathcal{M}$. Conversely, if $T \in \mathcal{M}$ then $\operatorname{supp} T \subseteq H$. In fact, if $\lambda_G(x)$ is a w^*-limit of operators of the form $u \cdot T$, $u \in A(G)$, then $\lambda_G(x) \in \mathcal{M}$ since $\mathcal{M}$ is invariant, and hence $x \in H$ by the definition of H. Now $\operatorname{supp} T \subseteq H$ implies $T \in VN_H(G)$ by Theorem 3.4.6. □

A W^*-subalgebra $\mathcal{M}$ of $VN(G)$ is said to be *invariant* if for any $T \in \mathcal{M}$ and $u \in A(G)$, $u \cdot T \in \mathcal{M}$, where $\langle u \cdot T, v\rangle = \langle T, uv\rangle$ for $v \in A(G)$. Clearly, $VN_H(G)$ is an invariant W^*-subalgebra for every closed subgroup H of G. However, Corollary 3.4.7 tells us that there are no other invariant W^*-subalgebras of $VN(G)$. The next theorem is the result announced prior to Proposition 3.4.4.

The next theorem is the second main result of this section.

THEOREM 3.4.8. *Let G be a locally compact group. There is a one-to-one correspondence between right invariant von Neumann subalgebras $\mathcal{B}$ of $\gamma(L^\infty(G))$ and invariant von Neumann subalgebras $\mathcal{M}$ of $VN(G)$ which is determined by*

$$\mathcal{M} = \mathcal{B}' \cap VN(G) \text{ and } \mathcal{B} = \mathcal{M}' \cap \gamma(L^\infty(G)).$$

PROOF. Let $\mathcal{B}$ be a right invariant von Neumann subalgebra of $\gamma(L^\infty(G))$ and let H be the closed subgroup of G defined by

$$H = \{x \in G : L_x f = f \text{ for all } f \in \gamma^{-1}(\mathcal{B})\}.$$

Then, by Proposition 3.3.4(ii), $\mathcal{B}' \cap VN(G) = VN_H(G)$.

Let $\mathcal{M} = VN_H(G)$. Then $\mathcal{B} = \mathcal{M}' \cap \gamma(L^\infty(G))$. Indeed, if $f \in \gamma^{-1}(\mathcal{B})$ then for all $x \in H$ and $h \in L^2(G)$,

$$[\lambda_G(x)\gamma(f)]h = \lambda_G(x)(fh) = (L_x f)(L_x h) = f(\lambda_G(x)h) = [\gamma(f)\lambda_G(x)]h,$$

so that $\gamma(f) \in \mathcal{M}'$. Conversely, if $f \in L^\infty(G)$ is such that $\gamma(f) \in \mathcal{M}'$, then for each fixed $x \in H$ and all $h \in L^2(G)$,

$$(L_{x^{-1}} f)h = L_{x^{-1}}(f\lambda_G(x)h) = L_{x^{-1}}(\gamma(f)\lambda_G(x)h) = L_{x^{-1}}(\lambda_G(x)fh) = fh$$

and therefore $L_{x^{-1}} f = f$ locally a.e. on G. Since this holds for all $x \in H$, $f \in \gamma^{-1}(\mathcal{B})$. Thus $\mathcal{B} = \mathcal{M}' \cap \gamma(L^\infty(G))$.

Let $\mathcal{M}$ be an invariant von Neumann subalgebra of $VN(G)$. By Corollary 3.4.7, $\mathcal{M} = VN_H(G)$, where $H = \{x \in G : \lambda_G(x) \in \mathcal{M}\}$. Let

$$\mathcal{B} = \{\gamma(f) : f \in L^\infty(G) \text{ such that } L_x f = f \text{ for all } x \in H\}.$$

Then $\mathcal{M} = \mathcal{B}' \cap VN(G)$. To see this, recall that since $\gamma^{-1}(\mathcal{B})$ is a right translation invariant w^*-closed subalgebra of $L^\infty(G)$, by Proposition 3.4.4 there exists a closed subgroup K of G such that $\mathcal{B}' \cap VN(G) = VN_K(G)$, where K is given by

$$K = \{x \in G : L_x f = f \text{ for all } f \in \gamma^{-1}(\mathcal{B})\}.$$

Then, by Proposition 3.4.4(i),

$$\gamma^{-1}(\mathcal{B}) = \{f \in L^\infty(G) : L_x f = f \text{ for all } x \in K\}.$$

On the other hand, by the definition of $\mathcal{B}$,

$$\gamma^{-1}(\mathcal{B}) = \{f \in L^\infty(G) : L_x f = f \text{ for all } x \in H\}.$$

This implies that $H = K$ and hence $\mathcal{B}' \cap VN(G) = VN_H(G) = \mathcal{M}$. This completes the proof of the theorem. □

REMARK 3.4.9. It should be pointed out that Takesaki and Tatsuuma [**271**] defined a notion of invariance for a von Neumann subalgebra $\mathcal{M}$ of $VN(G)$ in terms of the normal isomorphism $\pi : VN(G) \to VN(G)\bar{\otimes}VN(G)$ given by

$$\pi(T) = W^{-1}(T \otimes I)W, \ T \in VN(G),$$

where W is the unitary operator on $L^2(G \times G)$ defined by

$$Wf(x, y) = f(x, xy), \ f \in L^2(G \times G), \ x, y \in G.$$

More precisely, they called $\mathcal{M}$ invariant if $\pi(\mathcal{M}) \subseteq \mathcal{M}\bar{\otimes}\mathcal{M}$. They proved that $\mathcal{M}$ is invariant in this sense if and only if there exists a closed subgroup H of G such that $\mathcal{M} = VN_H(G)$. Corollary 3.4.7 above now shows that this notion of invariance is equivalent to the one we have given.

3.5. Invariant subalgebras of $A(G)$ and $B(G)$

In this section we are concerned with the structure of translation invariant $*$-subalgebras of $B(G)$. The first main result (Theorem 3.5.3) says that if such a subalgebra A is w^*-closed and separates the points of G, then A contains $A(G)$. This yields a description of the w^*-closed, invariant $*$-subalgebras of $B(G)$ when G is amenable: they turn out to be exactly the subalgebras of $B(G)$ of the form $B(G/N)$, where N is a closed normal subgroup of G and $B(G/N)$ is viewed as a subalgebra of $B(G)$ in the obvious way.

Let $P^+(G)$ denote the cone consisting of all $u \in P(G)$ such that $u(x) \geqslant 0$ for all $x \in G$, and view $P^+(G)$ as a subset of $M(G) = C_c(G)^*$, the space of all (not necessarily bounded) Radon measures on G. The following remarkable property of $P^+(G)$ is fundamental for all the results displayed in this section.

LEMMA 3.5.1. *Let $Q \subseteq P^+(G)$ be a subcone and suppose that Q is closed under pointwise multiplication and that Q separates the identity e from each other element of G. Then there exists a net $(u_\alpha)_\alpha$ in Q such that $u_\alpha dx \to \delta_e$ in the w^*-topology of $M(G)$.*

PROOF. It suffices to show that given a compact subset K of G with $e \in K$, an open neighbourhood U of e and $\epsilon > 0$, there exists $u \in Q$ such that

$$\int_U u(x)dx = 1 \quad \text{and} \quad \int_{K\setminus U} u(x)dx \leqslant \epsilon.$$

In fact, if $f \in C_c(G)$ is such that $\operatorname{supp} f \subseteq K$, then

$$\begin{aligned} |\langle \delta_e, f\rangle - \langle u dx, f\rangle| &\leqslant \left|\int_U u(x)(f(e) - f(x))dx\right| + \left|\int_{K\setminus U} f(x)u(x)dx\right| \\ &\leqslant \sup_{x\in U} |f(e) - f(x)| + \|f\|_\infty \epsilon. \end{aligned}$$

Let $Q_1 = \{u \in Q : u(e) = 1\}$. For $u \in Q_1$ and $0 < \delta < 1$, let

$$U_{u,\delta} = \{x \in G : u(x) \geqslant \delta\}.$$

Let K, U and ϵ be given. We claim that there exist $u \in Q_1$ and $0 < \delta < 1$ such that $K \cap U_{u,\delta} \subseteq U$. Towards a contradiction, assume that

$$K \cap U_{u,\delta} \cap (G\setminus U) \neq \varnothing$$

for every $u \in Q_1$ and $0 < \delta < 1$. Now, the family of sets

$$\{K \cap U_{u,\delta} \cap (G\setminus U) : u \in Q_1, 0 < \delta < 1\}$$

has the finite intersection property since obviously

$$U_{u,\delta} \cap U_{v,\eta} \supseteq U_{uv,\max\{\delta,\eta\}}$$

for any $\delta, \eta > 0$ and $u, v \in Q_1$. Since K is compact, it follows that there exists $x \in G\setminus U$ with $u(x) = 1$ for all $u \in Q_1$. However, this is a contradiction since Q_1 separates e from x. Consequently, there exist $v \in Q_1$ and $0 < \delta < 1$ such that $K \cap U_{v,\delta} \subseteq U$. Then

$$\int_{K\setminus U} v^n(x)dx \leqslant \delta^n |K\setminus U|$$

for all $n \in \mathbb{N}$. Fix any η with $\delta < \eta < 1$ and let $V = U_{v,\eta} \cap U$. Then

$$\int_U v^n(x)dx \geqslant \int_V v^n(x)dx \geqslant \eta^n |V|$$

and hence

$$\left(\int_U v^n(x)dx\right)^{-1} \int_{K\setminus U} v^n(x)dx \leqslant \left(\frac{\delta}{\eta}\right)^n \frac{|K\setminus U|}{|V|}$$

for all $n \in \mathbb{N}$. Since $(\delta/\eta)^n \to 0$, there exists $N \in \mathbb{N}$ such that

$$\left(\int_U v^n(x)dx\right)^{-1} \int_{K\setminus U} v^n(x)dx \leqslant \epsilon.$$

Let $u = \left(\int_U v^N(x)dx\right)^{-1} v^N \in Q$. Then

$$\int_U u(x)dx = 1 \quad \text{and} \quad \int_{K\setminus U} u(x)dx \leqslant \epsilon,$$

as required. □

Let M be a von Neumann algebra and A its predual. For $x \in A$, as before let $|x| \in A$ denote the absolute value of x and let $x^* \in A$ be defined by $\langle x^*, f\rangle = \overline{\langle f^*, x\rangle}$, $f \in M$. Recall that for $x \in A$ and $f \in M$ the products $f \cdot x \in A$ and $x \cdot f \in A$ are defined by $\langle f \cdot x, g\rangle = \langle gf, x\rangle$ and $\langle x \cdot f, g\rangle = \langle x, fg\rangle$, $g \in M$, respectively (see [**60**, 12.2.1]).

LEMMA 3.5.2. *Let X be a closed subspace of A. Suppose that X is M-invariant in the sense that $M \cdot X \subseteq X$ and $X \cdot M \subseteq X$. Then $|x| \in X$ and $x^* \in X$ for every $x \in X$.*

PROOF. The annihilator $X^\perp$ of X in M is a w^*-closed subspace and an ideal since X is M-invariant. Hence X is the predual of the von Neumann algebra $M/X^\perp$. Let $x \in X$. Then $x^* \in X$ since $\langle x^*, f\rangle = \overline{\langle f^*, x\rangle} = 0$ for $f \in X^\perp$. Let $|x| \in A$ and $|x|' \in X$ be the absolute value of x, viewed as a functional on M and as a functional on $M/X^\perp$, respectively. The characterization of the absolute value in [**60**, 12.2.9] shows that $|x| = |x|'$ and hence $|x| \in X$. □

A subset A of $B(G)$ is said to be *invariant* if $L_x u, R_x u \in A$ for all $u \in A$ and $x \in G$.

THEOREM 3.5.3. *Let G be a locally compact group and A a w^*-closed invariant subalgebra of $B(G)$. Suppose that A is closed under complex conjugation and separates the points of G. Then A contains $A(G)$.*

PROOF. Since the Dirac measures generate a w^*-dense subalgebra of $W^*(G)$, the enveloping von Neumann algebra of $C^*(G)$, and A is w^*-closed and invariant, it follows that A is $W^*(G)$-invariant. In the following, for $u \in B(G)$ we denote by $|u| \in P(G)$ the absolute value of u and by u^* the adjoint of u, viewed as a linear functional on $W^*(G)$. Let

$$|A| = \{|u| : u \in A\}, \ A_* = \{u^* : u \in A\} \text{ and } A_{sa} = \{u \in A : u = u^*\}.$$

Then, by Lemma 3.5.2, $|A| = A \cap P(G)$ and $A = A_*$. Hence

$$A_{sa} = \{u - v : u, v \in |A|\},$$

and therefore $|A|$ separates the points of G. Let $\mathrm{Re}|A| = \{\mathrm{Re}\, u : u \in |A|\}$. Since A is closed under conjugation, $\mathrm{Re}|A| \subseteq A \cap P(G)$. Moreover, $\mathrm{Re}|A|$ separates e from every $x \in G$, $x \neq e$. Now, $\mathrm{Re}|A|$ is a convex cone and therefore $Q = \{u^2 : u \in \mathrm{Re}|A|\}$ also separates e from every $x \in G$, $x \neq e$. As A is an algebra, Q is contained in A and closed under multiplication. By Lemma 3.5.1, there exists a net $(u_\alpha)_\alpha$ in Q such that

$$\langle u_\alpha, f\rangle = \int_G f(x) u_\alpha(x)dx \ \to \ \langle \delta_e, f\rangle = f(e)$$

for every $f \in C_c(G)$.

For $A(G) \subseteq A$, it suffices to show that $f * \widetilde{f} \in A$ for all $f \in C_c(G)$, where $\widetilde{f}(x) = \overline{f(x^{-1})}$. Fix $f \in C_c(G)$. Then $f \cdot u_\alpha \cdot f^* \in |A|$ by the $W^*(G)$-invariance of A. Since

$$\lim_\alpha \|f \cdot u_\alpha \cdot f^*\|_{B(G)} = \lim_\alpha (f \cdot u_\alpha \cdot f^*)(e) = \lim_\alpha \langle u_\alpha, f^* * f\rangle = (f^* * f)(e),$$

we can assume that $(f \cdot u_\alpha \cdot f^*)_\alpha$ is a bounded net in $B(G)$. Now, for all $g \in C_c(G)$,

$$\overline{\langle f * \widetilde{f}, g\rangle} = (f^* * g * f)(e) = \lim_\alpha \langle u_\alpha, f^* * g * f\rangle = \lim_\alpha \langle f \cdot u_\alpha \cdot f^*, g\rangle.$$

Thus $f * \widetilde{f} = \lim_\alpha \overline{f \cdot u_\alpha \cdot f^*}$ in the w^*-topology of $B(G)$. Since A is w^*-closed and invariant under complex conjugation, we conclude that $f * \widetilde{f} \in A$, as required. □

COROLLARY 3.5.4. *For a locally compact group G, the following conditions are equivalent.*

(i) *G is amenable.*
(ii) *The map $N \to B(G/N)$ is a bijection between the set of all closed normal subgroups of G and the set of all w^*-closed, translation invariant $*$-subalgebras A of $B(G)$ with $A \neq \{0\}$.*

PROOF. (i) $\Rightarrow$ (ii) It is clear that for any closed normal subgroup N of G, $B(G/N)$ is a w^*-closed, translation invariant $*$-subalgebras A of $B(G)$ and that $B(G/N_1) = B(G/N_2)$ implies that $N_1 = N_2$.

Now, let A be any w^*-closed, translation invariant $*$-subalgebra of $B(G)$ and let $N = \{x \in G : L_x u = u \text{ for all } u \in A\}$. Then N is a closed normal subgroup of G. Moreover, A viewed as a subalgebra of $B(G/N)$ separates the points of G/N. Theorem 3.5.3 implies that $A(G/N) \subseteq A$. Since G/N is amenable, $A(G/N)$ is w^*-dense in $B(G/N)$ and hence $A = B(G/N)$ since A is w^*-closed.

(ii) $\Rightarrow$ (i) Let A be the w^*-closure of $A(G)$ in $B(G)$. By (ii), $A = B(G/N)$ for some closed normal subgroup N of G. But $A(G)$ separates the points of G and so $N = \{e\}$ and $A = B(G)$. However, since $A = B_\lambda(G)$, this means that λ_G weakly contains every representation of G. This in turn implies that G is amenable. □

Let G and H be locally compact groups and let $\phi : H \to G$ be a continuous homomorphism. Let $j : B(G) \to B(H)$ denote the homomorphism defined by $j(u) = u \circ \phi$, $u \in B(G)$.

COROLLARY 3.5.5. *Suppose that ϕ is injective and H is amenable. Then $j(B(G))$ is w^*-dense in $B(H)$.*

PROOF. This follows from the preceding corollary since $j(B(G))$ is an invariant $*$-subalgebra of $B(H)$ which separates the points of H. □

In passing we give two immediate applications of Corollary 3.5.5.

EXAMPLE 3.5.6. (1) Let G_d be the group G with the discrete topology, and let $\phi : G_d \to G$ be the identity map. Suppose that G_d is amenable. Then Corollary 3.5.5 shows that $B(G)$ is w^*-dense in $B(G_d)$.

(2) Let $b(G)$ denote the Bohr compactification of G and let $AP(G)$ be the algebra of almost periodic functions on G. Recall that G is said to be maximally almost periodic if $AP(G)$ separates the points of G. Let $\phi : G \to b(G)$ be the canonical homomorphism. Since $j(C(b(G))) = AP(G)$, by Theorem 2.2.1(ii), $j(B(b(G))) = B(G) \cap AP(G)$. Consequently, $B(G) \cap AP(G)$ is w^*-dense in $B(G)$ provided that G is an amenable maximally almost periodic group.

Next we employ the methods that led to Theorem 3.5.3 to obtain a description of the norm closed, invariant $*$-subalgebras of $A(G)$.

THEOREM 3.5.7. *Let G be a locally compact group and let A be a norm-closed, translation invariant $*$-subalgebra of $A(G)$. If $A \neq \{0\}$, then $A = A(G/K)$ for some compact normal subgroup K of G.*

PROOF. Let $K = \{x \in G : L_x u = u \text{ for all } u \in A\}$. Then K is a closed normal subgroup of G and $A \subseteq A(G/K)$. Since $A \neq \{0\}$ and $A(G) \subseteq C_0(G)$, K must be compact. Moving to G/K, we can assume that $K = \{e\}$ and then have to show that $A = A(G)$. Observe next that A is $VN(G)$-invariant. Indeed, we have $u \cdot \lambda(x^{-1}) = L_x u \in A$ and $\lambda(x) \cdot u = R_x u \in A$ for all $u \in A$ and $x \in G$, and the set of all $T \in VN(G)$ satisfying $A \cdot T \subseteq A$ and $T \cdot A \subseteq A$ forms a w^*-closed subalgebra of $VN(G)$. Hence, by Lemma 3.5.2,

$$A \cap P(G) = \{|u| : u \in A\} \quad \text{and} \quad A = \{u^* : u \in A\}.$$

As we have seen in the proof of Theorem 3.5.3, for each $u \in A(G) \cap P^1(G)$, there exists a net $(u_\alpha)_\alpha$ in $A \cap P^1(G)$ such that $u_\alpha \to u$ in the w^*-topology of $B(G) = C^*(G)^*$. Since, as we shall see in Theorem 3.7.7, on the unit sphere of $B(G)$ the w^*-topology coincides with the multiplier topology, it follows that

$$\|u_\alpha v - uv\|_{A(G)} \to 0$$

for every $v \in A(G)$. Taking $v \in A$ gives $uA \subseteq A$ for all $u \in A(G) \cap P^1(G)$, and this implies that A is an ideal in $A(G)$. This in turn yields that $A = A(G)$. In fact, if A is a proper closed ideal of $A(G)$, then there exists $x \in G$ such that $u(x) = 0$ for all $u \in A$. This is a contradiction since A is translation invariant and $A \neq \{0\}$. □

We finish this section by giving an application of Theorem 3.5.7 to the structure of norm closed, invariant subalgebras of $B(G)$. Recall that for any unitary representation π of G, the space $A_\pi(G)$ is the closed linear subspace of $B(G)$ generated by the coefficients of π and that the spaces $A_\pi(G)$ are precisely the closed, invariant subspaces of $B(G)$ (Lemma 2.8.3(ii)). The set of all linear combinations of coefficients of all unitary representations of G which are disjoint from the left regular representation is of the form $A_\omega(G)$ for some representation ω, and then we have the Lebesgue decomposition $B(G) = A(G) \oplus_1 A_\omega(G)$ (Proposition 2.8.9(i)).

COROLLARY 3.5.8. *Let G be a locally compact group and let A be a norm closed, translation invariant $*$-subalgebra of $B(G)$ which separates the points of G. Then either $A(G) \subseteq A$ or $A \subseteq A_\omega$.*

PROOF. Recall first that $A = A_\pi$ for some unitary representation π of G. If $A \cap A(G) = \{0\}$ then π is disjoint from λ_G (Lemma 2.8.7) and so $A \subseteq A_\omega$. Thus suppose that $A \cap A(G) \neq \{0\}$. Then, by Theorem 3.5.7, $A \cap A(G) = A(G/K)$ for some compact normal subgroup K of G. We have to show that $K = \{e\}$.

Let $x, y \in K$ with $x \neq y$. Then there exists $u \in A$ such that $u(x) \neq u(y)$. Moreover, there exists $v \in A(G/K)$ such that $v(x)v(y) \neq 0$. Then $uv(x) \neq uv(y)$, and $uv \in A \cap A(G) = A(G/K)$ since $A(G)$ is an ideal in $B(G)$. This contradicts $x, y \in K$. □

3.6. Comparison of $A(G_1)\widehat{\otimes} A(G_2)$ and $A(G_1 \times G_2)$

Let G_1 and G_2 be locally compact groups. If $u_i \in A(G_i)$, $i = 1, 2$, then $u(x_1, x_2) = u_1(x_1)u_2(x_2)$, $x_i \in G_i$, defines an element of $A(G_1 \times G_2)$ with $\|u\| \leqslant \|u_1\| \cdot \|u_2\|$. Thus there exists a canonical contraction ϕ from the projective tensor product $A(G_1)\widehat{\otimes} A(G_2)$ into $A(G_1 \times G_2)$. Note that the range of ϕ is dense in $A(G_1 \times G_2)$. In fact, if $u_i = \xi_i * \check{\eta}_i$, $\xi_i, \eta_i \in L^2(G_i)$, $i = 1, 2$, then

$$\langle T, u_1 \times u_2 \rangle = \langle T(\xi_1 \otimes \xi_2), \eta_1 \otimes \eta_2 \rangle$$

for every $T \in VN(G_1 \times G_2)$. Since the linear span of the elementary tensors is dense in $L^2(G_1 \times G_2) = L^2(G_1) \otimes L^2(G_2)$, the claim follows.

If G_1 and G_2 are both abelian, then $A(G_i) = L^1(\widehat{G_i})$ and $A(G_1 \times G_2) = L^1(\widehat{G_1} \times \widehat{G_2})$, and $L^1(\widehat{G_1} \times \widehat{G_2})$ is isometrically isomorphic to $L^1(\widehat{G_1}) \hat{\otimes} L^1(\widehat{G_2})$. Consequently, in this case ϕ is an isometric isomorphism. The theme of this section is to treat the problem of when ϕ is a linear isomorphism for arbitrary locally compact groups G_1 and G_2.

For $n \in \mathbb{N}$, let ℓ_n^2 be $\mathbb{C}^n$ with the Euclidean scalar product and let M_n denote the algebra of all $n \times n$-matrices considered as operators on ℓ_n^2. The predual M_{n*} of M_n will be identified with $\ell_n^2 \hat{\otimes} \overline{\ell_n^2}$, where $\overline{\ell_n^2}$ denotes the complex conjugate space of ℓ_n^2 and the duality is given by

$$\langle \xi \otimes \overline{\eta}, T \rangle = \langle T\xi, \eta \rangle, \quad \xi, \eta \in \ell_n^2,\ T \in M_n.$$

Then, for $m, n \in \mathbb{N}$, there is canonical bilinear bijection

$$\phi_{m,n} : M_m \overline{\otimes} M_n \to (M_{m*} \otimes M_{n*})^*$$

given by, for $\xi_1, \eta_1 \in \ell_m^2$, $\xi_2, \eta_2 \in \ell_n^2$ and $T_1 \in M_m, T_2 \in M_n$,

$$\langle (\xi_1 \otimes \overline{\eta}_1) \otimes (\xi_2 \otimes \overline{\eta}_2), \phi_{m,n}(T_1 \otimes T_2) \rangle = \langle T_1\xi_1, \eta_1 \rangle \cdot \langle T_2\xi_2, \eta_2 \rangle.$$

LEMMA 3.6.1. *$\|\phi_{m,n}\| = 1$ and $\|\phi_{m,n}^{-1}\| = \min\{m, n\}$.*

PROOF. We identify $M_m \overline{\otimes} M_n$ with M_{mn}, the matrices acting as operators on $\ell_{mn}^2 = \ell_m^2 \otimes \ell_n^2$. The above formula implies that

$$\langle (\xi_1 \otimes \overline{\eta}_1) \otimes (\xi_2 \otimes \overline{\eta}_2), \phi_{m,n}(T) \rangle = \langle T(\xi_1 \otimes \xi_2), \eta_1 \otimes \eta_2 \rangle$$

for every $T \in M_m \overline{\otimes} M_n$, from which we easily conclude that $\|\phi_{m,n}\| = 1$.

Now let $N = \min\{m, n\}$ and let $\epsilon_1, \ldots, \epsilon_N$ be the standard basis of $\ell_N^2 \subseteq \ell_m^2, \ell_n^2$. Let $\omega = \sum_{i=1}^N \epsilon_i \otimes \epsilon_i \in \ell_m^2 \otimes \ell_n^2$, and define an operator T on $\ell_m^2 \otimes \ell_n^2$ by $T\xi = \langle \xi, \omega \rangle \omega$. Then $\|T\| \leqslant \|\omega\|^2$ and $\|T\omega\| = \|\omega\|^3$, so that $\|T\| = \|\omega\|^2$. On the other hand, $\|\omega\|^2 = N$.

By definition of $\phi_{m,n}$, we have

$$\begin{aligned}
\langle (\xi_1 \otimes \overline{\eta_1}) \otimes (\xi_2 \otimes \overline{\eta_2}), \phi_{m,n}(T) \rangle &= \langle \langle \xi_1 \otimes \xi_2, \omega \rangle \omega, \eta_1 \otimes \eta_2 \rangle \\
&= \sum_{i=1}^N \langle \xi_1, \epsilon_i \rangle \langle \xi_2, \epsilon_i \rangle \cdot \sum_{j=1}^N \langle \epsilon_j \otimes \eta_1 \rangle \langle \epsilon_j \otimes \eta_2 \rangle
\end{aligned}$$

for all $\xi_1, \xi_2 \in \ell_m^2$ and $\eta_1, \eta_2 \in \ell_n^2$. This equation implies

$$\begin{aligned}
|\langle (\xi_1 \otimes \overline{\eta_1}) \otimes (\xi_2 \otimes \overline{\eta_2}), \phi_{m,n}(T) \rangle| &\leqslant \|\xi_1\| \cdot \|\xi_2\| \cdot \|\eta_1\| \cdot \|\eta_2\| \\
&= \|\xi_1 \otimes \overline{\eta}_1\| \cdot \|\xi_2 \otimes \overline{\eta}_2\|
\end{aligned}$$

and hence $\|\phi_{m,n}(T)\| \leqslant 1$ by definition of the projective norm. Since $\|T\| = N$, it follows that $\|\phi_{m,n}^{-1}\| \geqslant N$.

To show that conversely $\|\phi_{m,n}^{-1}\| \leqslant N$, take any $T \in M_m \overline{\otimes} M_n$. Then there exist $\xi, \eta \in \ell_m^2 \otimes \ell_n^2$ such that $\|\xi\| = \|\eta\| = 1$ and $\langle T\xi, \eta \rangle = \|T\|$. Interpreting ξ and η as Hilbert-Schmidt operators, by definition of the tensor product of Hilbert spaces, there exist orthonormal systems $\{\xi_i^{(1)}\}$ and $\{\eta_i^{(1)}\}$ in ℓ_m^2 and $\{\xi_j^{(2)}\}$ and $\{\eta_j^{(2)}\}$ in ℓ_n^2 and $\lambda_i, \mu_j \in \mathbb{C}$, $1 \leqslant i, j \leqslant N$, such that $\sum_{i=1}^N |\lambda_i|^2 = \sum_{j=1}^N |\mu_j|^2 = 1$ and

$$\xi = \sum_{i=1}^N \lambda_i \left(\xi_i^{(1)} \otimes \xi_i^{(2)} \right) \quad \text{and} \quad \eta = \sum_{j=1}^N \mu_j \left(\eta_j^{(1)} \otimes \eta_j^{(2)} \right).$$

Then, using the Cauchy-Schwarz inequality, we get for the projective norm of u, $\|u\| \leqslant \sum_{i,j=1}^{N} |\lambda_i\mu_j| \leqslant N$. Since $\langle u, \phi_{m,n}(T)\rangle = \langle T\xi, \eta\rangle = \|T\|$, we conclude that $\|\phi_{m,n}(T)\| \geqslant \frac{1}{N}\|T\|$ and hence $\|\phi_{m,n}^{-1}\| \leqslant N$. □

THEOREM 3.6.2. *Let G_1 and G_2 be locally compact groups and and let*

$$\phi : A(G_1)\widehat{\otimes} A(G_2) \to A(G_1\times G_2)$$

be the canonical homomorphism. Then the following three conditions are equivalent.

(i) *ϕ is a surjective (not necessarily isometric) linear isomorphism.*
(ii) *ϕ is surjective.*
(ii) *At least one of G_1 and G_2 has an abelian subgroup of finite index.*

PROOF. (i) $\Rightarrow$ (ii) being trivial, suppose that ϕ is surjective. Then the dual map $\phi^* : VN(G_1\times G_2) \to [A(G_1)\widehat{\otimes} A(G_2)]^*$ satisfies $\|\phi^*(T)\| \geqslant c\|T\|$ for some constant $c > 0$. Let $n = \lfloor\frac{1}{c}\rfloor$ and $m = n+1$.

We are going to show that either $VN(G_1)$ or $VN(G_2)$ is of type $I_{\leqslant n}$, i.e., there are no direct summands of types $\mathrm{I}_k, k > n$, II or III. Assuming the contrary, it follows from [**270**, Chapter V, Proposition 1.35] that there exist pairwise equivalent orthogonal projections $P_1,\dots,P_m$ in $VN(G_1)$. Let $U_{ij} \in VN(G_1)$, $1 \leqslant i,j \leqslant m$, be partial isometries such that $U_{ij}^*U_{ij} = P_j$ and $U_{ij}^* = U_{ji}$ (and hence $U_{ij}U_{ij}^* = P_i$). The operators U_{ij} generate a subalgebra of $VN(G_1)$ which is isomorphic to M_m [**270**, Chapter IV, Proposition 1.8]. In the same way, we obtain a subalgebra of $VN(G_2)$. The canonical homomorphism

$$\Psi_m : M_m \overline{\otimes} M_m \to VN(G_1)\overline{\otimes} VN(G_2)$$

is injective and isometric. Indeed, Ψ is injective since $M_m\overline{\otimes} M_m$ is finite dimensional, and Ψ is an isometry by uniqueness of the C^*-norm. Now consider the following diagram:

$$\begin{array}{ccc} M_m\overline{\otimes} M_m & \xrightarrow{\ \phi_{m,m}\ } & [M_{m*}\widehat{\otimes} M_{m*}]^* \\ \Psi_m\big\downarrow & & \big\downarrow\Omega_m \\ VN(G_1)\overline{\otimes} VN(G_2) & \xrightarrow[\ \phi^*\ ]{} & [A(G_1)\widehat{\otimes} A(G_2)]^* \end{array}$$

This diagram commutes. To see this, let $u_i \in A(G_i)$ and $S_i \in M_m$, $i = 1,2$. Then

$$\begin{aligned} \langle\phi^*(\Psi_m(S_1\otimes S_2)), u_1\otimes u_2\rangle &= \langle\Psi_m(S_1\otimes S_2), u_1\times u_2\rangle \\ &= \langle S_1,u_1\rangle\langle S_2,u_2\rangle, \end{aligned}$$

and, on the other hand, if $u_i = \xi_i\otimes\overline{\eta_i}$, $i = 1,2$, then

$$\begin{aligned} \langle\Omega_m(\phi_{m,m}(S_1\otimes S_2)), u_1\otimes u_2\rangle &= \langle\phi_{m,m}(S_1\otimes S_2), \Omega_{m_*}(u_1\otimes u_2)\rangle \\ &= \langle\phi_{m,m}(S_1\otimes S_2), (\xi_1\otimes\overline{\eta_1})\otimes(\xi_2\otimes\overline{\eta_2})\rangle \\ &= \langle S_1,u_1\rangle\langle S_2,u_2\rangle. \end{aligned}$$

Since ϕ^* and Ω_m are both isometric, it follows that, for any $T \in M_m\widehat{\otimes} M_m$,

$$\|\phi_{m,n}(T)\| = \|\Omega_m(\phi_{m,n}(T))\| = \|\phi^*(\Psi_m(T))\| \geqslant c\,\|\Psi_m(T)\| = c\|T\|$$

and hence $\|\phi_{m,n}^{-1}\| \leqslant \frac{1}{c} < m$, contradicting Lemma 3.6.1. Thus one of $VN(G_1)$ or $VN(G_2)$ has to be of type $I_{\leqslant n}$. Theorem 1 of [**215**] (and its proof) now shows that G_1 or G_2 has an abelian subgroup of finite index.

(iii) $\Rightarrow$ (i) Suppose first that G_1, say, is abelian. Then the Fourier transform furnishes an isometric isomorphism between $A(G_1)$ and $L^1(\widehat{G_1})$. Let $L^1(\widehat{G_1}, A(G_2))$ denote the algebra of $A(G_2)$-valued Bochner integrable function on $\widehat{G_1}$ with convolution and the usual L^1-norm. Then $L^1(\widehat{G_1}, A(G_2))$ is isometrically isomorphic to the projective tensor product $L^1(\widehat{G_1})\,\widehat{\otimes}\,A(G_2)$. the isomorphism determined by assigning to $f\otimes u$, $f \in L^1(\widehat{G_1})$, $u \in A(G_2)$, the function $\gamma \to f(\gamma)u$ in $L^1(\widehat{G_1}, A(G_2))$.

Now, $A(G_1 \times G_2)$ is also isometrically isomorphic to $L^1(\widehat{G_1}, A(G_2))$. In fact, by [**270**, Chapter IV, Theorem 7.17] and standard facts, we have canonical isometric isomorphisms

$$\begin{aligned} L^1(\widehat{G_1}, A(G_2)) &= [L^\infty(\widehat{G_1})\,\overline{\otimes}\,VN(G_2)]_* \\ &= [VN(G_1)\,\overline{\otimes}\,VN(G_2)]_* \\ &= VN(G_1 \times G_2)_* = A(G_1 \times G_2). \end{aligned}$$

It is easily verified that the composition

$$A(G_1)\widehat{\otimes}A(G_2) \to L^1(\widehat{G_1}, A(G_2)) \to A(G_1 \times G_2)$$

of isometric isomorphisms equals the homomorphism ϕ. This shows (i) when G_1 is abelian.

Now assume that G_1 has an abelian subgroup H of finite index and let F be a representative system for the right cosets of H in G_1. Then every $u \in A(G_1)$ can be written as $u = \sum_{x\in F} \widetilde{R_x u|_H}$, and

$$\|u\| \leqslant \sum_{x\in F} \|\widetilde{R_x u|_H}\| \leqslant [G : H] \cdot \|u\|.$$

Thus every $w \in A(G_1)\,\widehat{\otimes}\,A(G_2)$ has a representation

$$w = \sum_{j=1}^{\infty} \left(\sum_{x\in F} \widetilde{R_x u_j|_H} \otimes v_j \right) = \sum_{x\in F} \left(\sum_{j=1}^{\infty} \widetilde{R_x u_j|_H} \otimes v_j \right)$$

such that $\|w\| = \sum_{j=1}^{\infty} \|u_j\| \cdot \|v_j\|$, and hence

$$\|w\| \leqslant \sum_{j=1}^{\infty} \left\| \sum_{x\in F} \widetilde{R_x u_j|_H} \right\| \cdot \|v_j\| \leqslant [G : H] \cdot \|w\|.$$

Of course, we have a similar representation and an analogous norm estimate for elements in $A(G_1\times G_2)$ in terms of elements of $A(H\times G_2)$. Since ϕ maps $A(H)\,\widehat{\otimes}\,A(G_2)$ onto $A(H \times G_2)$ and these spaces have the same codimension in $A(G_1)\,\widehat{\otimes}\,A(G_2)$ and $A(G_1 \times G_2)$, respectively, it follows that ϕ is surjective. □

COROLLARY 3.6.3. *The canonical homomorphism*

$$\phi : A(G_1)\,\widehat{\otimes}\,A(G_2) \to A(G_1 \times G_2)$$

is an isometric isomorphism if and only if at least one of G_1 *and* G_2 *is abelian.*

PROOF. Suppose first that G_1, say, is abelian. In this situation we have already seen in the first part of the proof of the implication (iii) $\Rightarrow$ (i) of Theorem 3.6.2 that ϕ is an isometric isomorphism.

Conversely, suppose that ϕ is an isometric isomorphism and, towards a contradiction, assume that none of G_1 and G_2 is abelian. Then (compare the proof of Theorem 3.6.2) $VN(G_i)$ contains a subalgebra isomorphic to M_2, $i = 1, 2$. Now, consider the above diagram with $m = 2$. Since Ψ_2 and Ω_2 are both isometric and the diagram is commutative, the norm of the map

$$\phi^*|_{\Omega_2((M_2\hat{\otimes}M_2)*)} = \Phi_2 \circ \phi_{2,2}^{-1} \circ \Omega_2^{-1}|_{\Omega_2((M_2\hat{\otimes}M_2)*)}$$

equals the norm of $\phi_{2,2}^{-1}$. Thus Lemma 3.6.1 implies that $\|\phi^*\| \geqslant 2$. This contradicts the hypothesis that $\|\phi^*\| = \|\phi\| = 1$. □

3.7. The w^*-topology and other topologies on $B(G)$

In this section we study various topologies on the Fourier-Stieltjes algebra $B(G)$ of a locally compact group G, with emphasis on the question of whether they coincide on the unit sphere of $B(G)$.

We define five topologies τ_{uc}, τ_{w^*}, τ_{bw^*}, τ_{nw^*} and τ_M on $B(G)$ by the statement that a net $(u_\alpha)_\alpha$ in $B(G)$ converges to $u \in B(G)$ with respect to

- τ_{uc} if $u_\alpha \to u$ uniformly on compact subsets of G;
- τ_{w^*} if $u_\alpha \to u$ in the w^*-topology $\sigma(B(G), C^*(G))$;
- τ_{bw^*} if $(u_\alpha)_\alpha$ is norm bounded and $u_\alpha \to u$ in τ_{w^*};
- τ_{nw^*} if $\|u_\alpha\| \to \|u\|$ and $u_\alpha \to u$ in τ_{w^*};
- τ_M if $\|(u_\alpha - u)v\| \to 0$ for all $v \in A(G)$.

We start the comparison of the topologies with a sequence of lemmas.

LEMMA 3.7.1. *The following hold.*

(i) *τ_M is stronger than τ_{uc}.*

(ii) *On norm bounded sets in $B(G)$, τ_{uc} is stronger than τ_{w^*}.*

PROOF. (i) Let K be any compact subset of G and choose $v \in A(G)$ such that $v = 1$ on K. Then, for every $x \in K$,

$$|u_\alpha(x) - u(x)| \leqslant \|(u_\alpha - u)v\|_\infty \leqslant \|(u_\alpha - u)v\|_{A(G)}.$$

(ii) Let $u_\alpha \to u$ in τ_{uc} and $\|u_\alpha\|, \|u\| \leqslant C < \infty$ for all α. Let $f \in C^*(G)$ and $\epsilon > 0$, and choose $g \in C_c(G)$ with $\|g - f\|_{C^*(G)} \leqslant \epsilon$. Then

$$\begin{aligned} |\langle u_\alpha, f\rangle - \langle u, f\rangle| &\leqslant |\langle u_\alpha - u, g\rangle| + |\langle u_\alpha, f - g\rangle| + |\langle u, g - f\rangle \\ &\leqslant \|(u_\alpha - u)|_{\operatorname{supp} g}\|_\infty \|g\|_\infty + 2C\|f - g\|_{C^*(G)}. \end{aligned}$$

Since $\|f - g\|_{C^*(G)} \leqslant \epsilon$ and $u_\alpha \to u$ uniformly on compact sets, it follows that $u_\alpha \to u$ in τ_{w^*}. □

The following lemma is a special case of [**211**, Lemma 8.2]. However, the proof given here is considerably simpler than the one given in [**211**].

LEMMA 3.7.2. *Let $(u_\alpha)_\alpha$ be a net in $B(G)$ such that $u_\alpha \to u \in B(G)$ in the topology τ_{nw^*}. Let $(e_\gamma)_\gamma$ be an approximate identity for $L^1(G)$, where every e_γ is of the form $e_\gamma = f_\gamma * f_\gamma^*$, $f_\gamma \in L^1(G)$, $f_\gamma \geqslant 0$ and $\|f_\gamma\|_1 = 1$. Then, given any $\epsilon > 0$, there exist indices α_0 and γ_0 such that*

$$\|e_{\gamma_0} * u - u\| \leqslant \epsilon \quad \text{and} \quad \|e_{\gamma_0} * u_\alpha - u_\alpha\| \leqslant \epsilon$$

for all $\alpha \geqslant \alpha_0$.

PROOF. Let $L^1(G)_e$ be $L^1(G)$ with an identity e adjoined if G is nondiscrete and equal to $L^1(G)$ if G is discrete. Then the enveloping C^*-algebra of $L^1(G)_e$ equals $C^*(G)_e$. Since $\|e_\gamma\|_{C^*(G)} \leqslant \|e_\gamma\|_1 = 1$, it follows that $0 \leqslant e_\gamma \leqslant e$ in the C^*-algebra sense and therefore $0 \leqslant (e - e_\gamma) * (e - e_\gamma) \leqslant e - e_\gamma$ in $C^*(G)_e$.

Since $\|u_\alpha\| \to \|u\|$, we may assume that $u \neq 0$. For any $f \in L^1(G)$ and $v \in B(G)$, as $\overline{e_\gamma} = e_\gamma$, we have

$$\begin{aligned} |\langle e_\gamma * v - v, f\rangle|^2 &= |\langle v, (\overline{e_\gamma} - e) * f\rangle|^2 \\ &\leqslant \|u\| \cdot |\langle |u|, (e_\gamma - e) * f * f^* * (e_\gamma - e)\rangle| \\ &\leqslant \|u\| \cdot \|f * f^*\|_{C^*(G)} |\langle |u|, (e - e_\gamma) * (e - e_\gamma)\rangle| \\ &\leqslant \|u\| \cdot \|f\|^2_{C^*(G)} \langle |u|, e - e_\gamma\rangle. \end{aligned}$$

To deduce this inequality, we have used the fact that if φ is a positive linear functional on a C^*-algebra A, then $\varphi(a^*ba) \leqslant \|b\|\varphi(a^*a)$ for all $a, b \in A$. Moreover, the last inequality holds because $|u|$ is a positive linear functional on $C^*(G)_e$.

Notice next that since $|u|$ is positive and $(e_\gamma)_\gamma$ is an approximate identity in $C^*(G)$, $\langle |u|, e_\gamma\rangle \to \langle |u|, e\rangle = \|u\|$. So, given $\epsilon > 0$, there exists γ_0 such that $\langle |u|, e - e_{\gamma_0}\rangle \leqslant \epsilon/\|u\|$. Since $u_\alpha \to u$ in τ_{nw^*} implies that $|u_\alpha| \to |u|$ [**68**], taking $v = u_\alpha$ in the above estimate and the supremum over all $f \in L^1(G)$ with $\|f\|_{C^*(G)} \leqslant 1$, it follows that

$$\|e_{\gamma_0} * u_\alpha - u_\alpha\|_{B(G)} \leqslant \|u_\alpha\|^{1/2} \langle |u_\alpha|, e - e_{\gamma_0}\rangle^{1/2} \to \|u\|^{1/2}\langle |u|, e - e_{\gamma_0}\rangle^{1/2} < \epsilon.$$

Thus there exists α_0 such that $\|u_\alpha\|^{1/2}\langle |u_\alpha|, e - e_{\gamma_0}\rangle^{1/2} \leqslant \epsilon$ for all $\alpha \geqslant \alpha_0$. This completes the proof of the lemma. □

LEMMA 3.7.3. *Let $f \in L^\infty(G)$ and $g \in L^1(G)$ and suppose that both f and g vanish outside some compact set K in G, and let $h = f * g$. Then the map $u \to h * u$ is continuous from $(B(G), \tau_{bw^*})$ to $(B(G), \tau_M)$.*

PROOF. Let $(u_\alpha)_\alpha$ be a net in $B(G)$ such that $u_\alpha \to u$ for some $u \in B(G)$ in the w^*-topology and suppose that $C = \sup \|u_\alpha\| < \infty$. Let $v \in A(G) \cap C_c(G)$ and let $\widetilde{K}$ be a compact subset of G containing $K^{-1} \cdot \operatorname{supp} v$. Then, for $p \in L^\infty(G)$ and $q \in L^1(G)$, we have

$$\langle (f * p)v, q\rangle = \langle p, f^* * (vq)\rangle = \langle 1_K p, f^* * (vq)\rangle = \langle [f * (1_K p)]v, q\rangle.$$

Since f and $\widetilde{1_K p}$ belong to $L^2(G)$, $f * \widetilde{1_K p} \in L^2(G) * \widetilde{L^2(G)} = A(G)$ and

$$\|(f * p)v\|_{A(G)} = \|(f * 1_K p)v\|_{A(G)} \leqslant \|f\|_2 \|\widetilde{1_K p}\|_2 \|v\|_{A(G)}.$$

Now take $p = g * (u_\alpha - u)$. Then

$$\|(f * g * (u_\alpha - u))v\| \leqslant \|f\|_2 \|[g * (u_\alpha - u)1_K]^\sim\|_2 \|v\|_{A(G)},$$

which tends to zero since $g * (u_\alpha - u) \to 0$ uniformly on K (compare [**59**, Lemma 13.5.1]). Thus we have seen that $\|[h * (u_\alpha - u)]v\|_{A(G)} \to 0$ for every $v \in A(G) \cap C_c(G)$.

Now let $w \in A(G)$ be arbitrary and let $\epsilon > 0$ be given and choose $v \in A(G) \cap C_c(G)$ such that $\|w - v\|_{A(G)} \leqslant \epsilon$. Notice that, by Lemma 2.1.13, $h * u_\alpha \in B(G)$ and

$$\|h * u_\alpha\|_{B(G)} \leqslant \|f * g\|_{C^*(G)} \|u_\alpha\|_{B(G)} \leqslant C\|f\|_1 \|g\|_1$$

for all α. Then, with $c = (C + \|u\|_{B(G)})\|f\|_1\|g\|_1$,

$$\begin{aligned} \|[(f * g) * (u_\alpha - u)]w\|_{B(G)} &\leqslant \|[(f * g) * (u_\alpha - u)]v\|_{B(G)} \\ &\quad + \|[(f * g) * (u_\alpha - u)](w - v)\|_{B(G)} \\ &\leqslant \|[(f * g) * (u_\alpha - u)]v\|_{B(G)} + \epsilon c. \end{aligned}$$

This completes the proof. □

The preceding three lemmas now lead to our first theorem.

THEOREM 3.7.4. *τ_{nw*} is stronger than τ_M. In particular, τ_{w*} and τ_M coincide on the unit sphere of $B(G)$.*

PROOF. Since, by Lemma 3.7.1, $\tau_M \supseteq \tau_{uc} \supseteq \tau_{w*}$ on norm bounded sets in $B(G)$, we only have to show that if $(u_\alpha)_\alpha$ is a net in $B(G)$ such that, for some $u \in B(G)$, $u_\alpha \to u$ in the w^*-topology and $\|u_\alpha\| \to \|u\|$, then $u_\alpha \to u$ in τ_M.

Let $(U_\gamma)_\gamma$ be a neighbourhood basis of the identity e of G. For each γ, choose an open, relatively compact neighbourhood V_γ of e such that $V_\gamma^{-1} = V_\gamma$ and $V_\gamma^2 \subseteq U_\gamma$, and let $f_\gamma = |V_\gamma|^{-1} 1_{V_\gamma}$ and $e_\gamma = f_\gamma * f_\gamma^*$. Then the net $(e_\gamma)_\gamma$ satisfies the hypotheses of Lemma 3.7.2. Hence, given $\epsilon > 0$, there exist α_0 and γ_0 such that

$$\|e_{\gamma_0} * u - u\|_{B(G)} \leqslant \epsilon \quad \text{and} \quad \|e_{\gamma_0} * u_\alpha - u_\alpha\|_{B(G)} \leqslant \epsilon$$

for all $\alpha \geqslant \alpha_0$. It follows that, for any $v \in A(G)$ and $\alpha \geqslant \alpha_0$,

$$\begin{aligned}\|(u_\alpha - u)v\|_{A(G)} &\leqslant \|(u_\alpha - e_{\gamma_0} * u_\alpha)v\|_{A(G)} + \|(e_{\gamma_0} * u - u)v\|_{A(G)} \\ &\quad + \|[e_{\gamma_0} * (u_\alpha - u)]v\|_{A(G)} \\ &\leqslant 2\epsilon\|v\|_{A(G)} + \|[e_{\gamma_0} * (u_\alpha - u)]v\|_{A(G)}.\end{aligned}$$

Now, since $u_\alpha \to u$ in τ_{w*} and the net $(u_\alpha)_\alpha$ is norm bounded, taking $h = e_{\gamma_0}$ in Lemma 3.7.3, we conclude that $\|[e_{\gamma_0} * (u_\alpha - u)]v\|_{A(G)} \leqslant \epsilon\|v\|_{A(G)}$ for all $\alpha \geqslant \alpha_1$ for some $\alpha_1 \geqslant \alpha_0$. This completes the proof of the theorem. □

COROLLARY 3.7.5. *τ_{nw*} is stronger than τ_{uc}. In particular, τ_{w*} and τ_{uc} coincide on the unit sphere of $B(G)$.*

PROOF. This is an immediate consequence of Theorem 3.7.4 and Lemma 3.7.1(i). □

COROLLARY 3.7.6. *Let K be a compact subset of G and*

$$A_K(G) = \{u \in A(G) : \operatorname{supp} u \subseteq K\}.$$

Then the norm topology and the w^*-topology coincide on the unit sphere of $A_K(G)$. In particular, if G is compact, then the w^*-topology and the norm topology agree on the unit sphere of $B(G)$.

PROOF. Let $(v_\alpha)_\alpha \subseteq A_K(G)$ and $v \in A_K(G)$ such that $\|v_\alpha\| = 1 = \|v\|$ for all α and $v = w^*\text{-}\lim_\alpha v_\alpha$. Choose $u \in A(G)$ with $u = 1$ on K. Then $v_\alpha - v = (v_\alpha - v)u$ and, by Theorem 3.7.4, $(v_\alpha - v)u \to 0$. □

We shall see below (Corollary 3.7.8) that the converse of the last assertion in Corollary 3.7.6 also holds. Next we list a number of topologies which coincide on the unit sphere.

THEOREM 3.7.7. *Let S denote the unit sphere of $B(G)$ and let $(u_\alpha)_\alpha$ be a net in S and $u \in S$. Then the following are equivalent.*

(1) *$u_\alpha \to u$ in the w^*-topology of $B(G)$.*
(2) *$u_\alpha \to u$ uniformly on compact subsets of G.*
(3) *For each $T \in C^*_\lambda(G)$, $u_\alpha \cdot T \to u \cdot T$ in the norm of $C^*_\lambda(G)$.*
(3') *For each $T \in VN(G)$, $u_\alpha \cdot T \to u \cdot T$ in the w^*-topology of $VN(G)$.*
(4) *For each $T \in C^*_\lambda(G)$ and $v \in A(G)$, $\|T(u_\alpha v) - T(uv)\|_{A(G)} \to 0$.*
(4') *For each $T \in C^*_\lambda(G)$ and $v \in A(G)$, $\langle T, u_\alpha v\rangle \to \langle T, uv\rangle$.*

(5) $\|u_\alpha v - uv\|_{A(G)} \to 0$ *for every* $v \in A(G)$.

(5') *For each* $T \in VN(G)$ *and* $v \in A(G)$, $\langle u_\alpha v, T\rangle \to \langle uv, T\rangle$.

PROOF. The equivalence of (1) and (2) and of (2) and (3) follows from Theorem 3.7.4. Clearly, (5) $\Rightarrow$ (3').

To show that (3') implies (1), let $f \in C_c(G)$ and choose $v \in A(G)$ with $v = 1$ on $\operatorname{supp} f$. Then $vf = f$ and

$$\begin{aligned}\langle u_\alpha, \lambda(f)\rangle &= \int_G u_\alpha(x) f(x)dx = \int_G u_\alpha(x)v(x)f(x)dx \\ &= \langle \lambda(u_\alpha f), v\rangle = \langle u_\alpha \cdot \lambda(f), v\rangle \to \langle u \cdot \lambda(f), v\rangle \\ &= \langle v \cdot \lambda(f), u\rangle = \langle \lambda(f), u\rangle.\end{aligned}$$

Thus $u_\alpha \to u$ in $\sigma(B(G), C_c(G))$, which implies (1) by the density of $C_c(G)$ in $C^*(G)$ and since the u_α and u are norm bounded.

(5) implies (3) since the operators with compact support are dense in $C^*_\lambda(G)$ and every $T \in C^*_\lambda(G)$ can be written as $T = v \cdot T$ for some $v \in A(G)$ with $v = 1$ on $\operatorname{supp} T$ and hence $u_\alpha \cdot T - u \cdot T = (u_\alpha v - uv) \cdot T$, which tends to 0 in $C^*_\lambda(G)$ since the net $(u_\alpha v - uv)_\alpha$ is norm bounded.

The implications (3) $\Rightarrow$ (1), (4') $\Rightarrow$ (1) and (5) $\Rightarrow$ (1) are all shown in the same manner as (3') $\Rightarrow$ (1). Finally, the remaining implications (5) $\Rightarrow$ (4), (4) $\Rightarrow$ (4') and (5) $\Rightarrow$ (5') are all evident. □

From Theorem 3.7.7 we deduce the following

COROLLARY 3.7.8. *Let* G *be amenable and suppose that the* w^**-topology and the norm topology agree on the unit sphere of* $B(G)$. *Then* G *must be compact.*

PROOF. Since G is amenable, $A(G)$ has a bounded approximate identity $(u_\alpha)_\alpha$ such that $\|u_\alpha\| = 1$ for all α (Theorem 2.8.2). Then $\|u_\alpha v - 1_G v\|_{A(G)} = \|u_\alpha v - v\|_{A(G)} \to 0$ for every $v \in A(G)$. Now the implication (5) $\Rightarrow$ (1) of Theorem 3.7.7 shows that $u_\alpha \to 1_G$ in the w^*-topology of $B(G)$. Since, by hypothesis, this topology coincides with the norm topology on the unit sphere, it follows that $1_G \in A(G)$, and hence G must be compact. □

We are now going to show that the amenability hypothesis in Corollary 3.7.8 can be dropped. However, the proof will have to employ a deep result from representation theory due to S.P. Wang. Actually, we will show a stronger result in that we replace the unit sphere of $B(G)$ by the smaller set $P^1_\lambda(G)$, where $P^1_\lambda(G) = B_\lambda(G) \cap P^1(G)$ is the set of all normalized continuous positive definite functions which are associated to representations of G that are weakly contained in the left regular representation.

LEMMA 3.7.9. *Let* H *be an open subgroup of* G. *If the identity map from* $(P^1_\lambda(G), w^*)$ *to* $(P^1_\lambda(G), \|\cdot\|)$ *is continuous, then the identity map from* $(P^1_\lambda(H), w^*)$ *to* $(P^1_\lambda(H), \|\cdot\|)$ *is also continuous.*

PROOF. For any $\varphi \in P_\lambda((H)$, the trivial extension $\widetilde{\varphi}$ of φ to G belongs to $P_\lambda(G)$. Indeed, $\widetilde{\varphi}$ is associated to the induced representations $\operatorname{ind}_H^G \pi_\varphi$ and $\pi_\varphi \prec \lambda_H$ implies

$$\operatorname{ind}_H^G \pi_\varphi \prec \operatorname{ind}_H^g \lambda_H \cong \lambda_G.$$

Now let $(\varphi_\alpha)_\alpha$ be a net in $P^1_\lambda(H)$ converging to $\varphi \in P_\lambda(H)$ in the w^*-topology. Then $\widetilde{\varphi}_\alpha \to \widetilde{\varphi}$ in the w^*-topology of $P_\lambda(G)$ and hence

$$\|\varphi_\alpha - \varphi\| = \|\widetilde{\varphi}_\alpha - \widetilde{\varphi}\| \to 0$$

by hypothesis. □

LEMMA 3.7.10. *Let $\varphi \in \mathrm{ex}(P^1_\lambda(G))$ and suppose that the identity map from $(\mathrm{ex}(P^1_\lambda(G)), w^*)$ to $(\mathrm{ex}(P^1_\lambda(G)), \|\cdot\|)$ is continuous at φ. Then π_φ is an isolated point of $\widehat{G}_r$.*

PROOF. Notice first that $\mathrm{ex}(P^1_\lambda(G)) \subseteq \mathrm{ex}(P^1(G))$ because if $\varphi \in P_\lambda(G)$ and $\psi \in P(G)$ are such that $c\varphi - \psi$ is positive definite for some $c \geqslant 0$, then $\psi \in P_\lambda(G)$. As mentioned in Chapter 1, if $\varphi_1, \varphi_2 \in P^1(G)$ and the associated GNS-representations are not equivalent, then $\|\varphi_1 - \varphi_2\| \geqslant 2$. By hypothesis, there exists a w^*-open subset U of $\mathrm{ex}(P^1_\lambda(G))$ such that

$$U \subseteq \{\psi \in \mathrm{ex}(P^1_\lambda(G)) : \|\psi - \varphi\| < 2\}.$$

It follows that $\pi_\varphi \cong \pi_\psi$ for all $\psi \in U$. Now the map $\phi : \psi \to \pi_\psi$ from $\mathrm{ex}(P^1_\lambda(G))$ onto $\widehat{G}_r$ is open (compare [**60**, Theorem 3.4.10]). Consequently, $\phi(U) = \{\pi_\varphi\}$ is open in $\widehat{G}_r$. □

THEOREM 3.7.11. *For any locally compact group G, the following conditions are equivalent.*

(i) *G is compact.*
(ii) *The w^*-topology and the norm topology coincide on the unit sphere of $B(G)$.*
(iii) *The w^*-topology and the norm topology agree on $P^1_\lambda(G)$.*

PROOF. The implications (i) $\Rightarrow$ (ii) and (ii) $\Rightarrow$ (iii) being clear, it only remains to prove that (iii) implies (i). Towards a contradiction, assume that G is noncompact. Then G contains a noncompact, σ-compact, open subgroup H. By Lemma 3.7.9, the w^*-topology and the norm topology agree on $P^1_\lambda(H)$. It now follows from Lemma 3.7.10 that $\widehat{H}_r$ is discrete. Finally, by [**283**, Theorem 7.6], a σ-compact locally compact group with discrete reduced dual is compact. This contradiction proves (iii) $\Rightarrow$ (i). □

When studying the w^*-topology on $B(G)$, the question of when $A(G)$ is w^*-closed in $B(G)$, arises naturally. It will turn out that for a large class of locally compact groups G, this forces G to be compact. As for Theorem 3.7.11, we will again have to utilize some profound result about dual spaces of locally compact groups, this time one due to Baggett [**7**].

LEMMA 3.7.12. *Let K be a compact normal subgroup of G. If $A(G)$ is w^*-closed in $B(G)$, then $A(G/K)$ is w^*-closed in $B(G/K)$.*

PROOF. Recall that $T_K : L^1(G) \to L^1(G/K)$ denotes the $*$-homomorphism defined by $T_K(f)(xK) = \int_K f(xk)dk$ for almost all $x \in G$, and that T_K extends to a $*$-homomorphism, also denoted T_K, from $C^*(G)$ onto $C^*(G/K)$. Let T_K^* denote the adjoint map. Then $T_K^*(B(G/K))$ consists precisely of those functions in $B(G)$ which are constant on cosets of K. Moreover, since K is compact,

$$T_K^*(A(G/K)) = A(G) \cap T_K^*(B(G/K)).$$

Now let $(u_\alpha)_\alpha$ be a net in $A(G/K)$ converging to some $u \in B(G/K)$ in the w^*-topology. Then, for every $f \in C^*(G)$,

$$\langle T_K^*(u_\alpha), f\rangle = \langle u_\alpha, T_K(f)\rangle \to \langle u, T_K(f)\rangle = \langle T_K^*(u), f\rangle.$$

Since $A(G)$ is w^*-closed in $B(G)$ and $T_K^*(u_\alpha) \in A(G)$, it follows that $T_K^*(u) \in A(G)$ and hence $T_K^*(u) \in A(G) \cap T_K^*(B(G/K))$. Consequently $u \in A(G/K)$. □

LEMMA 3.7.13. *If $A(G)$ is w^*-closed in $B(G)$ and H is an open subgroup of G, then $A(H)$ is w^*-closed in $B(H)$.*

PROOF. It suffices to show that the unit ball of $A(H)$ is w^*-closed in $B(H)$. Thus let $(u_\alpha)_\alpha$ be a net in $A(H)$ and $u \in B(G)$ such that

$$\|u_\alpha\| \leqslant 1, \|u\| \leqslant 1 \text{ and } u = w^* - \lim_\alpha u_\alpha.$$

Let $\widetilde{u}_\alpha$ and $\widetilde{u}$ denote the trivial extension of u_α and u to G, respectively. Then $\widetilde{u}_\alpha \in A(G)$ and $\widetilde{u} \in B(G)$ with norms $\leqslant 1$, and for every $f \in L^1(G)$,

$$\begin{aligned}\int_G \widetilde{u}_\alpha(x)f(x)dx &= \int_H u_\alpha(h)f(h)dh \\ &\to \int_H u(h)f(h)dh = \int_G \widetilde{u}(x)f(x)dx.\end{aligned}$$

Since $(\widetilde{u}_\alpha)_\alpha$ is a bounded net, it follows that $\widetilde{u}_\alpha \to \widetilde{u}$ in the w^*-topology of $B(G)$. Hence, by the hypothesis, $\widetilde{u} \in A(G)$ and so $u \in A(H)$. □

LEMMA 3.7.14. *Let G be an almost connected locally compact group. If $A(G)$ is w^*-closed in $B(G)$, then G is compact.*

PROOF. Since an almost connected locally compact group is a projective limit of Lie groups, there exists a compact normal subgroup K of G such that G/K is a Lie group. By Lemma 3.7.12, $A(G/K)$ is w^*-closed in $B(G/K)$. Therefore, we can assume that G is Lie group. Being a compactly generated Lie group, G is second countable and hence $A(G)$ is a separable Banach space. By the hypothesis,

$$A(G) = B_\lambda(G) = C_r^*(G)^*.$$

Now a C^*-algebra A with a separable dual Banach space A^* has a countable dual $\widehat{A}$. This appears to be folklore and can easily be seen as follows. Let S denote the set of all pure states of A. Then S is second countable, when equipped with the relative norm topology of A^*. Hence there exist $\varphi_n \in S$, $n \in \mathbb{N}$, such that

$$S = \bigcup_{n\in\mathbb{N}}\{\psi \in S : \|\psi - \varphi_n\| < 2\}.$$

Now, if $\varphi, \psi \in S$ are such that $\|\varphi - \psi\| < 2$, then the associated irreducible representations are equivalent [**60**, (2.12.1)]. This shows that $\widehat{A}$ is countable. It follows that $\widehat{G}_r$, the reduced dual of G, is countable. Finally, by [**7**, Theorem 2.5], a separable Lie group with countable reduced dual is compact. This shows that G is compact. □

COROLLARY 3.7.15. *Let G be any locally compact group and suppose that $A(G)$ is w^*-closed in $B(G)$. Then G contains a compact open subgroup.*

PROOF. Since G/G_0 is totally disconnected, there exists an open subgroup H of G such that H/G_0 is compact. By Lemma 3.7.13, $A(H)$ is w^*-closed in $B(H)$ and hence H is compact by Lemma 3.7.14. □

LEMMA 3.7.16. *If G is a discrete group and $A(G)$ is w^*-closed in $B(G)$, then G is finite.*

PROOF. Towards a contradiction, assume that G is infinite. Then G has a countable infinite subgroup H. By Lemma 3.7.13, $A(H)$ is w^*-closed in $B(H)$. As in the proof of Lemma 3.7.14, we now conclude that $\widehat{H}_r$ is countable. Applying [**7**, Theorem 2.5] again, it follows that H is finite, contradicting the choice of H. □

THEOREM 3.7.17. *Suppose that G contains an almost connected open normal subgroup. Then $A(G)$ is w^*-closed in $B(G)$ if and only if G is compact.*

PROOF. Let N be an almost connected open normal subgroup of G. Then $A(N)$ is w^*-closed in $B(N)$ (Lemma 3.7.13), and Corollary 3.7.15 implies that N is compact. Passing to G/N, we have that $A(G/N)$ is w^*-closed in $B(G/N)$. Since G/N is discrete, Lemma 3.7.16 shows that G/N must be finite. Hence G is compact. □

3.8. Notes and references

Host's idempotent theorem, which in particular identifies the idempotents in $B(G)$ as precisely the characteristic functions of open sets in the coset ring of G, was proved in [**129**] and has since become one of the major tools in the investigation of $A(G)$ and $B(G)$. As samples, we just mention the study of homomorphisms between Fourier algebras [**135**] and [**136**], the description of the closed ideals in $A(G)$ with bounded approximate identities in terms of their hulls [**85**], and amenability of $A(G)$ [**86**]. The idempotent theorem was earlier shown by Cohen [**33**] for abelian groups and was announced in the nonabelian case by Lefranc [**187**], who never published a proof. In contrast to Cohen's proof, which made substantial use of measure theory on the dual group of G, Host's proof is considerably simpler and based only on elementary facts from operator theory. Recently, Runde [**252**] obtained idempotent theorems of a similar flavour for representations on Banach spaces.

Wendel [**288**] proved that if G and H are two locally compact groups and the group algebras $L^1(G)$ and $L^1(H)$ are isometrically isomorphic, then G and H are topologically isomorphic. The same conclusion holds if there exists an isometric isomorphism between the measure algebras $M(G)$ and $M(H)$. In fact, as shown by Johnson [**140**], such an isomorphism maps $L^1(G)$ onto $L^1(H)$. For abelian groups these results are very special cases of, but certainly have been a motivation for, Walter's theorems presented in Section 3.2. Walter's viewpoint has been to consider $B(G)$ and $A(G)$ as dual objects for G and he perfectly succeeded in showing that both of them completely specify the underlying group G. Subsequently, several authors have investigated the problem of which maps between Fourier and Fourier-Stieltjes algebras ensure the groups to be topologically isomorphic. As a sample we mention that, using Walter's result [**280**], Arendt and deCanniere [**4**] have shown that the pointwise and the positive definite orderings in $B(G)$ and $A(G)$, respectively, determine G up to topological isomorphism. More precisely, the locally compact groups G and H are topologically isomorphic if (and only if) there exists a bijective linear mapping $\phi : B(G) \to B(H)$ such that $\phi(B(G)_+) = B(H)_+$ and $\phi(P(G)) = P(H)$, and similarly for Fourier algebras.

For locally compact abelian groups G and H, homomorphisms from $A(G)$ into $B(H)$ have been completely characterized by Cohen [**33**] in terms of continuous piecewise affine maps from sets in the open coset ring of H into G. Cohen's achievements have been extremely influential and inspired numerous subsequent investigations. As examples of contributions for general locally compact groups we mention [**135**], [**136**], [**138**]. The results we have presented in Section 3.3 are entirely due to Pham [**234**]. They provide satisfying descriptions of homomorphisms which are either positive or contractive. Moreover, in [**234**] completely bounded and completely contractive homomorphisms and isomorphisms between Fourier and Fourier-Stieltjes algebras have been studied. We did not touch these additional results because we will encounter operator space structures for the first time in Chapter 4.

The Galois type correspondence between closed subgroups of a locally compact group G and invariant von Neumann subalgebras of $L^\infty(G)$ and $VN(G)$ (Theorems 3.4.3 and 3.4.8), which might be considered as an extension of the Pontryagin duality theorem for locally compact abelian groups and the Tannaka-Krein duality theorem for compact groups, is due to Takesaki and Tatsuuma [**271**]. It is a continuation of previous investigations of the same authors on duality theorems for locally compact groups in terms of Hopf-von Neumann group algebras. Our approach, however, follows [**271**] only to some extent and appears to be considerably simpler. In particular, it entirely avoids the use of the Hopf-von Neumann algebra structure of $VN(G)$. The proof of Theorem 3.4.3 given here is taken from [**175**].

The material in Section 3.5 is taken from [**11**], where the structure of two-sided translation invariant w^*-closed $*$-subalgebras of $B(G)$ is studied. The main characterization of such subalgebras, when G is amenable, is parallel to that of w^*-closed right translation invariant $*$-subalgebras of $L^\infty(G)$, due to Takesaki and Tatsuuma [**271**], as presented in the preceding section. When G is non-amenable, no such structure theorem is known. It should also be pointed out that the problem of characterizing one-sided translation invariant w^*-closed subalgebras of $B(G)$ is open. Moreover, none of these problems seems to have been treated for the algebras $B_p(G)$, $1 < p < \infty$, formed by all continuous complex-valued functions on G which are multipliers of $A_p(G)$.

Corollary 3.6.3 was previously shown in [**221**]. Losert [**201**] mentions that the following generalization of Theorem 3.6.2 for Fourier spaces associated with unitary representations (compare Section 2.8) could be shown. For $i = 1, 2$, let G_i be a locally compact group and π_i a unitary representation. Then the spaces $A_{\pi_1}\widehat{\otimes}A_{\pi_2}$ and $A_{\pi_1\otimes\pi_2}(G_1 \times G_2)$ are linearly isomorphic if and only if at leat one of π_1 and π_2 is finite dimensional. Then, in addition, $A_{\pi_1}\widehat{\otimes}A_{\pi_2}$ and $A_{\pi_1\otimes\pi_2}(G_1 \times G_2)$ are isometrically isomorphic if and only if one of the representations is 1-dimensional.

The material of Section 3.7 up to Theorem 3.7.7 is taken from [**105**]. Theorem 3.7.11 improves on earlier results due to Derighetti [**50**] and McKennon [**211**] and actually verifies a conjecture of [**211**]. The remaining part of Section 3.7 appeared in [**10**]. Let H be a closed subgroup of the locally compact group G. In [**10**] it is also shown that the restriction map from $B(G)$ into $B(H)$ is w^*-w^*-continuous (if and) only if H is open in G. This result was in turn generalized in [**138**], where a systematic study of w^*-continuous homomorphisms between Fourier-Stieltjes algebras was conducted.

CHAPTER 4

Amenability Properties of $A(G)$ and $B(G)$

Amenable Banach algebras were introduced by B.E. Johnson. He showed the fundamental result that a locally compact group is amenable if and only if the group algebra $L^1(G)$ is amenable. We present a proof of the 'only if' part in Section 4.5. In particular, if G is abelian, then $A(G)$, being isometrically isomorphic to the L^1-algebra of the dual group $\widehat{G}$, is amenable. However, when G is nonabelian, then $A(G)$ need not be weakly amenable, even when G is compact.

In this chapter we will consider the completely bounded cohomology theory of the Fourier algebra $A(G)$ and of the Fourier-Stieltjes algebra $B(G)$. In Section 4.1 it is shown that $A(G)$, equipped with the operator space structure inherited from being embedded into $VN(G)^*$, is a completely contractive Banach algebra. Using this we establish in Section 4.2 the fundamental result, due to Ruan [**245**], that a locally compact group G is amenable precisely when $A(G)$ is operator amenable. Note that the natural operator space structure on $L^1(G)$ as predual of $L^\infty(G)$ is such that all bounded maps from $L^1(G)$ into any operator space are automatically completely bounded. Thus the notions of amenability and operator amenability coincide for $L^1(G)$.

Another classical result of B.E. Johnson states that for any locally compact group G, $L^1(G)$ is always weakly amenable. A simple proof, due to Despic and Ghahramani [**58**], is also included in Section 4.5. In Section 4.3 we are going to prove that $A(G)$ is operator weakly amenable for every locally compact group G. In contrast, it turns out that $A(G)$ is amenable if and only if G has an abelian subgroup of finite index (Section 4.5). The proof is based on the fact that G has to have an abelian subgroup of finite index if the antidiagonal $\Gamma_G = \{(x, x^{-1}) : x \in G\}$ belongs to the closed coset ring of $G \times G$. The proof of this latter fact depends on completely boundedness of the *flip map* $A(G) \to A(G)$, $u \to \check{u}$, where $\check{u}(x) = u(x^{-1})$ for $u \in A(G)$ and $x \in G$. It also involves the study of piecewise affine maps (Section 4.4) and the structure of ideals in $A(G)$ with bounded approximate identity.

4.1. $A(G)$ as a completely contractive Banach algebra

Let G be a locally compact group. The Fourier algebra $A(G)$ of G is endowed with the operator space structure which it inherits from being embedded into $VN(G)^* = A(G)^{**}$ (see Appendix A.3 and [**69**]).

LEMMA 4.1.1. *Let G and H be locally compact groups. Then the von Neumann algebras $VN(G) \overline{\otimes} VN(H)$ and $VN(G \times H)$ are isomorphic.*

PROOF. We first observe that the two Hilbert spaces $L^2(G) \overline{\otimes} L^2(H)$ and $L^2(G\times H)$ are isometrically isomorphic via the map ϕ determined by $f \otimes g \to f \times g$, where $(f \times g)(s,t) = f(s)g(t)$, $f \in L^2(G)$, $g \in L^2(H)$, $s \in G$ and $t \in H$. Thus $\mathcal{B}(L^2(G\times H))$ is isomorphic to $\mathcal{B}(L^2(G) \overline{\otimes} L^2(H))$, which contains $VN(G) \overline{\otimes} VN(G)$

as a $*$-subalgebra. Then ϕ is a unitary equivalence between the representations $\lambda_G \otimes \lambda_H$ and $\lambda_{G\times H}$. Indeed, for $x \in G$ and $y \in H$,

$$\begin{aligned}\lambda_{G\times H}(x,y)(\phi(f\otimes g))(s,t) &= \phi(f\otimes g)(x^{-1}s, y^{-1}t)\\ &= f(x^{-1}s)g(y^{-1}t)\\ &= \lambda_G(x)f(s)\lambda_H(y)g(t)\\ &= \phi(\lambda_G(x)f\otimes\lambda_H(y)g)(s,t).\end{aligned}$$

Consequently, ϕ induces an isomorphism between the von Neumann algebras

$$VN(G)\overline{\otimes}VN(H) = \lambda_G(G)''\overline{\otimes}\lambda_H(H)'' = (\lambda_G(G)\otimes\lambda_H(H))''$$

and $VN(G\times H) = \lambda_{G\times H}(G\times H)''$. □

LEMMA 4.1.2. *Let G and H be locally compact groups. Then*

$$A(G\times H) \cong A(G)\hat{\otimes} A(H).$$

Furthermore, the identity

$$A(G\times H) \cong A(G)\hat{\otimes} A(H)$$

is a (completely) isometric isomorphism.

PROOF. By Lemma 4.1.1, $VN(G\times H) \cong VN(G)\overline{\otimes}VN(H)$ and, by Proposition A.3.8,

$$A(G)\hat{\otimes} A(H) \cong (VN(G)\overline{\otimes}VN(H))_*.$$

Consequently, $A(G\times H) = VN(G\times H)_* \cong A(G)\hat{\otimes} A(H)$. □

In the sequel, let W denote the unitary operator on $L^2(G\times G)$ defined by

$$(Wf)(s,t) = f(s,st),\quad f\in L^2(G\times G),\ s,t\in G.$$

THEOREM 4.1.3. *For any locally compact group G, the Fourier algebra $A(G)$ is a completely contractive Banach algebra.*

PROOF. With I denoting the identity operator on $L^2(G)$, W satisfies the equation

$$W^*(\lambda_G(s)\otimes I)W = \lambda_G(s)\otimes\lambda_G(s) \tag{4.1}$$

for all $s\in G$. To prove (4.1) it suffices to show that

$$\langle W^*(\lambda_G(s)\otimes I)W\xi_1,\xi_2\rangle = \langle(\lambda_G(s)\otimes\lambda_G(s))\xi_1,\xi_2\rangle,$$

where $\xi_j = f_j\otimes g_j$, $f_j, g_j\in L^2(G)$, $j=1,2$. Now, observing that

$$(\lambda_G(s)\otimes I)\xi(x,y) = \xi(s^{-1}x,y)$$

for every $\xi \in L^2(G \times G)$, we get

$$\begin{aligned}
\langle W^*(\lambda_G(s) \otimes I)W\xi_1, \xi_2\rangle &= (\langle \lambda_G(s) \otimes I)W\xi_1, W\xi_2\rangle \\
&= \int_G \int_G W\xi_1(s^{-1}x, y)\, \overline{W\xi_2(x, y)}\, dxdy \\
&= \int_G \int_G \xi_1(s^{-1}x, s^{-1}xy)\, \overline{\xi_2(x, xy)}\, dxdy \\
&= \int_G \int_G f_1(s^{-1}x) g_1(s^{-1}xy)\, \overline{f_2(x) g_2(xy)}\, dydx \\
&= \int_G \int_G f_1(s^{-1}x)\, \overline{f_2(x)} g_1(s^{-1}y)\, \overline{g_2(y)}\, dydx \\
&= \langle (\lambda_G(s) \otimes \lambda_G(s))\xi_1, \xi_2\rangle,
\end{aligned}$$

as required. It follows that the map $\gamma : T \to W^*(T \otimes I)W$ is a normal isomorphism of $VN(G)$ into the tensor product $VN(G)\overline{\otimes}VN(G)$. Moreover, γ is a co-multiplication of $VN(G)$ in the sense that, denoting by i the identity mapping on $VN(G)$, the diagram

$$\begin{array}{ccc}
VN(G) & \xrightarrow{\quad\gamma\quad} & VN(G)\overline{\otimes}VN(G) \\
\gamma \downarrow & & \downarrow \gamma\otimes i \\
VN(G)\overline{\otimes}VN(G) & \xrightarrow[i\otimes\gamma]{} & VN(G)\overline{\otimes}VN(G)\overline{\otimes}VN(G)
\end{array}$$

is commutative. Indeed, since the operators $\lambda_G(s)$, $s \in G$, generate $VN(G)$, we only have to show that

$$[(\gamma \otimes i) \circ \gamma](\lambda_G(s)) = [(i \otimes \gamma) \circ \gamma](\lambda_G(s))$$

for all $s \in G$. But this is clear since both sides are equal to the operator $\lambda_G(s) \otimes \lambda_G(s) \otimes \lambda_G(s)$.

Now γ is a w^*-homomorphism, i.e., a w^*-w^*-continuous homomorphism, from $VN(G)$ into $VN(G)\overline{\otimes}VN(G)$. Hence γ is completely contractive. By Lemma 4.1.1,

$$\gamma(VN(G)) \subseteq VN(G)\overline{\otimes}VN(G) \cong VN(G \times G).$$

Thus the dual map $\gamma^* : VN(G \times G)^* \to VN(G)^*$ is also completely contractive [**70**, Proposition 3.2.2]. Since γ is w^*-w^*-continuous, it follows that

$$\gamma_* = \gamma^*|_{A(G\times G)} \to A(G)$$

is completely contractive as well. Identifying $A(G)\widehat{\otimes}A(G)$ with $A(G \times G)$ via $u \otimes v \to u \times v$, we may now deduce that $\gamma_*(u \otimes v) = uv$ for all $u, v \in A(G)$. Indeed, for each $s \in G$, we have

$$\begin{aligned}
\langle \gamma_*(u \otimes v), \lambda_G(s)\rangle &= \langle \gamma^*((u \otimes v), \lambda_G(s)\rangle = \langle u \otimes v, \gamma(\lambda_G(s))\rangle \\
&= \langle u \otimes v, \lambda_G(s) \otimes \lambda_G(s)\rangle = \langle u, \lambda_G(s)\rangle \cdot \langle v, \lambda_G(s)\rangle \\
&= u(s)v(s) = \langle uv, \lambda_G(s)\rangle,
\end{aligned}$$

and since the linear span of the set $\{\lambda_G(s) : s \in G\}$ is w^*-dense in $VN(G)$, it follows that $\gamma_*(u\otimes v) = uv$. Consequently, $A(G)$ is a completely contractive Banach algebra. □

REMARK 4.1.4. It should be noted that two different projective tensor products have been considered in this paper. The projective tensor product $A(G)\widehat{\otimes}A(H)$ of Fourier algebra $A(G)$ and $A(H)$ considered in Section 3.6 in the "Banach space projective tensor product", and the projective tensor product $\widehat{\otimes}$ considered in Lemma 4.1.2 should be "operator space projective tensor product". They are different unless either G or H is an abelian group, i.e., either $A(G) = L^1(\widehat{G})$ or $A(G) = L^1(\widehat{H})$.

4.2. Operator amenability of $A(G)$

In this section, we will show that the Fourier algebra $A(G)$ is operator amenable exactly when G is an amenable locally compact group. We start by providing a sufficient condition for operator amenability of $A(G)$.

LEMMA 4.2.1. *Let G be an amenable locally compact group. Suppose that there exists a bounded net $(w_\alpha)_\alpha$ in $B(G \times G)$ such that, for every $u \in A(G)$,*

(i) $\|u \cdot w_\alpha - w_\alpha \cdot u\| \to 0$;
(ii) $\|m(w_\alpha)u - u\| \to 0$.

Then $A(G)$ is operator amenable.

PROOF. Since G is amenable, by Theorem 2.8.5 $A(G)$ has a bounded approximate identity, $(u_\beta)_\beta$ say. Now consider the elements $v_{\alpha,\beta} = w_\alpha(u_\beta \otimes u_\beta)$ of $A(G)\widehat{\otimes}A(G) = A(G \times G)$. Clearly, the net $(v_{\alpha,\beta})_{(\alpha,\beta)}$ is bounded and $(u_\beta \otimes u_\beta)_\beta$ is a bounded approximate identity for $A(G \times G)$. Moreover, $(v_{\alpha,\beta})_{(\alpha,\beta)}$ is an approximate diagonal for $A(G)$. In fact, for each $u \in A(G)$,

$$u \cdot v_{\alpha,\beta} - v_{\alpha,\beta} \cdot u = [u \cdot w_\alpha - w_\alpha \cdot u](u_\beta \otimes u_\beta) \to 0$$

by (i), and by (ii),

$$m(v_{\alpha,\beta})u - u = m(w_\alpha)[u_\beta^2 u - u] + m(w_\alpha)u - u \to 0.$$

Since $A(G)$ is completely contractive by Theorem 4.1.3, Proposition A.4.2 yields that $A(G)$ is operator amenable. □

For $\xi \in L^2(G)$, let $w_\xi : G \times G \to \mathbb{C}$ be defined by

$$w_\xi(s,t) = \langle \lambda_G(s)\rho_G(t)\xi, \xi \rangle, \quad s,t \in G.$$

Then $w_\xi \in B(G \times G)$. Also, let $V \in \mathcal{B}(L^2(G))$ be defined by

$$V\xi(s) = \Delta(s)^{-1/2}\xi(s^{-1}), \quad \xi \in L^2(G), s \in G.$$

Then V is self-adjoint and unitary. Furthermore, for all $s \in G$,

$$\rho_G(s) = V^*\lambda_G(s)V.$$

Let W be as in the preceding section.

LEMMA 4.2.2. $(V^* \otimes I)W^* = W(V^* \otimes I)$.

PROOF. If $\xi \in L^2(G \times G)$, we have for almost all $(s,t) \in G \times G$,

$$\begin{aligned}(V^* \otimes I)W^*\xi(s,t) &= \Delta(s)^{-1/2}(W^*\xi)(s^{-1},t) \\ &= \Delta(s)^{-1/2}\xi(s^{-1},st) \\ &= (V^* \otimes I)\xi(s,st) \\ &= W(V^* \otimes I)\xi(s,t),\end{aligned}$$

as was to be shown. □

LEMMA 4.2.3. *Let G be a locally compact group and suppose that $(\xi_\alpha)_\alpha$ is a net of unit vectors in $L^2(G)$ such that*

(i) $\|W(\xi_\alpha \otimes \eta) - (\xi_\alpha \otimes \eta)\| \to 0$ *for all* $\eta \in L^2(G)$;
(ii) $\|\lambda_G(s)\rho_G(s)\xi_\alpha - \xi_\alpha\| \to 0$ *uniformly on compact subsets of* G.

Then the net $(w_{\xi_\alpha})_\alpha$ in $B(G \times G)$ satisfies conditions (i) *and* (ii) *in Lemma* 4.2.1.

PROOF. Let $\lambda = \lambda_G$ and $\rho = \rho_G$. To establish conditions (i) and (ii) of Lemma 4.2.1, it suffices to consider $u \in A(G)$ of the form $u(s) = \langle \lambda(s)\eta, \eta \rangle$, where $\eta \in L^2(G)$. Then, since $\rho(s) = V^*\lambda(s)V$ and using Lemma 4.2.2, for all $s, t \in G$,

$$\begin{aligned}
(w_{\xi_\alpha} \cdot u)(s,t) &= \langle \lambda(s)\rho(t)\xi_\alpha, \xi_\alpha \rangle \langle \lambda(t)\eta, \eta \rangle \\
&= \langle \lambda(s)V^*\lambda(t)V\xi_\alpha, \xi_\alpha \rangle \langle \lambda(t)\eta, \eta \rangle \\
&= \langle (\lambda(s) \otimes I)(V^*\lambda(t)V \otimes \lambda(t))(\xi_\alpha \otimes \eta), \xi_\alpha \otimes \eta \rangle \\
&= \langle (\lambda(s) \otimes I)(V^* \otimes I)W^*(\lambda(t) \otimes I)W(V \otimes I)(\xi_\alpha \otimes \eta), \xi_\alpha \otimes \eta \rangle \\
&= \langle (\lambda(s) \otimes \lambda(s))(V^*\lambda(t)V \otimes I)W^*(\xi_\alpha \otimes \eta), W^*(\xi_\alpha \otimes \eta) \rangle.
\end{aligned}$$

In addition, we have

$$\begin{aligned}
(u \cdot w_{\xi_\alpha})(s,t) &= \langle \lambda(s)\eta, \eta \rangle \langle \lambda(s)\rho(t)\xi_\alpha, \xi_\alpha \rangle \\
&= \langle \lambda(s)\eta, \eta \rangle \langle \lambda(s)V^*\lambda(t)V\xi_\alpha, \xi_\alpha \rangle \\
&= \langle (\lambda(s)V^*\lambda(t)V \otimes \lambda(s))(\xi_\alpha \otimes \eta), \xi_\alpha \otimes \eta \rangle.
\end{aligned}$$

Combining the preceding two equations, we get

$$\begin{aligned}
(u \cdot w_{\xi_\alpha} - w_{\xi_\alpha} \cdot u)(s,t) &= \langle (\lambda(s)V^*\lambda(t)V \otimes \lambda(s))(\xi_\alpha \otimes \eta), \xi_\alpha \otimes \eta \rangle \\
&\quad - \langle (\lambda(s)V^*\lambda(t)V \otimes \lambda(s))W^*(\xi_\alpha \otimes \eta), W^*(\xi_\alpha \otimes \eta) \rangle \\
&= \langle (\lambda(s)V^*\lambda(t) \otimes \lambda(s))(\xi_\alpha \otimes \eta - W^*(\xi_\alpha \otimes \eta), \xi_\alpha \otimes \eta \rangle \\
&\quad + \langle (\lambda(s)V^*\lambda(t)V \otimes \lambda(s))W^*(\xi_\alpha \otimes \eta - W^*(\xi_\alpha \otimes \eta) \rangle.
\end{aligned}$$

Consequently,

$$\|u \cdot w_{\xi_\alpha} - w_{\xi_\alpha} \cdot u\|_{B(G \times G)} \leqslant 2\|\eta\| \cdot \|\xi_\alpha \otimes \eta - W^*(\xi_\alpha \otimes \eta)\|,$$

which tends to 0 by hypothesis (i). Hence condition (i) of Lemma 4.2.1 holds for the net $(w_{\xi_\alpha})_\alpha$.

To show that condition (ii) of Lemma 4.2.1 is also satisfied, we may assume that $\eta \in C_c(G)$. Using the facts that V is unitary and that W commutes with $I \otimes V^*\lambda(s)V = I \otimes \rho(s)$, we have

$$\begin{aligned}
m(w_{\xi_\alpha})u(s) &= w_{\xi_\alpha}(s,s)u(s) \\
&= \langle \lambda(s)V^*\lambda(s)V\xi_\alpha, \xi_\alpha \rangle \langle \lambda(s)\eta, \eta \rangle \\
&= \langle (\lambda(s) \otimes \lambda(s))(I \otimes (V^*\lambda(s)V))(\eta \otimes \xi_\alpha), \eta \otimes \xi_\alpha \rangle \\
&= \langle W^*(\lambda(s) \otimes I)W(I \otimes V^*\lambda(s)V)(\eta \otimes \xi_\alpha), \eta \otimes \xi_\alpha \rangle \\
&= \langle W^*(\lambda(s) \otimes I)(I \otimes V^*\lambda(s)V)W(\eta \otimes \xi_\alpha), \eta \otimes \xi_\alpha \rangle \\
&= \langle W^*(I \otimes V^*)(\lambda(s) \otimes \lambda(s))(I \otimes V)W(\eta \otimes \xi_\alpha), \eta \otimes \xi_\alpha \rangle \\
&= \langle W^*(I \otimes V^*)W^*(\lambda(s) \otimes I)W(I \otimes V)W(\eta \otimes \xi_\alpha), \eta \otimes \xi_\alpha \rangle \\
&= \langle (\lambda(s) \otimes I)(I \otimes V)W(I \otimes V)W(\eta \otimes \xi_\alpha), W(I \otimes V)W(\eta \otimes \xi_\alpha) \rangle.
\end{aligned}$$

Let $w_\alpha = W(I\otimes V)W\eta\otimes\xi_\alpha$ and $v_\alpha = m(w_{\xi_\alpha})u - u$. Then

$$\begin{aligned} v_\alpha(s) &= (\langle\lambda(s)V^*\lambda(s)V\xi_\alpha,\xi_\alpha\rangle - 1)\langle\lambda(s)\eta,\eta\rangle \\ &= \langle(\lambda(s)\otimes I)\omega_\alpha,\omega_\alpha\rangle - \langle(\lambda(s)\otimes I)(\eta\otimes\xi_\alpha,\eta\otimes\xi_\alpha)\rangle \\ &= \langle(\lambda(s)\otimes I)(\omega_\alpha - (\eta\otimes\xi_\alpha)),\omega_\alpha\rangle \\ &\quad + \langle(\lambda(s)\otimes I)(\eta\otimes\xi_\alpha),\omega_\alpha - (\eta\otimes\xi_\alpha\rangle \end{aligned}$$

for all $s\in G$. This implies

$$\|v_\alpha\| \leqslant 2\|\eta\|\cdot\|\omega_\alpha - (\eta\otimes\xi_\alpha)\|.$$

Let $C = \operatorname{supp}\eta$. Since $\|\lambda(s)\rho(s)\xi_\alpha - \xi_\alpha\| \to 0$ uniformly on C, it follows that

$$\begin{aligned} \|\omega_\alpha - (\eta\otimes\xi_\alpha)\|^2 &= \int_G\int_G |\omega_\alpha(s,t) - (\eta\otimes\xi_\alpha)(s,t)|^2 dtds \\ &= \int_G\int_G |\eta(s)(\xi_\alpha(sts^{-1})\Delta(s)^{-1/2} - \xi_\alpha(t)|^2 dtds \\ &= \int_C |\eta(s)|\cdot\|\lambda(s)\rho(s)\xi_\alpha - \xi_\alpha\|_2^2 ds, \end{aligned}$$

which converges to 0. This implies

$$\|m(w_{\xi_\alpha})u - u\| = \|v_\alpha\| \to 0.$$

So condition (ii) of Lemma 4.2.1 is satisfied. □

For any compact subset C of G and $\epsilon > 0$, by Proposition 1.8.15 there exists $g_{C,\epsilon}\in L^1(G)$ such that $g_{C,\epsilon}\geqslant 0$, $\|g_{C,\epsilon}\|_1 = 1$ and $\|L_s g_{C,\epsilon} - g_{C,\epsilon}\|_1 \leqslant \epsilon$ for all $s\in C$. Fix such a $g_{C,\epsilon}$, and for any relatively compact neighbourhood U of e, let $f_U = |U|^{-1}1_U$. Then let

$$f_{U,C,\epsilon}(t) = \int_G g_{C,\epsilon}(s)\tau_s f_U(t)ds,$$

where

$$\tau_s f(t) = \Delta(s)f(s^{-1}ts).$$

LEMMA 4.2.4. *The function $f_{U,C,\epsilon}$ defined above has the following properties.*

(i) *$f_{U,C,\epsilon}\geqslant 0$ and $\|f_{U,C,\epsilon}\|_1 = 1$;*
(ii) *$\|\tau_s f_{U,C,\epsilon} - f_{U,C,\epsilon}\|_1 \leqslant \epsilon$ for all $s\in C$.*

PROOF. The assertions in (i) are clear. To prove (ii), observe first that

$$\begin{aligned} \tau_s f_{U,C,\epsilon}(t) &= \int_G g_{C,\epsilon}(w)f_U(w^{-1}(s^{-1}ts)w)dw \\ &= \int_G g_{C,\epsilon}(s^{-1}w)f_U(w^{-1}tw)dw. \end{aligned}$$

This implies that

$$\begin{aligned} \|\tau_s f_{U,C,\epsilon} - f_{U,C,\epsilon}\|_1 &\leqslant \int_G\int_G |g_{C,\epsilon}(s^{-1}w) - g_{C,\epsilon}(w)|f_U(w^{-1}tw)\Delta(w)dwdt \\ &= \int_G\int_G |g_{C,\epsilon}(s^{-1}w) - g_{C,\epsilon}(w)|f_U(t)dwdt \\ &= \int_G |g_{C,\epsilon}(s^{-1}w) - g_{C,\epsilon}(w)|dw \\ &= \|L_s g_{C,\epsilon} - g_{C,\epsilon}\|_1 \leqslant \epsilon \end{aligned}$$

for all $s\in C$. □

LEMMA 4.2.5. *Let* $\xi_{U,C,\epsilon} = \sqrt{f_{U,C,\epsilon}} \in L^2(G)$. *Then* $\xi_{U,C,\epsilon} \geqslant 0$ *and* $\|\xi_{U,C,\epsilon}\|_2 = 1$ *and*

(i) $\|\lambda_G(s)\rho_G(s)\xi_{U,C,\epsilon} - \xi_{U,C,\epsilon}\|_2^2 \leqslant \epsilon$ *for all* $s \in C$.

(ii) *For each finite subset* F *of* $L^2(G)$, *compact subset* C *of* G *and* $\epsilon > 0$, *there exists a neighbourhood* U_0 *of* e *such that*

$$\|W(\xi_{U,C,\epsilon} \otimes \eta) - (\xi_{U,C,\epsilon} \otimes \eta)\|_2^2 \leqslant \epsilon$$

for all neighbourhoods U *of* e *such that* $U \subseteq U_0$ *and all* $\eta \in F$.

PROOF. (i) follows from Lemma 4.2.4. Indeed, letting $\xi = \xi_{U,C,\epsilon}$,

$$\begin{aligned}
\|\lambda_G(s)\rho_G(s)\xi - \xi\|_2^2 &= \int_G \left|\Delta(s)^{1/2}\xi(s^{-1}ts) - \xi(t)\right|^2 dt \\
&\leqslant \int_G \left|\Delta(s)\xi(s^{-1}ts)^2 - \xi(t)^2\right| dt \\
&= \int_G |\tau_s f_{U,C,\epsilon}(t) - f_{U,C,\epsilon}(t)| dt \\
&= \|\tau_s f_{U,C,\epsilon} - f_{U,C,\epsilon}\|_1 \leqslant \epsilon
\end{aligned}$$

for all $s \in C$.

(ii) Retaining the notation of (i), for $\eta \in L^2(G)$ we have

$$\begin{aligned}
&\|W(\xi \otimes \eta) - (\xi \otimes \eta)\|_2^2 \\
= &\int_G \int_G |\xi(s)\eta(st) - \xi(s)\eta(t)|^2 ds dt \\
= &\int_G \int_G f_{U,C,\epsilon}(s)|\eta(st) - \eta(t)|^2 ds dt \\
= &\int_G \int_G \int_G g_{C,\epsilon}(w) f_U(w^{-1}sw)\Delta(w)|\eta(wsw^{-1}t) - \eta(t)|^2 ds dt dw.
\end{aligned}$$

Since $\|L_s\eta - \eta\|_2 \to 0$ as $s \to e$, there exists an open neighbourhood U_0 of e such that $\int_G |\eta(st) - \eta(t)|^2 dt \leqslant \epsilon^{1/2}$ for all $s \in V_0$ and $\eta \in F$. Let U_0 be an open neighbourhood of e such that $wsw^{-1} \in V_0$ for all $s \in U_0$ and $w \in \operatorname{supp} g_{C,\epsilon}$. Then, by the above,

$$\begin{aligned}
\|W(\xi \otimes \eta) - (\xi \otimes \eta)\|_2^2 &\leqslant \epsilon \int_G \int_G g_{C,\epsilon}(w) f_U(w^{-1}sw)\Delta(w) ds dw \\
&= \epsilon \|g_{C,\epsilon}\|_1 \|f_U\|_1 = \epsilon,
\end{aligned}$$

whenever $U \subseteq U_0$. This proves (ii). □

Let A denote the collection of all quadruples $\alpha = (C, F, \epsilon, U)$, where C is a compact subset of G, F a finite subset of $L^2(G)$ and $\epsilon > 0$, and U is an open neighbourhood of e in G such that the functions $\xi_{U,C,\epsilon}$ satisfy (i) and (ii) of Lemma 4.2.5 We order A by letting

$$\alpha = (C, F, \epsilon, U) \leqslant \alpha' = (C', F', \epsilon', U')$$

if and only if $C \subseteq C'$, $F \subseteq F'$, $\epsilon' \leqslant \epsilon$ and $U' \subseteq U$.

COROLLARY 4.2.6. *For* $\alpha \in A$, *let* $\xi_\alpha = \xi_{U,C,\epsilon}$ *be as in Lemma 4.2.5. Then*

(i) $\|\lambda_G(s)\rho_G(s)\xi_\alpha - \xi_\alpha\|_2 \to 0$ *uniformly on compact subsets of* G.

(ii) $\|W(\xi_\alpha \otimes \eta) - (\xi_\alpha \otimes \eta)\|_2 \to 0$ *for all* $\eta \in L^2(G)$.

PROOF. (i) is clear. For (ii), let $\eta \in L^2(G)$, $\delta > 0$ and a compact subset C_0 of G be given. By Lemma 4.2.5(ii), there exists a neighbourhood U_0 of e in G such that

$$\|W(\xi_{U,C_0,\delta} \otimes \eta) - (\xi_{U,C_0,\delta} \otimes \eta)\|_2 \leqslant \delta$$

for all neighbourhoods U of e such that $U \subseteq U_0$. Let $\alpha_0 = (C_0, \{\eta\}, \delta^2, U_0) \in A$. Then, for all $\alpha \geqslant \alpha_0$,

$$\|\lambda_G(s)\rho_G(s)\xi_\alpha - \xi_\alpha\|_2 \leqslant \sqrt{\epsilon} \leqslant \delta$$

for all $s \in C_0$ and also

$$\|W(\xi_\alpha \otimes \eta) - (\xi_\alpha \otimes \eta)\|_2 \leqslant \sqrt{\epsilon} \leqslant \delta.$$

This shows (ii). □

We are now able to present the main result of this section.

THEOREM 4.2.7. *Let G be a locally compact group. Then $A(G)$ is operator amenable if and only if G is amenable.*

PROOF. We only need to show that if G is amenable, then $A(G)$ is operator amenable. By Lemma 4.2.1, it suffices to find a net $(w_\alpha)_\alpha$ in $B(G \times G)$ satisfying the conditions (i) and (ii) of that lemma. However, if A is as defined prior to Corollary 4.2.5 and for $\alpha \in A$, ξ_α is as in Corollary 4.2.6, then the net $(w_\alpha)_\alpha$ in $B(G \times G)$, defined by

$$w_\alpha(s,t) = \langle \lambda_G(s)\rho_G(t)\xi_\alpha, \xi_\alpha \rangle, \quad s,t \in G,$$

satisfies conditions (i) and (ii) of Lemma 4.2.1. □

A locally compact group G is called an SIN-group (group with small invariant neighbourhoods) if every neighbourhood U of the identity contains a neighbourhood V of the identity such that $x^{-1}Vx = V$ for all $x \in G$. Clearly, compact groups and locally compact groups with open centres are SIN-groups. Note that SIN-groups are unimodular. Before concluding this section, we consider amenable SIN-groups. For such groups G, operator amenability of $A(G)$ can be established by simpler arguments, which we include for readers with a special interest in SIN-groups. The proof, however, uses the fact that closed subgroups of G are sets of synthesis (compare Section 6.1).

LEMMA 4.2.8. *Let G be an* SIN*-group and let $D_G = \{(s,s) : s \in G\}$ be the diagonal of $G \times G$. Let $\mathcal{V}$ be a neighbourhood basis of e in G consisting of relatively compact, conjugation invariant sets. Then there exists a bounded net $(u_V)_{V \in \mathcal{V}}$ in $B(G \times G)$ such that $u_V(s,s) = 1$ for all $s \in G$ and $u_V v \to 0$ for each $v \in I(D_G)$.*

PROOF. We can define a unitary representation π of $G \times G$ on $L^2(G)$ by

$$\pi(s,t)\xi = \lambda_G(s)\rho_G(t)\xi, \quad s,t \in G, \ \xi \in L^2(G).$$

For $V \in \mathcal{V}$, define the function u_V on $G \times G$ by

$$u_V(s,t) = |V|^{-1}\langle \pi(s,t)1_V, 1_V \rangle.$$

Then, for all $(s,t) \in G \times G$,

$$u_V(s,t) = \frac{1}{|V|}\int_G 1_V(s^{-1}xt)1_V(x)dx = \frac{|sVt^{-1} \cap V|}{|V|}$$

and hence, in particular, $u_V(s,s) = 1$ for all $s \in G$. Clearly, u_V is a continuous positive definite function and therefore $\|u_V\| = u_V(e) = 1$. If $v \in j(D_G)$, then there exists $V \in \mathcal{V}$ such that $V \cap \operatorname{supp} v = \varnothing$. Now $j(D_G)$ is dense in $I(D_G)$ since the subgroup D_G of $G \times G$ is a set of synthesis for $A(G \times G)$ (Theorem 6.1.9). Since $u_V v = 0$ for all $u \in j(D_G)$ and all small enough V, it follows that $u_V v \to 0$ for every $v \in I(D_G)$. □

THEOREM 4.2.9. *If G is an amenable* SIN*-group, then $A(G)$ is operator amenable.*

PROOF. Let $m : A(G \times G) \to A(G)$ denote the map defined by $m(v)(s) = v(s,s)$ for $v \in A(G \times G)$ and $s \in G$. By Lemma 4.2.1 it suffices to show that there exists a bounded net $(w_\alpha)_\alpha$ in $A(G \times G)$ such that

(i) $\|u \cdot w_\alpha - w_\alpha \cdot u\| \to 0$ for every $u \in A(G)$;
(ii) $\|m(w_\alpha)u - u\| \to 0$ for all $u \in A(G)$.

To show (ii), let $(v_\beta)_{\beta \in B}$ be a bounded approximate identity for $A(G)$ (Theorem 2.7.2) and let $(u_V)_{V \in \mathcal{V}}$ be the net constructed in Lemma 4.2.8. Let $A = B \times \mathcal{V}^B$ be the product directed set, and for $\alpha = (\beta, (V_\gamma)_{\gamma \in B}) \in A$, define $w_\alpha : G \times G \to \mathbb{C}$ by

$$w_\alpha(s,t) = u_{V_\beta}(s,t) v_\beta(s) v_\beta(t), \quad s,t \in G.$$

Then $(w_\alpha)_\alpha$ is a bounded net in $A(G \times G)$. Since $m(w_\alpha)(s) = u_V(s,s) v_\beta(s)^2 = v_\beta(s)^2$, the net $(m(w_\alpha))_{\alpha \in A}$ forms an approximate identity for $A(G)$.

To see (i), let $u \in A(G)$. Then

$$u \cdot (v_\beta \times v_\beta) - (v_\beta \times v_\beta) \cdot u \in I(D_G)$$

for all β. Moreover, for all $s,t \in G$,

$$\begin{aligned}(u \cdot w_\alpha - w_\alpha \cdot u)(s,t) &= u_{v_\beta}(s,t) v_\beta(s) v_\beta(t)[u(s) - u(t)] \\ &= u_{v_\beta}[u \cdot (v_\beta \times v_\beta) - (v_\beta \times v_\beta) \cdot u](s,t)\end{aligned}$$

and hence

$$\lim_\alpha \|u \cdot w_\alpha - w_\alpha \cdot u\| = \lim_\beta \lim_{v_\beta} \|u_{v_\beta}[u \cdot (v_\beta \times v_\beta) - (v_\beta \times v_\beta) \cdot u]\| = 0.$$

So (i) is satisfied. □

LEMMA 4.2.10. *Let H be a closed subgroup of G. Then $A(G)/I(H)$ is completely isometrically isomorphic to $A(H)$.*

PROOF. We know already that the map $u + I(H) \to u|_H$ is an isometric isomorphism from $A(G)/I(H)$ onto $A(H)$ (Corollary 2.7.4). Moreover,

$$I(H)^\perp = VN_H(G) = \{T \in VN(G) : \operatorname{supp} T \subseteq H\}$$

and $VN_H(G)$ is a von Neumann subalgebra of $VN(G)$ which is $*$-isomorphic to $VN(H)$ (Chapter 2). It now follows from [**16**] that $(A(G)/I(H))^*$ is completely isometrically isomorphic to $VN_H(G)$ and hence to $VN(H)$. Consequently $A(G)/I(H)$ is completely isometrically isomorphic to $A(H)$. □

PROPOSITION 4.2.11. *Let H be an open subgroup of G. Then $1_H \cdot A(G) = \{u \in A(G) : u = 0$ on $G \backslash H\}$ is completely isometrically isomorphic to $A(H)$.*

PROOF. For $u \in A(H)$, let $\widetilde{u}$ be defined by $\widetilde{u}(x) = u(x)$ for $x \in H$ and $\widetilde{u} = 0$ otherwise. Then the map $\phi : u \to \widetilde{u}$ is an isometric isomorphism of $A(H)$ onto $1_H \cdot A(G)$.

Let $n \in \mathbb{N}$ and $[u_{ij}] \in M_n(A(H))$ with $\|[u_{ij}]\|_n = 1$. It follows from Lemma 4.2.10 that $\|[\widetilde{u_{ij}} + I(H)]\|_n = 1$. Then, given $\epsilon > 0$, there exists $[v_{ij}] \in M_n(A(G))$ such that $\|[v_{ij} + I(H)]\| = \|[\widetilde{u_{ij}} + I(H)]\|$ and $\|[v_{ij}]\| \leqslant 1 + \epsilon$. Now, since $1_H \in B(G)$, $\|1_H\| = 1$ and hence the *cb*-norm of the multiplication operator M_{1_H} equals 1,

$$\|\phi_n[u_{ij}]\| = \|[\widetilde{u}_{ij}]\|_n = \|[1_H v_{ij}]\|_n \leqslant \|M_{1_H}\|_{cb}\|[v_{ij}]\| \leqslant 1 + \epsilon.$$

Since $\epsilon > 0$ was arbitrary, it follows that $\|\phi_n\| \leqslant 1$, and because this holds for all n, we get $\|\phi\|_{cb} \leqslant 1$.

Finally, the inverse map $\phi^{-1} : 1_H A(G) \to A(H)$ is simply the restriction to $1_H A(G)$ of the quotient map $A(G) \to A(G)/I(H)$ composed with the complete isometry $A(G)/I(H) \to A(H)$ (Lemma 4.2.10). Thus $\|\phi^{-1}\|_{cb} \leqslant 1$ and consequently $\|\phi\|_{cb} = 1$. □

COROLLARY 4.2.12. *For a locally compact group G, the following are equivalent.*

(i) *$A(G)$ has a nonzero closed ideal which is operator amenable.*
(ii) *G has an open amenable subgroup.*

PROOF. Suppose first that there exists an open amenable subgroup H of G and consider the closed ideal $I = 1_G \cdot A(G)$ of $A(G)$. By Proposition 4.2.11, I is completely isometrically isomorphic to $A(H)$. On the other hand, $A(H)$ is operator amenable since H is amenable (Theorem 4.2.7). Thus (ii) $\Rightarrow$ (i).

Conversely, let I be a nonzero closed ideal of $A(G)$ which is operator amenable. Then I has a bounded approximate identity [Theorem A.2.8]. It follows from Theorem 2.7.9 that G has an amenable open subgroup. □

We now present a class of locally compact groups G for which even $B(G)$ is operator amenable.

EXAMPLE 4.2.13. Let G be a semidirect product $G = N \rtimes K$, where N is a locally compact abelian group and K is a compact group and both are second countable. Suppose that $\widehat{G} = \widehat{K} \circ q \bigcup \{\pi_j : j \in \mathbb{N}\}$, where each π_j is a subrepresentation of the regular representation of G. Then

$$B(G) = A(K) \circ q \oplus A(G)$$

by Proposition 2.6.1. Then $B(G)$ is operator amenable. This follows from Theorem A.4.4 since $A(G)$ and $B(G)/A(G) = A(K)$ are both operator amenable by Ruan's theorem. Examples of such group are the p-adic motion groups.

4.3. Operator weak amenability of $A(G)$

If π is a continuous unitary representation of G on the Hilbert space $H(\pi)$, let $A_\pi(G)$ denote the closed linear span in $B(G)$ of the set of all coefficient functions $\langle \pi(\cdot)\xi, \eta \rangle$, $\xi, \eta \in H(\pi)$. The dual space $A_\pi(G)^*$ is canonically isomorphic to $VN_\pi(G)$, the von Neumann subalgebra of $\mathcal{B}(H(\pi))$ generated by the operators $\pi(x), x \in G$.

Let π and σ be two *disjoint representations* of G, that is, there are no subrepresentations π' of π and σ' of σ such that π' and σ' are unitarily equivalent. Then, denoting by $\oplus_p$ the ℓ^p-direct sum,

$$A_{\pi\oplus\sigma}(G) = A_\pi(G) \oplus_1 A_\sigma(G)$$

and

$$VN_{\pi\oplus\sigma}(G) = VN_\pi(G) \oplus_\infty VN_\sigma(G),$$

respectively.

LEMMA 4.3.1. *Let G be a noncompact locally compact group. Then the four representations*

$$1_G \times 1_G, 1_G \times \lambda_G, \lambda_G \times 1_G \text{ and } \lambda_G \times \lambda_G$$

of $G \times G$ are pairwise disjoint.

PROOF. Since G is noncompact, there cannot be a nonzero fixed vector ξ for λ_G in $L^2(G)$, because any such ξ would define a nonzero constant function in $A(G)$. Hence none of the representations $1_G \times \lambda_G$, $\lambda_G \times 1_G$ and $\lambda_G \times \lambda_G$ can have a nonzero vector in its Hilbert space which is fixed under the action of all elements of $G \times G$. Consequently, all three representations are disjoint from $1_G \times 1_G$.

Now $(1_G \times \lambda_G)(s, e)\xi = \xi$ for all $\xi \in (1_G \times \lambda_G)H = \mathbb{C} \otimes L^2(G)$ and $s \in G$. But there is no nonzero fixed vector for $\pi(G \times \{e\})$, where π is either $\lambda_G \times 1_G$ or $\lambda_G \times \lambda_G$. Hence $1_G \times \lambda_G$ is disjoint from $\lambda_G \times 1_G$ and $\lambda_G \times \lambda_G$. Similarly, $\lambda_G \times 1_G$ and $\lambda_G \times \lambda_G$ are disjoint. □

Let A_e denote the unitization of a Banach algebra A.

LEMMA 4.3.2. *Let G be a noncompact locally compact group. Then we have the following completely isometric identifications*

$$\begin{aligned} A(G)_e \widehat{\otimes} A(G) &\cong A_{(\lambda_G \times \lambda_G)\oplus(1_G \times \lambda_G)} \\ &= \operatorname{span}\{u, 1_G \times v : u \in A(G \times G), v \in A(G)\} \end{aligned}$$

and

$$\begin{aligned} A(G)_e \widehat{\otimes} A(G)_e &\cong A_{(\lambda_G \times \lambda_G)\oplus(1_G \times \lambda_G)\oplus(\lambda_G \times 1_G)\oplus(1_G \times 1_G)} \\ &= \operatorname{span}\{u, 1_G \times v, w \times 1_G, 1_{G\times G} : u \in A(G \times G), v, w \in A(G)\}, \end{aligned}$$

which are implemented by algebra isomorphisms.

PROOF. Since $A(G)^* = VN(G) = VN_{\lambda_G}(G)$ and $A(G)_e^* = VN_{\lambda_G \times 1_G}(G)$,

$$\begin{aligned} (A(G)_e \widehat{\otimes} A(G))^* &\cong VN_{\lambda_G \times 1_G}(G \times G) \overline{\otimes} VN(G) \\ &= (VN(G) \oplus_\infty VN_{1_G}(G)) \overline{\otimes} VN(G) \\ &= (VN(G) \overline{\otimes} VN(G)) \oplus_\infty (VN_{1_G}(G) \overline{\otimes} VN(G)) \\ &= VN(G \times G) \oplus_\infty VN_{1_G \times \lambda_G}(G \times G) \\ &= VN_{(\lambda_G \times \lambda_G)\oplus(1_G \times \lambda_G)}(G \times G) \\ &= \left(A_{(\lambda_G \times \lambda_G)\oplus(1_G \times \lambda_G)}\right)^*. \end{aligned}$$

It is clear that the resulting identification of the spaces $A(G)_e \widehat{\otimes} A(G)$ and $A_{(\lambda_G \times \lambda_G)\oplus(1_G \times \lambda_G)}(G \times G)$ is implemented by an algebra isomorphism. The second equality in the first statement follows from Lemma 4.3.1.

The proof of the second assertion is similar. □

THEOREM 4.3.3. *Let G be an arbitrary locally compact group. Then $A(G)$ is operator weakly amenable.*

PROOF. If G is compact, then $A(G)$ is even operator amenable by Theorem 4.2.7 and hence operator weakly amenable. Thus we may assume that G is non-compact. Consider the following direct sum

$$\pi = (\lambda_G \times \lambda_G) \oplus (1_G \times \lambda_G) \oplus (\lambda_G \oplus 1_G) \oplus (1_G \times 1_G)$$

of pairwise disjoint representations of $G \times G$ (Lemma 4.3.1). Let $A = A_\pi(G \times G) \subseteq B(G \times G)$. Then, by Lemma 4.3.2

$$A = \operatorname{span}\{u, 1_G \times v, v \times 1_G, 1_{G\times G} : u \in A(G \times G), v \in A(G)\}.$$

In particular, $A_\pi(G \times G)$ is a subalgebra of $B(G \times G)$. In the identification of $A(G)_e \hat{\otimes} A(G)_e$ in Lemma 4.3.2, the multiplication map $m : A(G)_e \hat{\otimes} A(G)_e \to A(G)_e$ corresponds to the map $R : A_\pi(G \times G) \to A_{\lambda_G} \bigoplus 1_G(G)$, which restricts functions to the diagonal $\Delta_G = \{(s,s) : s \in G\}$. Let $K_1 = \ker R$ and $K_0 = K_1 \cap (A(G) \hat{\otimes} A(G))$. Then

$$K_1 = \operatorname{span}\{u - 1_G \times R(u), u - R(u) \times 1_G : u \in A_\pi(G \times G)\}$$

and

$$K_0 = \ker R \cap A(G \times G) = \{u \in A(G \times G) : u|_{\Delta_G} = 0\}.$$

Since the closed subgroup Δ_G of $G \times G$ is a set of spectral synthesis for $A(G \times G)$ (Theorem 6.1.9), K_0 is the only closed ideal of $A(G \times G)$ with hull equal to Δ_G. In particular, $\overline{K_0^2} = K_0$. As $A(G \times G)$ is an ideal in $A_\pi(G \times G)$, it follows that

$$(A(G) \hat{\otimes} A(G))K_1 = A(G \times G)K_1 \subseteq A(G \times G) \cap K_1 = K_0.$$

On the other hand, using again that Δ_G is a set of spectral synthesis,

$$K_0 = \overline{A(G \times G)K_0} \subseteq \overline{A(G \times G)K_1} = \overline{(A(G) \hat{\otimes} A(G))K_1}.$$

Combining the preceding two equations with $K_0 = \overline{K_0^2}$, we obtain

$$\overline{K_0^2} = \overline{(A(G) \hat{\otimes} A(G))K_1}.$$

Now, $A(G)$ is completely contractive and $\overline{A(G)^2} = A(G)$ because $A(G)$ is a Tauberian algebra. This shows that the hypotheses of Theorem A.5.4 are satisfied and hence $A(G)$ is operator weakly amenable. □

4.4. The flip map and the antidiagonal

This section serves as a preparation for the next one. We study completely boundedness of the so-called flip map of the Fourier algebra $A(G)$ and provide a criterion for when the antidiagonal subgroup of $G \times G$ belongs to the coset ring. This latter goal is achieved by employing piecewise affine maps.

Let $j_* : B(G) \to B(G)$ be defined by $j_*(u) = \check{u}$, where as before, $\check{u}(x) = u(x^{-1})$, $x \in G$. Then j_* is a linear isometry. In fact, let $u(x) = \langle \pi(x)\xi, \eta \rangle$ for some unitary representation π of G and $\xi, \eta \in H(\pi)$ with $\|u\| = \|\xi\| \cdot \|\eta\|$. Then $\check{u}(x) = \overline{\langle \pi(x)\eta, \xi \rangle}$ and hence $\|\check{u}\| \leqslant \|\xi\| \cdot \|\eta\|$ and

$$\|u\| = \|(\check{u})\check{}\| \leqslant \|\check{u}\| \leqslant \|u\|.$$

Hence $j_* : A(G) \to A(G)$, the *flip map*, is a linear isometry.

PROPOSITION 4.4.1. *Let G be a locally compact group and let $j_* : A(G) \to A(G)$ be defined by $j_*(u) = \check{u}$, where $\check{u}(x) = u(x^{-1})$, $x \in G$. Then j_* is an isometry, and j_* is completely bounded if and only if G contains an abelian subgroup of finite index.*

PROOF. For any Banach space E, let $\overline{E}$ denote the complex conjugate space of E, i.e., the space E with scalar multiplication $\lambda \cdot x = \overline{\lambda} x$, $x \in E$, $\lambda \in \mathbb{C}$. Then the map $u \to \overline{u}$ is a complete isometry from $A(G)$ onto $\overline{A(G)}$. Therefore, if j_* is completely bounded, then so is the map

$$\overline{j_*} : A(G) \to \overline{A(G)}, \quad u \to \overline{\check{u}},$$

with $\|\overline{j_*}\|_{cb} = \|j_*\|_{cb}$. The adjoint of $\overline{j_*}$ is the map

$$\overline{j} : \overline{VN(G)} \to VN(G), \quad T \to T^*,$$

which then also is completely bounded by [**70**, Proposition 3.2.2], and satisfies $\|\overline{j}\|_{cb} = \|\overline{j_*}\|_{cb}$.

We now show that this forces $VN(G)$ to be subhomogeneous, that is, the irreducible representations of $VN(G)$ are finite dimensional and their degrees are bounded. Towards a contradiction, assume that this is not true. Then, for each $n \in \mathbb{N}$, $VN(G)$ contains the full matrix algebra M_n as a $*$-subalgebra [**261**, Lemma 9.3]. For each n, let $j_n : M_n \to M_n$ stand for taking the transpose of an $n \times n$-matrix. Since entrywise conjugation of matrices is a complete isometry, it follows from [**70**, Proposition 2.2.7] that

$$n = \|j_n\|_{cb} = \|\overline{j}|_{M_n}\|_{cb} \leqslant \|\overline{j}\|_{cb} = \|\overline{j_*}\|_{cb} = \|j_*\|_{cb}$$

for all $n \in \mathbb{N}$, which is a contradiction. So there exists $n \in \mathbb{N}$ such that every irreducible representation of $VN(G)$ has dimension n or less.

For $k \in \mathbb{N}$, let P_k denote the standard polynomial in k noncommuting variables, that is,

$$P_k(a_1, \ldots, a_k) = \sum (-1)^\sigma a_{\sigma(1)} a_{\sigma(2)} \cdots a_{\sigma(k)},$$

where the sum is taken over all permutations σ of $\{1, \ldots, k\}$ and $(-1)^\sigma$ denotes the signature of σ. Then, by a result of Amitsur and Levitzky [**3**], $VN(G)$ satisfies the polynomial identity $P_{2n} = 0$. Since $L^1(G)$ is isomorphically contained in $VN(G)$, $L^1(G)$ also satisfies $P_{2n} = 0$. Now, Theorem 1 of [**215**] implies that G has an abelian subgroup of finite index. □

LEMMA 4.4.2. *A subset C of a group G is a coset if and only if for any $r, s, t \in C$, $rs^{-1}t \in C$. Moreover, in this case $H = C^{-1}C$ is a subgroup of G and $C = sH$ for every $s \in C$.*

PROOF. Necessity of the condition is trivial. So suppose the condition is satisfied. If $s, t \in H$, then $s = s_1^{-1}s_2$ and $t = t_1^{-1}t_2$, where $s_j, t_j \in C$, $j = 1, 2$, and hence

$$st = s_1^{-1}(s_2 t_1^{-1} t_2) \in C^{-1}C \quad \text{and} \quad s^{-1} = s_2^{-1}s_1 \in C^{-1}C.$$

Consequently, H is a subgroup of G. Furthermore, if $s \in C$ and $t \in H$, where $t = t_1^{-1}t_2$, $t_1, t_2 \in C$, then $st = st_1^{-1}t_2 \in C$ and so $sH \subseteq C$. Conversely, $C = ss^{-1}C \subseteq sH$. □

Let G and H be groups and $Y \subseteq H$ a coset of some subgroup of H. A map $\alpha : Y \to G$ is called *affine* if

$$\alpha(rs^{-1}t) = \alpha(r)\alpha(s)^{-1}\alpha(t)$$

for all $r, s, t \in Y$. In this case, it follows from Lemma 4.4.2 that $\alpha(Y)$ is also a coset. Hence, if $s \in Y$, then the map

$$\beta : s^{-1}Y \to \alpha(s)^{-1}\alpha(Y), \ t \to \alpha(s)^{-1}\alpha(st), \ t \in s^{-1}Y,$$

is a homomorphism between the subgroups $s^{-1}Y$ of H and $\alpha(s)^{-1}\alpha(Y)$ of G. We now introduce the less known notion of piecewise affine map.

DEFINITION 4.4.3. Let G and H be groups and $Y \subseteq H$. A map $\alpha : Y \to G$ is called *piecewise affine* if there are sets Y_i, $i = 1, \ldots, n$ in $\mathcal{R}(H)$, the coset ring of H (see Appendix A.1), with the following properties:

(i) $Y = \bigcup_{i=1}^{n} Y_i$, and the sets Y_i are pairwise disjoint;
(ii) each Y_i is contained in a coset K_i on which there exists an affine map $\alpha_i : K_i \to G$ such that $\alpha_i|_{Y_i} = \alpha|_{Y_i}$, $1 \leqslant i \leqslant n$.

LEMMA 4.4.4. *Let $Y \subseteq H$ and $\alpha : Y \to G$ a map. If the graph*

$$\Gamma_\alpha = \{(s, \alpha(s)) : s \in Y\}$$

is a coset in $H \times G$, then Y is a coset in H and α is an affine map.

PROOF. Let $r, s, t \in Y$. Then, since Γ_α is a coset,

$$(r, \alpha(r))(s, \alpha(s))^{-1}(t, \alpha(t)) = (rs^{-1}t, \alpha(r)\alpha(s)^{-1}\alpha(t)) \in \Gamma_\alpha,$$

which implies that Y is a coset (Lemma 4.4.2). On the other hand, since Γ_α is a graph, $(rs^{-1}t, \alpha(rs^{-1}t)) \in \Gamma_\alpha$ and therefore $\alpha(r)\alpha(s)^{-1}\alpha(t) = \alpha(rs^{-1}t)$, which shows that α is affine. □

LEMMA 4.4.5. *Let G and H be groups, $Y \subseteq H$ and $\alpha : Y \to G$ a map. If $\Gamma_\alpha \in \mathcal{R}(H \times G)$, then α is piecewise affine.*

PROOF. Since $\Gamma_\alpha \in \mathcal{R}(H \times G)$, there exists a finite collection $\mathcal{S}$ of subgroups of $H \times G$ such that Γ_α belongs to the smallest ring of subsets of $H \times G$ generated by cosets of elements of $\mathcal{S}$. We may assume that $\mathcal{S}$ is closed under forming intersections. We may further assume that if $S_1, S_2 \in \mathcal{S}$ and $S_1 \subseteq S_2$, then $[S_2 : S_1] = \infty$. Now Γ_α can be written as a disjoint union $\Gamma_\alpha = \bigcup_{j=1}^{n} E_j$, where each E_j is of the form $E_j = L_j \backslash \bigcup_{i=1}^{m_j} M_{ji}$ and L_j and M_{ji} are cosets of subgroups in $\mathcal{S}$ and $M_{ji} \subseteq L_j$ for all j and i.

We claim that each L_j is a graph. Towards a contradiction, assume that for some j, there are elements (s, t_1) and (s, t_2) of L_j with $t_1 \neq t_2$. Let $t = t_1^{-1}t_2$ and let e denote the identity of H. Then, for any element $(s, \alpha(s))$ of E_j,

$$(s, \alpha(s)t) = (s, \alpha(s))(e, t) = (s, \alpha(s))(s, t_1^{-1})(s, t_2) \in L_j L_j^{-1} L_j = L_j$$

because L_j is a coset. Since $E_j \subseteq \Gamma_\alpha$ is a graph, it follows that $(s, \alpha(s)t) \in M_{ji}$ for some i. Thus

$$(s, \alpha(s)) = (s, \alpha(s)t)(e, t)^{-1} \in M_{ji}(e, t)^{-1}.$$

Hence $E_j \subseteq \bigcup_{i=1}^{m_j} M_{ji}(e, t)^{-1}$ and therefore

$$L_j \subseteq \textstyle\bigcup_{i=1}^{m_j}(M_{ji} \cup M_{ji}(e, t)^{-1}).$$

Hence the subgroup $L_i^{-1}L_i$ can be covered by finitely many cosets of subgroups which are of infinite index in $L_i^{-1}L_i$, which is impossible (see Lemma A.1.1). Thus L_i is a graph, and we may write

$$L_i = \{(s, \alpha_i(s)) : s \in K_i = q(L_i)\},$$

where $q : H \times G \to H$ is the projection $(h, g) \to h$. Now, if we let $Y_i = q(E_i)$ and $N_{ij} = q(M_{ij})$, we see that

$$Y_i = K_i \backslash \textstyle\bigcup_{j=1}^{m_i} N_{ij}$$

since $q|_{L_i}$ has inverse $s \to (s, \alpha_i(s))$. We also see from Lemma 4.4.4 that $\alpha_i : K_i \to G$ is an affine map and $\alpha_i|_{Y_i} = \alpha|_{Y_i}$. This completes the proof. □

LEMMA 4.4.6. *Let G and H be discrete groups, $Y \subseteq H$ and $\alpha : Y \to G$ a piecewise affine map. Then the map $\phi_\alpha : A(G) \to B(H)$ defined by*

$$\phi_\alpha(u)(s) = \begin{cases} u(\alpha(s)) & : \text{ if } s \in Y \\ 0 & : \text{ otherwise,} \end{cases}$$

is a completely bounded homomorphism.

PROOF. We shall build up the proof in stages, beginning with homomorphisms, then passing to affine maps and finally to piecewise affine maps.

(i) Suppose first that Y is a subgroup of H and α is a homomorphism. Then the map

$$\phi_\alpha^Y : A(G) \to B(Y), \quad u \to \phi_\alpha(u)|_Y$$

is an isometric homomorphism whose adjoint $(\phi_\alpha^Y)^* : W^*(Y) \to VN(G)$ is the $*$-homomorphism such that

$$(\phi_\alpha^Y)^*(\omega_Y(y)) = \lambda_G(\alpha(y)), \quad y \in Y.$$

Here $W^*(Y) = B(Y)^* = C^*(Y)^{**}$, the enveloping von Neumann algebra of $C^*(Y)$, and for $y \in Y$, $\omega_Y(y)$ is the element of $W^*(Y)$ defined by $\langle \omega_Y(y), u \rangle = u(y)$ for all $u \in B(Y)$.

Since Y is a subgroup of H, there is a canonical isometric embedding of $B(Y)$ into $B(H)$. In fact, for $v \in B(Y)$, the trivial extension $\widetilde{v}$ of v to all of H satisfies $\|\widetilde{v}\|_{B(H)} = \|v\|_{B(Y)}$. Now, define $m_Y : B(H) \to B(H)$ by $m_Y(u) = 1_Y u \in B(H)$. Then $m_Y(B(H)) = \{\widetilde{v} : v \in B(Y)\}$ and $m_Y^* : B(H)^* \to B(H)^*$ satisfies

$$m_Y^*(I) = \omega_H(e) \quad \text{and} \quad \|m_Y^*\|_{cb} = \|m_Y\|_{cb} = \|1_Y\|_{cb} = 1.$$

Hence by [**70**, Corollary 5.1.2], m_Y^* is completely positive.

The adjoint of the map $v \to \widetilde{v}$ is a $*$-homomorphism

$$\psi : \overline{\operatorname{span}\{\omega_H(y) : y \in Y\}}^{w^*} \quad \to \quad W^*(Y)$$

such that $\psi(\omega_H(y)) = \omega_Y(y)$ for all $y \in Y$. Hence it follows that $\phi_\alpha^* = \left(\phi_\alpha^Y\right)^* \circ \psi \circ m_Y^*$ is completely positive and contractive, and therefore ϕ_α is also completely contractive.

(ii) Next we suppose that Y is a coset and α is affine. Fix an element $y \in Y$ and let $\beta : y^{-1}Y \to \alpha(y)^{-1}\alpha(Y) \subseteq G$ be the map defined by

$$\beta(t) = \alpha(y)^{-1}\alpha(yt), \ t \in y^{-1}Y.$$

Then β is a homomorphism between the two subgroups $y^{-1}Y$ of H and $\alpha(y)^{-1}\alpha(Y)$ of G (compare the remark after Lemma 4.4.2). It is easily verified that ϕ_α can be written as a composition $\phi_\alpha = L_y \circ \phi_\beta \circ L_{\alpha(y)^{-1}}$, where $L_y : B(H) \to B(H)$ and $L_{\alpha(y)^{-1}} : A(G) \to A(G)$ are the left translation operators

$$L_y v(h) = v(y^{-1}h), \ h \in H, \quad \text{and} \quad L_{\alpha(y)^{-1}} u(t) = u(\alpha(y)t), \ t \in G,$$

respectively. Now the adjoints of these translation operators are multiplications given by the unitaries $\lambda_G(\alpha(y)^{-1})^*$ on $VN(G)$ and $\omega_H(y)^*$ on $W^*(H)$, respectively. Since both are complete isometries, it follows that ϕ_α is a complete contraction.

(iii) Finally, suppose that $\alpha : Y \to G$ is a piecewise affine map. Then Y can be written as a union $Y = \bigcup_{i=1}^n Y_i$ of pairwise disjoint sets $Y_i \in \mathcal{R}(H)$ with the property that Y_i is contained in a coset K_i and there exists an affine map $\alpha_i : K_i \to G$ such that $\alpha_i|_{Y_i} = \alpha|_{Y_i}$, $1 \leqslant i \leqslant n$. For each i, let $\phi_{\alpha_i} : A(G) \to B(H)$ be the algebra homomorphism given by

$$\phi_{\alpha_i}(u)(x) = \begin{cases} u(\alpha_i(x)) & : \ if\ x \in K_i \\ 0 & : \ otherwise. \end{cases}$$

By part (ii), we know that ϕ_{α_i} is completely bounded. Now, for every $u \in A(G)$,

$$\phi_\alpha(u) = \sum_{i=1}^{n} 1_{Y_i}\phi_{\alpha_i}(u).$$

Indeed, if $x \in G\backslash Y$, then $\phi_\alpha(u)(x) = 0$ and $1_{Y_i}(x) = 0$ for all i, and if $x \in Y$, then $x \in Y_j$ for exactly one $j \in \{1, \ldots, n\}$ and

$$\phi_{\alpha_j}(u)(x) = u(\alpha_j(x)) = u(\alpha(x)),$$

whereas $1_{Y_i}(x) = 0$ for $i \neq j$. Since $Y_i \in \mathcal{R}(H)$, $1_{Y_i} \in B(H)$ and the map $\omega_i : B(H) \to B(H)$ given by $\omega_i(v) = 1_{Y_i}v$, $v \in B(H)$, is completely bounded since $B(H)$ is a completely contractive Banach algebra.

Since the completely bounded maps form a linear space, it follows that $\phi_\alpha = \sum_{i=1}^{n} \omega_i \circ \phi_{\alpha_i}$ is completely bounded. This finishes the proof of the lemma. □

THEOREM 4.4.7. *Let G be a group and let $\Gamma_G = \{(x, x^{-1}) : x \in G\}$. Then Γ_G belongs to the coset ring $\mathcal{R}(G \times G)$ if and only if G has an abelian subgroup of finite index*

PROOF. Let G have an abelian subgroup, A say, of finite index and let $\{x_1, \ldots, x_n\}$ be a representative system for the left cosets of A in G. For $j = 1, \ldots, n$, set

$$A_j = \{(a, x_j a^{-1} x_j^{-1}) : a \in A\}.$$

Then A_j is a subgroup of $G \times G$ because A is abelian. Obviously,

$$\Gamma_G = (x_1, x_1^{-1})A_1 \cup \ldots \cup (x_n, x_n^{-1})A_n,$$

so that Γ_G is a finite union of left cosets of subgroups of $G \times G$, whence $\Gamma_G \in \mathcal{R}(G \times G)$.

Conversely, suppose that $\Gamma_G \in \mathcal{R}(G \times G)$. Since Γ_G is the graph of the map $x \to x^{-1}$, Lemma 4.4.5 implies that $x \to x^{-1}$ is a piecewise affine map. It then follows from Lemma 4.4.6 that the flip map $j_* : A(G) \to A(G), u \to \check{u}$ is completely bounded. Finally, Proposition 4.4.1 ensures that G has an abelian subgroup of finite index. □

4.5. Amenability and weak amenability of $A(G)$ and of $L^1(G)$

We remind the reader that a Banach algebra A is called *amenable* if the first cohomology group $H^1(A, E^*)$ is trivial for each A-bimodule E, that is, every bounded derivation from A into E^* is inner, and that A is said to be *weakly amenable* if $H^1(A, A^*) = \{0\}$ (compare Appendix A.2). In this section we study amenability and weak amenability of the Fourier $A(G)$ of a locally compact group G. Amenability turns out to be very restrictive in that for any locally compact group G, $A(G)$ is amenable (if and) only if G has an abelian subgroup of finite index. Essential ingredients of the proof are the results on the flip map and the antidiagonal obtained in the last section. Weak amenability appears to be a less accessible property. A sufficient condition is that the connected component of the identity of G be abelian. The question of whether the converse holds remains open.

LEMMA 4.5.1. *Let $(\mu_\alpha)_\alpha$ be a net in $M(G)$ such that, for some $\mu \in M(G)$, $\mu_\alpha \to \mu$ in the w^*-topology $\sigma(M(G), C_0(G))$ and $\|\mu_\alpha\| \to \|\mu\|$. Then for each $\epsilon > 0$, there exists a compact subset C of G such that $\int_{G\setminus C} d(|\mu_\alpha| + |\mu|)(x) < \epsilon$ eventually.*

PROOF. Given $\epsilon > 0$. Let $\delta = \epsilon/4$ and choose $f \in C_c(G)$ such that $\|f\|_\infty \leqslant 1$ and $|\langle f, \mu\rangle - \|u\|| < \delta$. There exists α_0 such that

$$|\langle f, \mu_\alpha\rangle - \langle f, \mu\rangle| < \delta \quad \text{and} \quad |\|\mu_\alpha\| - \|\mu\|| < \delta$$

for all $\alpha \geqslant \alpha_0$. Denoting by C the support of f, we then have

$$\begin{aligned} \|\mu\| &\leqslant |\langle f, \mu\rangle + \delta| \leqslant |\langle f, \mu - \mu_\alpha\rangle| + |\langle f, \mu_\alpha\rangle| + \delta \\ &\leqslant 2\delta + \int_C d|\mu_\alpha|(x) \end{aligned}$$

and hence

$$\int_{G\setminus C} d|\mu_\alpha|(x) = \|\mu_\alpha\| - \int_C d|\mu_\alpha|(x) \leqslant \|\mu\| + \delta - \int_{G\setminus C} d|\mu_\alpha|(x) \leqslant 3\delta$$

for all $\alpha \geqslant \alpha_0$. Thus

$$\int_{G\setminus C} d(|\mu_\alpha| + |\mu|)(x) < 4\delta$$

for all $\alpha \geqslant \alpha_0$, as was to be shown. □

REMARK 4.5.2. Let $(\mu_\alpha)_\alpha$ be a net in $M(G)$ such that $\mu_\alpha \to \mu$ in the w^*-topology and $|\mu_\alpha|(G) \to |\mu|(G)$. Then $|\mu_\alpha| \to |\mu|$ in the w^*-topology. To see this, after passing to a subnet, we may assume that $|\mu_\alpha| \to \nu$ for some $\nu \in M(G)$ since the unit ball of $M(G)$ is w^*-compact. Let $f \in C_c(G), f \geqslant 0$. Then for any $g \in C_c(G)$ with $|g| \leqslant f$, $\langle |\mu_\alpha|, f\rangle \geqslant |\langle \mu_\alpha, g\rangle| \to |\langle \mu, g\rangle|$. This implies

$$\langle \nu, f\rangle \geqslant \sup\{|\langle \mu, g\rangle| : g \in C_c(G), |g| \leqslant f\} = \langle |\mu|, f\rangle.$$

Consequently $\nu - |\mu|$ is positive. Since $(\nu - |\mu|)(G) = 0$, we conclude that $\nu = |\mu|$.

LEMMA 4.5.3. *Let $(\mu_\alpha)_\alpha$ be a net in $M(G)$ such that, for some $\mu \in M(G)$, $\mu_\alpha \to \mu$ in the w^*-topology and $\|\mu_\alpha\| \to \|\mu\|$. Then*

$$\|(\mu_\alpha - \mu) * f\|_1 \to 0 \quad \text{and} \quad \|f * (\mu_\alpha - \mu)\|_1 \to 0$$

for each $f \in L^1(G)$.

PROOF. We shall prove the first norm convergence, the proof of the second one being analogous. Since $\|\mu_\alpha\| \to \|\mu\|$, after passing to a subnet if necessary, we may assume that the net $(\|\mu_\alpha\|)_\alpha$ is bounded. Moreover, since $\|\mu_\alpha * f\|_1 \leqslant \|\mu_\alpha\| \cdot \|f\|_1$, we may assume that $f \in C_c(G)$ and then, in addition, that $0 \leqslant f \leqslant 1$. For any compact subset K of G we have

$$\|(\mu_\alpha - \mu) * f\|_1 \leqslant \int_K |(\mu_\alpha - \mu) * f|(x)dx + \int_{G\setminus K} (|\mu_\alpha| * f + |\mu| * f)(x)dx.$$

Now let $g_\alpha = |\mu_\alpha| * f$ and $g = |\mu| * f$. Then $g_\alpha, g \in L^1(G)$ and

$$\|g_\alpha\|_1 = \|\mu_\alpha\| \cdot \|f\|_1 \to \|\mu\| \cdot \|f\|_1 = \|g\|_1.$$

By the preceding remark, $g_\alpha \to g$ and $|\mu_\alpha| \to |\mu|$ in the w^*-topology. Therefore, given $\epsilon > 0$, by Lemma 4.5.1 there exist a compact subset K of G and α_0 such that

$$\int_{G\setminus K} (|\mu_\alpha| * f + |\mu| * f)(x)dx < \epsilon$$

for all $\alpha \geqslant \alpha_0$. On the other hand, since the map $y \to L_y f$ from G into $(C_0(G), \|\cdot\|_\infty)$ is continuous, the set $\{L_y f : y \in K\}$ is compact in $C_0(G)$ and hence there exists α_1 such that $\int_K |(\mu_\alpha - \mu) * f|(x)dx < \epsilon$ for all $\alpha \geqslant \alpha_1$. It follows that $\|(\mu_\alpha - \mu) * f\|_1 \leqslant 2\epsilon$ for $\alpha \geqslant \alpha_0, \alpha_1$. Since $\epsilon > 0$ was arbitrary, this shows that $\|(\mu_\alpha - \mu) * f\|_1 \to 0$. □

By the *strict topology* on $M(G)$ we mean the locally convex topology defined by the collection of seminorms $\sigma_f(\mu) = \|f * \mu\| + \|\mu * f\|$, $f \in L^1(G)$.

LEMMA 4.5.4. *Let G be a locally compact group, E a pseudo-unital $L^1(G)$-bimodule and $D : L^1(G) \to E^*$ a bounded derivation. Then there exists a unique derivation $\widetilde{D} : M(G) \to E^*$ which extends D and is continuous with respect to the strict topology on $M(G)$ and the w^*-topology on E^*. In particular, $\widetilde{D}$ is uniquely determined by its values on the set $\{\delta_x : x \in G\}$.*

PROOF. We first define a left module action of $M(G)$ on E. Fix $u \in E$ and let $f \in L^1(G)$ and $v \in E$ such that $u = f \cdot v$. For $\mu \in M(G)$, set $\mu \cdot u = (\mu * f) \cdot v$. This action is well defined. Indeed, if $g \in L^1(G)$ and $w \in E$ are such that also $u = g \cdot w$, then for any bounded approximate identity $(h_\alpha)_\alpha$ of $L^1(G)$,

$$(\mu * f) \cdot v = \lim_\alpha \mu * (h_\alpha * f) \cdot v = \lim_\alpha \mu * (h_\alpha * g) \cdot v = (\mu * g) \cdot w.$$

It is clear that this defines a left module action of $M(G)$ on E. Similarly, one defines a right module action of $M(G)$ on E. We claim that for each $\mu \in M(G)$, w^*-$\lim_\alpha (D(\mu * h_\alpha) - \mu \cdot D(h_\alpha))$ exists. To see this, let $u \in E$ and choose $f \in L^1(G)$ and $v \in E$ such that $u = v \cdot f$. Then, for any $\mu \in M(G)$,

$$\begin{aligned}
\langle u, (D(\mu * h_\alpha) - \mu \cdot D(h_\alpha)\rangle &= \langle v \cdot f, D(\mu * h_\alpha) - \mu \cdot D(h_\alpha)\rangle \\
&= \langle v, f \cdot D(\mu * h_\alpha) - (f * \mu) \cdot D(h_\alpha)\rangle \\
&= \langle v, D(f * \mu * h_\alpha) - D(f) \cdot (\mu * h_\alpha)\rangle \\
&\quad -\langle v, D(f * \mu * h_\alpha) + D(f * \mu) \cdot h_\alpha\rangle \\
&= \langle h_\alpha \cdot v, D(f * \mu)\rangle - \langle(\mu * h_\alpha) \cdot v, D(f)\rangle \\
&\to \langle v, D(f * \mu)\rangle - \langle \mu \cdot v, D(f)\rangle.
\end{aligned}$$

Therefore, we can define $\widetilde{D} : M(G) \to E^*$ by setting

$$\widetilde{D}(\mu) = w^* - \lim_\alpha (D(\mu * h_\alpha) - \mu \cdot D(h_\alpha)).$$

Then $\widetilde{D}$ extends D since

$$\widetilde{D}(f) = w^* - \lim_\alpha (D(f * h_\alpha) - f \cdot D(h_\alpha)) = w^* - \lim_\alpha D(f) \cdot h_\alpha = D(f)$$

for every $f \in L^1(G)$. Furthermore, for $\mu \in M(G)$ and $f \in L^1(G)$, we have

$$\begin{aligned}
\widetilde{D}(\mu) \cdot f &= w^* - \lim_\alpha (D(\mu * h_\alpha) \cdot f - \mu \cdot D(h_\alpha) \cdot f) \\
&= w^* - \lim_\alpha (D(\mu * h_\alpha * f) - (\mu * h_\alpha) \cdot D(f) \\
&\quad -w^* - \lim_\alpha \mu \cdot D(h_\alpha * \mu) + (\mu * h_\alpha) \cdot D(f) \\
&= D(\mu * f) - \mu \cdot D(f).
\end{aligned}$$

It remains to show that $\widetilde{D}$ is a derivation. Clearly, using the definition of the strict topology on $M(G)$, we have $\mu * h_\alpha \to \mu$ in the strict topology for each $\mu \in M(G)$. Hence, for $\mu, \nu \in M(G)$,

$$\begin{aligned}
\widetilde{D}(\mu * \nu) &= w^* - \lim_\alpha (w^* - \lim_\beta D((\mu * h_\alpha) * (\nu * h_\beta)) \\
&= w^* - \lim_\alpha (w^* - \lim_\beta (\mu * h_\alpha) \cdot D(\nu * h_\beta)) + D(\mu * h_\alpha) \cdot (\nu * h_\beta)) \\
&= \mu \cdot \widetilde{D}(\nu) + \widetilde{D}(\mu) \cdot \nu.
\end{aligned}$$

The final statement follows from the fact that if $\mu \in M(G)$ is such that $\mu \geqslant 0$ and $\|\mu\| = 1$, then there is a net $(\mu_\alpha)_\alpha$ in $M(G)$ such that $\mu_\alpha \to \mu$ in the w^*-topology, where each μ_α is of the form $\sum_{j=1}^n \lambda_j \delta_{x_j}$, where $x_j \in G$, $\lambda_j \geqslant 0$ and $\sum_{j=1}^n \lambda_j = 1$. □

THEOREM 4.5.5. *Let G be an amenable locally compact group. Then $L^1(G)$ is amenable.*

PROOF. Let E be an $L^1(G)$-bimodule. By Lemma A.2.6, we may assume that E is pseudo-unital. Let $D : L^1(G) \to E^*$ be a derivation and let $\widetilde{D}$ be the extension of D to $M(G)$ (Lemma 4.5.4). We shall show that $\widetilde{D}$ is inner using Day's fixed point theorem (Theorem 1.8.7). We define an affine action of G on K by

$$x \cdot \varphi = \delta_x \cdot \varphi \cdot \delta_{x^{-1}} + \widetilde{D}(\delta_x) \cdot \delta_{x^{-1}}, \quad x \in G, \varphi \in E^*.$$

This defines a group action of G on E^* since

$$\begin{aligned}
(xy) \cdot \varphi &= \delta_{xy} \cdot \varphi \cdot \delta_{(xy)^{-1}} + \widetilde{D}(\delta_{xy}) \cdot \delta_{(xy)^{-1}} \\
&= \delta_x \cdot (\delta_y \cdot \varphi \cdot \delta_{y^{-1}}) \cdot \delta_{x^{-1}} + \delta_x \cdot \widetilde{D}(\delta_y) + (\widehat{D}(\delta_x) \cdot \delta_y) \cdot \delta_{y^{-1}} * \delta_{x^{-1}} \\
&= \delta_x \cdot (\delta_y \cdot \varphi \cdot \delta_{y^{-1}} + \widetilde{D}(\delta_y)) \cdot \delta_{x^{-1}} + \widetilde{D}(\delta_x) \cdot \delta_{x^{-1}} \\
&= x \cdot (y \cdot \varphi)
\end{aligned}$$

for $x, y \in G$ and $\varphi \in E^*$. This action of G on E^* is affine, that is,

$$x \cdot (\lambda \varphi_1 + (1 - \lambda)\varphi_2) = \lambda x \cdot \varphi_1 + (1 - \lambda) x \cdot \varphi_2$$

for $\varphi_1, \varphi_2 \in E^*$ and $x \in G$. Using Lemma 4.5.4, it is easy to see that the map $\varphi \to x \cdot \varphi$ from E^* into E^* is w^*-w^*-continuous for fixed $x \in G$.

Now fix $x_0 \in G$ and let $\varphi_0 = \widetilde{D}(\delta_{x_0}) \cdot \delta_{x_0^{-1}}$. Since $\|x \cdot \varphi_0\| \leqslant \|\varphi_0\| + \|\widetilde{D}\|$, $\mathrm{co}\{x \cdot \varphi_0 : x \in G\}$, the convex hull of the G-orbit of φ_0, is bounded. Let K denote its w^*-closure. Then K is a w^*-compact G-invariant subset of E^*. It is also easy to check that the map $x \to x \cdot \varphi_0$ from G into K is continuous for the w^*-topology on K. Therefore, by Day's fixed point theorem for separately continuous actions (Theorem 1.8.7), there exists $\varphi \in K$ such that

$$\varphi = x \cdot \varphi = \delta_x \cdot \varphi \cdot \delta_{x^{-1}} + \widetilde{D}(\delta_x) \cdot \delta_{x^{-1}}$$

for all $x \in G$. Equivalently, $\widetilde{D}(\delta_x) = \varphi \cdot \delta_x - \delta_x \cdot \varphi$ for all $x \in G$, so that $\widetilde{D}$ is implemented by $-\varphi$. □

THEOREM 4.5.6. *Let G be any locally compact group. Then $L^1(G)$ is weakly amenable.*

PROOF. By Lemma 4.5.4 it suffices to show that every bounded derivation from $M(G)$ into $L^\infty(G)$, which is continuous with respect to the strict topology on $M(G)$ and the w^*-topology on $L^\infty(G)$, is an inner derivation. For $x, y \in G$, we have

$$\delta_{x^{-1}} \cdot D(\delta_x) = \delta_{x^{-1}} \cdot D(\delta_{xy^{-1}} * \delta_y) \tag{4.2}$$

$$= \delta_{x^{-1}} \cdot [\delta_{(xy^{-1})^{-1}} \cdot D(\delta_{xy^{-1}}] \cdot \delta_y + \delta_{y^{-1}} \cdot D(\delta_y). \tag{4.3}$$

For $\psi \in L^\infty(G)$, let $\mathrm{Re}(\psi)$ and $\mathrm{Im}(\psi)$ denote the real and imaginary part of ψ, respectively. Put

$$S = \{\mathrm{Re}(\delta_{x^{-1}} \cdot D(\delta_x)) : x \in G\} \subseteq L^\infty_{\mathbb{R}}(G).$$

Then S is bounded above by the constant function $\|D\|$. Since $L^\infty_{\mathbb{R}}(G)$ is a complete vector lattice, $\varphi_1 = \sup(S)$ exists in $L^\infty_{\mathbb{R}}(G)$. Moreover, it is not difficult to verify that

$$\sup(\delta_{x^{-1}} \cdot S \cdot \delta_x) = \delta_{x^{-1}} \cdot \sup(S) \cdot \delta_x, \quad x \in G, \tag{4.4}$$

and

$$\sup(\psi + S) = \psi + \sup(S), \quad \psi \in L^\infty_{\mathbb{R}}(G). \tag{4.5}$$

Taking $\sup_{x\in G}$ of the real parts in (4.2) and using (4.4) and (4.5), we obtain

$$\varphi_1 = \delta_{y^{-1}} \cdot \varphi_1 \cdot \delta_y + \delta_{y^{-1}} \cdot \mathrm{Re}(D(\delta_y)),$$

or equivalently

$$\mathrm{Re}(D(\delta_y)) = \delta_y \cdot \varphi_1 - \varphi_1 \cdot \delta_y, \quad y \in G.$$

Similarly, considering the imaginary parts and taking $\sup_{x\in G}$ in (4.2), we obtain $\varphi_2 \in L^\infty_{\mathbb{R}}(G)$ such that

$$\mathrm{Im}(D(\delta_y)) = \delta_y \cdot \varphi_2 - \varphi_2 \cdot \delta_y$$

for all $y \in G$. Thus, setting $\varphi = \varphi_1 + i\varphi_2$, we get

$$D(\delta_y) = \delta_y \cdot \varphi - \varphi \cdot \delta_y, \quad y \in G.$$

Since every positive measure in $M(G)$ is the strict limit of convex combinations of point masses and D is continuous with respect to the strict topology on $M(G)$ and the w^*-topology on $L^\infty(G)$, it follows that

$$D(\mu) = \mu \cdot \varphi - \varphi \cdot \mu$$

for all $\mu \in M(G)$, as required. □

LEMMA 4.5.7. *Let G be a locally compact group and H a closed subgroup of finite index. Then $A(G)$ is amenable if and only if $A(H)$ is amenable.*

PROOF. Since $A(H)$ is isomorphic to a quotient of $A(G)$, amenability of $A(G)$ implies amenability of $A(H)$. To see the converse, choose representatives $x_1 = e, x_2, \ldots, x_m$ for the left cosets of H in G. Define a map

$$\phi : \oplus_{j=1}^m A(H) \to A(G),\ \oplus_{j=1}^m u_j \to \sum_{j=1}^m L_{x_j} u_j.$$

Since the functions $L_{x_j} u_j$ are supported on disjoint open sets, ϕ is a homomorphism, and it is clearly continuous and surjective. Since $A(H)$ is amenable, so is $\oplus_{j=1}^m A(H)$ and hence also $A(G)$. □

COROLLARY 4.5.8. *Suppose that G has an abelian subgroup of finite index. Then $A(G)$ is amenable.*

PROOF. This follows from Theorem 4.5.5 and Lemma 4.5.7 observing that if H a closed abelian subgroup of finite index in G, then $A(H) = L^1(\widehat{H})$ and $L^1(\widehat{H})$ is amenable (Theorem 4.5.5). □

THEOREM 4.5.9. *For a locally compact group G, the following conditions are equivalent.*

(i) *$A(G)$ is amenable.*
(ii) *G has an abelian subgroup of finite index.*

PROOF. (ii) ⇒ (i) is Corollary 4.5.8.

(i) ⇒ (ii) Since $A(G)$ is amenable, by [**117**, Proposition VII.2.15] the kernel of the homomorphism

$$\Delta : A(G)\widehat{\otimes} A(G) \to A(G), \quad f \otimes g \to fg$$

has a bounded approximate identity, $(u_\alpha)_\alpha$ say. For each α, there are sequences $(f_{\alpha,j})_j$ and $(g_{\alpha,j})_j$ in $A(G)$ such that

$$\sum_{j=1}^{\infty} \|f_{\alpha,j}\|_{A(G)}\|g_{\alpha,j}\|_{A(G)} < \infty \quad \text{and} \quad u_\alpha = \sum_{j=1}^{\infty} f_{\alpha,j} \otimes g_{\alpha,j}.$$

Since the map $f \to \check{f}$ of $A(G)$ is an isometry, we can define a bounded net $(v_\alpha)_\alpha$ in $A(G)\widehat{\otimes} A(G)$ by

$$v_\alpha = \sum_{j=1}^{\infty} f_{\alpha,j} \otimes \check{g}_{\alpha,j}.$$

It is immediate that $(v_\alpha)_\alpha$ is a bounded approximate identity for the kernel of the homomorphism

$$\Gamma : A(G)\widehat{\otimes} A(G) \to A(G), \quad f \otimes g \to f\check{g}.$$

For $f, g \in A(G)$, the function $(x, y) \to f(x)g(y)$ belongs to $A(G \times G)$, and this defines a contractive homomorphism $\psi : A(G)\widehat{\otimes} A(G) \to A(G \times G)$. Let

$$I = \{u \in A(G \times G) : \psi(v_\alpha)u \to u\}.$$

Then I is a closed ideal of $A(G\times G)$ containing $\psi(\ker \Gamma)$, and $(\psi(v_\alpha))_\alpha$ is a bounded approximate identity for I. Since G, and hence $G \times G$, is amenable, Theorem 4.4.7 applies and shows that I is of the form

$$I = I(E) = \{u \in A(G \times G) : u|_E = 0\}$$

for some $E \in \mathcal{R}_c(G \times G)$. We claim that $E = \Gamma_G$. Since $\psi(v_\alpha)(x, x^{-1}) = 0$ for all α and all $x \in G$, we have $\Gamma_G \subseteq E$. On the other hand, if $(x, y) \notin \Gamma_G$ then, by regularity of $A(G \times G)$ (Proposition 2.3.2), there exists $w \in A(G \times G)$ such that $w|_\Gamma = 0$ and $w(x, y) \neq 0$. Then $w(x, y)(v_\alpha(x, y) - 1) \to 0$, and thus $v_\alpha(x, y) \to 1$ and $(x, y) \notin E$. This shows that $\Gamma_G = E \in \mathcal{R}_c(G \times G)$. From Theorem 4.4.7, we conclude that G has an abelian subgroup of finite index. □

It has been known for a long time that the measure algebra of a locally compact abelian group is amenable only if the group is discrete. Exploiting this fact and the preceding theorem, amenability of the Fourier-Stieltjes algebra can now been completely characterized in terms of the group structure.

COROLLARY 4.5.10. *The following are equivalent for a locally compact group G.*

(i) *The Fourier-Stieltjes algebra $B(G)$ is amenable.*
(ii) *G is compact and has an abelian subgroup of finite index.*

PROOF. (i) $\Rightarrow$ (ii) Since $A(G)$ is a translation invariant ideal of $B(G)$, it follows that $A(G)$ is an invariant subspace of $B(G)$, regarded as a predual of the von Neumann algebra $W^*(G)$. Hence there exists a central projection Z in $W^*(G)$ such that $A(G) = Z \cdot B(G)$ [**270**, p.123, Theorem 2.7], where $\langle u \cdot Z, T\rangle = \langle u, ZT\rangle$ for $u \in B(G)$ and $T \in W^*(G)$. Consequently, $B(G) = A(G) \oplus (I - Z) \cdot B(G)$ and hence $A(G)$ is complemented in $B(G)$. Since $B(G)$ is amenable, it follows that $A(G)$ is also amenable [Appendix A.2] and therefore G has an abelian closed subgroup A of finite index by Theorem 4.5.9. Then A is open in G and hence the restriction map from $B(G)$ to $B(A)$ is surjective. By a standard fact of amenable Banach algebras, $B(A)$ is also amenable. Now, $B(A)$ is isometrically isomorphic to $M(\widehat{A})$, the measure algebra of the dual group $\widehat{A}$ of A. However, amenability of $M(\widehat{A})$ forces $\widehat{A}$ to be discrete [**25**]. So A and hence G must be compact.

(ii) $\Rightarrow$ (i) If G is compact and has an abelian subgroup of finite index, then $B(G) = A(G)$ and $A(G)$ is amenable by Theorem 4.5.9. □

We now turn to weak amenability of Fourier algebras.

LEMMA 4.5.11. *Let G be a locally compact group and H a closed subgroup of G.*

(i) *If $A(G)$ is weakly amenable, then so is $A(H)$.*
(ii) *If H is open and $A(H)$ is weakly amenable, then so is $A(G)$.*

PROOF. Since $A(H)$ is a quotient of $A(G)$, (i) follows immediately from the basic hereditary properties of weak amenability for commutative Banach algebras [Appendix A.2].

For (ii), let X be a symmetric Banach $A(G)$-bimodule and let $D : A(G) \to X$ be a derivation. Let $u \in A(G) \cap C_c(G)$ and choose $x_1, \ldots, x_n \in G$ such that $\operatorname{supp} u \subseteq \bigcup_{j=1}^n x_j H$. For each $1 \leqslant j \leqslant n$, let $A_j = 1_{x_j H} A(G)$. Then A_j is a subalgebra of $A(G)$, which is isometrically isomorphic to $A(H)$. Since $A(H)$ is weakly amenable, it follows that $D|_{A_j} = 0$ for all j. In particular, $D(1_{x_j H} u) = 0$ for all j and hence $D(u) = 0$ since $u = \sum_{j=1}^n 1_{x_j H} u$. Since $A(G) \cap C_c(G)$ is dense in $A(G)$, we obtain that $D(u) = 0$ for all $u \in A(G)$. □

LEMMA 4.5.12. *Let G be a locally compact group such that G_0, the connected component of the identity of G, is abelian, and let K be a compact normal subgroup of G such that G/K is a Lie group. Then $A(G/K)$ is weakly amenable.*

PROOF. Let $q : G \to G/K$ denote the quotient homomorphism. Then the connected component of the identity of G/K equals $\overline{q(G_0)}$ (see [**125**]). Thus, since G_0 is abelian, so is $(G/K)_0$. In particular, $A((G/K)_0)$ is amenable (Theorem 4.5.9) and hence weakly amenable. Since G/K is a Lie group, $(G/K)_0$ is open in G/K. It now follows from Lemma 4.5.11(ii) that $A(G/K)$ is weakly amenable. □

THEOREM 4.5.13. *Let G be a locally compact group such that G_0 is abelian. Then $A(G)$ is weakly amenable.*

PROOF. We shall first treat the case where G/G_0 is compact. Let X be a symmetric Banach $A(G)$-bimodule and let $D : A(G) \to X$ be a derivation. Let $u \in A(G)$ and $\epsilon > 0$. We are going to show that $\|D(u)\| \leqslant \epsilon$. Since $\epsilon > 0$ is arbitrary, this is enough to conclude that $D(u) = 0$.

Since the right translation map $x \to R_x(u)$ from G into $A(G)$ is continuous, there is a neighbourhood U of e in G such that

$$\|u - R_x(u)\| < \frac{\epsilon}{1 + \|D\|}$$

for all $x \in U$. Since an almost connected group is a projective limit of Lie groups, there exists a compact normal subgroup K of G such that $K \subseteq U$ and G/K is a Lie group. Let

$$p_K(u) = \int_K R_k(u)dk,$$

where the integral is a Bochner integral with respect to normalized Haar measure on K. Then

$$\|u - p_K(u)\| \leqslant \int_K \|u - R_k(u)\|dk \leqslant \frac{\epsilon}{1 + \|D\|}.$$

Now, $p_K(u) \in A(G : K)$, the subalgebra of $A(G)$ consisting of those functions which are constant on cosets of K. Since $A(G : K)$ is isometrically isomorphic to $A(G/K)$ and $A(G/K)$ is weakly amenable, we obtain that $D|_{A(G:K)} = 0$. Thus

$$\|D(u)\| = \|D(u) - D(p_K(u))\| \leqslant \|D\| \cdot \|u - p_K(u)\| \leqslant \epsilon,$$

as was to be shown.

Now drop the hypothesis that G/G_0 is compact and note that G contains an open, almost connected subgroup H. We have shown so far that $A(H)$ is weakly amenable. By Lemma 4.5.11(ii), this implies that $A(G)$ is weakly amenable. □

PROPOSITION 4.5.14. *Let G be a Lie group such that $A(G)$ is weakly amenable. Then every compact subgroup of G has an abelian subgroup of finite index.*

PROOF. Let K be a compact subgroup of G. Since $A(G)$ is weakly amenable, $A(K)$ and $A(K_0)$ are also weakly amenable. But K_0 is a compact connected Lie group and therefore K_0 has to be abelian (Theorem 4.5.9). Finally, K_0 has finite index in K since K is compact and K_0 is open in K. □

COROLLARY 4.5.15. *For an almost connected semisimple Lie group G, the following conditions are equivalent.*

(i) *$A(G)$ is amenable.*
(ii) *G is amenable and $A(G)$ is weakly amenable.*
(iii) *G is finite.*

PROOF. The implications (i) ⇒ (ii) and (iii) ⇒ (i) are obvious. For (ii) ⇒ (iii), note that an amenable connected semisimple Lie group is compact. So G_0 is compact and hence abelian by Proposition 4.5.14. Since G_0 is semisimple, it has to be trivial, and consequently G is finite. □

4.6. Notes and references

The study of amenability and weak amenability and of their operator space variants for Banach algebras associated with locally compact groups started with Johnson's work some forty years ago. Johnson [**141**] proved that the L^1-algebra of a locally compact group G is amenable precisely when G is amenable. In contrast, $L^1(G)$ is weakly amenable for any locally compact group (Theorem 4.5.6). The completely different short proof presented here was found by Despic and Ghahramani [**58**]. Thus, if G is abelian, $A(G)$ is amenable (hence weakly amenable) since $A(G) \cong L^1(\widehat{G})$. For general locally compact groups, however, the problem of when the Fourier algebra is weakly amenable or even amenable, turned out to be more intricate.

Amenability of $A(G)$ proved to be a very restrictive property. The difficult implication of Theorem 4.4.7, stating that G must have an abelian subgroup of finite index whenever $A(G)$ is amenable, is due to Forrest and Runde [**86**]. As to weak amenability, the fact that $A(G)$ is weakly amenable provided that G_0, the connected component of the identity of G, is abelian ([**13**, Theorem 4.5]), was also established by Forrest and Runde [**86**].

Already twenty years earlier, Johnson [**143**] proved the surprising fact that $A(SO(3))$ fails to be weakly amenable by constructing a nonzero derivation from $A(SO(3))$ to $VN(SO(3))$. That the same conclusion holds for every nonabelian compact connected Lie group, was observed by Plymen [**238**]. See also [**89**].

Actually, there is strong evidence that this conjecture is true. Recently, Choi and Ghandehari [**29**], exploiting representation theory, have explicitly constructed nonzero derivations from $A(G)$ to $VN(G)$ for such important examples as the $ax+b$-group and the 3-dimensional reduced Heisenberg group. Moreover, these authors succeeded to show that a connected and simply connected Lie group G, for which $A(G)$ is weakly amenable, has to be solvable.

The main result of this chapter, Theorem 4.2.7, is due to Ruan [**245**] where he introduced the operator space structure of the Fourier algebra $A(G)$. This has been an important breakthrough in the study of the subject during the last twenty years. The short and elegant proof for the special case of an SIN-group (Theorem 4.2.9) was found by Spronk [**262**] using the deep fact, established by Takesaki and Tatsuuma [**272**] that closed subgroups of a locally compact group G are sets of spectral synthesis for $A(G)$ (see Theorem 6.1.9).

Theorem 4.2.7 is the operator space analogue of Johnson's result [**141**] that the L^1-group algebra of a locally amenable group is amenable (compare Theorem 4.5.5).

In contrast, the Fourier algebra $A(G)$ is always weakly operator amenable (Theorem 4.3.3). This result was proved by Spronk [**262**]. We conclude these notes by mentioning two further open problems.

(1) Is it true that $B(G)$ is weakly amenable if and only if G is compact and G_0 is abelian (Question 2.10 in [**264**])?

(2) When is $B(G)$ operator amenable? (Question 3.8 in [**264**])?

CHAPTER 5

Multiplier Algebras of Fourier Algebras

An important object associated to any (nonunital) commutative Banach algebra A is the multiplier algebra $M(A)$ of A, that is, the algebra of all bounded linear maps $T : A \to A$ satisfying the equation $T(ab) = aT(b)$ for all $a, b \in A$. When A is faithful, then the map $a \to T_a$, where $T_a(b) = ab$ for $b \in A$, is a continuous embedding of A into $M(A)$.

Let G be a locally compact group. Then $M(A(G))$ consists of all bounded continuous functions u on G such that $uA(G) \subseteq A(G)$, and since $A(G)$ is an ideal in $B(G)$, $B(G)$ embeds continuously into $M(A(G))$. If G is abelian, then as shown by Wendel [**288**], $M(L^1(G)) = M(G)$, and hence $M(A(G)) = B(G)$. It is not difficult to see that this holds true, more generally, when G is amenable. One of the profound achievements in abstract harmonic analysis has been that the converse holds, that is $M(A(G)) = B(G)$ forces G to be amenable. This was shown by Nebbia [**219**] for discrete groups G and by Losert [**201**] for nondiscrete G. We present these results in Sections 5.2 and 5.3, respectively.

A challenging problem arising in this context is to find elements of $M(A(G))\setminus B(G)$ if G is nonamenable. It turned out that there is a certain subalgebra of $M(A(G))$, the algebra $M_{cb}(A(G))$ of all so-called completely bounded multipliers, which is much more accessible than the whole algebra $M(A(G))$. Sections 5.4 and 5.5 are devoted to exploring $M_{cb}(A(G))$ and its close relationship with uniformly bounded, nonunitary representations of the group G.

We have seen in Section 2.7 that $A(G)$ possesses a bounded approximate identity if and only if G is amenable. In view of the importance of the existence of an approximate identity, the question for which nonamenable groups G the Fourier algebra has an approximate identity bounded in a norm weaker than the $A(G)$-norm, has become a major issue in the past. For the class of those connected real Lie groups, the Levi factor of which has finite centre, there is now a complete characterization of the groups G for which $A(G)$ admits a completely bounded approximate identity (compare the references in Section 5.8). In Section 5.6 we show that if Γ is a lattice in a second countable locally compact group G, then $A(G)$ has a *cb*-norm bounded approximate identity if (and only if) $A(\Gamma)$ does so. This is applied in Section 5.7 with $G = SL(2,\mathbb{R})$ and $\Gamma = \mathbb{F}_2$, the free group on two generators, to deduce the existence of such an approximate identity for $A(SL(2,\mathbb{R}))$ after establishing the existence for $A(\mathbb{F}_2)$.

5.1. Multipliers of $A(G)$

To start with, we remind the reader that if A is a faithful commutative Banach algebra, then a mapping $T : A \to A$ is a *multiplier* of A if $T(ab) = aT(b)$ for all $a, b \in A$. If $T \in M(A)$, the algebra of all multipliers of A, then there is a unique bounded continuous function $\widehat{T}$ on the spectrum $\sigma(A)$ of A such that

$$\widehat{T(a)}(\varphi) = \widehat{T}(\varphi)\widehat{a}(\varphi)$$

for all $a \in A$ and $\varphi \in \sigma(A)$. Thus $\widehat{T}\widehat{A} \subseteq \widehat{A}$ and conversely, if A is semisimple, then every bounded continuous function f on $\Delta(A)$ with the property that $f\widehat{A} \subseteq \widehat{A}$ defines a multiplier T_f of A by letting, for $a \in A$, $T_f(a)$ be the unique element of A satisfying $\widehat{T_f(a)}(\varphi) = f(\varphi)\widehat{a}(\varphi)$ for all $\varphi \in \sigma(A)$.

Turning to the Fourier algebra $A(G)$ of a locally compact group G and identifying $\sigma(A(G))$ with G, so that the Gelfand homomorphism of $A(G)$ is simply the identity mapping, we can then define a multiplier of $A(G)$ to be a bounded continuous function f on G with the property that $fA(G) \subseteq A(G)$. The multiplier norm of f is given by

$$\|f\|_{M(A(G))} = \sup\{\|fu\|_{A(G)} : u \in A(G), \|u\|_{A(G)} \leqslant 1\}.$$

With this norm, $M(A(G))$ is a semisimple commutative Banach algebra and $M(A(G))$ contains the Fourier-Stieltjes algebra $B(G)$ because $A(G)$ is an ideal in $B(G)$. However, the norm on $B(G)$ need not coincide with the multiplier norm.

Before attacking the problem of when $M(A(G)) = B(G)$, we collect a number of basic results about $M(A(G))$ which will be used in either this or later sections.

LEMMA 5.1.1. *Let H be a closed subgroup of a locally compact group G. If $u \in M(A(G))$, then $u|_H \in M(A(H))$ and*

$$\|u|_H\|_{M(A(H))} \leqslant \|u\|_{M(A(G))}.$$

PROOF. Given $v \in A(H)$, there exists $w \in A(G)$ with $w|_H = v$. So $(u|_H)v = (uw)|_H$ and $uw \in A(G)$ since $u \in M(A(G))$. This shows that $(u|_H)A(H) \subseteq A(H)$, that is, $u|_H \in M(A(H))$. Since

$$\begin{aligned}
\|(u|_H)v\|_{A(H)} &\leqslant \inf\{\|uw\|_{A(G)} : w \in A(G), w|_H = v\} \\
&\leqslant \|u\|_{M(A(G))} \cdot \inf\{\|w\|_{A(G)} : w \in A(G), w|_H = v\} \\
&= \|u\|_{M(A(G))}\|v\|_{A(H)},
\end{aligned}$$

it follows that $\|u|_H\|_{M(A(H))} \leqslant \|u\|_{M(A(G))}$. □

PROPOSITION 5.1.2. *Let f be a continuous function on a locally compact group G. Then the following four conditions are equivalent.*

(i) *f is a multiplier of $A(G)$.*
(ii) *There exists a (unique) σ-weakly continuous operator M_f on $VN(G)$ such that $M_f(\lambda_G(x)) = f(x)\lambda_G(x)$ for all $x \in G$.*
(iii) *f is bounded on G and there exists a constant $c > 0$ such that $\|\lambda_G(fg)\| \leqslant c\,\|\lambda_G(g)\|$ for all $g \in L^1(G)$.*
(iv) *$fv \in B_\lambda(G)$ for every $v \in B_\lambda(G)$.*

PROOF. (i) ⇒ (ii) Let $m_f : A(G) \to A(G)$ be the mapping $u \to fu$. The adjoint operator M_f of m_f is σ-weakly continuous and by the duality between $VN(G)$ and $A(G)$, $\langle \lambda_G(x), u\rangle = u(x)$ and therefore

$$\langle M_f(\lambda_G(x)), u\rangle = \langle \lambda_G(x), fu\rangle = f(x)u(x) = f(x)\langle \lambda_G(x), u\rangle$$

for all $u \in A(G)$ and $x \in G$. Thus $M_f(\lambda_G(x)) = f(x)\lambda_G(x)$ for all $x \in G$. Since the linear span of the set $\{\lambda_G(x) : x \in G\}$ is σ-weakly dense in $VN(G)$, M_f is the only σ-weakly continuous operator on $VN(G)$ satisfying this formula.

(ii) $\Rightarrow$ (iii) Clearly, for all $x \in G$,

$$|f(x)| = \|f(x)\lambda_G(x)\|_{VN(G)} = \|M_f(\lambda_G(x))\|_{VN(G)}.$$

Since a σ-weakly continuous operator on $VN(G)$ is norm bounded, this shows that f is bounded on G. Thus, if $g \in L^1(G)$ then $fg \in L^1(G)$ and hence

$$\begin{aligned} \langle M_f(\lambda_G(g)), u\rangle &= \int_G g(x)\langle M_f(\lambda_G(x)), u\rangle dx \\ &= \int_G g(x)f(x)\langle \lambda_G(x), u\rangle dx \\ &= \langle \lambda_G(fg), u\rangle \end{aligned}$$

for all $u \in A(G)$. It follows that $\|\lambda_G(fg)\| \leqslant \|M_f\| \cdot \|\lambda_G(g)\|$, so that (iii) holds.

(iii) $\Rightarrow$ (iv) Let $v \in B_\lambda(G)$ and $g \in L^1(G)$. Since $fg \in L^1(G)$, the duality between $C^*_\lambda(G)$ and $B_\lambda(G)$ yields

$$\begin{aligned} \left|\int_G f(x)g(x)v(x)dx\right| &= |\langle \lambda_G(fg), v\rangle| \leqslant \|\lambda_G(fg)\| \cdot \|v\|_{B_\lambda(G)} \\ &\leqslant c\,\|v\|_{B_\lambda(G)}\|\lambda_G(g)\|. \end{aligned}$$

Thus $\lambda_G(g) \to \int_G f(x)g(x)v(x)dx$ is well defined on $\lambda_G(L^1(G))$ and extends to a bounded linear functional on $C^*_\lambda(G)$. Hence there exists $w \in B_\lambda(G)$ with

$$\int_G g(x)f(x)v(x)dx = \int_G g(x)w(x)dx, \quad g \in L^1(G).$$

This implies that $fv = w \in B_\lambda(G)$.

(iv) $\Rightarrow$ (i) Suppose that $fv \in B_\lambda(G)$ for all $v \in B_\lambda(G)$. It follows form the closed graph theorem that the map

$$\overline{m}_f : B_\lambda(G) \to B_\lambda(G), \quad v \to fv,$$

is bounded. For $v \in B_\lambda(G) \cap C_c(G)$, we have

$$fv \in B_\lambda(G) \cap C_c(G) = A(G) \cap C_c(G).$$

Since $A(G) \cap C_c(G)$ is dense in $A(G)$ and the $A(G)$-norm and the $B_\lambda(G)$-norm coincide on $A(G)$, we conclude that $fv \in A(G)$ for every $v \in A(G)$. So f is a multiplier of $A(G)$. □

REMARK 5.1.3. Let $f \in M(A(G))$. Then, according to Proposition 5.1.2, we can associate with f operators $m_f, \overline{m}_f, M_f$, and $\overline{M}_f$ on $A(G), B_\lambda(G), VN(G)$, and $C^*_\lambda(G)$, respectively. Specifically, $m_f(u) = fu$ for $u \in A(G)$, $\overline{m}_f$ is the extension of m_f to $B_\lambda(G)$ given by the same formula, M_f is the adjoint of m_f, and $\overline{M}_f$ is the restriction of M_f to $C^*_\lambda(G)$. It follows from the proof of Proposition 5.1.2 that $(\overline{M}_f)^* = \overline{m}_f$. Thus

$$\|m_f\| \leqslant \|\overline{m}_f\| = \|\overline{M}_f\| \leqslant \|M_f\| = \|m_f\|,$$

and so all four norms are equal to $\|f\|_{M(A(G))}$.

LEMMA 5.1.4. *The unit ball of $M(A(G))$ is $\sigma(L^\infty(G), L^1(G))$-closed in $L^\infty(G)$.*

PROOF. Let $(u_\alpha)_\alpha$ be a net in the unit ball of $M(A(G))$ which converges to some $u \in L^\infty(G)$ in the $\sigma(L^\infty(G), L^1(G))$-topology. Let $f \in L^1(G)$. Then

$$\|\lambda_G(u_\alpha f)\| = \|M_{u_\alpha}(\lambda_G(f))\| \leqslant \|\lambda_G(f)\|,$$

for all α and, for all $\xi, \eta \in L^2(G)$,

$$\begin{aligned}\lim_{\alpha}\langle \lambda_G(u_\alpha f)\xi, \eta\rangle &= \lim_{\alpha}\int_G u_\alpha(x) f(x)(\xi * \widetilde{\eta})(x^{-1})\,dx \\ &= \int_G u(x) f(x)(\xi * \widetilde{\eta})(x^{-1})\,dx \\ &= \langle \lambda_G(uf)\xi, \eta\rangle.\end{aligned}$$

Hence we also have $\|\lambda_G(uf)\| \leqslant \|\lambda_G(f)\|$. The proof of Proposition 5.1.2, (iii) $\Rightarrow$ (iv), shows that for any $v \in B_\lambda(G)$, uv is equal to a function in $B_\lambda(G)$ locally almost everywhere. In particular, u is locally almost everywhere equal to a continuous function w on G. Thus, by (iv) $\Rightarrow$ (i) of Proposition 5.1.2, $w \in M(A(G))$. Moreover, $\|w\|_{M(A(G))} = \|\overline{m}_w\| \leqslant 1$. This proves that the unit ball of $M(A(G))$ is $\sigma(L^\infty(G), L^1(G))$-closed. □

Recall that $B(G)$ is the dual space of $C^*(G)$ and that, as we shall see later, $M(A(G)) = B(G)$ for amenable groups G. We next observe that $M(A(G))$ is a dual Banach space for any locally compact group G.

PROPOSITION 5.1.5. *Let G be an arbitrary locally compact group. If E is the completion of $L^1(G)$ with respect to the norm*

$$\|f\|_E = \sup\left\{\left|\int_G f(x)u(x)dx\right| : u \in M(A(G)), \|u\|_{M(A(G))} \leqslant 1\right\},$$

then $E^* = M(A(G))$. More precisely, every bounded linear functional ϕ on E is of the form

$$\phi(f) = \int_G f(x)u(x)dx, \quad f \in L^1(G),$$

for some $u \in M(A(G))$, and then $\|\phi\| = \|u\|_{M(A(G))}$.

PROOF. It is clear that for every $u \in M(A(G))$, the functional

$$f \to \int_G f(x)u(x)dx, \quad f \in L^1(G),$$

extends to a bounded linear functional ϕ_u on E and that $\|\phi_u\| \leqslant \|u\|_{M(A(G))}$. Conversely, let $\phi \in E^*$ with $\|\phi\| = 1$. Since the restriction of ϕ to $L^1(G)$ is a bounded linear functional on $L^1(G)$, there exists $v \in L^\infty(G)$ such that

$$\phi(f) = \int_G f(x)v(x)dx, \quad f \in L^1(G).$$

Since $|\int_G f(x)v(x)dx| \leqslant 1$ for all $f \in L^1(G)$ with $\|f\|_E \leqslant 1$, it follows from the Hahn-Banach theorem and the definition of $\|\cdot\|_E$ that v is contained in the $\sigma(L^\infty, L^1)$-closure of the unit ball of $M(A(G))$. Therefore, by Lemma 5.1.4, v is locally almost everywhere equal to some $u \in M(A(G))$, and $\|u\|_{M(A(G))} \leqslant 1$. This finishes the proof. □

We now proceed to show that if G is an amenable locally compact group, then $M(A(G)) = B(G)$ and the multiplier norm and the $B(G)$-norm coincide. For $v \in B(G)$, let $m_v(u) = vu$ for $u \in A(G)$.

LEMMA 5.1.6. *If $v \in B(G)$, then $\|m_v\|_{M(A(G))} \leqslant \|v\|_{B(G)}$ and equality holds if G is amenable.*

PROOF. Notice first that $m_v \in M(A(G))$ since $A(G)$ is an ideal in $B(G)$ and

$$\|m_v\|_{M(A(G))} = \sup\{\|vu\|_{A(G)} : u \in A(G)_1\} \leqslant \|v\|_{B(G)}.$$

Now assume that G is amenable. Since $B(G) = C^*(G)^*$ and $C_c(G)$ is dense in $C^*(G)$, given $\epsilon > 0$ there exists $f \in C_c(G)$ with $\|f\|_{C^*} = 1$ and $|\langle v, f\rangle| \geqslant \|v\|_{B(G)} - \epsilon$. As G is amenable, there exists $u \in A(G)$ such that $u = 1$ on $\operatorname{supp} f$ and $\|u\|_{A(G)} \leqslant 1 + \epsilon$. It follows that

$$\|v\|_{B(G)} - \epsilon \leqslant |\langle v, f\rangle| = |\langle v, uf\rangle| = |\langle uv, f\rangle| \leqslant \|uv\|_{A(G)},$$

and since $\|u\|_{A(G)} \leqslant 1 + \epsilon$, this implies that

$$\|m_v\|_{M(A(G))} \geqslant \frac{\|v\|_{B(G)} - \epsilon}{1 + \epsilon}.$$

Since $\epsilon > 0$ is arbitrary, we conclude that $\|m_v\|_{M(A(G))} \geqslant \|v\|_{B(G)}$. □

LEMMA 5.1.7. *Let G be an amenable locally compact group and let v be a multiplier of $A(G)$. Then $v \in B(G)$.*

PROOF. By the closed graph theorem, there exists $c > 0$ such that $\|vu\|_{A(G)} \leqslant c\,\|u\|_{A(G)}$ for all $u \in A(G)$. Let $s_1, \ldots, s_n \in G$ and $\epsilon > 0$. Since G is amenable, there exists $u \in A(G)$ such that $u(s_j) = 1$ for $j = 1, \ldots, n$ and $\|u\|_{A(G)} \leqslant 1 + \epsilon$. For any $c_1, \ldots, c_n \in \mathbb{C}$ this implies

$$\begin{aligned} \left|\sum_{j=1}^n c_j v(s_j)\right| &= \left|\sum_{j=1}^n c_j (vu)(s_j)\right| = \left|\left\langle \sum_{j=1}^n c_j \lambda(s_j), vu\right\rangle\right| \\ &\leqslant \|vu\|_{A(G)} \left\|\lambda\left(\sum_{j=1}^n c_j \delta_{s_j}\right)\right\|_{VN(G)} \\ &\leqslant c(1+\epsilon) \left\|\lambda\left(\sum_{j=1}^n c_j \delta_{s_j}\right)\right\|_{VN(G)}. \end{aligned}$$

Since $\epsilon > 0$ is arbitrary,

$$\left|\sum_{j=1}^n c_j v(s_j)\right| \leqslant c \left\|\lambda\left(\sum_{j=1}^n c_j \delta_{s_j}\right)\right\|_{VN(G)}.$$

Let i denote the identity map $G_d \to G$, where G_d denotes the group G with the discrete topology. Since the finite linear combinations of Dirac functions are dense in $C^*(G_d)$, this inequality shows that $v \in B_{\lambda \circ i}(G_d)$. Since v is continuous, it follows that $v \in B_\lambda(G)$. □

The following theorem is an immediate consequence of Lemma 5.1.6 and Lemma 5.1.7

THEOREM 5.1.8. *Let G be an amenable locally compact group. Then*

$$M(A(G)) = B(G) \quad \text{and} \quad \|v\|_{B(G)} = \|v\|_{M(A(G))}$$

for every $v \in B(G)$.

PROPOSITION 5.1.9. *Let $u \in B(G)$ be such that M_u is an isometry of $A(G)$. Then $u = \alpha\chi$ for some character χ of G and $\alpha \in \mathbb{C}$ with $|\alpha| = 1$.*

PROOF. Observe first that $u(x) \neq 0$ for every $x \in G$. Indeed, for any $v \in A(G)$ and $x \in G$, we have

$$\|(L_x u)v\| = \|L_x(uL_{x^{-1}}v)\| = \|uL_{x^{-1}}v\| = \|v\|,$$

so that $L_x u$ is also an isometric multiplier. Therefore, we only have to verify that $u(e) \neq 0$. For a contradiction, assume that $u(e) = 0$ and let $v \in P^1(G) \cap A(G)$. Then, given any $0 < \epsilon < 1$, by Corollary 2.5.12 there exists $w \in P^1(G) \cap A(G)$ such that $\|uvw\| \leqslant \epsilon$. This contradicts $1 = \|vw\| = \|uvw\|$.

We claim next that $M_u : A(G) \to A(G)$ is surjective. Since M_u is an isometry and a multiplier, $I = M_u(A(G))$ is a closed ideal of $A(G)$. Since $u(x) \neq 0$ for all $x \in G$ and $A(G)(x) \neq \{0\}$, the hull of I is empty and hence $I = A(G)$. Now, as M_u is surjective, $M_{u^{-1}}$ is also a multiplier and then $\|u^{-1}\| = \|M_{u^{-1}}\| = 1$ since $\|M_u\| = 1$. Now $\|u\|_\infty \leqslant 1$ and $\|u^{-1}\|_\infty \leqslant 1$ imply that $|u(x)| = 1$ for every $x \in G$.

Finally, there exist a unitary representation π of G and $\xi, \eta \in \mathcal{H}(\pi)$ such that ξ is cyclic, $\|\xi\| = \|\eta\| = 1$ and $u(x) = \langle \pi(x)\xi, \eta \rangle$ for all $x \in G$. It follows that

$$1 = |u(x)| = |\langle \pi(x)\xi, \eta \rangle| \leqslant \|\xi\| \cdot \|\eta\| = \|u\| = 1.$$

Thus $\pi(x)\xi \in \mathbb{C}\eta$ for all $x \in G$, whence $\mathcal{H}(\pi)$ is 1-dimensional. Thus $\xi = \alpha\eta$ for some $\alpha \in \mathbb{T}$ and then $u(x) = \alpha\chi(x)$ for some character χ of G. □

The converse to Theorem 5.1.8 also holds, that is, the condition that every multiplier of the Fourier algebra of a locally compact group G is given by an element of $B(G)$, forces G to be amenable. This result, which was established by Nebbia [**219**] for discrete groups and by Losert [**201**] for nondiscrete groups, is no doubt one of the highlights in the theory of multipliers of Fourier algebras and its proof requires fairly sophisticated arguments. Currently, there seems to be no approach covering both cases simultaneously. Even though some of Nebbia's tools are valid for nondiscrete groups as well, the essential steps require discreteness of G. On the other hand, Losert's treatment supposes G to be nondiscrete. As a consequence of this situation, the two cases are presented separately, the discrete one in the next section and the nondiscrete one in Section 5.3. In both cases we follow closely the original articles.

Slightly more general, the final result will be the following theorem.

THEOREM 5.1.10. *For a locally compact group G, the following statements are equivalent.*

(i) *G is amenable.*
(ii) $M(A(G)) = B(G)$.
(iii) *$A(G)$ is closed in $M(A(G))$.*

REMARK 5.1.11. (1) Condition (iii) means that there exists a constant $c > 0$ such that, for all $u \in A(G)$,

$$\sup\{\|uv\|_{A(G)} : v \in A(G)_1\} \geqslant c\,\|u\|_{A(G)}.$$

By the open mapping theorem, this in turn is equivalent to $A(G)$ being closed in $M(A(G))$.

(2) If G is amenable, $A(G)$ has a bounded approximate identity of norm bound 1, so that in (1) the constant c can be chosen to be equal to 1.

(3) To prove Theorem 5.1.10, it only remains to show the implication (iii) $\Rightarrow$ (i). In fact, (i) $\Rightarrow$ (ii) holds by Theorem 5.1.8 and (ii) $\Rightarrow$ (iii) follows since condition (ii) implies that the norm on $B(G)$ is equivalent to the multiplier norm.

We close this section with a fairly simple result which might support the expectation that Theorem 5.1.10 is true.

PROPOSITION 5.1.12. *Let G be a nonamenable locally compact group and suppose that $A(G)$ admits a sequential approximate identity . Then $M(A(G))$ properly contains $B(G)$.*

PROOF. Let $(u_n)_n$ be a sequence in $A(G)$ such that $\|u_n v - v\| \to 0$ for every $v \in A(G)$. Then, since G is not amenable, it follows from Theorem 2.7.2 that $\sup_{n\in\mathbb{N}} \|u_n\| = \infty$. On the other hand, as $u_n v \to v$ for every $v \in A(G)$, the sequence of multipliers M_{u_n} of $A(G)$ is pointwise bounded. The uniform boundedness principle now asserts that the sequence of multiplier norms $\|M_{u_n}\|$, $n \in \mathbb{N}$, is bounded. This shows that on $B(G) \subseteq M(A(G))$ the $B(G)$-norm and the multiplier norm are not equivalent. Thus $B(G)$ is not closed in $M(A(G))$ and, in particular, not equal to $M(A(G))$. □

There are several nonamenable locally compact groups, for instance $SL(2,\mathbb{R})$, $SL(2,\mathbb{Z})$ and free groups (see Section 5.7), the Fourier algebras of which possess a sequential (multiplier bounded) approximate identity.

THEOREM 5.1.13. *For any locally compact group G, we have $UC(\widehat{G}) = A(G) \cdot VN(G)$ if and only if G is amenable.*

PROOF. Assume first that G is amenable and view $VN(G)$ as an $A(G)$-Banach module. Since $A(G)$ has a bounded approximate identity (Theorem 2.7.2), it follows from the Cohen's factorization theorem (see [**126**, p.268]) that $A(G) \cdot VN(G)$ is closed in $VN(G)$ and hence equal to $UC(\widehat{G})$.

To attack the much more complicated reverse implication, let Φ denote the map from the projective tensor product $A(G)\widehat{\otimes} VN(G)$ into $UC(\widehat{G})$ defined by $\Phi(u \otimes T) = u \cdot T$, $u \in A(G)$, $T \in VN(G)$. By hypothesis, Φ is surjective and hence the algebras $A(G)\widehat{\otimes} VN(G)$ and $UC(\widehat{G})$ are topologically isomorphic. Thus the adjoint map $\Phi^* : UC(\widehat{G})^* \to [A(G)\widehat{\otimes} VN(G)]^*$ is bounded below, that is, there exists a constant $c > 0$ such that $\|\Phi^*(\varphi)\| \geqslant c\|\varphi\|$ for all $\varphi \in UC(\widehat{G})^*$.

Now, view $A(G)$ as a subspace of $VN(G)^*$ and, for $u \in A(G)$, let $r(u)$ denote the restriction of the associated functional to the subspace $UC(\widehat{G})$. Then, for each $T \in VN(G)$ and $v \in A(G)$,

$$\langle \Phi^*(r(u)), v \otimes T\rangle = \langle r(u), \Phi(v \otimes T)\rangle = \langle u, v \cdot T\rangle.$$

We claim that r is an isometry. To see this, note that $C^*_\lambda(G)$ is ultraweakly (or weak*) dense in $VN(G)$, and hence the unit ball $C^*_\lambda(G)_1$ is ultraweakly dense in the unit ball $VN(G)_1$. Since $C^*_\lambda(G) \subseteq UC(\widehat{G})$, it follows that $\|r(u)\| = \|u\|$ for all $u \in A(G)$. So

$$\|\Phi^* \circ r(u)\| \geqslant c\,\|r(u)\| = c\|u\|$$

and therefore $\Phi^* \circ r(A(G))$ is closed in $[A(G)\widehat{\otimes} VN(G)]^*$.

Recall next that $[A(G)\widehat{\otimes} VN(G)]^*$ is canonically isometrically isomorphic to $\mathcal{B}(A(G), VN(G)^*)$, the isomorphism Ψ being given by, for $\varphi \in [A(G)\widehat{\otimes} VN(G)]^*$,

$$\langle \Psi(\varphi)(u), T\rangle = \langle \varphi, u \otimes T\rangle, \ u \in A(G), T \in VN(G).$$

Let $\Omega = \Psi \circ \Phi^* \circ r : A(G) \to \mathcal{B}(A(G), VN(G)^*)$. Then $\Omega(A(G))$ is closed in $\mathcal{B}(A(G), VN(G)^*)$ by what we have seen above. Moreover, for $u, v \in A(G)$ and

$T \in VN(G)$,

$$\langle \Omega(v)(u), T \rangle = \langle \Phi^*(r(u)), u \otimes T \rangle = \langle vu, T \rangle.$$

Thus Ω equals the multiplication operator on $A(G)$ defined by v. It follows that $A(G)$ is closed in $M(A(G))$, and consequently G must be amenable by Theorem 5.1.10. □

5.2. $M(A(G)) = B(G)$ implies amenability of G: The discrete case

If A and B are commutative Banach algebras and B is a C-module for some subalgebra C of A, then one can introduce the notion of a C-multiplier from A into B. However, we refrain from doing so and instead restrict to two special cases which play a role in the sequel. We start with an arbitrary locally compact group G. Later, G will be assumed to be discrete.

DEFINITION 5.2.1. A bounded linear operator ϕ from $C_0(G)$ into $VN(G)$ is called a *multiplier of* $C_0(G)$ *into* $VN(G)$ if $\phi(uf) = u \cdot \phi(f)$ for all $u \in A(G)$ and $f \in C_0(G)$. Recall that the module action of $A(G)$ on $VN(G)$ is defined by $\langle u \cdot T, v \rangle = \langle T, uv \rangle$ for $u, v \in A(G)$ and $T \in VN(G)$. Let $M(C_0(G), VN(G))$ denote the linear space of all such multipliers. It is clear that equipped with the operator norm, $M(C_0(G), VN(G))$ is a Banach space.

DEFINITION 5.2.2. A bounded linear operator Γ from $A(G)$ into the space $M(G)$ of bounded regular Borel measures on G is called a *multiplier of* $A(G)$ *into* $M(G)$ if $\Gamma(uv) = u\Gamma(v)$ holds for all $u, v \in A(G)$. Endowed with the operator norm, the space $M(A(G), M(G))$ is a Banach space. We denote by $M(A(G), L^1(G))$ the closed subspace of $M(A(G), M(G))$ which consists of all Γ such that $\Gamma(u) \in L^1(G)$ for all $u \in A(G)$.

The following simple lemma shows that the spaces of multipliers defined above are isomorphic.

LEMMA 5.2.3. *The spaces* $M(A(G), M(G))$ *and* $M(C_0(G), VN(G))$ *are isometrically isomorphic, the isomorphism* Φ *being given by*

$$\langle f, \Gamma(u) \rangle = \langle \Phi(\Gamma)(f), u \rangle$$

for $u \in A(G), f \in C_0(G)$ *and* $\Gamma \in M(A(G), M(G))$.

PROOF. For $\Gamma \in M(A(G), M(G))$, $f \in C_0(G)$ and $u \in A(G)$, we have

$$|\langle \Phi(\Gamma)(f), u \rangle| = |\langle f, \Gamma(u) \rangle| \leqslant \|\Gamma\| \cdot \|u\| \cdot \|f\|.$$

Thus $\Phi(\Gamma)(f) \in VN(G)$, and $\Phi(\Gamma)$ is a bounded linear operator from $C_0(G)$ into $VN(G)$. Moreover, for $v \in A(G)$,

$$\begin{aligned} \langle \Phi(\Gamma)(vf), u \rangle &= \langle vf, \Gamma(u) \rangle = \langle f, v\Gamma(u) \rangle \\ &= \langle \Phi(\Gamma)(f), uv \rangle = \langle v \cdot \Phi(\Gamma)(f), u \rangle. \end{aligned}$$

This shows that $\Phi(\Gamma) \in M(C_0(G), VN(G))$. Given $T \in M(C_0(G), VN(G))$, for each u, the assignment $f \to \langle T(f), u \rangle$ defines an element $\Gamma(u)$ of $M(G)$ and

$$\begin{aligned} \|\Gamma(u)\| &= \sup\{|\langle f, \Gamma(u) \rangle| : f \in C_0(G), \|f\|_\infty \leqslant 1\} \\ &= \sup\{|\langle T(f), u \rangle| : f \in C_0(G), \|f\|_\infty \leqslant 1\} \\ &\leqslant \|u\| \cdot \sup\{\|T(f)\| : f \in C_0(G), \|f\|_\infty \leqslant 1\}. \end{aligned}$$

Furthermore,

$$\begin{aligned}\langle f, \Gamma(uv)\rangle &= \langle T(f), uv\rangle = \langle u \cdot T(f), v\rangle = \langle f, \Gamma(uv)\rangle \\ &= \langle T(uf), v\rangle = \langle \Gamma(v), uf\rangle = \langle u\Gamma(v), f\rangle.\end{aligned}$$

Thus Γ is an element of $M(A(G), M(G))$ satisfying $\Phi(\Gamma) = T$ and $\|\Gamma\| \leqslant \|T\|$. Combining these facts shows that Φ is an isometric isomorphism. □

If $\mu \in M(G)$, then $\Gamma_\mu : A(G) \to M(G)$, $u \to \mu u$ defines an element of $M(A(G), M(G))$ and, using that $\|u\|_\infty \leqslant \|u\|_{A(G)}$ for $u \in A(G)$, it is easily seen that $\|\Gamma_\mu\| \leqslant \|\mu\|$. Conversely, every $\Gamma \in M(A(G)), M(G))$ is given by a Radon measure which, however, in general is unbounded. Indeed, let $T = \Phi(\Gamma) \in M(C_0(G), VN(G))$ and $u \in A(G) \cap C_c(G)$, and choose $v \in A(G)$ such that $v = 1$ on $\operatorname{supp} u$. Then

$$\begin{aligned}|\langle T, u\rangle| &= |\langle T, uv\rangle| = |\langle u \cdot T, v\rangle| = |\langle u, \Gamma(v)\rangle| \\ &\leqslant \|u\|_\infty \|v\|_{A(G)} \|\Gamma\|.\end{aligned}$$

We shall see in Theorem 5.2.5 below that every $\Gamma \in M(A(G), M(G))$ is given by a bounded Radon measure if and only if G is amenable.

Let E be the subspace of $C_0(G)$ consisting of all functions f of the form $f(x) = \sum_{j=1}^\infty u_j(x)g_j(x)$, where $u_j \in A(G)$ and $g_j \in C_0(G)$ are such that $\sum_{j=1}^\infty \|u_j\|_{A(G)}\|g_j\|_\infty < \infty$. We define a norm on E by setting

$$\|f\|_E = \inf\left\{\sum_{j=1}^\infty \|u_j\|_{A(G)}\|g_j\|_\infty : f(x) = \sum_{j=1}^\infty u_j(x)g_j(x) \text{ for all } x \in G\right\},$$

where the infimum is taken over all such representations of f. It is easy to verify that, with this norm, E is a Banach space.

LEMMA 5.2.4. *The space $M(C_0(G), VN(G))$ is isometrically isomorphic to the dual space of E, the duality being given by the formula*

$$\langle \phi_T, f\rangle = \sum_{j=1}^\infty \langle T(g_j), u_j\rangle$$

for $T \in M(C_0(G), VN(G))$ and $f = \sum_{j=1}^\infty u_j g_j$.

PROOF. Let $f = \sum_{j=1}^\infty u_j g_j \in E$, so that $\sum_{j=1}^\infty \|u_j\|_{A(G)}\|g_j\|_\infty < \infty$. We first show that if $f = 0$ and $T \in M(C_0(G), VN(G))$, then $\sum_{j=1}^\infty \langle T(g_j), u_j\rangle = 0$. For this, let $\epsilon > 0$ be given and choose $N \in \mathbb{N}$ such that $\sum_{j=N+1}^\infty \|u_j\|_{A(G)}\|g_j\|_\infty \leqslant \epsilon$. For any compact and symmetric subset C of G with $|C| > 0$, let $v_C = |C|^{-1}(1_{C^2} * 1_C) \in A(G)$. Then $0 \leqslant v_C \leqslant 1$ and $v_C|_C = 1$. Therefore, the functions v_C form an approximate identity for $C_0(G)$. Thus there exists $v \in A(G)$ such that

$$\sum_{j=1}^N \|g_j - vg_j\|_\infty \|u_j\|_{A(G)} \leqslant \epsilon.$$

We next observe that $\sum_{j=1}^\infty \langle T(vg_j), u_j\rangle = 0$. Indeed, since

$$\langle T(vg_j), u_j\rangle = \langle vT(g_j), u_j\rangle = \langle T(g_j), vu_j\rangle = \langle u_jT(g_j), v\rangle = \langle T(u_jg_j), v\rangle,$$

we get, by continuity of T,

$$\sum_{j=1}^\infty \langle T(vg_j), u_j\rangle = \sum_{j=1}^\infty \langle T(u_jg_j), v\rangle = \langle T, f\rangle = 0.$$

It now follows that

$$\begin{aligned}\left|\sum_{j=1}^{\infty}\langle T(g_j), u_j\rangle\right| &= \left|\sum_{j=1}^{\infty}\langle T(g_j - vg_j), u_j\rangle\right| \\ &\leqslant \|T\| \cdot \sum_{j=1}^{N} \|g_j - vg_j\|_\infty \|u_j\|_{A(G)} \\ &\quad + \|T\|(1 + \|v\|_\infty) \cdot \sum_{j=N+1}^{\infty} \|g_j\|_\infty \|u_j\|_{A(G)} \\ &\leqslant \epsilon \cdot \|T\|(2 + \|v\|_\infty).\end{aligned}$$

Since $\epsilon > 0$ was arbitrary, it follows that $\sum_{j=1}^{\infty}\langle T(g_j), u_j\rangle = 0$.

This shows that the value $\sum_{j=1}^{\infty}\langle T(g_j), u_j\rangle$ does not depend on the representation of $f \in E$. To each $T \in M(C_0(G), VN(G))$, we can therefore associate a linear functional F_T on E by setting

$$\langle F_T, f\rangle = \sum_{j=1}^{\infty}\langle T(g_j), u_j\rangle, \quad f = \sum_{j=1}^{\infty} u_j g_j \in E.$$

It is clear that F_T is bounded and $\|F_T\| \leqslant \|T\|$ since

$$|\langle F_T, f\rangle| \leqslant \|T\| \cdot \sum_{j=1}^{\infty} \|u_j\|_{A(G)} \|g_j\|_\infty$$

for any representation $f = \sum_{j=1}^{\infty} u_j g_j$ of f. To see the converse $\|F_T\| \geqslant \|T\|$, we simply choose $g \in C_0(G)$ with $\|g\|_\infty \leqslant 1$ and $\|T(g)\| \geqslant \|T\| - \epsilon$ and then $v \in A(G)$ with $\|v\|_{A(G)} \leqslant 1$ and $|\langle T(g), v\rangle| \geqslant \|T(g)\| - \epsilon$. Then

$$|\langle F_T, vg\rangle| = |\langle T(g), v\rangle| \geqslant \|T(g)\| - \epsilon \geqslant \|T\| - 2\epsilon.$$

Thus $T \to F_T$ is an embedding of $M(C_0(G), VN(G))$ into E^*.

Finally, every $F \in E^*$ arises in this manner. In fact, since

$$|\langle F, ug\rangle| \leqslant \|F\| \cdot \|ug\|_E \leqslant \|F\| \cdot \|u\|_{A(G)} \|g\|_\infty.$$

for each $g \in C_0(G)$ and $u \in A(G)$, the assignment $u \to \langle F, ug\rangle$ defines an element $T(g) \in A(G)^* = VN(G)$. Moreover,

$$\langle T(vg), u\rangle = \langle F, uvg\rangle = \langle T(g), uv\rangle = \langle v \cdot T(g), u\rangle$$

for all $u, v \in A(G)$. Consequently, $T \in M(C_0(G), VN(G))$ and $F = F_T$. □

THEOREM 5.2.5. *For any locally compact group G, the following four conditions are equivalent.*

(i) *G is amenable.*
(ii) *For every $f \in C_0(G)$ there exist $u \in A(G)$ and $g \in C_0(G)$ such that $f = ug$.*
(iii) *$M(A(G), M(G)) = M(G)$.*
(iv) *$M(A(G), L^1(G)) = L^1(G)$.*

PROOF. (i) $\Rightarrow$ (ii) Since $A(G)$ has a bounded approximate identity and $C_0(G)$ is a Banach $A(G)$-module, $A(G)C_0(G)$ is closed in $C_0(G)$ by the of Cohen-Hewitt factorization theorem [**126**, p.268]. On the other hand, $A(G)C_c(G) = C_c(G)$ since for any compact subset C of G there exists $v \in A(G)$ such that $v = 1$ on C. So $A(G)C_0(G)$ is dense in $C_0(G)$, and hence (ii) follows.

(ii) $\Rightarrow$ (iii) If (ii) holds, then the space E defined above coincides with $C_0(G)$ and Lemma 5.2.3 shows that $M(A(G), M(G)) = M(G)$ using that $M(A(G), M(G))$ is isometrically isomorphic to $M(C_0(G), VN(G))$ (Lemma 5.2.3).

(iii) $\Rightarrow$ (iv) If T is an $A(G)$-multiplier from $A(G)$ into $L^1(G)$, then by (iii) there exists $\mu \in M(G)$ such that $T(u) = u\mu$ and $u\mu \in L^1(G)$ for all $u \in A(G)$. However, this clearly forces μ to be absolutely continuous with respect to Haar measure.

(iv) $\Rightarrow$ (i) By Proposition 1.8.20, it suffices to prove that $\|\lambda(f)\| = \|f\|_1$ for every nonnegative function f in $L^1(G)$. To that end, observe first that, for any $f \in L^1(G)$, $u \in A(G)$ and $\xi \in L^2(G)$,

$$\begin{aligned} \|\lambda(uf)\xi\|_2 &= \|(uf) * \xi\|_2 \leqslant \| \, |uf| * |\xi| \, \|_2 \\ &\leqslant \| \, \|u\|_\infty |f| * |\xi| \, \|_2 \\ &\leqslant \|u\|_\infty \|\lambda(|f|)\| \cdot \|\xi\|_2. \end{aligned}$$

Thus $\|u \cdot \lambda(f)\| = \|\lambda(uf)\| \leqslant \|u\|_\infty \|\lambda(|f|)\|$ and hence $\|u \cdot \lambda(f)\| \leqslant \|u\|_\infty \|\lambda(f)\|$ for any nonnegative $f \in L^1(G)$ and all $u \in A(G)$.

Now recall that the two multiplier algebras $M(A(G), M(G))$ and $M(C_0(G), VN(G))$ are isometrically isomorphic and that an ismomorphism is given by the multiplier $T_f : C_0(G) \to VN(G)$, where for $g \in C_0(G)$, $T_f(g)$ is given by $\langle T_f(g), u\rangle = \langle M_f(u), g\rangle$, that is, $\langle uf, g\rangle = \langle T_f(g), u\rangle = \langle u \cdot \lambda(f), g\rangle$. Then $T_f(u) = u \cdot \lambda(f)$ for $u \in A(G)$ since

$$\begin{aligned} \langle T_f(u), v\rangle &= \langle M_f(v), u\rangle = \langle fv, u\rangle = \langle f, uv\rangle \\ &= \langle \lambda(f), uv\rangle = \langle u \cdot \lambda(f), v\rangle. \end{aligned}$$

Since $A(G)$ is uniformly dense in $C_0(G)$, it follows that

$$\begin{aligned} \|M_f\| &= \sup\{\|T_f(u)\| : u \in C_0(G), \|u\|_\infty \leqslant 1\} \\ &= \sup\{\|u \cdot \lambda(f)\| : u \in A(G), \|u\|_\infty \leqslant 1\} \end{aligned}$$

and hence, whenever $f \geqslant 0$, $\|M_f\| \leqslant \|\lambda(f)\|$. On the other hand, since $M(A(G), L^1(G)) = L^1(G)$, by the closed graph theorem there exists a constant $c > 0$ such that $\|f\|_1 \leqslant c\|M_f\|$ for all $f \in L^1(G)$ and hence $\|f\|_1 \leqslant c\|\lambda(f)\|$ whenever $f \geqslant 0$. Now, since $f \geqslant 0$,

$$\begin{aligned} \|f * f^*\|_1 &= \int_G \int_G f(y) f^*(y^{-1}x) dy dx \\ &= \int_G \int_G f(y) f(x^{-1}y) \Delta(x^{-1}y) dx dy \\ &= \int_G \int_G f(y) \left(\int_G f(xy)\Delta(y) dx \right) dy = \|f\|_1^2. \end{aligned}$$

Thus, for any nonnegative f in $L^1(G)$,

$$\|f\|_1 = \|f * f^*\|_1^{1/2} \leqslant c^{1/2} \lambda(f * f^*)^{1/2} = c^{1/2} \|\lambda(f)\| \leqslant c^{1/2} \|f\|_1.$$

By induction we obtain $\|f\|_1 \leqslant c^{1/2n} \|\lambda(f)\|$ for every $n \in \mathbb{N}$, and hence

$$\|f\|_1 \leqslant \lim_{n \to \infty} c^{1/2n} \|\lambda(f)\| = \|\lambda(f)\|.$$

So we have seen that $\|f\|_1 = \|\lambda(f)\|$ for every nonnegative $f \in L^1(G)$, and this implies that G is amenable, as pointed out above. $\square$

From now on we only consider discrete groups G. Then $M(G) = l^1(G)$ and $M(A(G), l^1(G))$ can be identified with the space of all functions f on G such that $fu \in l^1(G)$ for all $u \in A(G)$. Indeed, let $T \in M(A(G), l^1(G))$ be given and for every $x \in G$, choose $u_x \in A(G)$ with $u_x(x) \neq 0$. Define $f : G \to \mathbb{C}$ by $f(x) = T(u_x)(x)u_x(x)$, $x \in G$. Then, for any $v \in A(G)$ and $x \in G$,

$$u_x(x)f(x)v(x) = T(u_x)(x)v(x) = T(u_x v)(x) = u_x(x)T(v)(x)$$

and hence $T(v)(x) = f(x)v(x)$ since $u_x(x) \neq 0$.

Define L to be the completion of the space of finitely supported functions on G with the norm $\|f\|_L = [(f * f^*)^2(e)]^{1/4}$, where the power is meant as convolution power. The following lemma (actually, a somewhat more general version of it) was shown in [**235**].

LEMMA 5.2.6. *Let G be a discrete group and let f be a function on G with finite support F. Then there exists a function u on G, which is of absolute value one on F, such that*

$$\|uf\|_L \leqslant 2\sqrt{2}\,\|f\|_2.$$

PROOF. Let $F = \{x_1, \ldots, x_n\}$ and $f = \sum_{j=1}^n c_j \delta_{x_j}$, and suppose first that $c_j \in \mathbb{R}$ for all j. Let $r_1, \ldots, r_n$ be different Rademacher functions on the interval $[0,1]$ and for each $t \in [0,1]$, define a function u_t on G by

$$u_t = \sum_{j=1}^n r_j(t)\delta_{x_j}.$$

Then, by definition of the norm $\|\cdot\|_L$ and of u_t,

$$\begin{aligned}
\int_0^1 \|u_t f\|_L^4 dt &= \int_0^1 [(u_t f) * (u_t f)^*]^2(e) dt \\
&= \int_0^1 \left(\sum_{j_1,\ldots,j_4=1}^n \prod_{i=1}^4 c_{j_i} r_{j_i}(t) \delta_e(x_{j_1} x_{j_2}^{-1} x_{j_3} x_{j_4}^{-1}) \right) dt \\
&= \sum_{j_1,\ldots,j_4=1}^n \left(\prod_{i=1}^4 c_{j_i} \right) \delta_e(x_{j_1} x_{j_2}^{-1} x_{j_3} x_{j_4}^{-1}) \int_0^1 \prod_{i=1}^4 r_{j_i}(t)\, dt.
\end{aligned}$$

Let Q denote the set of all $q = (q_1, \ldots, q_4)$ in $\{1, \ldots, n\}^4$ such that

$$\int_0^1 \left(\prod_{i=1}^4 r_{j_i}(t) \right) dt \neq 0.$$

Since the products of Rademacher functions form an orthonormal system, for each $q \in Q$ there exists a permutation σ_q of $q = \{q_1, \ldots, q_4\}$ such that $\sigma_q(q) =$

$(p_1, p_2, p_1, p_2) \in Q$. Thus, for every $q \in Q$, we have $\prod_{i=1}^4 c_{q_i} \geqslant 0$. This implies that

$$\begin{aligned}
\int_0^1 \|u_t f\|_L^4 dt &= \sum_{q\in Q}\left(\prod_{i=1}^4 c_{q_i}\right)\delta_e(x_{q_1}x_{q_2}^{-1}x_{q_3}x_{q_4}^{-1}) \leqslant \sum_{q\in Q}\left(\prod_{i=1}^4 c_{q_i}\right) \\
&= \sum_{q\in Q}\left(\prod_{i=1}^4 c_{q_i}\right)\cdot\int_0^1\prod_{i=1}^4 r_{q_i}(t)dt \\
&= \sum_{j_1,\dots,j_4=1}^{n}\left(\prod_{i=1}^4 c_{j_i}\right)\cdot\int_0^1\prod_{i=1}^4 r_{j_i}(t)dt \\
&\leqslant \int_0^1\left|\sum_{j_1,\dots,j_4=1}^{n}\left(\prod_{i=1}^4 c_{j_i}\right)\left(\prod_{i=1}^4 r_{j_i}(t)\right)\right| dt \\
&\leqslant \int_0^1\left|\sum_{j=1}^{n} c_j r_j(t)\right|^4 dt.
\end{aligned}$$

By a well-known result about Rademacher functions [**295**, Theorem (8.4)], it follows that

$$\int_0^1 \|u_t f\|_L^4 dt \leqslant 4\left(\sum_{j=1}^n c_j^2\right)^2 = 4\,\|f\|_2^4.$$

Now allow f to be complex-valued, say $f = f_1 + if_2$, where f_1 and f_2 are real-valued. Then the preceding inequality yields

$$\begin{aligned}
\int_0^1 \|u_t f\|_L^4 dt &\leqslant \int_0^1 (\|u_t f_1\|_L + \|u_t f_2\|_L)^4\, dt \\
&\leqslant \int_0^1 (2\cdot\max\{\|u_t f_1\|_L, \|u_t f_2\|_L)^4\, dt \\
&\leqslant 2^4\cdot 4\cdot\max\{\|f_1\|_2^4, \|f_2\|_2^4\} \\
&\leqslant 2^6\|f\|_2^4.
\end{aligned}$$

This implies that there exists at least one $t_0 \in [0, 1]$ such that

$$\|u_{t_0} f\|_L \leqslant 2\sqrt{2}\,\|f\|_2.$$

Since u_{t_0} is of absolute value one on F, the proof is complete. □

LEMMA 5.2.7. *Let G be a discrete group. Then $M(\ell^\infty(G), B(G)) = \ell^2(G)$.*

PROOF. Note first that, since G is discrete, $\ell^2(G)B(G) \subseteq A(G)$ and hence $\ell^2(G) \subseteq M(\ell^\infty(G), B(G))$. To prove the reverse inclusion, let $v \in M(\ell^\infty(G), B(G))$ and for any finite subset F of G, let $v_F = 1_F v \in A(G)$. Then, denoting by $|\cdot|$ the multiplier norm on $M(\ell^\infty(G), B(G))$,

$$\|v_F\|_{A(G)} = \|1_F v\|_{B(G)} \leqslant |v|\cdot\|1_F\|_\infty = |v|.$$

By Lemma 5.2.6, there exists a function u of absolute value one on F such that

$$\|uv_F\|_L \leqslant 2\sqrt{2}\,\|v_F\|_2.$$

Now, for any finitely supported function f on G,

$$\|f\|_2 \leqslant \|f\|_{A(G)}^{1/3}\|f\|_L^{2/3}.$$

It follows that for some function u of absolute value one,

$$\begin{aligned}\|v_F\|_2 &= \|uv_F\|_2 \leqslant \|uv_F\|_{A(G)}^{1/3}\|uv_F\|_L^{2/3} \\ &\leqslant (2\sqrt{2})^{2/3}\|uv_F\|_{A(G)}^{2/3}\|v_F\|_2^{2/3}\end{aligned}$$

and therefore

$$\|v_F\|_2 \leqslant 8\,\|uv_F\|_{A(G)} = 8\,\|(1_F u)v\|_{B(G)} \leqslant 8\,|v|.$$

Since F is an arbitrary finite subset of G, we conclude that $v \in \ell^2(G)$. □

THEOREM 5.2.8. *Let G be a discrete group. Then $M(A(G)) = B(G)$ if and only if G is amenable.*

PROOF. We only have to show that if $M(A(G)) = B(G)$, then G is amenable. Towards a contradiction, suppose that G fails to be amenable. Then, by Theorem 5.2.5, (iv) $\Rightarrow$ (i), there exists $f \in M(A(G), l^1(G))$ such that $f \notin l^1(G)$. Let $g = |f|^{1/2}$; then $g \notin l^2(G)$, but $g \in M(A(G), l^2(G))$. Indeed, for each $v \in A(G)$, we have

$$\sum_{x\in G} |g(x)v(x)|^2 \leqslant \|v\|_\infty \cdot \sum_{x\in G} |f(x)v(x)| < \infty.$$

If now $hg \in B(G)$ for all $h \in l^\infty(G)$, then $g \in l^2(G)$ by Lemma 5.2.7 So there exists $h \in l^\infty(G)$ such that $hg \notin B(G)$. However,

$$(hg)A(G) = h(gA(G)) \subseteq h\,l^2(G) \subseteq l^2(G) \subseteq A(G).$$

This shows that $hg \in M(A(G))$ even though $hg \notin B(G)$. Consequently, G has to be amenable. □

LEMMA 5.2.9. *Let G be a discrete group, and suppose that $A(G)$ is closed in $M(A(G))$. Then $\ell^2(G)$ is closed in $M(A(G), \ell^2(G))$.*

PROOF. Since $A(G)$ is closed in $M(A(G))$, the original norm and the multiplier norm of $A(G)$ are equivalent. Thus there exists a constant $c > 0$ such that, for every $u \in A(G)$,

$$\sup\{\|uv\|_{A(G)} : v \in A(G), \|v\|_{A(G)} \leqslant 1\} \geqslant c\|u\|_{A(G)}.$$

It follows that $\ell^2(G)$ is closed in $M(A(G), \ell^2(G))$ once we have shown the existence of a constant $C > 0$ such that $\|uv\|_2 \geqslant C\|u\|_2$ for all $u \in \ell^2(G)$ and all $v \in A(G)$ wirh $\|v\|_{A(G)} \leqslant 1$. By Lemma 5.2.7, $\ell^2(G) = M(\ell^\infty(G), B(G))$. Thus there exists a constant $d > 0$ such that $\|uw\|_{A(G)} \geqslant 2d\|u\|_2$ for all $u \in \ell^2(G)$ and all $w \in \ell^\infty(G)$ with $\|w\|_\infty \leqslant 1$.

Now, fix $u \in \ell^2(G)$ and choose $w \in \ell^\infty(G)$ with $\|w\|_\infty \leqslant 1$ and $\|uw\|_{A(G)} \geqslant d\|u\|_2$. Note that c and d do not depend on u and w. For any $v \in A(G)$ with $\|v\|_{A(G)} \leqslant 1$, it follows that

$$\|uvw\|_{A(G)} \geqslant c\|uw\|_{A(G)} \geqslant cd\|u\|_2.$$

Now, for $f \in \ell^2(G)$, $f = f * \delta_e$ and hence $\|f\|_{A(G)} \leqslant \|f\|_2$. Therefore

$$\|uvw\|_{A(G)} \leqslant \|uvw\|_2 \leqslant \|uv\|_2,$$

and hence $\|uv\|_2 \geqslant cd\|u\|_2$, as required. □

For $1 \leqslant p < \infty$, let $M(A(G), L^p(G))$ denote the space of all multipliers from $A(G)$ into $L^p(G)$, that is, the space all bounded linear operators $T : A(G) \to L^p(G)$ satisfying $T(uv) = uT(v)$ for all $u, v \in A(G)$. Equipped with the operator norm, $M(A(G), L^p(G))$ is a Banach space. Since $\|u\|_\infty \leqslant \|u\|_{A(G)}$ for every $u \in A(G)$, the map $f \to M_f$, where $M_f(u) = fu$, from $L^p(G)$ into $M(A(G), L^p(G))$ is norm decreasing.

LEMMA 5.2.10. *Let $1 < p < \infty$ and suppose that on $L^p(G) \subseteq M(A(G), L^p(G))$ the $\|\cdot\|_p$-norm and the multiplier norm are equivalent. Then the same is true for $L^1(G) \subseteq M(A(G), L^1(G))$, and hence $L^1(G)$ is closed in $M(A(G), L^1(G))$.*

PROOF. By hypothesis, there exists a constant $c > 0$ such that

$$\|g\|_p \leqslant c \cdot \sup\{\|gv\|_p : v \in A(G), \|v\|_{A(G)} \leqslant 1\}.$$

Let $f \in L^1(G)$ and set $g = |f|^{1/p} \in L^p(G)$. Choose $v \in A(G)$ with $\|gv\|_p \geqslant \frac{1}{2c}\|g\|_p$ and $\|v\|_{A(G)} \leqslant 1$. Then

$$\begin{aligned} \|f\|_1 &= \|g\|_p^p \leqslant (2c)^p \|gv\|_p^p = (2c)^p \|fv^p\|_1 \\ &\leqslant (2c)^p \sup\{\|fu\|_1 : u \in A(G), \|u\|_{A(G)} \leqslant 1\}. \end{aligned}$$

This proves the first statement of the lemma. In particular, it follows that the multiplier norm on $L^1(G) \subseteq M(A(G), L^1(G))$ is a complete norm, and this in turn implies that $L^1(G)$ is closed in $M(A(G), L^1(G))$. □

5.3. $M(A(G)) = B(G)$ implies amenability of G: The nondiscrete case

This section is devoted to prove Theorem 5.1.10 for nondiscrete locally compact groups. The proof is fairly intricate and technical and requires several new ideas. The exposition is based on [**201**] and some handwritten notes provided by Losert.

LEMMA 5.3.1. *Let $T \in C^*_\lambda(G)$ and let K be a compact subset of G, and consider $L^2(K)$ as a closed subspace of $L^2(G)$. Then the operator*

$$T : L^2(K) \to L^2(G), \quad g \to T(g),$$

is compact.

PROOF. Of course, we can assume that K has positive measure $|K|$. Since the subspace of compact operators is closed in $\mathcal{B}(L^2(K), L^2(G))$ and $C_c(G)$ is dense in $C^*_\lambda(G)$, we can moreover assume that $T = \lambda_G(f)$, where $f \in C_c(G)$. If $g \in L^2(K)$ with $\|g\|_2 \leqslant |K|^{-1/2}$, then $\|g\|_1 \leqslant \|g\|_2 \cdot |K|^{1/2} \leqslant 1$ and hence the function

$$\lambda_G(f)g = f * g = \int_{K^{-1}} g(y^{-1}) f * \delta_{y^{-1}}\, dy \in L^2(G)$$

belongs to the closed absolutely convex hull of the set $A = \{f * \delta_a : a \in K\}$. Now, since the map $x \to f * \delta_x$ from G into $L^2(G)$ is continuous, A is compact and so is its absolutely convex hull. □

An interesting simple consequence of the proof of Lemma 5.3.1 is the well-known fact that $C^*_\lambda(G)$ is unital only if G is discrete. Indeed, if an operator $\lambda_G(f)$ is invertible in $C^*_\lambda(G)$, then the set

$$\{\lambda_G(f)g : g \in L^2(K), \|g\|_2 \leqslant |K|^{-1/2}\}$$

can be relatively compact only if $L^2(K)$ is finite dimensional.

In the following two lemmas, f denotes a fixed continuous function on $\mathbb{R}$ such that $0 \leqslant f \leqslant 1$, $f(t) = 0$ for $t \leqslant 1/4$ and $f(t) = 1$ for $t \geqslant 1/2$. Then, for any operator T on $L^2(G)$, the self-adjoint operator $f(T)$ on $L^2(G)$ is defined.

LEMMA 5.3.2. *Let $T \in VN(G)$ such that $-I \leqslant T \leqslant I$, $u \in A(G) \cap P^1(G)$ and $\delta > 0$ such that $\langle T, u\rangle > 1 - \delta^2/2$. Then the following hold.*

(i) *If $x \in G$ is such that $\langle \lambda_G(x^{-1})T\lambda_G(x), u\rangle > 1 - \delta^2/2$, then*

$$\langle f(T)\lambda_G(x)f(T), L_x u\rangle > 1 - 2\delta.$$

(ii) *Let $T_1, \ldots, T_n \in VN(G)$, $n \in \mathbb{N}$, be pairwise commuting operators such that $0 \leqslant T_j \leqslant I$ for all j and $\langle \sum_{j=1}^n T_j^2, u\rangle < (\delta^2/4n)^2$. Then the operator*

$$S = \left(\prod_{j=1}^n (I - T_j)\right) T \left(\prod_{j=1}^n (I - T_j)\right)$$

satisfies $\langle S, u\rangle > 1 - \delta^2$.

PROOF. (i) Since $u \in A(G) \cap P^1(G)$, there exists $\xi \in L^2(G)$ with $\|\xi\|_2 = 1$ and $u(y) = \langle \lambda_G(y)\xi, \xi\rangle$ for all $y \in G$. Since $2(1-t) \geqslant 1$ for $t \leqslant 1/2$, by definition of f we have $(1 - f(t))^2 \leqslant 2(1-t)$ for $t \leqslant 1$. As $-I \leqslant T \leqslant I$, this and the monotony of the functional calculus imply that

$$(I - f(T))^2 = (1 - f)^2(T) \leqslant 2(I - T).$$

Since $f(T)$ is self-adjoint, this in turn implies

$$\begin{aligned} \|\lambda_G(x)\xi - f(T)\lambda_G(x)\xi\|_2^2 &= \langle (I - f(T))^2\lambda_G(x)\xi, \lambda_G(x)\xi\rangle \\ &\leqslant 2\langle (I - T)\lambda_G(x)\xi, \lambda_G(x)\xi\rangle \\ &= 2(1 - \langle \lambda_G(x^{-1})T\lambda_G(x), u\rangle) < \delta^2. \end{aligned}$$

Because $\langle T, u\rangle > 1 - \delta^2/2$, the same inequality holds with e instead of x. It follows that

$$\begin{aligned} \langle f(T)\lambda_G(x)f(T), L_x u\rangle &= \langle f(T)\lambda_G(x)f(T)\xi, \lambda_G(x)\xi\rangle \\ &= \langle \lambda_G(x)f(T)\xi, f(T)\lambda_G(x)\xi\rangle \\ &= \langle \lambda_G(x)\xi, \lambda_G(x)\xi\rangle + \langle \xi, \lambda_G(x^{-1})f(T)\lambda_G(x)\xi - \xi\rangle \\ &\quad + \langle f(T)\xi - \xi, \lambda_G(x^{-1})f(T)\lambda_G(x)\xi\rangle \\ &\geqslant 1 - \|\lambda_G(x^{-1})f(T)\lambda_G(x)\xi - \xi\|_2 \\ &\quad - \|f(T)\xi - \xi\|_2 \cdot \|f(T)\lambda_G(x)\xi\|_2 \\ &> 1 - 2\delta. \end{aligned}$$

(ii) Let $R = \prod_{j=1}^n (I - T_j)$. Since the operators T_j are pairwise commuting,

$$I - R = \sum_{j=1}^n \left(T_j \cdot \prod_{i=1}^{j-1}(I - T_i)\right),$$

as is easily verified by induction on n. Since $0 \leqslant T_j \leqslant I$, $0 \leqslant \prod_{i=1}^{j-1}(I - T_i) \leqslant I$ for $j = 1, \ldots, n$, and hence

$$\begin{aligned} \|R\xi - \xi\|_2 &= \left\| \left(\sum_{j=1}^n T_j \prod_{i=1}^{j-1}(I - T_i)\right)\xi \right\|_2 \leqslant \sum_{j=1}^n \left\| \prod_{i=1}^{j-1}(I - T_i)\right\| \cdot \|T_j\xi\| \\ &\leqslant \sum_{j=1}^n \|T_j\xi\|_2. \end{aligned}$$

Now by hypothesis, for each j,

$$\|T_j\xi\|_2^2 = \langle T_j^2\xi, \xi\rangle = \langle T_j^2, u\rangle < (\delta^2/4n)^2$$

and therefore

$$\|R\xi - \xi\|_2 < \delta^2/4.$$

This in turn implies

$$\begin{aligned}
\langle S, u\rangle &= \langle S\xi, \xi\rangle = \langle RTR, u\rangle = \langle TR\xi, R\xi\rangle \\
&= \langle T\xi, \xi\rangle + \langle TR\xi, R\xi - \xi\rangle + \langle TR\xi - T\xi, \xi\rangle \\
&\geqslant \langle T\xi, \xi\rangle - \|T\| \cdot \|R\xi\|_2 \cdot \|R\xi - \xi\|_2 - \|T\| \cdot \|R\xi - \xi\|_2 \cdot \|\xi\|_2 \\
&\geqslant \langle T\xi, \xi\rangle - \frac{\delta^2}{4} - \frac{\delta^2}{4} = \langle T, u\rangle - \frac{\delta^2}{2} \\
&> 1 - \frac{\delta^2}{2} - \frac{\delta^2}{2} = 1 - \delta^2
\end{aligned}$$

since $\|\xi\|_2 \leqslant 1$, $\|T\| \leqslant 1$, $\|R\| \leqslant 1$ and $\langle T, u\rangle > 1 - \frac{\delta^2}{2}$. □

As a heuristic motivation for the next theorem and its proof we mention the following fact for a nondiscrete locally abelian group G and $u_1, \ldots, u_n \in A(G)$. Let $u_j = \widehat{f_j}$ with $f_j \in L^1(\widehat{G})$, so that $\|u_j\| = \|f_j\|_1$. Then, since $\widehat{G}$ is noncompact, one can find characters $\chi_1, \ldots, \chi_n \in \widehat{G}$ such that the functions $\delta_{\chi_1} * f_1, \ldots, \delta_{\chi_n} * f_n$ in $L^1(\widehat{G})$ have pairwise almost disjoint supports.

PROPOSITION 5.3.3. *Suppose that G is nondiscrete and let $u_j \in A(G) \cap P^1(G)$ and $x_j \in G$, $j = 1, \ldots, n$. Then, given $0 < \epsilon < 1$, there exist $w_j \in P^1(G)$, $j = 1, \ldots, n$, such that*

$$\left\| \sum_{j=1}^n \mu_j L_{x_j}(u_j w_j) \right\|_{A(G)} \geqslant (1 - \epsilon) \sum_{j=1}^n |\mu_j|$$

for any choice of $\mu_1, \ldots, \mu_n \in \mathbb{C}$.

PROOF. Suppose that there exist operators $S_i \in C^*_\lambda(G)$ and positive definite functions $w_i \in P^1(G)$, $i = 1, \ldots, n$, with the following properties:

(1) $\langle S_i, L_{x_i}(u_i w_i)\rangle \geqslant 1 - \epsilon$;

(2) the subspaces $S_i(L^2(G))$ are pairwise orthogonal, and so are the subspaces $S_i^*(L^2(G))$.

Let N_i denote the null space of S_i, $1 \leqslant i \leqslant n$. It follows from (2) and standard functional analysis arguments that the orthogonal complements $N_i^\perp$ are pairwise orthogonal. Since

$$\|S_i\| = \sup\{\|S_i\xi\| : \xi \in N_i^\perp, \|\xi_i\| \leqslant 1\},$$

(2) implies that $\|\sum_{i=1}^n \alpha_i S_i\| = \max_{1 \leqslant i \leqslant n} |\alpha_i| \cdot \|S_i\|$ for any $\alpha_1, \ldots, \alpha_n \in \mathbb{C}$. Consequently, for any choice of $\mu_1, \ldots, \mu_n \in \mathbb{C}$,

$$\left| \left\langle \sum_{i=1}^n \alpha_i S_i, \sum_{j=1}^n \mu_j L_{x_j}(u_j w_j) \right\rangle \right| \leqslant \left\| \sum_{j=1}^n \mu_j L_{x_j}(u_j w_j) \right\|,$$

whenever $|\alpha_i| \leqslant 1$ for all i. Now let $\alpha_i = |\mu_i|/\mu_i$ and $v_j = L_{x_j}(u_j w_j)$, $1 \leqslant i, j \leqslant n$, choose $\beta_{ij} \in \mathbb{C}$ with $|\beta_{ij}| = 1$ and $|\langle S_i, v_j\rangle| = \beta_{ij}\langle S_i, v_j\rangle$. Then, observing that

$\|v_j\| = \|u_j w_j\| = (u_j w_j)(e) = 1$, we have

$$\begin{aligned}\left\langle \sum_{i=1}^n \alpha_i S_i, \sum_{j=1}^n \mu_j v_j \right\rangle &= \sum_{j=1}^n \mu_j \left(\langle \alpha_j S_j, v_j \rangle + \sum_{i \neq j} \langle \alpha_i S_i, v_j \rangle \right) \\ &\geqslant \sum_{j=1}^n |\mu_j| \left(\langle S_j, v_j \rangle - \sum_{i \neq j} |\langle S_i, v_j \rangle| \right) \\ &\geqslant \sum_{j=1}^n |\mu_j| \left(2\langle S_j, v_j \rangle - \sum_{i=1}^n |\langle S_i, v_j \rangle| \right) \\ &= \sum_{j=1}^n |\mu_j| \left(2\langle S_j, v_j \rangle - \sum_{i=1}^n \beta_{ji} \langle S_i, v_j \rangle \right) \\ &\geqslant \sum_{j=1}^n |\mu_j| (2(1-\epsilon) - 1) \\ &= (1 - 2\epsilon) \sum_{j=1}^n |\mu_j|.\end{aligned}$$

This shows that

$$\left\| \sum_{j=1}^n \mu_j L_{x_j}(u_j w_j) \right\|_{A(G)} \geqslant (1 - 2\epsilon) \sum_{j=1}^n |\mu_j|,$$

the desired inequality.

The existence of operators S_i and functions w_i satisfying properties (1) and (2) above is now shown by induction on n, using Lemmas 5.3.1 and 5.3.2 and the function f fixed before Lemma 5.3.2.

Let $n = 1$ and let $(g_\alpha)_\alpha$ be an approximate identity of $L^1(G)$ such that, for each α, $g_\alpha \geqslant 0$, $\|g_\alpha\|_1 = 1$ and $g_\alpha^* = g_\alpha$. Then $\lambda(g_\alpha) \to I$ in $C_\lambda^*(G)$ and $\|\lambda(g_\alpha)\| \leqslant 1$ and $\lambda(g_\alpha)^* = \lambda(g_\alpha)$ for every α. For sufficiently large α, the operator $T = \lambda(g_\alpha)$ satisfies $-I \leqslant T \leqslant I$, $\langle T, u_1 \rangle \geqslant 1 - \epsilon^2/8$ and

$$\langle \lambda(x_1^{-1}) T \lambda(x_1), u_1 \rangle = \langle T, L_{x_1} R_{x_1} u_1 \rangle \geqslant 1 - \epsilon^2/8.$$

Now let $R_1 = f(T)$, $R_1' = f(2T)$, $S_1 = R_1 \lambda(x_1) R_1$ and $w_1 = 1_G$. Then applying Lemma 5.3.2(i) with $\delta = \epsilon/2$, it follows that $\langle S_1, L_{x_1}(u_1 w_1) \rangle \geqslant 1 - \epsilon$. Moreover, we have $0 \leqslant R_1, R_1' \leqslant I$ and $R_1 R_1' = R_1 = R_1' R_1$, since $1 \geqslant f \geqslant 0$ and $f(t)f(2t) = f(t)$ for all $t \in \mathbb{R}$.

To carry out the inductive step, suppose that $R_i, P_i \in C_\lambda^*(G)$ and $w_i \in P^1(G)$, $1 \leqslant i \leqslant n$, have been constructed such that $0 \leqslant R_i, P_i \leqslant I, R_i P_i = R_i, P_i P_j = 0$ for $i \neq j$ and the operators $S_i = R_i \lambda_G(x_i) R_i$ satisfy (1). Let $Q_i = f(2P_i)$, $1 \leqslant i \leqslant n$. Then

$$f(P_i) Q_i = f(P_i) = Q_i f(P_i) \quad \text{and} \quad Q_i Q_j = 0, \; i \neq j,$$

since $f(t)f(2t) = f(t)$ for all t. Now define operators $T_1, T_2 \in C_\lambda^*(G)$ by

$$T_1 = \sum_{i=1}^n Q_i^2 \quad \text{and} \quad T_2 = \lambda(x_{n+1}) T_1 \lambda(x_{n+1}^{-1}).$$

Since the module action of $A(G)$ on $VN(G)$ is continuous, we have $u_{n+1} \cdot T_k \in C_\lambda^*(G)$ for $k = 1, 2$.

By Lemma 5.3.1, for every non-empty compact subset K of G, the operators

$$h \to u_{n+1} \cdot T_1(h) \quad \text{and} \quad h \to u_{n+1} \cdot T_2(h)$$

from $L^2(K)$ into $L^2(G)$ are compact. Since G is nondiscrete, $L^2(K)$ is infinite dimensional, and hence there exists $h \in L^2(G)$ such that $\|h\|_2 = 1$ and

$$\langle u_{n+1} \cdot T_j(h), h\rangle < \left(\frac{\epsilon^2}{32n}\right)^2, \; j = 1, 2.$$

Let $w_{n+1} = \langle\lambda(\cdot)h, h\rangle \in P^1(G)$. Then

$$\langle T_j, u_{n+1}w_{n+1}\rangle < \left(\frac{\epsilon^2}{32n}\right)^2, \; j = 1, 2.$$

As in the case $n = 1$ above, we can see that there exists $T \in C^*_\lambda(G)$ such that $-I \leqslant T \leqslant I$, $\langle T, u_{n+1}w_{n+1}\rangle < 1 - \epsilon^2/16$ and

$$\langle\lambda(x_{n+1}^{-1})T\lambda(x_{n+1}), u_{n+1}w_{n+1}\rangle = \langle T, L_{x_{n+1}}R_{x_{n+1}}(u_{n+1}w_{n+1})\rangle > 1 - \frac{\epsilon^2}{16}.$$

Define operators T', R_{n+1} and R'_{n+1} by

$$T' = \prod_{i=1}^{n}(I - R'_i)T\prod_{i=1}^{n}(I - R'_i), \; R_{n+1} = f(T') \text{ and } R'_{n+1} = f(2T').$$

Since $(I - R'_i)R'_i = R'_i - R'_i = 0$, we have $T'R'_i = 0$ and hence

$$R_{n+1}R'_{n+1} = R'_{n+1}R_{n+1} = R_{n+1} \quad \text{and} \quad R'_iR'_{n+1} = R'_{n+1}R'_i = 0, \; 1 \leqslant i \leqslant n.$$

Taking $\delta = \epsilon/\sqrt{8}$, Lemma 5.3.2(ii) implies

$$\langle T', u_{n+1}w_{n+1}\rangle > 1 - \frac{\epsilon^2}{8} \quad \text{and} \quad \langle\lambda(x_{n+1}^{-1})T'\lambda(x_{n+1}), u_{n+1}w_{n+1}\rangle > 1 - \frac{\epsilon^2}{8}.$$

Finally, setting $S_{n+1} = R_{n+1}\lambda(x_{n+1})R_{n+1}$ and $\delta = \epsilon/2$, Lemma 5.3.2(i) gives

$$\langle S_{n+1}, L_{x_{n+1}}(u_{n+1}w_{n+1})\rangle > 1 - \epsilon.$$

This completes the inductive step and hence the proof of the lemma. □

COROLLARY 5.3.4. *Let G be nondiscrete and suppose that there exists a constant $c > 0$ such that*

$$\sup\{\|uv\|_{A(G)} : v \in A(G), \|v\|_{A(G)} \leqslant 1\} \geqslant c\,\|u\|_{A(G)}$$

for all $u \in A(G)$. Then, given $u \in A(G) \cap P^1(G)$ and $x_1, \ldots, x_n \in G$, there exists $v \in A(G) \cap P^1(G)$ such that

$$\sum_{j=1}^{n} \|v \cdot L_{x_j}u\|_{A(G)} \geqslant \frac{cn}{4}.$$

PROOF. By Proposition 5.3.3, taking $\mu_j = 1$ and $u_j = u$ for all j and $\epsilon = 1/2$, there exist $w_j \in P^1(G)$, $1 \leqslant j \leqslant n$, such that

$$\left\|\sum_{j=1}^{n} L_{x_j}(uw_j)\right\| \geqslant \frac{n}{2}.$$

By hypothesis, there exists $v \in A(G)$ with $\|v\| \leqslant 1$ and

$$\left\| v \cdot \sum_{j=1}^{n} L_{x_j}(uw_j) \right\| \geqslant c \cdot \left\| \sum_{j=1}^{n} L_{x_j}(uw_j) \right\| \geqslant \frac{cn}{2}.$$

Write $v = v_1 + iv_2$, where v_1 and v_2 are hermitean. Then $\|v_i\| \leqslant 1$, $i = 1, 2$, and

$$\left\| v_i \cdot \sum_{j=1}^{n} L_{x_j}(uw_j) \right\| \geqslant \frac{cn}{4}$$

for at least one i. Thus we can assume that v is hermitean and satisfies $\|v \cdot \sum_{j=1}^{n} L_{x_j}(uw_j)\| \geqslant cn/4$. Now, let $v = v^+ - v^-$, $v^+, v^- \in P(G)$. Then $\|v^+\| + \|v^-\| = \|v\| \leqslant 1$, and this implies that

$$\left\| v^+ \cdot \sum_{j=1}^{n} L_{x_j}(uw_j) \right\| \geqslant \frac{cn}{4} \|v^+\|$$

or

$$\left\| v^- \cdot \sum_{j=1}^{n} L_{x_j}(uw_j) \right\| \geqslant \frac{cn}{4} \|v^-\|.$$

It follows that at least one of the two elements $\|v^+\|^{-1}v^+$ and $\|v^-\|^{-1}v^-$ of $A(G) \cap P^1(G)$ has the desired property. □

LEMMA 5.3.5. *For real-valued functions $f, g \in L^2(G)$ we have*

$$\int_G |f(x)^2 - g(x)^2| dx \leqslant \left(\left(\int_G (f(x)^2 + g(x)^2) dx \right)^2 - 4 \left(\int_G f(x)g(x) dx \right)^2 \right)^{1/2}.$$

PROOF. In fact, since f and g are real-valued,

$$\begin{aligned} \int_G |f(x)^2 - g(x)^2| dx &\leqslant \|f - g\|_2 \|f + g\|_2 \\ &= \left(\|f\|_2^2 - 2\langle f, g\rangle + \|g\|_2^2 \right)^{1/2} \left(\|f\|_2^2 + 2\langle f, g\rangle + \|g\|_2^2 \right)^{1/2} \\ &= \left(\|f\|_2^4 + 2\|f\|_2^2 \|g\|_2^2 + \|g\|_2^4 - 4\langle f, g\rangle^2 \right)^{1/2}, \end{aligned}$$

which equals the right hand side of the stated inequality. □

In the sequel, $LUC(G)$ denotes the space of all left uniformly continuous functions on G. Recall that $LUC(G)$ consists of all those bounded continuous functions f on G for which the map $x \to L_x f$ from G into $L^\infty(G)$ is continuous. Every $T \in VN(G)$ defines an operator, also denoted T, on the tensor product $L^2(G) \otimes L^2(G)$ as follows. If $T = w^* - \lim_{n\to\infty} \sum_{i=1}^{n} \alpha_{ni}\lambda(x_{ni})$, $x_{ni} \in G$, $\alpha_{ni} \in \mathbb{C}$, then, for $\xi, \eta \in L^2(G)$,

$$T(\xi \otimes \eta) = \lim_{n\to\infty} \sum_{i=1}^{n} \alpha_{ni}\lambda(x_{ni})\xi \otimes \lambda(x_{ni})\eta.$$

Let m be a mean on $LUC(G)$, which is given by an integral as in the next lemma. Then m cannot be (left) invariant unless G is compact. The following lemma provides a bound for the lack of invariance.

LEMMA 5.3.6. *Let $0 < \epsilon < 1$ and let U, V, W be relatively compact neighbourhoods of e in G with $U^{-1}W \subseteq V$ and $|V| < (1+\epsilon)|W|$. Let $\xi, \eta \in L^2(G)$ and $T \in VN(G)$ such that $\|\xi\|_2 = \|\eta\|_2 = 1$ and $\|T\| \leqslant 1$ and $\operatorname{supp} T \subseteq U$. Define a mean m on $LUC(G)$ by $m(f) = \int_G f(x)|\xi(x)|^2 dt$. Then, for each $f \in LUC(G)$ and $a \in G$,*

$$\begin{aligned} |m(L_a f - f)| &\leqslant 2\sup\{\|L_y f - f\| : y \in V\} \\ &\quad + \|f\|_\infty (2 + 3\epsilon - |\langle T(\xi \otimes \eta), L_{a^{-1}}\xi \otimes \eta\rangle|^2). \end{aligned}$$

PROOF. Let $\delta = \sup\{\|L_y f - f\| : y \in V\}$. Since $\|\eta\|_2 = 1$, we get

$$\begin{aligned} m(f) &= \int_G f(x)|\xi(x)|^2 \|\eta\|_2^2 dx \\ &= \int_{G\times G} f(x)|\xi(x)\eta(y)|^2 d(x,y) \\ &= \int_{G\times G} f(x)|(\xi \otimes \eta)(x,y)|^2 d(x,y). \end{aligned}$$

This formula and the definition of δ imply

$$\left| m(f) - |V|^{-1} \int_V \int_{G\times G} f(t^{-1}y)|(\xi\otimes\eta)(x,y)|^2 d(x,y)dt \right| \leqslant \delta.$$

Now, this double integral equals

$$\frac{1}{|V|} \int_G f(y) \left(\int_{V\times G} |(\xi\otimes\eta)(ty,x)|^2 d(t,x) \right) dy.$$

Similarly, it is shown that

$$\left| m(L_a f) - \frac{1}{|W|} \int_G f(y) \left(\int_{W\times G} |(L_{a^{-1}}\xi \otimes \eta)(ty,x)|^2 d(t,x) \right) dy \right| \leqslant \delta.$$

Now, put

$$g(y) = \left(\int_{W\times G} |(L_{a^{-1}}\xi\otimes\eta)(ty,x)|^2 d(t,x) \right)^{1/2}$$

and

$$h(y) = \left(\int_{V\times G} |(\xi\otimes\eta)(ty,x)|^2 d(t,x) \right)^{1/2}.$$

Notice next that $\int_G h(y)^2 dy = |V|$ and $\int_G g(y)^2 dy = |W|$. Indeed,

$$\int_G h(y)^2 dy = \|\eta\|_2^2 \int_V \int_G |\xi(ty)|^2 dy\, dt = \|\xi\|_2^2 \cdot \|\eta\|_2^2 \cdot |V| = |V|,$$

and similarly for g. Since

$$\big| |V|^{-1} - |W|^{-1} \big| = -(|V|^{-1} - |W|^{-1}) < \epsilon |V|^{-1},$$

we conclude that

$$\begin{aligned}|m(L_af - f)| &\leqslant 2\delta + \left|\int_G f(y)\left(\frac{g(y)^2}{|V|} - \frac{h(y)^2}{|W|}\right)dy\right| \\ &\leqslant 2\delta + \|f\|_\infty |W|^{-1}\int_G |g(y)^2 - h(y)^2|dy \\ &\quad + \|f\|_\infty| \, |V|^{-1} - |W|^{-1}| \int_G h(y)^2 dy| \\ &\leqslant 2\delta + \|f\|_\infty\left(\epsilon + |W|^{-1}\int_G |g(y)^2 - h(y)^2|dy\right).\end{aligned}$$

To finish the proof, by the preceding inequality, it remains to show that

$$\int_G |g(y)^2 - h(y)^2|dy \leqslant |W|\left(2 + 2\epsilon - |\langle T(\xi\otimes\eta), L_{a^{-1}}\xi\otimes\eta\rangle|\right)^2.$$

To that end, we are going to find a lower bound for $\int_G g(y)h(y)dy$ and then apply Lemma 5.3.5. Notice first that, by definition of g and h,

$$g(y) = \Delta(y)^{-1/2}\left\|L_{a^{-1}}\xi\otimes\eta|_{Wy\times G}\right\|_2$$

and

$$h(y) = \Delta(y)^{-1/2}\left\|\xi\otimes\eta|_{Vy\times G}\right\|_2,$$

where Δ denotes the modular function of G. We claim that

$$\|(\xi\otimes\eta)|_{Vy\times G}\|_2 \geqslant \|T(\xi\otimes\eta)|_{Vy\times G}\|_2 \geqslant \|T(\xi\otimes\eta)|_{Wy\times G}]\|_2.$$

The first inequality is clear since $\|T\| \leqslant 1$. For the second we are going to show that

$$T[(\xi\otimes\eta)|_{Vy\times G}]|_{Wy\times G} = T(\xi\otimes\eta)|_{Wy\times G},$$

which requires some effort. Consider non-empty compact subsets C of $G\backslash Vy$ and D of Wy and functions $\sigma \in L^2(C)$ and $\tau \in L^2(D)$. Define $u \in A(G)$ by $u(x) = \langle\lambda(x)\sigma, \tau\rangle$, $x \in G$. If $u(x) \neq 0$, then

$$\varnothing \neq xC \cap D \subseteq x[(G\backslash V) \cap x^{-1}W]y$$

and hence $x \notin U$. Thus $\operatorname{supp} u \cap \operatorname{supp} T = \varnothing$ and hence $u \cdot T = 0$ since $\operatorname{supp} u$ is compact. It follows that $\langle T, uv\rangle = 0$ for all $v \in A(G)$ and therefore $\langle T(\sigma\otimes\varphi), \tau\otimes\psi\rangle = 0$ for all $\varphi, \psi \in L^2(G)$. Since C and D are arbitrary non-empty compact subsets of $G\backslash Vy$ and Wy, respectively, a simple approximation argument gives that $\langle T(\sigma\otimes\varphi), \tau\otimes\psi\rangle = 0$ for all $\varphi, \psi \in L^2(G)$ whenever $\sigma \in L^2(G\backslash Vy)$ and $\tau \in L^2(Wy)$. This in turn implies that $T(\sigma\otimes\varphi) = 0$ almost everywhere on $Wy \times G$ for every $\varphi \in L^2(G)$ whenever $\sigma \in L^2(G\backslash Vy)$. Returning to ξ and η, we therefore obtain

$$T(\xi\otimes\eta)|_{Wy\times G} = T[(\xi\otimes\eta)|_{Vy\times G}]|_{Wy\times G}.$$

We can now estimate

$$\begin{aligned}\int_G g(y)h(y)dy &= \int_G \|L_{a^{-1}}\xi \otimes \eta|_{Wy\times G}\|_2 \cdot \|\xi\otimes\eta|_{Vy\times G}\|_2 \\ &\geqslant \int_G \|L_{a^{-1}}\xi \otimes \eta|_{G\times Wy}\|_2 \cdot \|T(\xi\otimes\eta)|_{Wy\times G}\|_2 \\ &= \int_G \Delta(y)^{-1}\int_{Wy\times G} |(L_{a^{-1}}\xi\otimes\eta)(T(\xi\otimes\eta))(t,x)|\, d(t,x)\, dy \\ &= \int_G\int_{W\times G} |(L_{a^{-1}}\xi\otimes\eta)(T(\xi\otimes\eta))(t,x)|\, d(ty,x)dy \\ &\geqslant |W|\cdot |\langle T(\xi\otimes\eta), L_{a^{-1}}\xi\otimes\eta\rangle|.\end{aligned}$$

This inequality combined with Lemma 5.3.5 yields

$$\begin{aligned}\left(\int_G |g(y)^2 - h(y)^2|dy\right)^2 &\leqslant (|V|+|W|)^2 - 4\left(\int_G g(y)h(y)dy\right)^2 \\ &\leqslant |W|^2(2+\epsilon)^2 - 4|W|^2\cdot|\langle T(\xi\otimes\eta), L_{a^{-1}}\xi\otimes\eta\rangle|^2 \\ &\leqslant 4|W|^2(1+2\epsilon - |\langle T(\xi\otimes\eta), L_{a^{-1}}\xi\otimes\eta\rangle|^2).\end{aligned}$$

Now using $1+s \leqslant (1+s/2)^2$ and setting $s = 2\epsilon - |\langle T(\xi\otimes\eta), L_{a^{-1}}\xi\otimes\eta\rangle|^2$, we conclude that

$$\int_G |g(y)^2 - h(y)^2|dy \leqslant |W|\left(2+2\epsilon - |\langle T(\xi\otimes\eta), L_{a^{-1}}\xi\otimes\eta\rangle|^2\right),$$

which completes the proof of the lemma. □

Let $\mathcal{M}$ denote the set of all means on $LUC(G)$, the space of all left uniformly continuous functions on G. For any mean m on either $LUC(G)$ or $L^\infty(G)$ and $x \in G$, we put

$$d(m,x) = \sup\{|m(L_x f - f)| : f \in LUC(G), \|f\|_\infty \leqslant 1\}.$$

LEMMA 5.3.7. *Suppose that there exists $c > 0$ with the following property. For any finitely many (not necessarily distinct) $y_1,\dots,y_q \in G$, $q \in \mathbb{N}$, there exists $m \in \mathcal{M}$ such that $\sum_{j=1}^q d(m,y_j) \leqslant cq$. Then, given $x_1,\dots,x_n \in G$, $n \in \mathbb{N}$, there exists $m \in \mathcal{M}$ such that $d(m,x_j) \leqslant c$ for $j = 1,\dots,n$.*

PROOF. To each $m \in \mathcal{M}$ we associate the n-tuple $(d(m,x_j))_{j=1}^n \in \mathbb{R}^n$. Let C denote the convex hull of the set of all these n-tuples. We claim that for each $\epsilon > 0$, there exists $(t_j)_{j=1}^n \in C$ with $\|(t_j)_{j=1}^n\|_\infty \leqslant c+\epsilon$.

Towards a contradiction, assume that for some $\epsilon > 0$ no such n-tuple $(t_j)_{j=1}^n$ exists. Then, by the separation theorem for convex sets, there exists $(s_j)_{j=1}^n \in \mathbb{R}^n$ such that $\|(s_j)_{j=1}^n\|_1 = 1$ and $\sum_{j=1}^n s_j t_j \geqslant c+\epsilon$ for all $(t_j)_{j=1}^n \in C$. Since $C \subseteq [0,\infty)^n$, we can assume that $s_j \geqslant 0$ for all j. Moreover, since C is bounded, a straightforward approximation argument shows that we can assume that $s_j = p_j/q \in \mathbb{Q}$ for $j = 1,\dots,n$, where $\sum_{j=1}^n p_j = q$ and $\sum_{j=1}^n s_j t_j \geqslant c+\epsilon/2$ for all $(t_j)_{j=1}^n \in C$. Now define $y_1,\dots,y_q \in G$ by

$$y_1 = \ldots = y_{p_1} = x_1, \quad y_{p_1+1} = \ldots = y_{p_1+p_2} = x_2, \ldots,$$

$$y_{1+\sum_{j=1}^{n-1} p_j} = \ldots = y_q = x_n.$$

Now, by hypothesis, there exists $m \in \mathcal{M}$ with $\sum_{i=1}^{q} d(m, y_i) \leqslant cq$. However, by definition of $(s_j)_{j=1}^{n}$,

$$\frac{1}{q}\sum_{i=1}^{q} d(m, y_i) = \frac{1}{q}\sum_{j=1}^{n} p_j d(m, x_j) = \sum_{j=1}^{n} s_j d(m, x_j) \geqslant c + \frac{\epsilon}{2}.$$

This contradiction shows that for each $\epsilon > 0$ there exists $(t_j)_{j=1}^{n} \in C$ with $\|(t_j)_{j=1}^{n}\|_\infty \leqslant c + \epsilon$.

Since C is the convex hull of all vectors $d(m, x_j)_{j=1}^{n}$, $m \in \mathcal{M}$, there exist $m_k \in \mathcal{M}$ and $\alpha_k \geqslant 0$, $k = 1, \ldots, r$, such that

$$t_j = \sum_{k=1}^{r} \alpha_k d(m_k, x_j), \ j = 1, \ldots, n, \quad \text{and} \ \sum_{k=1}^{r} \alpha_k = 1.$$

Finally, put $m = \sum_{k=1}^{r} \alpha_k m_k$. Then $m \in \mathcal{M}$ and

$$\begin{aligned} d(m, x_j) &\leqslant \sup\left\{\sum_{k=1}^{r} \alpha_k |m_k(L_{x_j} f - f)| : f \in LUC(G), \|f\|_\infty \leqslant 1\right\} \\ &\leqslant \sum_{k=1}^{r} \sup\left\{\alpha_k |m_k(L_{x_j} f - f)| : f \in LUC(G), \|f\|_\infty \leqslant 1\right\} \\ &= \sum_{k=1}^{r} \alpha_k d(m_k, x_j) = t_j \leqslant c + \epsilon \end{aligned}$$

for all j. Since $\epsilon > 0$ is arbitrary, it follows from the compactness of $\mathcal{M}$ that there exists $m \in \mathcal{M}$ with $d(m, x_j) \leqslant c$ for $j = 1, \ldots, n$. □

PROPOSITION 5.3.8. *Suppose that c satisfies the hypothesis of Lemma 5.3.7. Then, given $x_1, \ldots, x_n \in G$ and $\epsilon > 0$, there exists a mean $\tilde{m}$ on $L^\infty(G)$ such that*

$$d(\tilde{m}, x_j) \leqslant c + \epsilon, \ 1 \leqslant j \leqslant n.$$

PROOF. Let m be any mean on $LUC(G)$ and choose $u \in L^1(G)$ such that $u \geqslant 0$ and $\|u\|_1 = 1$. Since $u * L^\infty(G) \subseteq LUC(G)$, we can define $\tilde{m}$ on $L^\infty(G)$ by $\tilde{m}(f) = m(u * f)$, $f \in L^\infty(G)$. Then $\tilde{m}$ is a mean on $L^\infty(G)$ and

$$\tilde{m}(L_{x_j} f - f) = m((u * \delta_{x_j} - u) * f), \quad 1 \leqslant j \leqslant n.$$

Let $\delta = \epsilon/2$. There exists $v \in L^1(G)$ with $v \geqslant 0$ and $\|v\|_1 = 1$ such that

$$\|(u * \delta_{x_j} - u) * v - (u * \delta_{x_j} - u)\|_1 < \delta$$

for $j = 1, \ldots, n$. Moreover, there exist $y_i \in G$ and $\alpha_i \in \mathbb{R}$, $i = 1, \ldots, r$, such that $\alpha_i \geqslant 0$, $\sum_{i=1}^{r} \alpha_i = 1$ and

$$\left\| \left(u - \sum_{i=1}^{r} \alpha_i \delta_{y_i}\right) * (\delta_{x_j} * v - v) \right\|_1 \leqslant \delta$$

for $j = 1, \ldots, n$. Then, for $f \in L^\infty(G)$ with $\|f\|_\infty \leqslant 1$ and all $j = 1, \ldots, n$, it follows that

$$\begin{aligned}
|\tilde{m}(L_{x_j} f - f)| &= |m((u * \delta_{x_j} - u) * f)| \\
&\leqslant |m((u * \delta_{x_j} - u) * v * f)| + \|(u * \delta_{x_j} - u) * (\delta_e - v) * f\|_\infty \\
&\leqslant |m((u * \delta_{x_j} - u) * v * f)| + \|(u * \delta_{x_j} - u) * (\delta_e - v)\|_1 \\
&\leqslant \delta + |m((u * \delta_{x_j} - u) * v * f)| \\
&\leqslant \epsilon + \left| m\left(\left(\sum_{i=1}^r \alpha_i \delta_{y_i}\right) * (\delta_{x_j} * v - v) * f\right)\right| \\
&\leqslant \epsilon + \sum_{i=1}^r \alpha_i |m(\delta_{y_i} * (\delta_{x_j} * v - v) * f)| \\
&= \epsilon + \sum_{i=1}^r \alpha_i |m((\delta_{y_i x_j y_i^{-1}} - \delta_e) * \delta_{y_i} * v * f)| \\
&\leqslant \epsilon + \sum_{i=1}^r \alpha_i d(m, y_i x_j y_i^{-1}).
\end{aligned}$$

Applying Lemma 5.3.6, taking the elements $y_i x_j y_i^{-1}$, $1 \leqslant j \leqslant n$, $1 \leqslant i \leqslant r$, we find $m \in \mathcal{M}$ such that $d(m, y_i x_j y_i^{-1}) \leqslant c$ for all $1 \leqslant j \leqslant n$ and $1 \leqslant i \leqslant r$. Thus $|\tilde{m}(L_{x_j} f - f)| \leqslant c + \epsilon$ for $j = 1, \ldots, n$, as required. □

Employing all the preceding results, we can now complete the proof of Theorem 5.1.10 by showing the following

THEOREM 5.3.9. *Let G be a nondiscrete locally compact group and suppose that $A(G)$ is closed in $M(A(G))$. Then G is amenable.*

PROOF. Since $A(G)$ is closed in $M(A(G))$, the original norm and the multiplier norm on $A(G)$ are equivalent. So there exists a constant $C > 0$ such that

$$\sup\{\|uv\|_{A(G)} : v \in A(G), \|v\|_{A(G)} \leqslant 1\} \geqslant C\,\|u\|_{A(G)}$$

for all $u \in A(G)$. By Proposition 5.3.10, it suffices to find a constant $c < 2$ with the following property: for any finitely many $x_1, \ldots, x_n \in G$, there exists a mean m on $LUC(G)$ with $\sum_{j=1}^n d(m, x_j) \leqslant cn$. Let

$$\epsilon = C^2/200 \quad \text{and} \quad c = 2 + 3\epsilon - C^2/64 < 2.$$

For each relatively compact neighbourhood V of e in G we are going to construct a mean m_V on $LUC(G)$ such that

$$\sum_{j=1}^n |m_V(L_{x_j^{-1}} f - f)| \leqslant 2 \cdot \sup\{\|L_y f - f\|_\infty : y \in V\} + cn\|f\|_\infty$$

for all $f \in LUC(G)$. Let $\mathcal{V}$ be a neighbourhood basis of the identity consisting of compact sets. Since $y \to L_y f$ is continuous with respect to the $\|\cdot\|_\infty$-norm, taking then a w^*-cluster point of the net $(m_V)_{V \in \mathcal{V}}$, we will obtain a mean m on $LUC(G)$ with the desired properties.

Thus fix $x_1, \ldots, x_n \in G$ and a neighbourhood V of e. We choose neighbourhoods U and W of e such that $U^{-1}W \subseteq V$ and $|W| \geqslant |V|/(1+\epsilon)$ and

$u \in P^1(G) \cap A(G)$ with $\operatorname{supp} u \subseteq U$. Then, by Corollary 5.3.4, there exists $v \in P^1(G) \cap A(G)$ such that

$$\sum_{j=1}^{n} \|vL_{x_j}u\|_{A(G)} > \frac{cn}{4}.$$

Let $c_j = \|vL_{x_j}u\|_{A(G)}$, $1 \leqslant j \leqslant n$, and let $\xi, \eta \in L^2(G)$ such that $u(x) = \langle \lambda(x)\xi, \xi\rangle$ and $v(x) = \langle \lambda(x)\eta, \eta\rangle$ for all $x \in G$. For each j there exists $S_j \in VN(G)$ such that $\|S_j\| \leqslant 1$ and $\langle S_j, vL_{x_j}u\rangle = c_j$. Clearly,

$$\langle L_{x_j^{-1}} S_J, (L_{x_j^{-1}}v)u\rangle = \langle S_j, vL_{x_j}u\rangle.$$

Now define $w \in A(G)$ by

$$w(x) = |W|^{-1}|xW \cap V| = |W|^{-1}(1_V * \check{1}_W)(x).$$

Then $w = 1$ on U and $\|w\|_{A(G)} \leqslant |W|^{-1/2}|V|^{1/2} \leqslant 1 + \epsilon$.

For $1 \leqslant j \leqslant n$, let $T_j = (1+\epsilon)^{-1} w \cdot (L_{x_j^{-1}} S_j)$. Then $\|T_j\| \leqslant 1$ and, since $\operatorname{supp} u \subseteq U$ and $w = 1$ on U,

$$\begin{aligned}
\langle T_j(\eta \otimes \xi), L_{x_j^{-1}}\eta \otimes \xi\rangle &= \langle T_j, (L_{x_j^{-1}}v)u\rangle \\
&= \frac{1}{1+\epsilon}\langle w \cdot (L_{x_j^{-1}}S_j), (L_{x_j^{-1}}v)u\rangle \\
&= \frac{1}{1+\epsilon}\langle L_{x_j^{-1}}S_j, (L_{x_j^{-1}}v)u\rangle \\
&= \frac{1}{1+\epsilon}\langle S_j, v(L_{x_j^{-1}}u)\rangle \\
&= \frac{c_j}{1+\epsilon} \geqslant \frac{c_j}{2}.
\end{aligned}$$

We now define a mean m_V on $LUC(G)$ by

$$m_V(f) = \int_G f(x)|\eta(x)|^2 dx, \quad f \in LUC(G).$$

It follows from Lemma 5.3.6 that

$$|m_V(L_{x_j}f - f)| \leqslant 2 \sup\{\|L_y f - f\| : y \in V\} + \|f\|_\infty \left(2 + 3\epsilon - \frac{c_j}{4}\right)$$

for all $f \in LUC(G)$. Since $\sum_{j=1}^n c_j \geqslant cn/8$,

$$n\sum_{j=1}^{n} c_j^2 \geqslant \left(\sum_{j=1}^{n} c_j\right)^2 \geqslant \frac{n^2c^2}{64}$$

and therefore

$$\begin{aligned}
\sum_{j=1}^{n} |m_V(L_{x_j}f - f)| &\leqslant 2 \sup\{\|L_y f - f\| : y \in V\} + n\|f\|_\infty \left(2 + 3\epsilon - \frac{c^2}{256}\right) \\
&= 2 \sup\{\|L_y f - f\| : y \in V\} + dn\|f\|_\infty.
\end{aligned}$$

This completes the proof of the theorem. □

PROPOSITION 5.3.10. *Suppose that c has the property of Lemma 5.3.7. If $c < 2$, then G is amenable.*

PROOF. We are going to show that given $x_1, \ldots, x_n \in G$ and $\epsilon > 0$, there exists a mean $\widetilde{m}$ on $L^\infty(G)$ such that $d(\widetilde{m}, x_j) \leqslant c + 2\epsilon$ for $j = 1, \ldots, n$. It then follows from [**236**, Proposition 4.21] that G is amenable.

Let m be any mean on $LUC(G)$. Choose $u \in L^1(G)$ such that $u \geqslant 0$ and $\|u\|_1 = 1$, and define $\widetilde{m}$ on $L^\infty(G)$ by $\widetilde{m}(f) = m(u * f)$, $f \in L^\infty(G)$. Then $\widetilde{m}$ is a mean on $L^\infty(G)$ and

$$\widetilde{m}(L_{x_j} f - f) = m((u * \delta_{x_j} - u) * f), \quad 1 \leqslant j \leqslant n.$$

There exists $v \in L^1(G)$ with $v \geqslant 0$ and $\|v\|_1 = 1$ such that

$$\left\| \left(u - \sum_{i=1}^{r} \alpha_i \delta_{y_i} \right) * (\delta_{x_j} * v - v) \right\|_1 \leqslant \epsilon$$

for $j = 1, \ldots, n$. Using these two inequalities, it follows for $f \in L^\infty(G)$ with $\|f\|_1 \leqslant 1$ and $j = 1, \ldots, n$,

$$\begin{aligned}
|\widetilde{m}(L_{x_j} f - f)| &= |m((u * \delta_{x_j} - u) * f| \\
&\leqslant |m((u * \delta_{x_j} - u) * v * f)| + \|(u * \delta_{x_j} - u) * (\delta_e - v) * f\|_\infty \\
&\leqslant |m((u * \delta_{x_j} - u) * v * f)| + \|(u * \delta_{x_j} - u) * (\delta_e - v)\|_1 \\
&\leqslant \epsilon + |m((u * \delta_{x_j} - u) * v * f)| \\
&\leqslant 2\epsilon + \left| m\left(\left(\sum_{i=1}^{r} \alpha_i \delta_{y_i} \right) * (\delta_{x_j} * v - v) * f \right) \right| \\
&\leqslant 2\epsilon + \sum_{i=1}^{r} \alpha_i |m(\delta_{y_i} * (\delta_{x_j} * v - v) * f)| \\
&= 2\epsilon + \sum_{i=1}^{r} \alpha_i |m((\delta_{y_i x_j y_i^{-1}} - \delta_e) * \delta_{y_i} * v * f)| \\
&\leqslant 2\epsilon + \sum_{i=1}^{r} \alpha_i d(m, y_i x_j y_i^{-1}).
\end{aligned}$$

Now, by Lemma 5.3.7, there exists $m \in \mathcal{M}$ such that $d(m, y_i x_j y_i^{-1}) \leqslant c$ for all $1 \leqslant j \leqslant n$ and $1 \leqslant i \leqslant r$. Thus $|\widetilde{m}(L_{x_j} f - f)| \leqslant c + 2\epsilon$ for $j = 1, \ldots, n$, as required. □

5.4. Completely bounded multipliers

Let G be a locally compact group. We have seen in Theorem 5.1.10 that every multiplier of the Fourier algebra $A(G)$ is given by an element of $B(G)$ if and only if the group G is amenable. Of course, there is genuine interest in knowing how in the nonamenable case elements of $M(A(G)) \backslash B(G)$ can arise. Note that the proofs of Nebbia and of Losert for discrete and nondiscrete groups, respectively, are highly nonconstructive. It turned out that there is a subalgebra of $M(A(G))$, the algebra of all so-called completely bounded multipliers, which is much easier accessible than $M(A(G))$ itself. In this section we study these completely bounded multipliers, and in the subsequent section we show that they are closely related to uniformly bounded representations of G in Hilbert spaces.

DEFINITION 5.4.1. Let G be a locally compact group. A multiplier u of $A(G)$ is called *completely bounded* if the associated σ-weakly continuous map M_u :

$VN(G) \to VN(G)$ satisfying $M_u(\lambda_G(x)) = u(x)\lambda_G(x)$ for all $x \in G$ is completely bounded. The space of completely bounded multipliers of $A(G)$ is denoted $M_{cb}(A(G))$.

The proof of the following simple fact is left to the reader.

LEMMA 5.4.2. *The space $M_{cb}(A(G))$ of completely bounded multipliers of $A(G)$ is a Banach space with the norm*

$$\|u\|_{M_{cb}(A(G))} = \|M_u\|_{cb}.$$

We now first present a theorem which identifies those elements of $M(A(G))$ which are completely bounded. This result will prove extremely useful in the sequel.

THEOREM 5.4.3. *Let G be a locally compact group and $K = SU(2)$, and let f be a multiplier of $A(G)$. Then the following conditions are equivalent.*

(i) *f is a completely bounded multiplier of $A(G)$.*
(ii) *For every locally compact group H, $f \times 1_H$ is a multiplier of $A(G \times H)$.*
(iii) *$f \times 1_K$ is a multiplier of $A(G \times K)$.*

Moreover, if any one of these conditions is satisfied, then

$$\|f\|_{M_{cb}(A(G))} = \|f \times 1_K\|_{M(A(G\times K))} = \sup_H \|f \times 1_H\|_{M(A(G\times H))},$$

where the supremum is taken over all locally compact groups H.

PROOF. (i) $\Rightarrow$ (ii) Let $f \in M_{cb}(A(G))$ and let H be a closed subgroup of G. Then M_f is completely bounded and hence there exists a σ-weakly continuous operator $\widetilde{M}_f$ on $VN(G \times H) = VN(G)\widehat{\otimes}VN(H)$ such that

$$\widetilde{M}_f(T \otimes S) = M_f(T) \otimes S, \quad T \in VN(G),\, S \in VN(H).$$

In particular, for $x \in G$ and $y \in H$,

$$\widetilde{M}_f(\lambda_G(x) \otimes \lambda_H(y)) = f(x)\lambda_G(x) \otimes \lambda_H(y).$$

By Proposition 5.1.2, (ii) $\Rightarrow$ (i), this means that $\widetilde{M}_f \in M(A(G \times H))$. Moreover,

$$\|f \times 1_H\|_{M(A(G\times H))} = \|\widetilde{M}_f\| \leqslant \|M_f\|_{cb} = \|f\|_{M_{cb}(A(G))}$$

(Lemma 5.4.2).

The implication (ii) $\Rightarrow$ (iii) being trivial, suppose that (iii) holds. It is well known that for each $n \in \mathbb{N}$, the group $K = SU(2)$ has exactly one n-dimensional irreducible representation. Therefore, $VN(K)$ is isometrically isomorphic to the l^∞-direct sum $\oplus_{n=1}^\infty M_n(\mathbb{C})$.

Since $f \times 1_K \in M(A(G \times K))$, $M_{f\times 1_K}$ is a σ-weakly continuous operator on $VN(G)\widehat{\otimes}VN(K)$ satisfying

$$M_{f\times 1_K}(\lambda_G(x) \otimes \lambda_K(y)) = f(x)\lambda_G(x) \otimes \lambda_K(y)$$

for all $x \in G$ and $y \in K$. Since the sets $\{\lambda_G(x) : x \in G\}$ and $\{\lambda_K(y) : y \in K\}$ are total in $VN(G)$ and $VN(K)$), respectively, two consecutive approximation arguments yield

$$M_{f\times 1_K}(T \otimes S) = M_f(T) \otimes S, \quad T \in VN(G),\, S \in VN(K).$$

By restricting $M_{f\times 1_K}$ to each of the components of the direct sum decomposition

$$VN(G)\hat{\otimes}VN(K) = \bigoplus_{n=1}^{\infty} VN(G)\otimes M_n(\mathbb{C}),$$

and denoting by i_n the embedding of $M_n(\mathbb{C})$ into $VN(K)$, we recognize that $\|M_f\otimes i_n\| \leqslant \|M_{f\times 1_K}\|$ for all $n\in\mathbb{N}$. This shows that M_f is completely bounded and also that $\|M_f\|_{cb} \leqslant \|M_{f\times 1_K}\|$.

Concerning the last statement of the theorem, we have already seen that, for any locally compact group H, $\|f\times 1_H\|_{M(A(G\times H))} \leqslant \|f\|_{M_{cb}(A(G))}$. Combining the last two inequalities, we get

$$\begin{aligned}\sup_H \|f\times 1_H\|_{M(A(G\times H))} &\leqslant \|f\|_{M_{cb}(A(G))} = \|M_f\|_{cb}\\ &\leqslant \|M_{f\times 1_K}\| = \|f\times 1_K\|_{M(A(G\times K))}\\ &\leqslant \sup_H \|f\times 1_H\|_{M(A(G\times H))}.\end{aligned}$$

This completes the proof of the theorem. □

We have seen in Section 4.1 that $M(A(G))$ is a dual Banach space. In passing we apply Theorem 5.4.3 to show that the same is true of $M_{cb}(A(G))$.

LEMMA 5.4.4. *The unit ball of $M_{cb}(A(G))$ is $\sigma(L^\infty(G), L^1(G))$-closed in $L^\infty(G)$.*

PROOF. Let $(f_\alpha)_\alpha$ be a net in the unit ball of $M_{cb}(A(G))$ which converges to some $f\in L^\infty(G)$ in the $\sigma(L^\infty(G), L^1(G))$-topology, and let $K = SU(2)$. Then, by Theorem 5.4.3, $(f_\alpha\times 1_K)_\alpha$ is a net in the unit ball of $M(A(G\times K))$ converging to $f\times 1_K$ in the $\sigma(L^\infty(G\times K), L^1(G\times K))$-topology. Now, by Lemma 5.1.4, the unit ball of $M(A(G\times K))$ is closed in $L^\infty(G\times K)$ in the $\sigma(L^\infty(G\times K), L^1(G\times K))$-topology. Therefore we can assume that $f\times 1_K\in M(A(G\times K))$, and this implies that $f\in M_{cb}(A(G))$ (Theorem 5.4.3). □

The proof of the following proposition uses the preceding lemma and is similar to the proof of Proposition 5.1.5.

PROPOSITION 5.4.5. *If E_0 is the completion of $L^1(G)$ with respect to the norm*

$$\|f\|_{E_0} = \sup\left\{\left|\int_G f(x)u(x)dx\right| : u\in M_{cb}(A(G)), \|u\|_{M_{cb}(A(G))}\leqslant 1\right\},$$

then $E_0^* = M_{cb}(A(G))$. More precisely, every bounded linear functional ψ on E_0 is of the form

$$\psi(f) = \int_G f(x)u(x)dx, \quad f\in L^1(G),$$

for some $u\in M_{cb}(A(G))$, and then $\|\phi\| = \|u\|_{M_{cb}(A(G))}$.

The next theorem provides the link between completely bounded multipliers and uniformly bounded representations which will be established in Section 5.5.

THEOREM 5.4.6. *Let G be a locally compact group and u a continuous function on G. Then the following conditions are equivalent.*

(i) *u is a completely bounded multiplier of $A(G)$.*
(ii) *There exist a Hilbert space $\mathcal{H}$ and bounded continuous maps ξ and η from G into $\mathcal{H}$ such that, for all $x, y\in G$,*

$$u(y^{-1}x) = \langle \xi(x), \eta(y)\rangle.$$

Moreover, if these conditions are satisfied, then

$$\|u\|_{M_{cb}(A(G))} = inf\{\|\xi\|_\infty \|\eta\|_\infty\},$$

where the infimum is taken over all such pairs (ξ, η).

PROOF. The proof of (i) $\Rightarrow$ (ii) rests on a representation theorem for completely bounded maps on unital C^*-algebras [**233**, Theorem 7.4]. Given the completely bounded map $M_u : VN(G) \to VN(G) \subseteq \mathcal{B}(L^2(G))$, there exist a Hilbert space $\mathcal{H}$, a $*$-representation $\pi : VN(G) \to \mathcal{B}(\mathcal{H})$ and two bounded linear operators T_1 and T_2 from $L^2(G)$ to $\mathcal{H}$ such that $M_u(S) = T_2^*\pi(S)T_1$ for all $S \in VN(G)$ and $\|u\|_{M_{cb}(AG))} = \|M_u\|_{cb} = \|T_1\| \cdot \|T_2\|$. Let σ denote the unitary representation of G associated to the $*$-representation $\pi \circ \lambda_G$ of $L^1(G)$. We claim that

$$M_u(\lambda_G(x)) = u(x)\lambda_G(x) = T_2^*\sigma(x)T_1$$

for all $x \in G$. Fix $x \in G$ and let $\mathcal{V}$ denote the collection of all compact neighbourhoods of x in G, ordered by $V_1 \geqslant V_2$ if $V_1 \subseteq V_2$. For each $V \in \mathcal{V}$, choose a continuous nonnegative function f_V with supp $f_V \subseteq V$ and $\int_G f_V(x)dx = 1$. Then the net $(\lambda_G(f_V))_{V \in \mathcal{V}}$ converges σ-strongly to $\lambda_G(x)$ in $VN(G)$, and $(\pi \circ \lambda_G(f_V))_{V \in \mathcal{V}}$ converges σ-strongly to $\sigma(x)$ in $\mathcal{B}(\mathcal{H})$. Since $M_u(\lambda_G(f_V)) = T_2^*\pi(\lambda_G(f_V))T_1$ for every V and M_u is σ-weakly continuous, we conclude that $M_u(\lambda_G(x)) = T_2^*\sigma(x)T_1$.

Now, fix any $f_0 \in L^2(G)$ with $\|f_0\|_2 = 1$ and define maps ξ and η from G into H by

$$\xi(x) = \sigma(x)T_1\lambda_G(x^{-1})f_0 \quad \text{and} \quad \eta(x) = \sigma(x)T_2\lambda_G(x^{-1})f_0,$$

$x \in G$. Then ξ and η are bounded and continuous, and for $x, y \in G$, we have

$$\begin{aligned} \langle \xi(x), \eta(y) \rangle &= \langle \sigma(x)T_1\lambda_G(x^{-1})f_0, \sigma(y)T_2\lambda_G(y^{-1})f_0 \rangle \\ &= \langle T_2^*\sigma(y^{-1}x)T_1\lambda_G(x^{-1})f_0, \lambda_G(y^{-1})f_0 \rangle \\ &= u(y^{-1}x)\langle \lambda_G(y^{-1}x)\lambda_G(x^{-1})f_0, \lambda_G(y^{-1})f_0 \rangle \\ &= u(y^{-1}x). \end{aligned}$$

Moreover, $\|\xi\|_\infty \|\eta\|_\infty \leqslant \|T_1\| \cdot \|T_2\| = \|u\|_{M_{cb}(A(G))}$.

(ii) $\Rightarrow$ (i) We show that, given any $v \in A(G)$, we have $uv \in A(G)$ and $\|uv\|_{A(G)} \leqslant \|\xi\|_\infty \|\eta\|_\infty$. There exist $f, g \in L^2(G)$ such that $v(x) = \langle \lambda_G(x)f, g \rangle$ for all $x \in G$ and $\|v\|_{A(G)} = \|f\|_2 \|g\|_2$. Choose an orthonormal basis $(e_\alpha)_{\alpha \in A}$ of H and, for each $\alpha \in A$, define functions f_α and g_α in $L^2(G)$ by

$$f_\alpha(x) = \langle \xi(x^{-1}), e_\alpha \rangle f(x) \quad \text{and} \quad g_\alpha(x) = \langle \eta(x^{-1}), e_\alpha \rangle g(x).$$

The series $\sum_{\alpha \in A} \|f_\alpha\|_2^2$ and $\sum_{\alpha \in A} \|g_\alpha\|_2^2$ converge. More precisely,

$$\begin{aligned} \sum_{\alpha \in A} \|f_\alpha\|_2^2 &= \int_G |f(x)|^2 \left(\sum_{\alpha \in A} |\langle \xi(x^{-1}, e_\alpha \rangle|^2 \right) dx \\ &= \int_G |f(x)|^2 \|\xi(x^{-1})\|^2 dx \\ &\leqslant \|\xi\|_\infty^2 \|f\|_2^2, \end{aligned}$$

and similarly $\sum_{\alpha\in A}\|g_\alpha\|_2^2 \leqslant \|\eta\|_\infty^2\|g\|_2^2$. These inequalities imply

$$\begin{aligned}\sum_{\alpha\in A}\|\langle\lambda_G(\cdot)f_\alpha, g_\alpha\rangle\|_{A(G)} &\leqslant \sum_{\alpha\in A}\|f_\alpha\|_2\|g_\alpha\|_2\\ &\leqslant \left(\sum_{\alpha\in A}\|f_\alpha\|_2^2\right)^{1/2}\left(\sum_{\alpha\in A}\|g_\alpha\|_2^2\right)^{1/2}\\ &\leqslant \|\xi\|_\infty\|\eta\|_\infty\|f\|_2\|g\|_2\\ &= \|\xi\|_\infty\|\eta\|_\infty\|v\|_{A(G)}.\end{aligned}$$

Now, convergence in $A(G)$ being verified, we get for each $x \in G$, by Parseval's equation,

$$\begin{aligned}\sum_{\alpha\in A}\langle\lambda_G(x)f_\alpha, g_\alpha\rangle &= \sum_{\alpha\in A}\int_G f_\alpha(x^{-1}y)\overline{g_\alpha(y)}\,dy\\ &= \sum_{\alpha\in A}\int_G \langle\xi(y^{-1}x), e_\alpha\rangle f(x^{-1}y)\overline{\langle\eta(y^{-1}), e_\alpha\rangle g(y)}\,dy\\ &= \int_G f(x^{-1}y)\overline{g(y)}\left(\sum_{\alpha\in A}\langle\xi(y^{-1}x), e_\alpha\rangle\langle e_\alpha, \eta(y^{-1})\rangle\right)dy\\ &= \int_G f(x^{-1}y)\overline{g(y)}\langle\xi(y^{-1}x), \eta(y^{-1})\rangle\,dy\\ &= u(x)\langle\lambda_G(x)f, g\rangle = u(x)v(x).\end{aligned}$$

Thus $uv \in A(G)$ and $\|uv\|_{A(G)} \leqslant \|\xi\|_\infty\|\eta\|_\infty\|v\|_{A(G)}$ and hence, by Theorem 5.4.3, $\|u\|_{M_{cb}(AG))} \leqslant \|\xi\|_\infty\|\eta\|_\infty$.

Since we have seen in the proof of the implication (i) $\Rightarrow$ (ii) that there exists a representation of u as in (ii) with $\|\xi\|_\infty\|\eta\|_\infty \leqslant \|u\|_{M_{cb}(AG))}$, it follows that

$$\|u\|_{M_{cb}(AG))} = \inf\|\xi\|_\infty\|\eta\|_\infty,$$

where the infimum is taken over all such pairs (ξ, η). ☐

COROLLARY 5.4.7. *Let G_d be the group G with the discrete topology. Then*

$$M_{cb}(A(G)) = M_{cb}(A(G_d)) \cap C(G).$$

PROOF. Let $j : G_d \to G$ be the identity mapping. If $u \in M_{cb}(A(G))$ and ξ, η are as in (ii) of Theorem 5.4.6, then

$$(u \circ j)(y^{-1}x) = \langle\xi \circ j(x), \eta \circ j(y)\rangle$$

for all $x, y \in G_d$, and hence the implication (ii) $\Rightarrow$ (i) of Theorem 5.4.6 shows that $u \circ j \in M_{cb}(A(G_d))$.

Conversely, let $u \in M_{cb}(A(G_d)) \cap C(G)$ and, as in Theorem 5.4.3, let $K = SO(2)$ and $f = u \times 1_K$. Then f is continuous on $G \times K$ and a multiplier of $A(G_d \times K)$. It follows that for any $v \in A(G \times K)$,

$$vf \in A(G_d \times K) \cap C(G \times K) = A(G \times K).$$

Consequently, f is a multiplier of $A(G \times K)$ and hence $u \in M_{cb}A(G))$ by Theorem 5.4.3. ☐

The preceding corollary is of course reminiscent of the fact that for any locally compact group G, $B(G) = B(G_d) \cap C(G)$. The next corollary is the analogue for completely bounded multipliers of Lemma 5.1.1.

COROLLARY 5.4.8. *Let H be a closed subgroup of G and let $u \in M_{cb}(A(G))$. Then $u|_H \in M_{cb}(A(H))$ and*

$$\|u|_H\|_{M_{cb}(A(H))} \leqslant \|u\|_{M_{cb}(A(G))}.$$

PROOF. Simply apply Lemma 5.1.1 and use the equivalence of (i) and (ii) in Theorem 5.4.6. □

We conclude this section with two results about completely positive multipliers.

PROPOSITION 5.4.9. *Let u be a continuous function on a locally compact group G. Then the following conditions are equivalent.*

(i) *u is a completely positive multiplier of $A(G)$.*
(ii) *u is a positive definite function.*

PROOF. (i) ⇒ (ii) Let $n \in \mathbb{N}$, $x_1, \ldots, x_n \in G$ and $\alpha_1, \ldots, \alpha_n \in \mathbb{C}$. Choose a unit vector $\xi \in L^2(G)$ and put $\xi_j = \alpha_j \lambda_G(x_j^{-1})\xi$, $1 \leqslant j \leqslant n$. Then, since by hypothesis $M_u = m_u^*$ is n-positive,

$$\begin{aligned}\sum_{i,j=1}^{n} \overline{\alpha_j}\alpha_i u(x_j^{-1}x_i) &= \sum_{i,j=1}^{n} u(x_j^{-1}x_i)\langle \alpha_i\xi, \alpha_j\xi\rangle \\ &= \sum_{i,j=1}^{n} u(x_j^{-1}x_i)\langle \lambda_G(x_j^{-1}x_i)\xi_i, \xi_j\rangle \\ &= \sum_{i,j=1}^{n} \langle M_u(\lambda_G(x_j)^*\lambda_G(x_i))\xi_i, \xi_j\rangle \geqslant 0.\end{aligned}$$

(ii) ⇒ (i) Since $u \in P(G)$, u is a multiplier of $A(G)$. To prove that M_u is completely positive, it is sufficient to show that for any finitely many $T_1, \ldots, T_n \in VN(G)$ and $\xi_1, \ldots, \xi_n \in L^2(G)$,

$$\sum_{i,j=1}^{n} \langle M_u(T_j^*T_i)\xi_i, \xi_j\rangle \geqslant 0.$$

Since the set $\{\lambda_G(x) : x \in G\}$ spans a σ-strongly dense subset of $VN(G)$, it is enough to consider $T_i \in \operatorname{span}(\lambda_G(G))$. Then we find $x_1, \ldots, x_r \in G$ such that each T_i, $1 \leqslant i \leqslant n$, is of the form $T_i = \sum_{k=1}^{r} \alpha_{ik}\lambda_G(x_k)$, $\alpha_{ik} \in \mathbb{C}$, $1 \leqslant k \leqslant r$. Let

$$\eta_k = \lambda_G(x_k)\left(\sum_{i=1}^{n} \alpha_{ik}\xi_i\right) \in L^2(G), \quad 1 \leqslant k \leqslant r.$$

Then, for $1 \leqslant i, j \leqslant n$,

$$\langle M_u(T_j^*T_i)\xi_i, \xi_j\rangle = \sum_{k,l=1}^{r} \overline{\alpha}_{jl}\alpha_{ik}u(x_l^{-1}x_k)\langle \lambda_G(x_l^{-1}x_k)\xi_i, \xi_j\rangle$$

and hence, by definition of η_k,

$$\sum_{i,j=1}^{n} \langle M_u(T_j^*T_i)\xi_i, \xi_j\rangle = \sum_{k,l=1}^{r} u(x_l^{-1}x_k)\langle \eta_k, \eta_l\rangle.$$

Since the $(r \times r)$-matrices $A = (u(x_l^{-1}x_k))_{1 \leqslant k,l \leqslant r}$ and $B = (\langle \eta_k, \eta_l \rangle)_{1 \leqslant k,l \leqslant r}$ are positive definite, it follows that

$$\sum_{k,l=1}^{r} u(x_l^{-1}x_k)\langle \eta_k, \eta_l \rangle = \operatorname{trace}(AB) \geqslant 0.$$

This proves (i). □

PROPOSITION 5.4.10. *Let G be an amenable locally compact group and let u be a positive multiplier of $A(G)$. Then u is completely positive.*

PROOF. Since M_u is positive, we have $\langle M_u(T^* * T)\xi, \xi \rangle \geqslant 0$ for all $T \in VN(G)$ and $\xi \in L^2(G)$. Thus, taking $T = \lambda_G(f)$ and $\xi = g$, where $f, g \in C_c(G)$,

$$\begin{aligned} \int_G u(x)(f^* * f)(x)(g * \widetilde{g})(x^{-1})dx &= \int_G u(x)(f^* * f)(x)\langle \lambda_G(x)g, g\rangle dx \\ &= \int_G (f^* * f)(x)\langle M_u(\lambda_G(x))g, g\rangle dx \\ &= \langle M_u(\lambda_G(f^* * f))g, g\rangle, \end{aligned}$$

which is $\geqslant 0$. Since G is amenable, there exists a net $(g_\alpha)_\alpha$ in $C_c(G)$ such that $(g_\alpha * \widetilde{g}_\alpha)(y) \to 1$ uniformly on compact subsets of G. It follows that

$$\int_G u(x)(f^* * f)(x)dx \geqslant 0$$

for all $f \in C_c(G)$, and this shows that u is positive definite. □

COROLLARY 5.4.11. *Let G and H be locally compact groups.*

(i) *$B(G)$ is contained in $M_{cb}(A(G))$ and for any $u \in B(G)$,*

$$\|u\|_{M_{cb}(A(G))} \leqslant \|u\|_{B(G)}.$$

(ii) *If G is amenable, then*

$$M(A(G)) = M_{cb}(A(G)) = B(G),$$

and the three corresponding norms agree.

(iii) *If $u \in M_{cb}(A(G))$ and $v \in M_{cb}(A(H)$, then $u \times v \in M_{cb}(A(G \times H))$.*

PROOF. (i) Let $K = SU(2)$ and choose a representation π of G such that u is a coordinate function of π. It is clear that $u \times 1_K \in B(G \times K)$ and, identifying $\mathcal{H}(\pi)$ with $\mathcal{H}(\pi \times 1_K)$ in the usual manner, it is easily seen that $\|u \times 1_K\|_{B(G \times K)} = \|u\|_{B(G)}$. Because $A(G \times K)$ is an ideal in $B(G \times K)$,

$$\|u \times 1_K\|_{M(A(G \times K))} \leqslant \|u \times 1_K\|_{B(G \times K)}.$$

Now (i) follows from Theorem 5.4.3.

(ii) If G is amenable, then $M(A(G)) = B(G)$ with the same norm (Theorem 5.1.8). So (ii) follows from (i).

(iii) Again, let $K = SU(2)$. By Theorem 5.4.3, $u \times 1_H \times 1_K$ and $1_G \times v \times 1_K$ both belong to $M(A(G \times H \times K))$, and hence so does their product $u \times v \times 1_K$. Using Theorem 5.4.3, (iii) $\Rightarrow$ (i), we conclude that $u \times v \in M_{cb}(A(G \times H))$. □

REMARK 5.4.12. Let G be a nonamenable locally compact group and suppose that $A(G)$ has an approximate identity which is bounded in the $M_{cb}(A(G))$-norm. Then $M_{cb}(A(G)) \backslash B(G) \neq \varnothing$. In fact, if $(u_\alpha)_\alpha$ is such an approximate identity, then $(\|u_\alpha\|_{A(G)})_\alpha$ is unbounded because G is nonamenable. It follows that the two norms $\|\cdot\|_{M_{cb}(A(G))}$ and $\|\cdot\|_{B(G)}$ on $B(G)$ are not equivalent and therefore $B(G)$ cannot be closed in $M_{cb}(A(G))$.

COROLLARY 5.4.13. *Let G be a locally compact group. For a subset E of G, the following are equivalent.*

(i) *$1_E \in B(G)$ with $\|1_E\|_{B(G)} = 1$.*
(ii) *$1_E \in M_{cb}(A(G))$ with $\|1_E\|_{M_{cb}(A(G))} = 1$.*
(iii) *E is a coset of an open subgroup of G.*

PROOF. (ii) $\Rightarrow$ (iii) Clearly, E is open in G. Replacing E by some translate if necessary, we can assume that $e \in E$. We are going to show that then E is a subgroup of G. By Theorem 5.4.6, there exist a Hilbert space H and bounded continuous maps $\xi, \eta : G \to H$ such that $1 = \|1_E\|_{M_{cb}(A(G))} = \|\xi\|_\infty \|\eta\|_\infty$ and

$$1_E(y^{-1}x) = \langle \xi(x), \eta(y) \rangle, \quad x, y \in G.$$

Obviously, we can further assume that $\|\xi\|_\infty = \|\eta\|_\infty = 1$. The Cauchy-Schwarz inequality implies that, for all $x, y \in G$, $y^{-1}x \in E$ if and only if $\eta(y) = \xi(x)$. In particular, as $e \in E$, $\xi(e) = \eta(e)$. Thus, with $\xi = \xi(e)$,

$$E = \{x \in G : \xi(x) = \xi\} = \{y \in G : \eta(y^{-1}) = \xi\}.$$

Hence, if $x, y \in E$, then

$$1_E(yx) = \langle \xi(x), \eta(y^{-1}) \rangle = \langle \xi, \xi \rangle = 1,$$

so that $yx \in E$. This shows that E is a subsemigroup of G. Finally, $x \in E$ implies $x^{-1} \in E$. Indeed, if $x \in E$ then $e \in x^{-1}E$ and $\|1_{x^{-1}E}\|_{M_{cb}(A(G))} = 1$. Therefore, by the foregoing, $x^{-1}E$ is a subsemigroup of G and hence $x^{-1} = x(x^{-1}x^{-1}) \in x(x^{-1}E) = E$. Thus E is a subgroup of G. □

5.5. Uniformly bounded representations and multipliers

In this section we investigate a class of representations of locally compact groups in Hilbert spaces, the coefficient functions of which turn out to be completely bounded multipliers. Moreover, if such a representation is not similar to a unitary representation, then its coefficient functions provide multipliers which are not in $B(G)$.

DEFINITION 5.5.1. A strongly continuous representation π of a locally compact group G in a Hilbert space is called *uniformly bounded* if

$$\|\pi\| := \sup_{x \in G} \|\pi(x)\| < \infty.$$

Two uniformly bounded representations π_1 in $\mathcal{H}(\pi_1)$ and π_2 in $\mathcal{H}(\pi_2)$ are said to be *similar* if there exists a bounded linear operator $T : \mathcal{H}(\pi_1) \to \mathcal{H}(\pi_2)$ with bounded inverse $T^{-1} : \mathcal{H}(\pi_2) \to \mathcal{H}(\pi_1)$ such that, for all $x \in G$,

$$\pi_2(x) = T\pi_1(x)T^{-1}.$$

We start with a result which the reader might expect in view of what we have said above and the fact that $M(A(G)) = B(G)$ when G is amenable.

THEOREM 5.5.2. *Let G be an amenable locally compact group. Then every uniformly bounded representation of G is similar to a unitary representation.*

PROOF. Let π be a uniformly bounded representation of G on the Hilbert space $H(\pi)$. For $\xi, \eta \in H(\pi)$, define $f_{\xi,\eta} : G \to \mathbb{C}$ by

$$f_{\xi,\eta}(x) = \langle \pi(x^{-1})\xi, \pi(x^{-1})\eta \rangle, \quad x \in G.$$

For $x, y \in G$, we then have

$$\begin{aligned} |f_{\xi,\eta}(x) - f_{\xi,\eta}(y)| &= |\langle \pi(x^{-1})\xi, \pi(x^{-1})\eta \rangle - \langle \pi(y^{-1})\xi, \pi(y^{-1})\eta \rangle| \\ &\leqslant |\langle \pi(x^{-1})\xi, \pi(x^{-1})\eta - \pi(y^{-1})\eta \rangle| \\ &\quad + |\langle \pi(x^{-1})\xi - \pi(y^{-1})\xi, \pi(y^{-1})\eta \rangle| \\ &\leqslant \|\pi\| \cdot \|\xi\| \cdot \|\pi(x^{-1})\eta - \pi(y^{-1})\eta\| \\ &\quad + \|\pi(x^{-1})\xi - \pi(y^{-1})\xi\| \cdot \|\pi\| \cdot \|\eta\|. \end{aligned}$$

Since π is strongly continuous, this shows that $f_{\xi,\eta}$ is a bounded continuous function.

Let m be a left invariant mean on $C^b(G)$ and define $\beta : \mathcal{H}(\pi) \times \mathcal{H}(\pi) \to \mathbb{C}$ by

$$\beta(\xi, \eta) = \langle m, f_{\xi,\eta} \rangle, \quad \xi, \eta \in \mathcal{H}(\pi).$$

Clearly, β is a positive semidefinite sesquilinear form on $\mathcal{H}(\pi)$. Let $|\xi|_0 = \beta(\xi, \xi)^{1/2}$, $\xi \in \mathcal{H}(\pi)$. Then $|\xi|_0 \leqslant \|\pi\| \cdot \|\xi\|$ and conversely

$$\|\xi\|^2 \leqslant \|\pi(x)\|^2 f_{\xi,\xi}(x) \leqslant \|\pi\|^2 f_{\xi,\xi}(x)$$

for all $x \in G$, whence $\|\xi\| \leqslant \|\pi\| \cdot |\xi|_0$. Thus $|\cdot|_0$ and $\|\cdot\|$ are equivalent norms. Since β is bounded, there exists $S \in \mathcal{B}(\mathcal{H}(\pi))$ such that

$$\beta(\xi, \eta) = \langle S(\xi), \eta \rangle, \quad \xi, \eta \in \mathcal{H}(\pi).$$

Because β is positive definite, S is a positive operator. Let $T = S^{1/2}$, then T is invertible. In fact, the equivalence of the norms $\|\cdot\|$ and $|\cdot|_0$ implies that $T(\mathcal{H}(\pi))$ is closed in $\mathcal{H}(\pi)$ and that $T^{-1} : T(\mathcal{H}(\pi)) \to \mathcal{H}(\pi)$ is bounded. Moreover, if $\eta \in \mathcal{H}(\pi)$ is such that $\eta \perp T(\mathcal{H}(\pi))$, then $0 = \langle \eta, T(T(\eta)) \rangle = \|T(\eta)\|^2$, so that $\eta = 0$. Finally, the representation $x \to \tau(x) = T \circ \pi(x) \circ T^{-1}$ is unitary. Indeed, since $f_{\pi(x)\xi', \pi(x)\eta'} = \delta_x * f_{\xi',\eta'}$ for all $\xi', \eta' \in \mathcal{H}(\pi)$ and $x \in G$, we have

$$\begin{aligned} \langle \tau(x)\xi, \tau(x)\eta \rangle &= \beta(\pi(x)T^{-1}\xi, \pi(x)T^{-1}\eta) \\ &= \langle m, f_{\pi(x)T^{-1}\xi, \pi(x)T^{-1}\eta} \rangle \\ &= \langle m, \delta_x * f_{T^{-1}\xi, T^{-1}\eta} \rangle \\ &= \langle m, f_{T^{-1}\xi, T^{-1}\eta} \rangle \\ &= \beta(T^{-1}\xi, T^{-1}\eta) \\ &= \langle \xi, \eta \rangle \end{aligned}$$

for all $x \in G$ and $\xi, \eta \in \mathcal{H}(\pi)$. □

LEMMA 5.5.3. *Let G be a locally compact group and π a strongly continuous uniformly bounded representation of G on the Hilbert space $\mathcal{H}(\pi)$. Then the representation $\pi \otimes \lambda_G$ of G on $\mathcal{H}(\pi) \overline{\otimes} L^2(G)$ is similar to the unitary representation $x \to I \otimes \lambda_G(x)$, where I denotes the identity operator on $\mathcal{H}(\pi)$. More precisely, there exists an invertible operator $T \in \mathcal{B}(\mathcal{H}(\pi) \overline{\otimes} L^2(G))$ such that*

$$(\pi \otimes \lambda_G)(x) = T((I \otimes \lambda(x)))T^{-1}, \quad x \in G,$$

and $\|T\| = \|T^{-1}\| = \|\pi\|$.

PROOF. The Hilbert spaces $\mathcal{H}(\pi)\otimes L^2(G)$ and $L^2(G,\mathcal{H}(\pi))$ are isometrically isomorphic. Define $T\in\mathcal{B}(L^2(G,\mathcal{H}(\pi))$ by $(T\xi)(x)=\pi(x)(\xi(x))$ for $\xi\in L^2(G,\mathcal{H}(\pi))$ and $x\in G$. Since

$$\|T\xi\|^2=\int_G\|\pi(x)(\xi(x))\|_2^2\leqslant\|\pi\|\cdot\|\xi\|_2^2$$

for $\xi\in L^2(G,\mathcal{H}(\pi))$, we have $\|T\|\leqslant\|\pi\|$. Conversely, given $\epsilon>0$, choose $\xi_0\in\mathcal{H}(\pi)$ with $\|\xi_0\|=1$, $x_0\in G$ and a relatively compact neighbourhood V of x_0 such that $\|\pi(x)\xi_0\|\geqslant\|\pi\|-\epsilon$ for all $x\in V$ and define ξ by $\xi(x)=|V|^{-1}1_V(x)\xi_0$. It then follows that

$$\|T(\xi)\|^2=|V|^{-1}\int_V\|\pi(x)\xi_0\|_2^2dx\geqslant(\|\pi\|-\epsilon)^2.$$

Since $\epsilon>0$ is arbitrary, $\|T\|\geqslant\|\pi\|$. The inverse of T is given by $T^{-1}(\xi)(x)=\pi(x^{-1})(\xi(x))$, and then an analogous argument shows that $\|T^{-1}\|=\|\pi\|$. We now have, for $\xi\in L^2(G,\mathcal{H}(\pi))$ and $x,y\in G$,

$$\begin{aligned}T(I\otimes\lambda_G(x))T^{-1}(\xi)(y)&=\pi(y)(T^{-1}(\xi)(x^{-1}y))\\&=\pi(y)(\pi(y^{-1}x)(\xi(x^{-1}y))=\pi(x)(\xi(x^{-1}y))\\&=[(\pi(x)\otimes\lambda_G(x))(\xi)](y).\end{aligned}$$

This completes the proof. □

Employing the preceding lemma and Theorem 5.4.6, we are now able to prove the first main result of this section.

THEOREM 5.5.4. *Let G be a locally compact group and π a strongly continuous uniformly bounded representation of G on the Hilbert space $\mathcal{H}(\pi)$. Then for $\xi,\eta\in\mathcal{H}(\pi)$, the coefficient function $u(x)=\langle\pi(x)\xi,\eta\rangle$, $x\in G$, is a completely bounded multiplier of $A(G)$ and*

$$\|u\|_{M_{cb}(A(G))}\leqslant\|\pi\|^2\cdot\|\xi\|\cdot\|\eta\|.$$

PROOF. Let $v\in A(G)$ and $\check{v}(x)=v(x^{-1})$, $x\in G$. There exist $f,g\in L^2(G)$ such that

$$v(x)=\langle\lambda_G(x)f,g\rangle=\check{v}(x^{-1})=(f*\widetilde{g})(x^{-1}),$$

where, as usual, $\widetilde{g}(y)=\overline{g(y^{-1})}$ for $y\in G$, and

$$\|v\|_{A(G)}=\|f\|_2\|g\|_2=\|\check{u}\|_{A(G)}.$$

Let $T\in\mathcal{B}(\mathcal{H}(\pi)\overline{\otimes}L^2(G))$ be as in Lemma 5.5.3 and put $\widetilde{\xi}=T^{-1}(\xi\otimes f)$ and $\widetilde{\eta}=T^*(\eta\otimes g)$. Then, for all $x\in G$,

$$\begin{aligned}u(x)v(x)&=\langle(\pi(x)\otimes\lambda_G(x))(\xi\otimes f),\eta\otimes g\rangle\\&=\langle T(I\otimes\lambda_G(x))T^{-1}(\xi\otimes f),\eta\otimes g\rangle\\&=\langle(I\otimes\lambda_G(x))\widetilde{\xi},\widetilde{\eta}\rangle.\end{aligned}$$

Thus uv is a coefficient function of a unitary representation. Therefore $uv\in B(G)$ and

$$\begin{aligned}\|uv\|_{B(G)}&\leqslant\|\widetilde{\xi}\|\cdot\|\widetilde{\eta}\|\leqslant\|T^{-1}\|\cdot\|\xi\otimes f\|\cdot\|T^*\|\cdot\|\eta\otimes g\|\\&\leqslant\|\pi\|^2\cdot\|\xi\|\cdot\|\eta\|\cdot\|f\|_2\cdot\|g\|_2\\&=\|\pi\|^2\cdot\|\xi\|\cdot\|\eta\|\cdot\|v\|_{A(G)}.\end{aligned}$$

This shows that $v \to uv$ is a bounded linear map from $A(G)$ into $B(G)$. Since $u(A(G) \cap C_c(G)) \subseteq B(G) \cap C_c(G) \subseteq A(G)$, it follows that even $uA(G) \subseteq A(G)$. Consequently, $u \in M(A(G))$ and $\|u\|_{M(A(G))} \leqslant \|\pi\|^2 \|\xi\| \cdot \|\eta\|$.

Let now H be an arbitrary locally compact group and define a representation $\widetilde{\pi}$ of $G \times H$ in $\mathcal{H}(\pi)$ by $\widetilde{\pi}(x, y) = \pi(x)$, $x \in G$, $y \in H$. Then

$$\langle \widetilde{\pi}(x, y)\xi, \eta \rangle = (u \times 1_H)(x, y),$$

and applying the first part of the proof with G replaced by $G \times H$ and π by $\widetilde{\pi}$, we obtain that $u \times 1_H \in M(A(G \times H))$ and

$$\|u \times 1_H\|_{M(A(G \times H))} \leqslant \|\pi\|^2 \|\xi\| \cdot \|\eta\|.$$

Theorem 5.4.3 and Theorem 5.4.6 yield that $u \in M_{cb}(A(G))$ and $\|u\|_{M_{cb}(A(G))} \leqslant \|\pi\|^2 \cdot \|\xi\| \cdot \|\eta\|$. □

Now the question arises whether Theorem 5.5.4 gives us with multipliers which are not in $B(G)$. The next theorem provides an affirmative answer, at least when the representation π is cyclic. Recall that a representation $\pi : G \to \mathcal{B}(\mathcal{H}(\pi))$ is called cyclic if there exists a vector $\xi \in \mathcal{H}(\pi)$ such that the span of $\{\pi(x)\xi : x \in G\}$ is dense in $\mathcal{H}(\pi)$. Unfortunately, the proof of the following theorem builds on heavy machinery, namely the positive solution to the similarity problem for cyclic representations of C^*-algebras (see [**112**]).

THEOREM 5.5.5. *Let π be a strongly continuous uniformly bounded representation of a locally compact group G on a Hilbert space $\mathcal{H}(\pi)$. Moreover, assume that π is cyclic. Then the following conditions are equivalent.*

(i) *All the coefficients of π are contained in $B(G)$.*
(ii) *π is similar to a unitary representation.*

PROOF. (ii) $\Rightarrow$ (i) is trivial because the sets of coefficient functions of two similar representations coincide.

Conversely, suppose that (i) holds and consider the sesquilinear map $s : \mathcal{H}(\pi) \times \mathcal{H}(\pi) \to B(G)$ defined by

$$s(\xi, \eta)(x) = \langle \pi(x)\xi, \eta \rangle, \quad \xi, \eta \in \mathcal{H}(\pi).$$

We first show that s is separately continuous. For fixed $\eta \in \mathcal{H}(\pi)$, the map $\xi \to s(\xi, \eta)$ from $\mathcal{H}(\pi)$ into $\mathcal{B}(\mathcal{H}(\pi))$ has a closed graph. Indeed, if $(\xi_n)_n$ is a sequence in $\mathcal{H}(\pi)$ such that

$$\|\xi_n - \xi\| \to 0 \quad \text{and} \quad \|s(\xi_n, \eta) - u\|_{B(G)} \to 0$$

for some $\xi \in \mathcal{H}(\pi)$ and $u \in B(G)$, then

$$\langle \pi(x)\xi, \eta \rangle = \lim_{n\to\infty} \langle \pi(x)\xi_n, \eta \rangle = \lim_{n\to\infty} s(\xi_n, \eta)(x) = u(x)$$

for all $x \in G$. Thus $s(\xi, \eta) = u$ and hence s is continuous in the first variable by the closed graph theorem. The same argument applies to the second variable. Now, being separately continuous, s is a bounded bilinear map.

Since π is strongly continuous and uniformly bounded, we can associate to each $f \in L^1(G)$ an operator

$$\widetilde{\pi}(f) = \int_G f(x)\pi(x)dx \in \mathcal{B}(\mathcal{H}(\pi)),$$

where the integral converges in the strong operator topology.

It is easy to verify that $\tilde{\pi}$ is a bounded representation of the convolution algebra $L^1(G)$ and that $\|\tilde{\pi}\| = \|\pi\|$. Since

$$\langle \tilde{\pi}(f)\xi, \eta\rangle = \int_G f(x)\langle \pi(x)\xi, \eta\rangle dx$$

for all $\xi, \eta \in \mathcal{H}(\pi)$ and

$$\|\langle \pi(\cdot)\xi, \eta\rangle\|_{B(G)} \leqslant \|s\| \cdot \|\xi\| \cdot \|\eta\|,$$

it follows from the duality between $C^*(G)$ and $B(G)$ that

$$\|\tilde{\pi}(f)\| \leqslant \|s\| \cdot \|f\|_{C^*(G)}, \ f \in L^1(G).$$

Thus $\tilde{\pi}$ extends to a bounded representation, also denoted $\tilde{\pi}$, of $C^*(G)$ on $\mathcal{H}(\pi)$. Since π is cyclic, to is $\tilde{\pi}$. Therefore, by the solution to the similarity problem [**112**], there exists an invertible operator $T \in \mathcal{B}(\mathcal{H}(\pi))$ such that

$$f \to \tilde{\rho}(f) = T\,\tilde{\pi}(f)T^{-1}$$

is a $*$-representation of $C^*(G)$. Now, let

$$\rho(x) = T\pi(x)T^{-1}, \quad x \in G.$$

Then ρ is a strongly continuous representation of G and

$$\tilde{\rho}(f) = \int_G f(x)\rho(x)dx, \quad f \in L^1(G).$$

Since $\|\tilde{\rho}(f)\| \leqslant \|f\|_{C^*(G)} \leqslant \|f\|_1$ for all $f \in L^1(G)$, it follows that $\|\rho(x)\| \leqslant 1$ for all $x \in G$. Indeed, if $\xi, \eta \in \mathcal{H}(\pi)$ are of norm one, then $f \to \langle \tilde{\rho}(f)\xi, \eta\rangle$ is a bounded linear functional of norm at most one and hence

$$\int_G f(x)\langle \rho(x)\xi, \eta\rangle dx = \int_G f(x)\varphi(x)dx$$

for some $\varphi \in L^\infty(G)$ with $\|\varphi\|_\infty \leqslant 1$, which in turn implies that $|\langle \rho(x)\xi, \eta\rangle| \leqslant 1$ for all $x \in G$. Consequently, ρ is a unitary representation. This completes the proof of (i) $\Rightarrow$ (ii). □

It follows from Theorems 5.5.4 and 5.5.5 that if G possesses a uniformly bounded cyclic representation which is not similar to a unitary representation, then

$$M_{cb}(A(G))\backslash B(G) \neq \varnothing.$$

Actually, it is sufficient that some quotient group of G admits such a representation.

COROLLARY 5.5.6. *Let G be a locally compact group and N a closed normal subgroup of G. If G/N admits a uniformly bounded cyclic representation which is not similar to a unitary representation, then $M_{cb}(A(G))\backslash B(G) \neq \varnothing$.*

PROOF. Let ρ be a uniformly bounded representation of G/N which is not similar to a unitary representation and let $\pi = \rho \circ q$, where $q : G \to G/N$ is the quotient homomorphism. Then π is a uniformly bounded representation which is not similar to a unitary representation, as required. □

5.6. Multiplier bounded approximate identities in $A(G)$

We remind the reader that the Fourier algebra $A(G)$ of a locally compact group G has a bounded approximate identity if and only if G is amenable (Theorem 2.7.2). It is an extremely challenging and difficult problem to identify those groups G for which $A(G)$ admits an approximate identity which is bounded in the *cb*-multiplier norm.

Suppose that H is a closed subgroup of G and that $(u_\alpha)_\alpha$ is an approximate identity for $A(G)$ such that $\|u_\alpha\|_{M_{cb}(A(G))} \leqslant c < \infty$ for all α. Then $\|u_\alpha|_H\|_{M_{cb}(A(H))} \leqslant c$ and since $A(H) = A(G)|_H$, it follows that $(u_\alpha|_H)_\alpha$ is an approximate identity for $A(H)$ bounded in the *cb*-multiplier norm. In this section we are going to show that the converse is true at least when H is a lattice in G (and G is second countable). This result will be applied in Section 5.7 in case $G = SL(2, \mathbb{R})$ and $H = \mathbb{F}_2$, the free group on two generators.

LEMMA 5.6.1. *Let G be a locally compact group and let $c \geqslant 1$. Then the following three conditions are equivalent.*

(i) *There exists a net $(w_\alpha)_\alpha$ in $A(G)$ such that $\sup_\alpha \|w_\alpha\|_{M_{cb}(A(G))} \leqslant c$ and $w_\alpha \to 1_G$ in the $\sigma(L^\infty, L^1)$-topology.*
(ii) *There exists a net $(v_\alpha)_\alpha$ in $A(G)$ such that $\sup_\alpha \|v_\alpha\|_{M_{cb}(A(G))} \leqslant c$ and $v_\alpha \to 1$ uniformly on compact subsets of G.*
(iii) *There exists an approximate identity $(u_\alpha)_\alpha$ in $A(G)$ such that, for all α, $\|u_\alpha\|_{M_{cb}(A(G))} \leqslant c$.*

PROOF. (i) $\Rightarrow$ (ii) Let the net $(w_\alpha)_\alpha$ satisfy (i) and choose $f \in C_c(G)$ with $f \geqslant 0$ and $\|f\|_1 = 1$. For each α, put

$$v_\alpha(x) = (f * w_\alpha)(x) = \int_G f(xy) w_\alpha(y^{-1}) dy, \quad x \in G.$$

Let $K \subseteq G$ be compact. Then the functions $L_{x^{-1}} f, x \in K$, form a compact subset of $L^1(G)$. Since $w_\alpha \to 1_G$ in the $\sigma(L^\infty, L^1)$-topology and

$$\sup_\alpha \|w_\alpha\|_\infty \leqslant \sup_\alpha \|w_\alpha\|_{M_{cb}(A(G))} \leqslant c,$$

the convergence is uniform on compact subsets of $L^1(G)$. Thus

$$v_\alpha(x) = \langle \check{w}_\alpha, L_{x^{-1}} f\rangle \ \to \ \langle 1_G, L_{x^{-1}} f\rangle = (f * 1_G)(x) = 1$$

uniformly on K. Moreover, since $f \in C_c(G), f \geqslant 0$ and $\|f\|_1 = 1$, v_α is contained in the $\sigma(L^\infty, L^1)$-closed convex hull of left translates of w_α. Because the unit ball of $M_{cb}(A(G))$ is $\sigma(L^\infty(G), L^1(G))$-closed (Lemma 5.4.4), it follows that

$$\|v_\alpha\|_{M_{cb}(A(G))} \leqslant \|w_\alpha\|_{M_{cb}(A(G))} \leqslant c.$$

So the net $(v_\alpha)_\alpha$ satisfies the conditions in (ii).

(ii) $\Rightarrow$ (iii) Suppose that $(v_\alpha)_\alpha$ is a net as in (ii). Again, choose $f \in C_c(G)$ with $f \geqslant 0$ and $\|f\|_1 = 1$ and put $u_\alpha = f * v_\alpha$. Then, as in the proof of (i) $\Rightarrow$ (ii), we have $\|u_\alpha\|_{M_{cb}(A(G))} \leqslant c$ for all α. Let $w \in A(G) \cap C_c(G)$ and set

$$K = \operatorname{supp} f, \ L = \operatorname{supp} w \quad \text{and} \quad h_\alpha = v_\alpha 1_{K^{-1}L}.$$

Then, for each $x \in L$,

$$u_\alpha(x) = \int_K f(y) v_\alpha(y^{-1}x) dy = (f * h_\alpha)(x),$$

and similarly, for $x \in L$,

$$(f * 1_G)(x) = (f * 1_{K^{-1}L})(x).$$

By hypothesis, $v_\alpha \to 1$ uniformly on the compact sets and hence $h_\alpha \to 1$ uniformly on the compact set $K^{-1}L$. Since h_α vanishes outside of $K^{-1}L$, it follows that $h_\alpha \to 1_{K^{-1}L}$ in $L^2(G)$. This implies that $f * h_\alpha \to f * 1_{K^{-1}L}$ in $A(G)$ and, since $w \in A(G)$, this in turn yields that

$$\|(f * h_\alpha)w - (f * 1_{K^{-1}L})w\|_{A(G)} \to 0.$$

By the above equations and since $L = \operatorname{supp} w$, we have

$$u_\alpha w = (f * v_\alpha)w \quad \to \quad (f * 1_G)w = w$$

in $A(G)$. Finally, since $A(G) \cap C_c(G)$ is dense in $A(G)$ and

$$\sup_\alpha \|u_\alpha\|_{M(A(G))} \leqslant \sup_\alpha \|u_\alpha\|_{M_{cb}(A(G))} < \infty,$$

a simple triangle inequality argument shows that $\|u_\alpha w - w\|_{A(G)} \to 0$ for every $w \in A(G)$.

Conversely, (iii) $\Rightarrow$ (ii) $\Rightarrow$ (i). Indeed, (iii) $\Rightarrow$ (ii) since the $A(G)$-norm dominates the L^∞-norm and since, given any compact subset K of G, there exists $u \in A(G)$ with $u|_K = 1$. Clearly, (ii) $\Rightarrow$ (i) because $C_c(G)$ is dense in $L^1(G)$. $\square$

COROLLARY 5.6.2. *Let G be a locally compact group and let $c \geqslant 1$. Suppose that G satisfies the equivalent conditions of Lemma* 5.6.1. *Then there exists a net $(u_\beta)_\beta$ in $A(G) \cap C_c(G)$ with the following properties.*

(1) $\|u_\beta\|_{M_{cb}(A(G))} \leqslant c$ *for all* β.
(2) $\|u_\beta u - u\|_{A(G)} \to 0$ *for every* $u \in A(G)$.
(3) $u_\beta \to 1$ *uniformly on compact subsets of* G.

PROOF. Let $(v_\alpha)_\alpha$ be a net in $A(G)$ satisfying condition (ii) in Lemma 5.6.1. Choose $f \in C_c(G)$ with $f \geqslant 0$ and $\|f\|_1 = 1$, and for each α let $u_\alpha = f * \widetilde{v_\alpha}$. Then in the proof of the implication (ii) $\Rightarrow$ (iii) of Lemma 5.6.1 the net $(u_\alpha)_\alpha$ was shown to be an approximate identity for $A(G)$ such that $\|u_\alpha\|_{M_{cb}(A(G))} \leqslant c$ for all α.

Let K be a compact subset of G containing the identity and let $C = K^{-1} \operatorname{supp} f$. Then, for each $x \in K$, $f(xy) = 0$ for all $y \in G \backslash C$ and hence

$$\begin{aligned} |u_\alpha(x) - 1| &= \left| \int_C f(xy)(v_\alpha(x^{-1}) - 1)dy \right| \\ &\leqslant \sup_{y \in C} |v_\alpha(y^{-1}) - 1| \cdot \int_C f(xy)dy \\ &\leqslant \sup_{y \in C} |v_\alpha(y^{-1}) - 1|. \end{aligned}$$

Since, by condition (ii), $v_\alpha \to 1$ uniformly on the compact set C^{-1}, it follows that $u_\alpha \to 1$ uniformly on K.

Thus the net $(u_\alpha)_\alpha$ has the properties (1), (2) and (3), but u_α may not have compact support. Therefore, for each α and $n \in \mathbb{N}$, choose $u_{\alpha,n} \in A(G) \cap C_c(G)$ such that $\|u_{\alpha,n} - u_\alpha\|_{A(G)} \leqslant 1/n$. Then

$$\begin{aligned} \|u_{\alpha,n}\|_{M_{cb}(A(G))} &\leqslant \|u_{\alpha,n} - u_\alpha\|_{M_{cb}(A(G))} + \|u_\alpha\|_{M_{cb}(A(G))} \\ &\leqslant \|u_{\alpha,n} - u_\alpha\|_{A(G)} + \|u_\alpha\|_{M_{cb}(A(G))} \\ &\leqslant c + 1/n. \end{aligned}$$

For each α and n, we now define

$$v_{\alpha,n} = \frac{c}{c+1/n} u_{\alpha,n} \in A(G) \cap C_c(G).$$

It is straightforward to check that if the set of all $\beta = (\alpha, n)$ is given the product ordering, then the net $(v_\beta)_\beta$ in $A(G) \cap C_c(G)$ has properties (1), (2) and (3). □

A *lattice* Γ in a locally compact group G is a closed discrete subgroup for which G/Γ admits a finite G-invariant measure. A locally compact group G which contains a lattice is necessarily unimodular. In the following, G is a second countable locally compact group and Γ is a lattice in G. Then Γ is countable and there exists a Borel cross-section for the left cosets of Γ in G, that is, a Borel subset S of G with the property that each $x \in G$ has a unique decomposition $x = sa$ with $s \in S$ and $a \in \Gamma$. Equip Γ with counting measure and let ν be a left invariant measure on G/Γ such that $\nu(G/\Gamma) = 1$. Then

$$\int_G f(y)d\mu(y) = \int_{G/\Gamma} \left(\sum_{a\in\Gamma} f(xa) \right) d\nu(x\Gamma)$$

defines a Haar measure on G and $\mu(S) = \nu(G/\Gamma) = 1$. For every bounded function f on Γ, define $\widetilde{f}$ on G by

$$\widetilde{f} = 1_S * f * \check{1}_S.$$

Then $\widetilde{f}$ is a bounded continuous function on G since $1_S \in L^1(G)$, G is unimodular and $1_S * f$ is the bounded function on G given by

$$(1_S * f)(sa) = f(a), \quad s \in S, a \in \Gamma.$$

We now show that if $f \in A(\Gamma)$ (respectively, $f \in M_{cb}(A(\Gamma))$), then $\widetilde{f}$ has the corresponding property. The *cb*-statement builds on Theorem 5.4.6.

PROPOSITION 5.6.3. *Let f be a bounded function on Γ and let $\widetilde{f}$ be as above.*

(i) *If $f \in A(\Gamma)$, then $\widetilde{f} \in A(G)$ and*

$$\|\widetilde{f}\|_{A(G)} \leqslant \|f\|_{A(\Gamma)}.$$

(ii) *If $f \in M_{cb}(A(\Gamma))$, then $\widetilde{f} \in M_{cb}(A(G))$ and*

$$\|\widetilde{f}\|_{M_{cb}(A(G))} \leqslant \|f\|_{M_{cb}(A(\Gamma))}.$$

PROOF. (i) There exist $g, h \in l^2(\Gamma)$ such that

$$f = g * \check{h} \quad \text{and} \quad \|f\|_{A(\Gamma)} = \|g\|_2 \|h\|_2.$$

Let $g_1 = 1_S * g$ and $h_1 = 1_S * h$. Then $g_1, h_1 \in L^2(G)$ and

$$\widetilde{f} = 1_S * f * \check{1}_S = (1_s * g) * (\check{h} * \check{1}_S) = g_1 * \check{h}_1 \in A(G).$$

Moreover,

$$\|\widetilde{f}\|_{A(G)} \leqslant \|g_1\|_2 \|h_1\|_2 = \|g\|_2 \|h\|_2 = \|f\|_{A(\Gamma)}.$$

This proves (i).

(ii) Every $y \in G$ has a unique decomposition $y = \sigma(y)\gamma(y)$ with $\sigma(y) \in S$ and $\gamma(y) \in \Gamma$. For each $x \in G$, define $\sigma_x : S \to S$ by $\sigma_x(s) = \sigma(xs), s \in S$, and let $l_x : G/\Gamma \to G/\Gamma$ be the left translation $y\Gamma \to xy\Gamma$. Then, with $q : S \to G/\Gamma$ the map $s \to s\Gamma$,

$$q(\sigma_x(s)) = q(xs\gamma(xy)^{-1}) = xs\Gamma = l_x(q(s))$$

for all $s \in S$. Since q and l_x are Borel isomorphisms, so is $\sigma_x : S \to S$. Since the measure ν on G/Γ is left translation invariant, the above formula implies that $\mu|_S$ is σ_x-invariant for every $x \in G$.

We next rewrite the function $\widetilde{f} = 1_S * f * \check{1}_S$ in a suitable way. Since

$$(1_S * f)(sa) = f(a), \quad a \in \Gamma, s \in S,$$

we get, for each $x \in G$,

$$(1_S * f)(x) = (1_S * f)(\sigma(x)\gamma(x)) = f(\gamma(x))$$

and therefore

$$\begin{aligned} \widetilde{f}(x) &= \int_G (1_S * f)(y)1_S(x^{-1}y)d\mu(y) = \int_{xS} f(\gamma(y))d\mu(y) \\ &= \int_S f(\gamma(xs))d\mu(s). \end{aligned}$$

Now, since $xs = \sigma_x(s)\gamma(xs)$ and $ys = \sigma_y(s)\gamma(ys)$ for $x, y \in G$ and $s \in S$, we have

$$yx^{-1}\sigma_x(s) = ys(xs)^{-1}\sigma_x(s) = \sigma_y(s)\gamma(ys)\gamma(xs)^{-1}.$$

As $\sigma_x(s) \in S$ and $\gamma(ys)\gamma(xs)^{-1} \in \Gamma$, this means that $\gamma(yx^{-1}\sigma_x(s)) = \gamma(ys)\gamma(xs)^{-1}$ and therefore, since $\mu|_S$ is σ_x-invariant,

$$\begin{aligned} \widetilde{f}(yx^{-1}) &= \int_S f(\gamma(yx^{-1}s))d\mu(s) \\ &= \int_S f(\gamma(yx^{-1}\sigma_x(s))d\mu(s) \\ &= \int_S f(\gamma(ys)\gamma(xs)^{-1})d\mu(s) \end{aligned}$$

for all $x, y \in G$.

Since $f \in M_{cb}(A(\Gamma))$, by Theorem 5.4.6 there exist a Hilbert space $\mathcal{H}$ and bounded maps $\xi, \eta : \Gamma \to \mathcal{H}$ such that

$$\|f\|_{M_{cb}(A(\Gamma))} = \|\xi\|_\infty \|\eta\|_\infty \quad \text{and} \quad f(b^{-1}a) = \langle \xi(a), \eta(b) \rangle$$

for all $a, b \in \Gamma$. Define $\widetilde{\xi}, \widetilde{\eta} : G \to L^2(S, \mathcal{H}, \mu|_S)$ by

$$\widetilde{\xi}(x)(s) = \xi(\gamma(x^{-1}s)^{-1}) \quad \text{and} \quad \widetilde{\eta}(x)(s) = \eta(\gamma(x^{-1}s)^{-1}),$$

$x \in G, s \in S$. Then $\widetilde{\xi}$ and $\widetilde{\eta}$ are bounded Borel measurable maps from G into $\mathcal{H}$ satisfying

$$\sup_{x \in G} \|\widetilde{\xi}(x)\|_{L^2} \leqslant \|\xi\|_\infty \quad \text{and} \quad \sup_{x \in G} \|\widetilde{\eta}(x)\|_{L^2} \leqslant \|\eta\|_\infty.$$

Here we have used that $\mu(S) = 1$. Moreover, the above formula for $\widetilde{f}$ gives

$$\begin{aligned} \widetilde{f}(y^{-1}x) &= \int_S f(\gamma(y^{-1}s)\gamma(x^{-1}s)^{-1})d\mu(s) \\ &= \int_S \langle \xi(\gamma(x^{-1}s)^{-1}), \eta(\gamma(y^{-1}s)^{-1}) \rangle d\mu(s) \\ &= \int_S \langle \widetilde{\xi}(x)(s), \widetilde{\eta}(x)(s) \rangle d\mu(s) \\ &= \langle \widetilde{\xi}(x), \widetilde{\eta}(y) \rangle \end{aligned}$$

for all $x, y \in G$. Since $\widetilde{f}$ is continuous, Theorem 5.4.6, (iii) $\Rightarrow$ (i) implies that $\widetilde{f} \in M_{cb}(A(G))$ and

$$\|\widetilde{f}\|_{M_{cb}(A(G))} \leqslant \|\widetilde{\xi}\|_\infty \|\widetilde{\eta}\|_\infty \leqslant \|\xi\|_\infty \|\eta\|_\infty = \|f\|_{M_{cb}(A(\Gamma))}.$$

This finishes the proof of (ii) □

With the preceding proposition at hand, we can now deduce the main result of this section.

THEOREM 5.6.4. *Let Γ be a lattice in a second countable locally compact group G and let $c \geqslant 1$. Then the following are equivalent.*

(i) *$A(G)$ has an approximate identity $(u_\alpha)_\alpha$ such that $\|u_\alpha\|_{M_{cb}(A(G))} \leqslant c$ for all α.*

(ii) *$A(\Gamma)$ has an approximate identity $(v_\alpha)_\alpha$ such that $\|v_\alpha\|_{M_{cb}(A(\Gamma))} \leqslant c$ for all α.*

PROOF. As mentioned at the outset of this section, (i) $\Rightarrow$ (ii) holds for any closed subgroup of an arbitrary locally compact group.

(ii) $\Rightarrow$ (i) Let $(v_\alpha)_\alpha \subseteq A(\Gamma)$ be as in (ii), and as before let S be a Borel cross-section for the left cosets of Γ in G. For each α, let

$$u_\alpha = 1_S * v_\alpha * \check{1}_S \in A(G).$$

Since Γ is discrete and $v_\alpha \to 1$ pointwise on Γ, also $v_\alpha \to 1$ in the $\sigma(l^\infty(\Gamma), l^1(\Gamma))$-topology.

To each $g \in L^1(G)$, we now associate the function $T(g)$ on Γ defined by

$$T(g)(t) = \int_G \int_S g(x) 1_S(xyt^{-1}) dy dx, \quad t \in \Gamma.$$

Then $\sum_{t\in\Gamma} |T(g)(t)| \leqslant \|g\|_1$ and hence $T : g \to T(g)$ is a bounded linear map from $L^1(G)$ into $\ell^1(\Gamma)$. By the definition of T,

$$\langle 1_S * f * \check{1}_S, g\rangle = \langle f, T(g)\rangle$$

for all $f \in \ell^\infty(\Gamma)$ and $g \in L^1(G)$. Thus $1_S * f * \check{1}_S = T^*(f)$, that is, $f \to 1_S * f * \check{1}_S$ is the adjoint of the map T. It follows that $u_\alpha \to 1_G$ in the $\sigma(L^\infty(G), L^1(G))$-topology. Moreover, by Proposition 5.6.3(ii), we have

$$\|u_\alpha\|_{M_{cb}(A(G))} \leqslant \|v_\alpha\|_{M_{cb}(A(G))} \leqslant c$$

for all α. Now (i) follows from Lemma 5.6.1, (i) $\Rightarrow$ (iii). □

5.7. Examples: Free groups and $SL(2,\mathbb{R})$

Let G be a free group on finitely many generators $a_1, \ldots, a_N$. Any element x of G has a unique representation as a finite product of a_i and a_j^{-1}, $i, j = 1, \ldots, N$, which does not contain two adjacent inverse factors, $a_i a_i^{-1}$ or $a_j^{-1} a_j$. This representation is called the *word of* x. The number of factors in the word is called the *length of* x and denoted $|x|$. For $x, y \in G$, the word of xy can be found as follows. Take the words of x and y and delete the maximal number of products of the form $a_i a_i^{-1}$ of $a_i^{-1} a_i$. If this number of products is k, then the word for xy starts with the first $|x| - k$ factors in the word of x followed by the last $|y| - k$ factors in the word of y.

For $m \in \mathbb{N}_0$, let $E_m = \{x \in G : |x| = m\}$ and let 1_{E_m} denote the characteristic function of E_m. Note that the sets E_m are finite. Actually, $E_m = \{e\}$ and it is not difficult to show that the number of elements in E_m equals $2N(2N-1)^{m-1}$ for $m \geqslant 1$.

LEMMA 5.7.1. *Let G be a free group on finitely many generators and let $\alpha > 0$. Then the function $x \to e^{-\alpha|x|}$ on G is positive definite.*

PROOF. By Schoenberg's theorem (Section 1.2), it suffices to show that the function $x \to |x|$ on G is negative definite, that is, for any finitely many $x_1, \ldots, x_n \in G$ and $c_1, \ldots, c_n \in \mathbb{C}$ with $\sum_{j=1}^n c_j = 0$, we have

$$\sum_{i,j=1}^{n} c_i \overline{c_j}\, |x_i^{-1} x_j| \leqslant 0.$$

Let $A = \{a_1, \ldots, a_N\}$ be a set of free generators of G and let

$$M = \{(x,y) \in G \times G : x^{-1}y \in A\}$$

and

$$\overline{M} = \{(x,y) \in G \times G : (y,x) \in M\} = \{(x,y) \in G \times G : x^{-1}y \in A^{-1}\}.$$

Let H be a Hilbert space with orthonormal basis $\{v_{(x,y)} : (x,y) \in M\}$, and for $(x,y) \in \overline{M}$ put $v_{(x,y)} = -v_{(y,x)}$.

Now consider an arbitrary element x of G and let $x = x_1 \cdot \ldots \cdot x_n$ be the word of x, that is, $n = |x|$ and $x_j \in A \cup A^{-1}$ for each j. Put $y_0 = e$ and $y_k = x_1 \cdot \ldots \cdot x_k$ for $1 \leqslant k \leqslant n$. Then $y_{k-1}^{-1} y_k = x_k$ and hence $(y_{k-1}, y_k) \in M \cup \overline{M}$. Define a map $f : G \to \mathcal{H}$ by

$$f(x) = v_{(y_0,y_1)} + v_{(y_1,y_2)} + \ldots + v_{(y_{n-1},y_n)}.$$

Since $x = x_1 \cdot \ldots \cdot x_n$ is the word of x, $y_0, \ldots, y_n$ are distinct elements of G. Therefore the elements $v_{(y_{k-1},y_k)}$ of $\mathcal{H}$ are orthogonal and hence $\|f(x)\|^2 = n = |x|$. The main step now is to show that

$$\|f(y) - f(x)\|^2 = |y^{-1}x|, \quad x, y \in G.$$

Let $n = |x|$ and $x = x_1 \cdot \ldots \cdot x_n$ as before and let $m = |y|$ and $y = y_1 \cdot \ldots \cdot y_m$ be the word of y. Then

$$y^{-1}x = y_m^{-1} y_{m-1}^{-1} \cdot \ldots \cdot y_1^{-1} x_1 \cdot \ldots \cdot x_{n-1} x_n.$$

Let $l \in \mathbb{N}_0$ be the number for which $y_{l+1} \neq x_{l+1}$, but $y_j = x_j$ for $j \leqslant l$. Then $l \leqslant \min\{n, m\}$ and

$$y^{-1}x = y_m^{-1} \cdot \ldots \cdot y_{l+1}^{-1} x_{l+1} \cdot \ldots \cdot x_n$$

is the word of $y^{-1}x$. In particular, $|y^{-1}x| = m + n - 2l$. Now

$$f(x) = v_{(e,x_1)} + v_{(x_1,x_1x_2)} + \ldots + v_{(x_1\cdot\ldots\cdot x_{n-1},x_1\cdot\ldots\cdot x_n)}$$

and

$$f(y) = v_{(e,y_1)} + v_{(y_1,y_1y_2)} + \ldots + v_{(y_1\cdot\ldots\cdot y_{m-1},y_1\cdot\ldots\cdot y_m)}.$$

Clearly, the first l terms in the expression for $f(y)$ coincide with the first l terms in the expression for $f(x)$. Using that $v_{(s,t)} = -v_{(t,s)}$ for $(s,t) \in A \cup \overline{A}$, we get

$$\begin{aligned} f(y) - f(x) &= v_{(y_1\cdot\ldots\cdot y_l,y_1\cdot\ldots\cdot y_{l+1})} + \ldots + v_{(y_1\cdot\ldots\cdot y_{m-1},y_1\cdot\ldots\cdot y_m)} \\ &\quad + v_{(x_1\cdot\ldots\cdot x_{l+1},x_1\cdot\ldots\cdot x_l)} + \ldots + v_{(x_1\cdot\ldots\cdot x_n,x_1\cdot\ldots\cdot x_{n-1})}. \end{aligned}$$

Now the elements

$$y_1 \cdot \ldots \cdot y_m, \ldots, y_1 \cdot \ldots \cdot y_l = x_1 \cdot \ldots \cdot x_l, \ldots, x_1 \cdot \ldots \cdot x_n$$

of G are all distinct. So the $m+n-2l$ unit vectors in the decomposition of $f(y)-f(x)$ are orthogonal. It follows that

$$\|f(y)-f(x)\|^2 = m+n-2l = |y^{-1}x|,$$

as stated above.

Let now $x_1,\ldots,x_n \in G$ and $c_1,\ldots,c_n \in \mathbb{C}$ such that $\sum_{j=1}^n c_j = 0$. Then

$$\begin{aligned}
\sum_{i,j=1}^n c_i\bar{c}_j\,|x_i^{-1}x_j| &= \sum_{i,j=1}^n c_i\bar{c}_j\,\|f(x_i)-f(x_j)\|^2 \\
&= \sum_{i,j=1}^n c_i\bar{c}_j\,\left(\|f(x_i)\|^2+\|f(x_j)\|^2-2\langle f(x_i),f(x_j)\rangle\right) \\
&= \sum_{i=1}^n c_i\|f(x_i)\|^2\cdot\sum_{j=1}^n \bar{c}_j + \sum_{j=1}^n \bar{c}_j\|f(x_j)\|^2\cdot\sum_{i=1}^n c_i \\
&\quad -2\sum_{i,j=1}^n c_i\bar{c}_j\,\langle f(x_i),f(x_j)\rangle \\
&= -2\left\|\sum_{i=1}^n c_i f(x_i)\right\|^2 \leqslant 0.
\end{aligned}$$

This shows that the function $x\to|x|$ is negative definite and completes the proof. □

LEMMA 5.7.2. *Let k,l and m be nonnegative integers, and let f and g be functions on G with finite support contained in E_k and E_l, respectively. Then*

$$\|1_{E_m}(f*g)\|_2 \leqslant \|f\|_2\|g\|_2$$

if $|k-l|\leqslant m\leqslant k+l$ and $k+l-m$ is even and

$$\|1_{E_m}(f*g)\|_2 = 0$$

for all other values of m.

PROOF. We have, for all $x\in G$,

$$(f*g)(x) = \sum_{s,t\in G,st=x} f(s)g(t) = \sum_{|s|=k,|t|=l,st=x} f(s)g(t).$$

Since the word of st is obtained from the words of s and t by deleting an even number of factors, $|st|$ can only attain the values

$$|k-l|,\,|k-l|+2,\ldots,k+l-2,\,k+l.$$

Thus $\|1_{E_m}(f*g)\|_2^2 = 0$ if m is not one of these numbers. Assume now that $m=k+l$. If $|x|=m$, x can only in one way be written as a product $x=st$, where $|s|=k$ and $|t|=l$. In fact, s consists of the first k letters in the word of x, and t consists of the last l letters in the word of x. Therefore

$$\begin{aligned}
\|1_{E_m}(f*g)\|_2^2 &= \sum_{|s|=k,|t|=l,|st|=k+l} |f(s)|^2|g(t)|^2 \\
&\leqslant \sum_{|s|=k,|t|=l} |f(s)|^2|g(t)|^2 \leqslant \|f\|_2^2\|g\|_2^2.
\end{aligned}$$

Let now m be one of the numbers $k+l-2, k+l-4, \ldots, |k-l|$. Then $m = k+l-2p$, where $1 \leqslant p \leqslant \min\{k,l\}$. If $|x| = m$, $x = st$, where $|s| = k$ and $|t| = l$, then the word of x consists of the first $k-p$ letters of s and the last $l-p$ letters of t.

We define functions f' and g' with supports in E_{k-p} and E_{l-p}, respectively, by

$$f'(y) = \left(\sum_{|z|=p} |f(yz)|^2\right)^{1/2} \quad \text{if } |y| = k-p \quad \text{and} \quad f'(y) = 0 \text{ otherwise}$$

and

$$g'(y) = \left(\sum_{|z|=p} |f(z^{-1}y)|^2\right)^{1/2} \quad \text{if } |y| = l-p \quad \text{and} \quad g'(y) = 0 \text{ otherwise.}$$

Since any element $x \in G$ with $|x| = k$ can only in one way be written as $x = yx$, where $|y| = k-p$ and $|z| = l-p$, we have

$$\|f'\|_2^2 = \sum_{|y|=k-p} \left(\sum_{|z|=p} |f(yz)|^2\right) = \|f\|_2^2.$$

In the same way one gets $\|g'\|_2 = \|g\|_2$.

Let now $x \in G$ with $|x| = m$. Let s' consist of the first $k-p$ letters of x, and let t' consist of the last $l-p$ letters of x. Then $x = s't'$. Moreover, if $x = st$, where $|s| = k$ and $|t| = l$, then $s = s'z$ and $t = z^{-1}t'$ where $|z| = p$. Consequently,

$$\begin{aligned}
|(f * g)(x)| &= \left| \sum_{|s|=k,|t|=l,st=x} f(s)g(t) \right| \\
&= \left| \sum_{|z|=p,|s'z|=k,|z^{-1}t'|=l} f(s'z)g(z^{-1}t') \right| \\
&= \left| \sum_{|z|=p} f(s'z)g(z^{-1}t') \right| \\
&\leqslant \left(\sum_{|z|=p} |f(s'z)|^2\right)^{1/2} \left(\sum_{|z|=p} |g(z^{-1}t')|^2\right)^{1/2} \\
&= f'(s')g'(t').
\end{aligned}$$

Now $f'(s')g'(t') = (f' * g')(x)$ because $x = s't'$ is the unique splitting of x in a product of elements with $|s'| = k-p$ and $|t'| = l-p$. Thus

$$|f * g| 1_{E_m}(x) \leqslant (f' * g') 1_{E_m}(x)$$

for all $x \in G$. Since $(k-p)+(l-p) = m$, it follows from the first part of the proof that

$$\|1_{E_m}(f * g)\|_2 \leqslant \|1_{E_m}(f' * g')\|_2 \|f'\|_2 \|g'\|_2 = \|f\|_2 \|g\|_2.$$

This finishes the proof of the lemma. □

LEMMA 5.7.3. *Let f be a function on G with finite support. Then*

$$\|\lambda_G(f)\| \leqslant \sum_{n=0}^{\infty}(n+1)\|1_{E_n}f\|_2.$$

PROOF. Since $f = \sum_{n=0}^{\infty} 1_{E_n} f$, it is enough to show that when the support of f is contained in E_n, then $\|\lambda_G(f)\| \leqslant (n+1)\|f\|_2$. Thus let f be such a function and let $g \in \ell^2(G)$ be arbitrary and set $g_k = 1_{E_k} g$, $k \in \mathbb{N}_0$. Then g_k has support in E_k and $\|g\|_2^2 = \sum_{k=0}^{\infty} \|g_k\|_2^2$. Let

$$h = f * g = \sum_{k=0}^{\infty} f * g_k \quad \text{and} \quad h_m = 1_{E_m} h,\ m \in \mathbb{N}_0.$$

Note that $h \in \ell^2(G)$ and $\|h\|_2^2 = \sum_{m=0}^{\infty} \|h_m\|_2^2$. From Lemma 5.7.2, we get that

$$\|1_{E_m}(f * g_k)\|_2 \leqslant \|f\|_2 \|g_k\|_2,$$

if $|n-k| \leqslant m \leqslant n+k$ and $n+k-m$ is even, and $\|1_{E_M}(f * g_k)\|_2 = 0$ otherwise. Therefore

$$\begin{aligned}
\|h_m\|_2 &= \left\| \sum_{k=0}^{\infty} 1_{E_m}(f * g_k) \right\|_2 \leqslant \sum_{k=0}^{\infty} \|1_{E_m}(f * g_k)\|_2 \\
&\leqslant \|f\|_2 \left(\sum_{k=|m-n|,\, m+n-k \text{ even}}^{m+n} \|g_k\|_2 \right).
\end{aligned}$$

Writing $k = m+n-2l$, we get

$$\begin{aligned}
\|h_m\|_2 &\leqslant \|f\|_2 \left(\sum_{l=0}^{\min\{m,n\}} \|g_{m+n-2l}\|_2 \right) \\
&\leqslant \|f\|_2 \left(\sum_{l=0}^{\min\{m,n\}} \|g_{m+n-2l}\|_2^2 \right)^{1/2} \left(\sum_{l=0}^{\min\{m,n\}} 1_{\{l\}} \right)^{1/2} \\
&\leqslant (n+1)^{1/2} \|f\|_2 \left(\sum_{l=0}^{\min\{m,n\}} \|g_{m+n-2l}\|_2^2 \right)^{1/2}
\end{aligned}$$

and therefore

$$\begin{aligned}
\|h\|_2^2 &= \sum_{m=0}^{\infty} \|h_m\|_2^2 \\
&\leqslant (n+1)\|f\|_2^2 \sum_{m=0}^{\infty} \left(\sum_{l=0}^{\min\{m,n\}} \|g_{m+n-2l}\|_2^2 \right) \\
&= (n+1)\|f\|_2^2 \sum_{l=0}^{n} \left(\sum_{m=l}^{\infty} \|g_{m+n-2l}\|_2^2 \right) \\
&= (n+1)\|f\|_2^2 \sum_{l=0}^{n} \left(\sum_{k=n-l}^{\infty} \|g_k\|_2^2 \right) \\
&\leqslant (n+1)\|f\|_2^2 \sum_{l=0}^{\infty} \|g_l\|_2^2 \\
&= (n+1)^2 \|f\|_2^2 \|g\|_2^2.
\end{aligned}$$

This proves that $\|f * g\|_2 \leqslant (n+1)\|f\|_2\|g\|_2$ for every $g \in \ell^2(G)$, that is, $\|\lambda_G(f)\| \leqslant (n+1)\|f\|_2$. □

LEMMA 5.7.4. *Let f be a function on G with finite support. Then*

$$\|\lambda_G(f)\| \leqslant 2 \left(\sum_{x \in G} |f(x)|^2 (1+|x|)^4 \right)^{1/2}.$$

PROOF. It follows from Lemma 5.7.3 and the Cauchy-Schwarz inequality that

$$\begin{aligned}
\|\lambda_G(f)\| &\leqslant \sum_{n=0}^{\infty} (n+1)\|1_{E_n} f\|_2 = \sum_{n=0}^{\infty} \frac{1}{n+1} \left((n+1)^2 \|1_{E_n} f\|_2\right) \\
&\leqslant \left(\sum_{n=0}^{\infty} \left(\frac{1}{n+1} \right)^2 \right)^{1/2} \left(\sum_{n=0}^{\infty} (n+1)^4 \|1_{E_n} f\|_2^2 \right)^{1/2} \\
&= \sqrt{\frac{\pi^2}{6}} \left(\sum_{x \in G} |f(x)|^2 (1+|x|)^4 \right)^{1/2},
\end{aligned}$$

and hence the statement of the lemma since $\sqrt{\pi^2/6} \leqslant 2$. □

LEMMA 5.7.5. *Let u be a function on G for which*

$$\sup_{x \in G} |u(x)|(1+|x|)^2 < \infty.$$

Then u is a multiplier of $A(G)$ and

$$\|u\|_{M(A(G))} \leqslant 2 \sup_{x \in G} |u(x)| \cdot (1+|x|)^2.$$

PROOF. Let $F(G)$ denote the space of functions on G with finite support. For any $v \in F(G) \subseteq A(G)$ we have $uv \in F(G) \subseteq A(G)$ and, since $\lambda_G(F(G))$ is dense in

$VN(G) = A(G)^*$,

$$\begin{aligned}\|uv\|_{A(G)} &= \sup\{|\langle uv, \lambda_G(f)\rangle| : f \in F(G), \|\lambda_G(f)\| \leqslant 1\} \\ &= \sup\{|\langle v, \lambda_G(uf)\rangle| : f \in F(G), \|\lambda_G(f)\| \leqslant 1\} \\ &\leqslant \|v\|_{A(G)} \cdot \sup\{\|\lambda_G(uf)\| : f \in F(G), \|\lambda_G(f)\| \leqslant 1\}.\end{aligned}$$

Now, by Lemma 5.7.4,

$$\|\lambda_G(uf)\| \leqslant 2\left(\sum_{x\in G} |u(x)|^2|f(x)|^2(1+|x|)^4\right)^{1/2}.$$

Put $c = \sup_{x\in G} |u(x)| \cdot (1+|x|)^2$ and observe that $\|\lambda_G(f)\| \geqslant \|f * \delta_e\|_2 = \|f\|_2$. It follows that

$$\|\lambda_G(uf)\| \leqslant 2c\,\|f\|_2 \leqslant 2c\,\|\lambda_G(f)\|$$

for every $f \in F(G)$ and hence

$$\|uv\|_{A(G)} \leqslant 2c\,\|v\|_{A(G)}.$$

Since $F(G)$ is dense in $A(G)$, we conclude that $uA(G) \subseteq A(G)$ and $\|u\|_{M(A(G))} \leqslant 2c$, as stated. □

THEOREM 5.7.6. *Let G be a free group. Then there exists a net $(u_\alpha)_\alpha$ of functions in $A(G)$ with finite support such that for each $v \in A(G)$:*

(i) $\|u_\alpha v\|_{A(G)} \leqslant \|v\|_{A(G)}$ *for every* α.
(ii) $\|u_\alpha v - v\|_{A(G)} \to 0$.

PROOF. We assume first that G is finitely generated. By Lemma 5.7.1, the function u_γ, $\gamma > 0$, defined by $u_\gamma(x) = e^{-\gamma|x|}$, $x \in G$, is positive definite. For each $\gamma > 0$ and $n \in \mathbb{N}$, put

$$v_{\gamma,n}(x) = e^{-\gamma|x|} \text{ if } |x| \leqslant n \text{ and } v_{\gamma,n}(x) = 0 \text{ if } |x| > n.$$

Then $v_{\gamma,n} \in F(G)$ and by Lemma 5.7.5,

$$\|v_{\gamma,n} - u_\gamma\|_{M(A(G))} \leqslant 2 \sup_{|x|>n} e^{-\gamma|x|}(1+|x|)^2.$$

Since $e^{-\gamma m}(1+m)^2 \to 0$ as $m \to \infty$, it follows that $\|v_{\gamma,n} - u_\gamma\|_{M(A(G))} \to 0$ as $n \to \infty$ and, in particular,

$$\|v_{\gamma,n}\|_{M(A(G))} \to \|u_\gamma\|_{M(A(G))} = u_\gamma(e) = 1.$$

Thus, setting $u_{\gamma,n} = \|v_{\gamma,n}\|^{-1}_{M(A(G))} v_{\gamma,n}$, we have

$$\|u_{\gamma,n}\|_{M(A(G))} = 1 \text{ and } \|u_{\gamma,n} - u_\gamma\|_{M(A(G))} \to 0 \text{ as } n \to \infty.$$

Now let $v = \sum_{j=1}^r c_j\delta_{x_j} \in F(G)$ and let $n \geqslant \max\{|x_j| : 1 \leqslant j \leqslant r\}$. Then

$$\begin{aligned}\|u_\gamma v - v\|_{A(G)} &\leqslant \|(u_\gamma - u_{\gamma,n})v\|_{A(G)} + \|u_{\gamma,n}v - v\|_{A(G)} \\ &\leqslant \|u_\gamma - u_{\gamma,n}\|_{M(A(G))}\|v\|_{A(G)} \\ &\quad + \sum_{J=1}^r |c_j| \cdot \|u_{\gamma,n}(x_j)\delta_{x_j} - \delta_{x_j}\|_{A(G)} \\ &= \|u_\gamma - u_{\gamma,n}\|_{M(A(G))}\|v\|_{A(G)} + \sum_{j=1}^r |c_j| \cdot |e^{-\gamma|x_j|} - 1|.\end{aligned}$$

Since $\|u_{\gamma,n} - u_\gamma\|_{M(A(G))} \to 0$ for each γ as $n \to \infty$ and $e^{-\gamma|x|} \to 1$ as $\gamma \to 0$ for each x, it follows that $\|u_\gamma v - v\|_{A(G)} \to 0$ as $\gamma \to 0$ and therefore

$$\lim_{\gamma \to 0, n \to \infty} \|u_{\gamma,n} v - v\|_{A(G)} = 0.$$

Finally, since $F(G)$ is dense in $A(G)$ and $\|u_{\gamma,n}\|_{M(A(G))} = 1$, a simple triangle inequality argument shows that $u_{\gamma,n} v \to v$ as $\gamma \to 0$ and $n \to \infty$ for all $v \in A(G)$.

Now drop the hypothesis that G be finitely generated and let $\mathcal{H}$ denote the collection of all finitely generated subgroups of G. For $H \in \mathcal{H}$, the mapping

$$A(H) \to \widetilde{A(H)}, \ u \to \widetilde{u},$$

where $\widetilde{u}(x) = u(x)$ for $x \in H$ and $\widetilde{u}(x) = 0$ for $x \in G \backslash H$, is an isometric embedding of $A(H)$ into $A(G)$, and $\cup_{H \in \mathcal{H}} \widetilde{A(H)}$ is dense in $A(G)$. Moreover, by the first part of the proof, each $A(H)$ has an approximate identity with multiplier norm bounded by one, and for $u \in A(H)$,

$$\begin{aligned}
\|\widetilde{u}\|_{M(A(G))} &= \sup\{\|\widetilde{u}v\|_{A(G)} : v \in A(G), \|v\|_{A(G)} \leqslant 1\} \\
&\leqslant \sup\{\|\widetilde{u}\widetilde{w}\|_{A(G)} : w \in A(H), \|w\|_{A(H)} \leqslant 1\} \\
&= \sup\{\|uw\|_{A(H)} : w \in A(H), \|w\|_{A(H)} \leqslant 1\} \\
&= \|u\|_{M(A(H))}.
\end{aligned}$$

Using these facts, another simple triangle inequality argument completes the proof of the theorem. □

We can now apply the results of Section 4.6 to deduce from Theorem 5.7.6 that the Fourier algebra of $SL(2,\mathbb{R})$ has a multiplier bounded approximate identity.

COROLLARY 5.7.7. *$A(SL(2,\mathbb{R}))$ possesses an approximate identity which is bounded by one in the multiplier norm.*

PROOF. It is well-known that the two matrices

$$\begin{pmatrix} 1 & 2 \\ 0 & 1 \end{pmatrix} \quad \text{and} \quad \begin{pmatrix} 1 & 0 \\ 2 & 1 \end{pmatrix}$$

generate a closed subgroup H of $G = SL(2,\mathbb{R})$ which is isomorphic to $\mathbb{F}_2$. Moreover, G/H admits a finite G-invariant measure. Thus H is a lattice in $SL(2,\mathbb{R})$. By the preceding theorem, $A(H)$ has a approximate identity bounded by one in the multiplier norm. The statement now follows from Theorem 5.6.4. □

5.8. Notes and references

The fundamental Lemma 5.1.2, which gives various characterizations of multipliers of $A(G)$, can be found in [**27**]. The fact that $M(A(G)) = B(G)$ when G is amenable (Theorem 5.1.8) was independently shown in [**76**] and [**242**].

All the results about completely bounded multipliers and uniformly bounded representations given in Sections 5.4 and 5.5, are due to de Canniere and Haagerup [**27**], and our presentation follows very closely that of [**27**]. De Canniere and Haagerup were interested in when $M_{cb}(A(G))$, the space of completely bounded multipliers of $A(G)$, is strictly larger than $B(G)$. It was shown by Cowling [**37**] that the coefficient functions of uniformly bounded representations define multipliers of the Fourier algebra. In [**27**, Theorem 2.2] (Theorem 5.5.4) a slightly simplified proof was given, and it was moreover shown that these multipliers of $A(G)$

are even completely bounded. This raises the problem of characterizing those uniformly bounded representations which provide multipliers not belonging to $B(G)$. Within the class of cyclic representations, Theorem 5.5.5 [**27**, Theorem 2.3] identifies the ones in question as precisely those which are not similar to a unitary representation. The proof, however, is fairly difficult as it uses the solution, due to Haagerup [**112**], of the so-called similarity problem. As pointed out in [**27**, Section 2], many semisimple Lie groups (e.g. $SL(2,\mathbb{R}), SL(n,\mathbb{C})$ and $SU(n,1)$, $n \geqslant 2$) admit uniformly bounded irreducible representations which are not similar to unitary representations and thus lead to multipliers in $M_{cb}(A(G))\backslash B(G)$.

In [**111**], Haagerup proved that if G is a free group on at least two generators, then $C^*_\lambda(G)$, the reduced group C^*-algebra of G, has the metric approximation property, although it is not nuclear. As an application, he obtained that $A(G)$ has an approximate identity which is bounded by 1 in the multiplier norm.

CHAPTER 6

Spectral Synthesis and Ideal Theory

Let A be a regular and semisimple commutative Banach algebra with Gelfand spectrum $\sigma(A)$ and Gelfand representation $A \to C_0(\sigma(A))$, $a \to \widehat{a}$. We remind the reader that associated to any closed subset E of $\sigma(A)$ are two distinguished ideals of A, namely

$$I(E) = \{a \in A : \widehat{a} = 0 \text{ on } E\}$$

and

$$j(E) = \{a \in A : \widehat{a} \text{ has compact support disjoint from } E\}.$$

Then $I(E)$ and $j(E)$ are the largest and the smallest ideal of A with zero set E. The set E is called a *spectral set* or *set of synthesis* if $\overline{j(E)} = I(E)$, and one says that *spectral synthesis holds for* A if every closed subset of $\sigma(A)$ is a set of synthesis. Moreover, E is called a *Ditkin set* if $a \in \overline{aj(E)}$ for every $a \in I(E)$. The main purpose of this chapter is to study spectral sets and Ditkin sets and the spectral synthesis problem for Fourier algebras of locally compact groups.

A famous theorem of Malliavin [**207**] states that spectral synthesis fails for $A(G)$ whenever G is any nondiscrete abelian locally compact group. Using this and a deep theorem of Zelmanov [**293**] ensuring the existence of infinite abelian subgroups of infinite compact groups, we prove in Section 6.2 that for an arbitrary locally compact group G, under a mild additional hypothesis, spectral synthesis holds for $A(G)$ if and only if G is discrete.

One of the most interesting problems in the ideal theory of a commutative Banach algebra is to identify the closed ideals with bounded approximate identities. For Fourier algebras this problem is treated in Section 6.5. Improving on earlier results for abelian groups and employing operator space techniques, it is shown that for an amenable group G, the closed ideals in $A(G)$ with bounded approximate identities are in one-to-one correspondence with the closed sets in the coset ring of G. Moreover, for a closed subgroup H of G, the ideal $I(H)$ has an approximate identity with norm bounded by 2.

Suppose that G is amenable and let H be a closed normal subgroup of G. In Section 6.6 we prove the existence of a bijection $e : I \to e(I)$ between the set of all closed ideals of $A(G/H)$ and the set of all closed ideals of $A(G)$, which are invariant under translation by elements of H. Moreover, we show that a closed ideal I of $A(G/H)$ has a bounded approximate identity if and only if the same is true of $e(I)$.

After collecting some basic material and proving that closed subgroups are sets of synthesis (Section 6.1), we proceed in Sections 6.2 and 6.3 to establish various injection and projection theorems for spectral sets and Ditkin sets as well as their local variants. Given a closed subgroup H and a closed normal subgroup N of G, these results relate spectral sets and Ditkin sets for $A(H)$ and $A(G/N)$, respectively, to such sets for $A(G)$.

6.1. Sets of synthesis and Ditkin sets

Let G be a locally compact group and $A(G)$ the Fourier algebra of G. Recall that a closed subset E of $G = \sigma(A(G))$ is a *set of synthesis* (or *spectral set*) if $\overline{j(E)} = I(E)$ and E is a Ditkin set if $u \in \overline{uj(E)}$ for each $u \in I(E)$. Both notions admit local variants as follows. The set E is said to be a set of *local synthesis* (or *local spectral set*) if $I(E) \cap C_c(G) \subseteq \overline{j(E)}$, and E is called a *local Ditkin set* if $u \in \overline{uj(E)}$ for every $u \in I(E) \cap C_c(G)$. Clearly, every local Ditkin set is a local spectral set. Notice that $\varnothing$ is a set of synthesis since $A(G) \cap C_c(G)$ is dense in $A(G)$. On the other hand, $\varnothing$ is a Ditkin set if and only if $u \in \overline{uA(G)}$ for every $u \in A(G)$.

LEMMA 6.1.1. *Let E be an open and closed subset of G. Then E is a set of local synthesis. If, in addition, E has the property that $u \in \overline{uA(G)}$ for each $u \in I(E)$, then E is a set of synthesis.*

PROOF. Let $T \in VN(G)$ such that $\operatorname{supp} T \subseteq E$, and let $u \in I(E) \cap C_c(G)$. Choose $v \in A(G)$ such that $v = 1$ on $\operatorname{supp} u$. Since E is open in G, $\operatorname{supp} u \cap \operatorname{supp} T = \varnothing$ and hence $u \cdot T = 0$. Thus

$$\langle T, u\rangle = \langle T, uv\rangle = \langle u \cdot T, v\rangle = 0.$$

Now suppose that there exists a net $(v_\alpha)_\alpha$ in $A(G)$ such that $uv_\alpha \to u$. Then, as in the previous case, $u \cdot T = 0$ and hence

$$\langle T, u\rangle = \lim_\alpha \langle T, uv_\alpha\rangle = \lim_\alpha \langle u \cdot T, v_\alpha\rangle = 0,$$

as required. □

We shall introduce some notions which allow us to treat spectral set and Ditkin sets and their local variants to some extent simultaneously.

DEFINITION 6.1.2. Let X be an $A(G)$-invariant linear subspace of $VN(G)$. A closed subset E of G is called an *X-spectral set* or *set of X-synthesis* if for any $T \in X$ such that $\operatorname{supp} T \subseteq E$ we have $T \in I(E)^\perp$.

The set E is called an *X-Ditkin set* for $A(G)$ if for every $T \in X$ and $u \in I(E)$ there exists a net $(v_\alpha)_\alpha$ in $j(E)$ such that

$$\langle v_\alpha \cdot T, u\rangle \to \langle T, u\rangle.$$

REMARK 6.1.3. (a) We shall frequently use the following simple fact. Suppose that E is X-Ditkin. Then, given $T \in X$ and $u \in I(E)$, there exists $v \in j(E)$ such that $\langle T, u\rangle = \langle T, vu\rangle$. To see this, choose $v = 0$ when $\langle T, u\rangle = 0$, and if $\langle T, u\rangle \neq 0$, notice that $\{\langle T, vu\rangle : v \in j(E)\} = \mathbb{C}$.

(b) Every X-Ditkin set is an X-spectral set. Indeed, if $T \in X$ is such that $\operatorname{supp} T \subseteq E$ and $u \in I(E)$, then taking $v \in j(E)$ as in (a), it follows that $\langle T, u\rangle = \langle T, vu\rangle = 0$ since $vu \in j(E)$ and $\operatorname{supp} T \subseteq E$.

We next identify the proper choices of X to recover the original notions. For that, let $UC(\widehat{G})$ denote the closed linear span of

$$\{u \cdot T : T \in VN(G), u \in A(G)\}.$$

Then $UC(\widehat{G})$ is the norm closure of $UC_c(\widehat{G})$, the set of operators in $VN(G)$ with compact support. To explain the notation, we briefly mention that when G is abelian, $UC(\widehat{G})$ is precisely the C^*-algebra of bounded uniformly continuous functions on the dual group $\widehat{G}$ of G. Furthermore, $UC(\widehat{G})$ is a C^*-subalgebra of $VN(G)$ (see [**174**] and [**102**] for more details).

LEMMA 6.1.4. *Let E be a closed subset of G. Then*

(i) *E is of local synthesis if and only if E is of $UC_c(\widehat{G})$-synthesis.*
(ii) *E is of synthesis if and only if E is of $VN(G)$-synthesis.*

PROOF. We show (i), the proof of (ii) being similar (in fact, easier).

Suppose first that E is of local synthesis, and let $T \in UC_c(\widehat{G})$ such that $\operatorname{supp} T \subseteq E$. Choose $v \in A(G) \cap C_c(G)$ such that $v = 1$ in some open neighbourhood V of $\operatorname{supp} T$. Then $u - uv = 0$ on V for all $u \in A(G)$, and this implies $v \cdot T - T = 0$ (Proposition 2.5.3(ii)). Now, since $\operatorname{supp} T \subseteq E$, we have $\langle T, u\rangle = 0$ for all $u \in I(E) \cap C_c(G)$. It follows that, for $u \in I(E)$,

$$\langle T, u\rangle = \langle v \cdot T, u\rangle = \langle T, vu\rangle = 0,$$

since $vu \in I(E) \cap C_c(G)$.

Conversely, suppose that E is of $UC_c(\widehat{G})$-synthesis, and let $u \in I(E) \cap C_c(G)$. We have to show that $\langle T, u\rangle = 0$ whenever $T \in VN(G)$ annihilates $j(E)$. Choose $v \in A(G) \cap C_c(G)$ such that $v = 1$ on $\operatorname{supp} u$. Then $v \cdot T \in UC_c(\widehat{G})$, and hence $v \cdot T$ annihilates $I(E)$. Thus $\langle T, u\rangle = \langle T, vu\rangle = 0$. □

LEMMA 6.1.5. *Let E be a closed subset of G. Then*

(i) *E is a local Ditkin set if and only if E is $UC_c(\widehat{G})$-Ditkin.*
(ii) *E is a Ditkin set if and only if E is $VN(G)$-Ditkin.*

PROOF. Suppose first that E is a Ditkin set (respectively, a local Ditkin set), and let $u \in I(E)$ and $T \in VN(G)$ (respectively, $T \in UC_c(\widehat{G})$). If $T \in UC_c(\widehat{G})$, then choose $v \in A(G) \cap C_c(G)$ such that $v \cdot T = T$. Now, by hypothesis, there exists a net $(v_\alpha)_\alpha$ in $j(E)$ such that

$$v_\alpha u \to u \quad \text{and} \quad v_\alpha(uv) \to uv$$

in $A(G)$, respectively. It follows that

$$\langle v_\alpha \cdot T - T, u\rangle = \langle T, v_\alpha u - u\rangle \to 0$$

in the first case, whereas in the second case

$$\langle v_\alpha \cdot T - T, u\rangle = \langle v_\alpha \cdot (v \cdot T) - v \cdot T, u\rangle = \langle T, v_\alpha uv - uv\rangle \to 0.$$

Conversely, suppose that E is $UC_c(\widehat{G})$-Ditkin, and let $T \in VN(G)$ and $u \in I(E) \cap C_c(G)$. Choose $v \in A(G) \cap C_c(G)$ such that $v = 1$ on $\operatorname{supp} u$. Since $v \cdot T \in UC_c(\widehat{G})$, there exists a net $(v_\alpha)_\alpha$ in $j(E)$ such that

$$\langle v_\alpha \cdot (v \cdot T), u\rangle \to \langle v \cdot T, u\rangle,$$

and hence, since $vu = u$,

$$\langle T, v_\alpha u\rangle \to \langle T, u\rangle.$$

Thus $\langle T, u\rangle = 0$ whenever T annihilates $uj(E)$, as required. The proof that E is Ditkin if it is $VN(G)$-Ditkin is even simpler. □

PROPOSITION 6.1.6. *Let G be a locally compact satisfying $u \in \overline{uA(G)}$ for every $u \in A(G)$, and let E be a closed subset of G.*

(i) *If E is a $UC_c(\widehat{G})$-spectral set, then E is a spectral set.*
(ii) *If E is $UC_c(\widehat{G})$-Ditkin, then E is a Ditkin set.*

PROOF. Choose a net $(u_\alpha)_\alpha$ in $A(G) \cap C_c(G)$ such that $\|u_\alpha u - u\| \to 0$.

Suppose first that E is $UC_c(\widehat{G})$-spectral, and let $u \in I(E)$ and $T \in VN(G)$ such that supp $T \subseteq E$. Then, since $u_\alpha \cdot T \in UC_c(\widehat{G})$ and supp $u_\alpha \cdot T \subseteq E$, we have that $\langle u_\alpha \cdot T, u\rangle = 0$. Hence $\langle T, u\rangle = 0$ since $\langle T, u_\alpha u\rangle \to \langle T, u\rangle$.

Now, let E be $UC_c(\widehat{G})$-Ditkin and let $T \in VN(G)$ and $u \in I(E)$. We have to show that there exists $v \in j(E)$ such that $\langle T, u\rangle = \langle T, vu\rangle$. We can assume that $\langle T, u\rangle \neq 0$. Since $u_\alpha \cdot T \in UC_c(\widehat{G})$, for each α there is a $v_\alpha \in j(E)$ such that

$$\langle u_\alpha \cdot T, u\rangle = \langle u_\alpha \cdot T, v_\alpha u\rangle.$$

Since $\langle T, u_\alpha u\rangle \to \langle T, u\rangle$, $\langle T, u\rangle \neq 0$ and $u_\alpha v_\alpha \in j(E)$, we must have that $\{\langle T, vu\rangle : v \in j(E)\} = \mathbb{C}$, whence $\langle T, u\rangle = \langle T, vu\rangle$ for some $v \in j(E)$. □

Proposition 6.1.6 shows that if $A(G)$ has an approximate identity in the weakest possible sense, then local spectral sets (local Ditkin sets) are necessarily spectral sets (Ditkin sets).

REMARK 6.1.7. Let $X = (X_*)^*$ be a dual Banach space and let Y be a w^*-closed linear subspace of X and

$${}_\perp Y = \{u \in X_* : \langle f, u\rangle = 0 \text{ for all } f \in Y\}.$$

Then $Y = ({}_\perp Y)^\perp$. The inclusion $Y \subseteq ({}_\perp Y)^\perp$ being clear, suppose there exists $g \in ({}_\perp Y)^\perp \backslash Y$. Then, by the Hahn-Banach theorem, we find $u \in X_*$ such that $\langle u, f\rangle = 0$ for all $f \in Y$, but $\langle u, g\rangle \neq 0$. So $u \in {}_\perp Y$ and $g \notin ({}_\perp Y)^\perp$, a contradiction.

LEMMA 6.1.8. *Let E be a closed subset of the locally compact group G and let $VN_E(G)$ be the w^*-closure of the linear span of the set $\{\lambda_G(x) : x \in E\}$. Then the following are equivalent:*

(i) *E is a set of synthesis for $A(G)$.*
(ii) *For any $T \in VN(G)$,* supp *$T \subseteq E$ implies that $T \in VN_E(G)$.*

PROOF. We apply the above remark with $X = VN(G)$, $X_* = A(G)$ and $Y = VN_E(G)$. Then

$$I(E) = \{u \in A(G) : \langle u, \lambda_G(x)\rangle = 0 \text{ for all } x \in E\} = {}_\perp VN_E(G),$$

and hence $VN_E(G) = ({}_\perp VN_E(G))^\perp = I(E)^\perp$.

(i) ⇒ (ii) Let $T \in VN(G)$ be such that supp $T \subseteq E$. Then $\langle T, u\rangle = 0$ for all $u \in A(G) \cap C_c(G)$ with $\operatorname{supp} u \cap E = \varnothing$ (Proposition 2.5.3(ii)). It follows from (i) and the above remark that $T \in j(E)^\perp = J(E)^\perp = I(E)^\perp = VN_E(G)$.

(ii) ⇒ (i) Suppose that E fails to be a set of synthesis. Then $J(E)$ is a proper closed subspace of $I(E)$. Choose $u \in I(E) \backslash J(E)$. Then, by the Hahn-Banach theorem, there exists $T \in VN(G)$ such that $T \in J(E)^\perp$ and $\langle T, u\rangle \neq 0$. Proposition 2.5.3(ii) implies that $\operatorname{supp} T \subseteq E$, but $T \notin I(E)^\perp = VN_E(G)$. This contradicts (ii). □

The following theorem together with Theorem 6.1.12 already provides us with a wealth of sets of synthesis for $A(G)$.

THEOREM 6.1.9. *Let G be a locally compact group and H a closed subgroup of G. Then H is a set of synthesis for $A(G)$.*

PROOF. The statement follows from Lemma 6.1.8. □

COROLLARY 6.1.10. *Let H be a closed subgroup of G and $a \in G$. Then the coset aH is a set of synthesis.*

PROOF. We only have to note that, for any closed subset E of G, the map $u \to L_a u$ is an isometric algebra isomorphism mapping $I(E)$ and $j(E)$ onto $I(aE)$ and $j(aE)$, respectively. □

THEOREM 6.1.11. *Let G be a locally compact group and X an $A(G)$-invariant linear subspace of $VN(G)$. Suppose that E_1 and E_2 are closed subsets of G such that $E_1 \cap E_2$ is an X-Ditkin set. Then $E_1 \cup E_2$ is an X-spectral set if and only if both E_1 and E_2 are X-spectral sets.*

PROOF. Suppose first that E_1 and E_2 are X-spectral sets, and let $T \in X$ be such that $\operatorname{supp} T \subseteq E_1 \cup E_2$ and $u \in I(E_1 \cup E_2)$. Since $E_1 \cap E_2$ is X-Ditkin, there exists $v \in j(E_1 \cap E_2)$ such that $\langle T, u\rangle = \langle v \cdot T, u\rangle$. Since v has compact support disjoint from $E_1 \cap E_2$, there are compact sets F_1 and F_2 such that

$$\operatorname{supp}(v \cdot T) = F_1 \cup F_2 \quad \text{and} \quad F_j \subseteq E_j \backslash (E_1 \cap E_2),\ j = 1, 2.$$

Now there exist $v_j \in A(G) \cap C_c(G)$, $j = 1, 2$, such that $v_j = 1$ on a neighbourhood of F_j and $\operatorname{supp} v_1 \cap \operatorname{supp} v_2 = \varnothing$. Then $(v_1 + v_2)v \cdot T = v \cdot T$ since $v_1 + v_2 = 1$ on some neighbourhood of $F_1 \cup F_2$. Moreover,

$$\operatorname{supp}(v_j v) \cdot T \subseteq E_j \quad \text{and} \quad (v_j v) \cdot T \in X \ \ (j = 1, 2),$$

since X is $A(G)$-invariant. Since E_1 and E_2 are of X-synthesis, it follows that $\langle (v_j v) \cdot T, u\rangle = 0$ for $j = 1, 2$, an d hence $\langle v \cdot T, u\rangle = 0$, as was to be shown.

Conversely, suppose that $E_1 \cup E_2$ is of X-synthesis, and let $T \in X$ with $\operatorname{supp} T \subseteq E_1$ and $u \in I(E_1)$ be given. As above, since $E_1 \cap E_2$ is an X-Ditkin set, $\langle T, u\rangle = \langle T, vu\rangle$ for some $v \in j(E_1 \cap E_2)$. Since $\operatorname{supp}(v \cdot T)$ is a compact set contained in $E_1 \backslash E_2$, there exists $w \in A(G)$ so that $w = 1$ on a compact neighbourhood of $\operatorname{supp}(v \cdot T)$ and $w = 0$ on E_2. It follows that

$$wu \in I(E_1 \cup E_2) \quad \text{and} \quad v \cdot T = (wv) \cdot T.$$

Now, since $v \cdot T \in X$, $\operatorname{supp}(v \cdot T) \subseteq E_1 \cup E_2$ and $E_1 \cup E_2$ is of X-synthesis,

$$\langle T, u\rangle = \langle v \cdot T, u\rangle = \langle v \cdot T, wu\rangle = 0.$$

This shows that E_1 is an X-spectral set. In the same manner it is shown that E_2 is an X-spectral set. □

THEOREM 6.1.12. *Let G and X be as in Theorem 6.1.11, and let E and F be closed subsets of G.*

(i) *If both E and F are X-Ditkin sets, then $E \cup F$ is an X-Ditkin set.*
(ii) *If $E \cup F$ and $E \cap F$ are X-Ditkin sets, then E and F are X-Ditkin sets.*

PROOF. (i) Given $u \in I(E \cup F)$ and $T \in X$, there exist $v \in J(E)$ such that $\langle T, u\rangle = \langle T, uv\rangle$ and then $w \in J(F)$ such that $\langle T, uv\rangle = \langle T, (uv)w\rangle$. Thus $vw \in J(E \cup F)$ and

$$\langle T, u\rangle = \langle T, u(vw)\rangle,$$

as required.

(ii) Suppose that $E \cap F$ and $E \cup F$ are both X-Ditkin sets, and let $T \in X$ and $u \in I(E)$. Since $E \cap F$ is X-Ditkin, there exists $v \in j(E \cap F)$ such that

$\langle T, u\rangle = \langle T, uv\rangle$. Let $C = F \cap \mathrm{supp}(vu)$, a compact set disjoint from E. Thus there exists $w \in A(G) \cap C_c(G)$ such that $w = 0$ on a neighbourhood of E and $w = 1$ on C. So, in particular, $w \in J(E)$. Let $u' = vu - vwu$, then $u' \in I(E \cup F)$ since $w = 1$ on C. Since $E \cup F$ is an X-Ditkin set, there exists $v' \in J(E \cup F)$ such that $\langle T, u'\rangle = \langle T, u'v'\rangle$. It follows that $(v - vw)v' + vw \in J(E)$ and

$$\begin{aligned}\langle T, u\rangle &= \langle T, vu\rangle = \langle T, u'\rangle + \langle T, uvw\rangle \\ &= \langle T, u'v'\rangle + \langle T, uvw\rangle \\ &= \langle T, u((v - vw)v' + vw)\rangle.\end{aligned}$$

This proves that E is X-Ditkin, and similarly for F. □

The following corollary strengthens the second statement of Lemma 6.1.1.

COROLLARY 6.1.13. *Suppose that $A(G)$ has an approximate identity, and let E be an open and closed subset of G. Then E is a Ditkin set.*

PROOF. Since, by hypothesis, the empty set is a Ditkin set for $A(G)$, the claim follows by applying Theorem 6.1.12 to E and $F = G\backslash E$. □

6.2. Malliavin's theorem for $A(G)$

As mentioned in the introduction to this chapter, according to Malliavin's theorem, spectral synthesis holds for the Fourier algebra $A(G)$ of a locally compact abelian group G if and only if G is discrete. In this section, we extend this result to general locally compact groups.

LEMMA 6.2.1. *Let K be a compact normal subgroup of G and $q : G \to G/K$ the quotient homomorphism. Let E be a closed subset of G/K. If $q^{-1}(E)$ is a local spectral set for $A(G)$, then E is a local spectral set for $A(G/K)$.*

PROOF. Let $u \in I(E) \cap C_c(G/K)$ and $\epsilon > 0$ be given, and let $u_1 = u \circ q \in A(G)$. Then, since K is compact, $u_1 \in I(q^{-1}(E)) \cap C_c(G)$. Hence there exists $v_1 \in j(q^{-1}(E))$ such that $\|u_1 - v_1\|_{A(G)} \leqslant \epsilon$. Now define v on G/K by

$$v(xK) = \int_K v_1(xk)\, dk = \int_K R_k v_1(x)\, dk,$$

where dk denotes normalized Haar measure of K. Then $v \in A(G/K)$ and

$$\|u - v\|_{A(G/K)} = \left\| \int_K R_k(u_1 - v_1)\, dk \right\|_{A(G)} \leqslant \|u_1 - v_1\|_{A(G)} \leqslant \epsilon.$$

Then v has compact support since $C = \mathrm{supp}\, v_1$ is compact and $\mathrm{supp}\, v \subseteq q(C)$. Moreover, v vanishes in a neighbourhood of E. Indeed, since $C \cap q^{-1}(E) = \varnothing$, there exists a symmetric neighbourhood V of e in G such that $VC \cap q^{-1}(E) = \varnothing$ and hence $C \cap Vq^{-1}(E) = \varnothing$. Since $v_1 = 0$ on $Vq^{-1}(E)$, v vanishes on the neighbourhood $q(Vq^{-1}(E))$ of E. Thus $v \in j(E)$, as required. □

PROPOSITION 6.2.2. *Let G be a nontrivial connected locally compact group. Then local spectral synthesis fails for $A(G)$.*

PROOF. Towards a contradiction, assume that local spectral synthesis holds for $A(G)$. Since $G \neq \{e\}$, G has a proper compact subgroup K such that G/K is a Lie group. By Lemma 6.2.1, local spectral synthesis holds for $A(G/K)$. Now the nontrivial connected Lie group G/K contains a closed nondiscrete abelian subgroup

H (a one-parameter group). Then local spectral synthesis holds for $A(H)$ as well (Corollary 6.3.4). This contradicts Malliavin's theorem. □

THEOREM 6.2.3. *Let G be a locally compact group. Then*

(i) *Local spectral synthesis holds for $A(G)$ if and only if G is discrete.*
(ii) *Spectral synthesis holds for $A(G)$ if and only if G is discrete and $u \in \overline{uA(G)}$ for every $u \in A(G)$.*

PROOF. Suppose first that local spectral synthesis holds for $A(G)$. Then, G must be totally disconnected. Fix a compact open subgroup K of G. We have to show that K is finite. Towards a contradiction, assume that K is infinite. Then, by a theorem of Zelmanov [**293**, Theorem 2], K contains an infinite abelian closed subgroup H. Now local spectral synthesis holds for $A(H)$, and hence so does spectral synthesis since H is abelian. This contradicts Malliavin's theorem. Thus G is discrete. Conversely, if G is discrete, then local spectral synthesis holds for $A(G)$ by Lemma 6.1.1.

For (ii), notice first that if synthesis holds for $A(G)$, then G is discrete by (i) and $u \in \overline{uA(G)}$ for each $u \in A(G)$. Indeed, denoting by E the zero set of u, we have $h(uA(G)) = E$ and therefore $\overline{uA(G)} = I(E)$. Again, the converse follows from Lemma 6.1.1. □

6.3. Injection theorems for spectral sets and Ditkin sets

The subject of this section is to establish so-called injection theorems, which constitute an additional method to produce sets of synthesis and Ditkin sets.

REMARK 6.3.1. Let H be a closed subgroup of G and let $r : A(G) \to A(H)$ denote the restriction map (Section 2.3).

(1) For any $A(G)$-invariant linear subspace X of $VN(G)$ and closed subgroup H of G, let

$$X_H = r^{*-1}(X).$$

Then X_H is an $A(H)$-invariant linear subspace of $VN(H)$. Indeed, if $v \in A(H)$ and $S \in X_H$, then choosing $u \in A(G)$ with $r(u) = v$,

$$r^*(v \cdot S) = u \cdot r^*(S) \in X,$$

and hence $v \cdot S = r^{*-1}(u \cdot r^*(S)) \in r^{*-1}(X)$.

(2) If $S \in X_H$ and $\operatorname{supp} S \subseteq E$ for some closed subset E of H, then $\operatorname{supp} r^*(S) \subseteq E$. In fact, let $x \in \operatorname{supp} r^*(S)$. Then there exists a net $(u_\alpha)_\alpha$ in $A(G)$ such that

$$r^*(r(u_\alpha) \cdot S) = u_\alpha \cdot r^*(S) \;\to\; \lambda_G(x) = r^*(\lambda_H(x)).$$

Hence $r(u_\alpha) \cdot S \to \lambda_H(x)$.

LEMMA 6.3.2. *Let H be a closed subgroup of G. Then*

$$r^*(UC_c(\widehat{H})) = UC_c(\widehat{G}) \cap VN_H(G).$$

PROOF. First, let $T \in UC_c(\widehat{H})$. There exists $w \in A(H) \cap C_c(H)$ such that $w \cdot T = T$. Next, choose $u_1 \in A(G)$ with $u_1|_H = w$ and $u_2 \in A(G) \cap C_c(G)$ such that $u_2 = 1$ on the compact set supp w, and put $v = u_1u_2$. Then $v \in A(G) \cap C_c(G)$ and $v|_H = w$. For all $u \in A(G)$, it follows that

$$\begin{aligned}\langle r^*(T), u\rangle &= \langle r^*(w \cdot T), u\rangle = \langle w \cdot T, r(u)\rangle \\ &= \langle T, wr(u)\rangle = \langle T, r(vu)\rangle \\ &= \langle r^*(T), vu\rangle = \langle v \cdot r^*(T), u\rangle.\end{aligned}$$

Thus $r^*(T) = v \cdot r^*(T)$, which belongs to $UC_c(\widehat{G}) \cap VN_H(G)$.

Conversely, let $S \in UC_c(\widehat{G}) \cap VN_H(G)$ and let $T \in VN(H)$ such that $r^*(T) = S$. Since supp S is compact, there exists $v \in A(G) \cap C_c(G)$ such that $v = 1$ on a neighbourhood of supp S. Then $v \cdot S = S$ and $r(v) \cdot T \in UC_c(\widehat{H})$. Moreover, $r^*(r(v) \cdot T) = S$. Indeed, for all $u \in A(G)$,

$$\begin{aligned}\langle r^*(r(v) \cdot T), u\rangle &= \langle r(v) \cdot T, r(u)\rangle = \langle T, r(vu)\rangle \\ &= r^*(T), vu\rangle = \langle S, vu\rangle \\ &= \langle v \cdot S, u\rangle = \langle S, u\rangle.\end{aligned}$$

This shows that $UC_c(\widehat{G}) \cap VN_H(G) \subseteq r^*(UC_c(\widehat{H}))$. □

THEOREM 6.3.3. *Let G be a locally compact group and let X be an $A(G)$-invariant linear subspace of $VN(G)$. Let H be a closed subgroup of G and E a closed subset of H. Then E is an X-spectral set for $A(G)$ if and only if it is an X_H-spectral set for $A(H)$.*

PROOF. Suppose first that E is of X-synthesis and let $S \in X_H = r^{*-1}(X)$ be such that supp $S \subseteq E$. Then $r^*(S) \in X$ and supp $r^*(S) \subseteq E$ by Remark 6.3.1(2). Then, by hypothesis, for every $u \in I(E)$,

$$0 = \langle r^*(S), u\rangle = \langle S, r(u)\rangle.$$

Since $r(I(E)) = \{v \in A(H) : v|_E = 0\}$, it follows that E is of X_H-synthesis.

Conversely, suppose that E is of X-synthesis, and let $T \in X$ such that supp $T \subseteq E$. Since H is a set of synthesis for $A(G)$ by Theorem 6.1.9, T annihilates $I(H)$. So there exists a unique $S \in X_H$ such that $r^*(S) = T$. Clearly, supp $S \subseteq E$ also and hence, by hypothesis, $\langle T, u\rangle = \langle S, r(u)\rangle = 0$ for all $u \in I(E)$, as required. □

COROLLARY 6.3.4. *Let H be a closed subgroup of G and E a closed subset of H. Then E is a spectral set (local spectral set) for $A(G)$ if and only if E is a spectral set (local spectral set) for $A(H)$.*

PROOF. Since $VN(G)_H = r^{*-1}(VN(G)) = VN(H)$, the statement about spectral sets follows from Lemma 6.1.4 and Theorem 6.3.3, taking $X = VN(G)$. Turning to local spectral sets, recall from Lemma 6.3.2 that

$$UC_c(\widehat{H}) = r^{*-1}(UC_c(\widehat{G}) \cap VN_H(G)) = r^{*-1}(UC_c(\widehat{G})) = UC_c(\widehat{G})_H.$$

Now apply Theorem 6.3.3 with $X = UC_c(\widehat{G})$ and Lemma 6.1.4. □

PROPOSITION 6.3.5. *Let X be an $A(G)$-invariant linear subspace of $VN(G)$. Let H be a closed subgroup of G and E a closed subset of H. If E is an X-Ditkin set for $A(G)$, then E is an X_H-Ditkin set for $A(H)$.*

PROOF. It suffices to show that given $u \in A(H)$ such that $u|_E = 0$ and $S \in X_H$, there exists $v \in A(H)$ which vanishes in a neighbourhood of E in H and satisfies

$\langle S, u\rangle = \langle S, vu\rangle$. Choose $u_1 \in A(G)$ extending u. Since $r^*(S) \in X$ and E is X-Ditkin, there exists $v_1 \in A(G)$ such that $v_1 = 0$ in a neighbourhood of E in G and $\langle r^*(S), u_1\rangle = \langle r^*(S), v_1u_1\rangle$ (Remark 6.1.3). Then

$$\begin{aligned}\langle S, u\rangle &= \langle S, r(u_1)\rangle = \langle r^*(S), u_1\rangle = \langle r^*(S), v_1u_1\rangle \\ &= \langle S, r(v_1)u\rangle,\end{aligned}$$

and hence $v = r(v_1)$ has the required property. □

THEOREM 6.3.6. *Let G be an amenable locally compact group, H a closed subgroup of G and E a closed subset of H. If E is a Ditkin set (local Ditkin set) for $A(H)$, then it is a Ditkin set (local Ditkin set) for $A(G)$.*

PROOF. Let E be a Ditkin set for $A(H)$. We show that given $u \in I(E)$ and $\epsilon > 0$, there exists $v \in j(E)$ such that $\|u - vu\|_{A(G)} \leqslant C\,\epsilon$ for some constant $C > 0$. There exists $w_1 \in A(H)$ such that $w_1 = 0$ in a neighbourhood of E in H and $\|u|_H - w_1(u|_H)\| \leqslant \epsilon$. Choose $w_2 \in A(G)$ extending w_1. By Theorem 6.3.3, E is a set of synthesis for $A(G)$. So there exists $w_3 \in j(E)$ such that $\|w_2 - w_3\| \leqslant \epsilon/\|u\|$. Then

$$\begin{aligned}\|r(u - w_3u)\| &\leqslant \|r(u) - r(w_2u)\| + \|r(u)(r(w_2) - r(w_3))\| \\ &\leqslant \|r(u) - w_1r(u)\| + \|u\|\|w_2 - w_3\| \leqslant 2\epsilon.\end{aligned}$$

Now let $w_4 = u - uw_3 \in A(G)$. Since G is amenable, by Lemma 6.3.7 below there exists $w_5 \in j(H)$ such that

$$\|w_4 - w_4w_5\| \leqslant 3\,\|r(w_4)\| + \epsilon,$$

and hence $\|w_4 - w_4w_5\| \leqslant 7\epsilon$. Finally, let

$$v = w_3 + w_5 - w_3w_5 \in A(G).$$

Then v has compact support since both w_3 and w_5 have compact support. Moreover, v vanishes in a neighbourhood of E in G since w_3 does so and w_5 vanishes in a neighbourhood of H. Thus $v \in j(E)$ and

$$\|u - vu\|_{A(G)} = \|w_4 - w_5w_4\|_{A(G)} \leqslant 7\epsilon.$$

This shows that E is Ditkin set for $A(G)$. The proof of the local version is entirely analogous. □

LEMMA 6.3.7. *Let G be an amenable locally compact group and H a closed subgroup of G. Given $u \in A(G)$ and $\epsilon > 0$, there exists $v \in A(G)$ such that v vanishes in a neighbourhood of H and satisfies*

$$\|u - vu\| \leqslant 3\,\|u|_H\| + \epsilon.$$

PROOF. Since

$$\|u|_H\| = \inf\{\|u + w\| : w \in I(H)\},$$

we find $w \in I(H)$ such that

$$\|u - w\| \leqslant \|u|_H\| + \epsilon/4.$$

Now H is a set of synthesis (Theorem 6.1.9) and since G is amenable, the ideal $I(H)$ has an approximate identity of norm bound 2. Hence there exists $v \in j(H)$

with $\|v\| \leqslant 2$ and $\|w - vw\| \leqslant \epsilon/4$. It follows that

$$\begin{aligned} \|u - vu\| &\leqslant \|u - w\| + \|w - wv\| + \|v(w - u)\| \\ &\leqslant \|u|_H\| + \epsilon/2 + \epsilon/4 + 2(\|u|_H\| + \epsilon/4) \\ &= 3\|u|_H\| + \epsilon, \end{aligned}$$

as was to be shown. □

6.4. A projection theorem for local spectral sets

Let G be a locally compact abelian group, H a closed subgroup of G and $r : \widehat{G} \to \widehat{H}$ the restriction map. Then a closed subset E of $\widehat{H}$ is a local spectral set for $L^1(H)$ if and only if $r^{-1}(E)$ is a local spectral set for $L^1(G)$ [**241**, Section 7.3]. In this section we present the analogue, due to Lohoué [**196**], for Fourier algebras of general locally compact groups and their quotient groups. The proof of Theorem 6.4.6 below in the abelian case is essentially based on the Poisson formula on the dual group of G. Since such a nice dual object is not available for general locally compact groups, the approach has to be completely different and the arguments used are naturally much more complicated.

LEMMA 6.4.1. *Let $\gamma : G \to \mathbb{C}^\times$ be a continuous homomorphism from G into the multiplicative group of nonzero complex numbers.*

(i) *Then $\gamma u \in A(G)$ for every $u \in A(G) \cap C_c(G)$.*
(ii) *Given a compact subset K of G, there exists a constant $C(K)$ such that, for all $u \in A(G)$ with* $\operatorname{supp} u \subseteq K$,

$$\|\gamma u\| \leqslant C(K)\|u\|.$$

PROOF. Let K be a compact subset of G. Then we can find compact subsets K_1 and K_2 of positive measure such that the function

$$w = 1_{K_1K_2} * |K_2|^{-1}1_{K_2}$$

is identically 1 on K. If $u \in A(G)$ is such that $\operatorname{supp} u \subseteq K$, then $\gamma u = \gamma wu$. Therefore, (i) follows once we have recognized that $\gamma w \in A(G)$. For that, let

$$v_1 = \gamma \cdot 1_{K_1K_2} \quad \text{and} \quad v_2 = |K_2|^{-1}\gamma^{-1} \cdot 1_{K_2}.$$

Then $v_1, v_2 \in L^2(G)$ and hence $\gamma w = v_1 * \check{v}_2 \in A(G)$. Moreover,

$$\|v_1\|_2 \leqslant \sup_{x \in K_1K_2} |\gamma(x)| \cdot \|1_{K_1K_2}\|_2 = \frac{1}{|K_1K_2|^{1/2}} \sup_{x \in K_1K_2} |\gamma(x)|,$$

and, since γ is multiplicative,

$$\|v_2\|_2 \leqslant \sup_{x \in K_2} \frac{1}{|\gamma(x)|} \cdot \|1_{K_2}\|_2 = \frac{1}{|K_2|^{1/2}} \sup_{x \in K_2^{-1}} |\gamma(x)|.$$

Now, define $C(K)$ by

$$C(K) = |K_1K_2|^{-1/2}|K_2|^{-1/2} \cdot \sup_{x \in K_1K_2} |\gamma(x)| \cdot \sup_{x \in K_2^{-1}} |\gamma(x)| < \infty.$$

Then $\|\gamma w\| \leqslant \|v_1\|_2\|v_2\|_2 \leqslant C(K)$ and hence

$$\|\gamma u\| = \|(\gamma w)u\| \leqslant C(K)\|u\|.$$

This establishes (ii). □

Since, given any compact subset K of G, there exists $u \in A(G) \cap C_c(G)$ such that $u = 1$ on K, the statement of Lemma 6.4.1 can be rephrased by saying that γ belongs locally to $A(G)$. In particular, the modular function of G belongs locally to $A(G)$.

COROLLARY 6.4.2. *Let $S \in VN(G)$ with compact support, and let $\gamma : G \to \mathbb{C}^\times$ be a continuous homomorphism. Then there exists an operator, denoted γS, in $VN(G)$ such that $\langle \gamma S, u\rangle = \langle S, \gamma u\rangle$ for all $u \in A(G) \cap C_c(G)$. Moreover,* $\operatorname{supp}(\gamma S) = \operatorname{supp} S$.

PROOF. According to Lemma 6.4.1, we can define γS on $A(G) \cap C_c(G)$ by $\langle \gamma S, u\rangle = \langle S, \gamma u\rangle$. Fix a compact neighbourhood K of $\operatorname{supp} S$ and let $C(K)$ as given by Lemma 6.4.1. Choose $w \in A(G) \cap C_c(G)$ such that $w = 1$ on a neighbourhood of $\operatorname{supp} S$ and $\operatorname{supp} w \subseteq K$. Then $w \cdot S = S$ and, for any $u \in A(G) \cap C_c(G)$,

$$\begin{aligned} |\langle \gamma S, u\rangle| &= |\langle S, \gamma u\rangle| = |\langle w \cdot S, \gamma u\rangle| = |\langle S, \gamma(wu)\rangle| \\ &\leqslant \|S\| \cdot \|\gamma(wu)\| \leqslant \|S\| \cdot C(K) \cdot \|wu\| \\ &\leqslant C(K)\|S\| \cdot \|w\| \cdot \|u\|. \end{aligned}$$

Thus γS extends uniquely to a bounded linear operator on $A(G)$.

For the last statement, let $a \in \operatorname{supp} S$ and $u \in A(G)$ with $u \cdot (\gamma S) = 0$. Then, for all $v \in A(G) \cap C_c(G)$,

$$\langle u \cdot S, v\rangle = \langle u \cdot (\gamma S), \gamma^{-1} v\rangle = 0,$$

so that $u \cdot S = 0$ and hence $u(a) = 0$. This shows that $\operatorname{supp} S \subseteq \operatorname{supp}(\gamma S)$. The reverse inclusion follows by replacing γ with γ^{-1}. □

LEMMA 6.4.3. *Let N be a closed normal subgroup of G and K a compact subset of G. Then there exists a constant $c(K) > 0$ with the following property: If $u \in A(G)$ is such that* $\operatorname{supp} u \subseteq K$, *then $T_N(u) \in A(G/N)$ and*

$$\|T_N(u)\|_{A(G/N)} \leqslant c(K)\|u\|_{A(G)}.$$

PROOF. There exist $f, g \in L^2(G)$ such that $u = f * \check{g}$ and $\|u\| = \|f\|_2\|g\|_2$. Choose a compact symmetric neighbourhood V of e and set $v_1 = 1_{KV}, v_2 = |V|^{-1}1_V$ and $v = v_1 * v_2 \in A(G)$. Then $v = 1$ on K and, for every $x \in G$,

$$\begin{aligned} v(x)(f * \check{g})(x) &= \int_G v_1(xy)v_2(y)\,dy \cdot \int_G f(xz)g(z)\,dz \\ &= \int_G\int_G f(xyz)v_1(xy)g(yz)v_2(y)\,dzdy \\ &= \int_G\int_G (R_z f)v_1(xy)(R_z g \cdot v_2)\check{}(y^{-1})\,dydz \\ &= \int_G ((R_z f \cdot v_1) * (R_z g \cdot v_2)\check{})(x)\,dz. \end{aligned}$$

Since the map T_N commutes with vector-valued integration and T_N is a homomorphism from $L^1(G)$ onto $L^1(G/N)$, we get

$$\begin{aligned} T_N(v(f * \check{g}))(\dot{x}) &= \int_G T_N\left[(R_z f \cdot v_1) * (R_z g \cdot v_2)\check{}\right](\dot{x})\,dz \\ &= \int_G T_N(R_z f \cdot v_1) * T_N\left((R_z g \cdot v_2)\check{}\right)(\dot{x})\,dz, \end{aligned}$$

where of course in the last term $*$ denotes convolution on G/N.

We temporarily fix $z \in G$ and estimate the 2-norms of the functions $T_N(R_z f \cdot v_1)$ and $T_N((R_z g \cdot v_2)\check{})$. For the first one we get, using Weil's formula and the Cauchy-Schwarz inequality,

$$\begin{aligned}
\|T_N(R_z f \cdot v_1)\|_2^2 &= \int_{G/N} \left| \int_N R_z f(xn) v_1(xn)\, dn \right|^2 d\dot{x} \\
&\leqslant \int_{G/N} \left(\int_N |R_z f(xn) v_1(xn)| dn \right)^2 d\dot{x} \\
&\leqslant \int_{G/N} \left(\int_N |R_z f(xn)|^2 dn \cdot \int_N |v_1(xn)|^2 dn \right) d\dot{x} \\
&= |KV| \cdot \int_G |R_z f(x)|^2 dx = |KV| \cdot \|R_z f\|_2^2. \qquad (6.1)
\end{aligned}$$

To treat the second function above, we note that the support of $T_N(\Delta_G^{-1} \cdot R_z g \cdot v_2)$ is contained in VN/N, we obtain

$$\begin{aligned}
\|T_N((R_z g \cdot v_2)\check{})\|_2^2 &= \int_{G/N} \left| \Delta_{G/N}(\dot{x}) T_N(\Delta_G^{-1} \cdot R_z g \cdot v_2)(\dot{x}) \right|^2 d\dot{x} \\
&\leqslant \left(\sup_{\dot{x} \in q(V)} \Delta_{G/N}(\dot{x}) \right)^2 \cdot \int_{G/N} \left| T_N(\Delta_G^{-1} \cdot R_z g \cdot v_2)(\dot{x}) \right|^2 d\dot{x}
\end{aligned}$$

Observe next that, for every $x \in G$,

$$\begin{aligned}
\int_N |\Delta_G^{-1}(xn) 1_V(xn)|^2 dn &= \int_{x^{-1}V} \Delta_G^{-2}(xn) dn \\
&\leqslant \sup\{\Delta_G(t)^{-2} : t \in V\}.
\end{aligned}$$

Setting now

$$c_1 = \sup\{\Delta_{G/N}(\dot{x}) : \dot{x} \in q(V)\} \quad \text{and} \quad c_2 = \sup\{\Delta_G(t)^{-2} : t \in V\},$$

and applying Weil's formula and Cauchy-Schwarz again, we get

$$\begin{aligned}
\|T_N(\Delta_G^{-1} R_z g \cdot v_2\|_2^2 &\leqslant \int_{G/N} \left(\int_N |(\Delta_G^{-1} R_z g \cdot v_2)(xn)| dn \right)^2 d\dot{x} \\
&\leqslant \int_{G/N} \left(\|L_x R_z g \cdot v_2)|_N\|_2 \cdot \int_N |\Delta_G^{-1}(xn) 1_V xn) dn|^2 \right) d\dot{x} \\
&\leqslant \int_G |R_z g \cdot v_2(x)|^2 dx \cdot \sup_{x \in G} \left\{ \int_N |\Delta_G^{-1}(xn) 1_V(xn)|^2 dn \right\} \\
&= c_1 c_2 \|R_z g \cdot v_2\|_2^2.
\end{aligned}$$

Since $v = 1$ on K and $\operatorname{supp} u \subseteq K$, we can now estimate the norm of $T_N(u)$ as follows.

$$\begin{aligned}
\|T_N(u)\| &= \left\| \int_G [T_N(R_z f \cdot v_1) * T_N((R_z g \cdot v_2)\check{})]\, dz \right\| \\
&\leqslant \int_G \|T_N(R_z f \cdot v_1) * T_N((R_z g \cdot v_2)\check{})\| dz \\
&\leqslant \int_G \|T_N(R_z f \cdot v_1)\|_2 \cdot \|T_N((R_z g \cdot v_2)\check{}\|_2 dz.
\end{aligned}$$

Now, by (6.1) above

$$\|T_N(R_zf\cdot v_1)\|_2^2 \leqslant |KV|\cdot\|R_zf\cdot v_1\|_2^2,$$

and by definition of c_1 and c_2,

$$\|T_N(R_zf\cdot v_2)\check{}\|_2^2 \leqslant c_1c_2\cdot\|R_zg\cdot v_2\|_2^2.$$

It follows that

$$\begin{aligned}
\|T_N(u)\| &\leqslant |KV|^{1/2}(c_1c_2)^{1/2}\int_G \|R_zf\cdot v_1\|_2\|R_zg\cdot v_2\|_2\,dz \\
&\leqslant (|KV|c_1c_2)^{1/2}\left(\int_G \|R_zf\cdot v_1\|_2^2\,dz\right)^{1/2}\left(\int_G \|R_zg\cdot v_2\|_2^2\,dz\right)^{1/2} \\
&\leqslant (|KV|c_1c_2)^{1/2}\|f\|_2\|v_1\|_2\cdot\|g\|_2\|v_2\|_2 \\
&= |KV|\cdot|V|^{-1}c_1c_2\|u\|_{A(G)} \\
&\leqslant |K|\cdot c_1c_2\|u\|_{A(G)}.
\end{aligned}$$

Since this inequality holds for all compact symmetric neighbourhoods of e, we can let $V\to\{e\}$. Then $c_1\to\Delta_{G/N}(\dot{e})=1$ and $c_2\to 1$. Consequently,

$$\|T_N(u)\| \leqslant |K|\cdot\|u\|_{A(G)}.$$

□

LEMMA 6.4.4. *Let $S\in VN(G)$ with compact support and let $f,\varphi,\psi\in A(G)\cap C_c(G)$. Then*

$$(f*\check{S})(x)(\varphi*\check{\psi})(x) = \int_G [(fR_z\varphi)*(R_z\psi\cdot S)\check{}](x)dz$$

for all $x\in G$.

PROOF. Since right translation on $A(G)$ is continuous, the map

$$z\to(L_xf)R_z(L_x\varphi)R_z\psi$$

from G into $A(G)$ is continuous with compact support. Therefore,

$$u(x)=\int_G (L_xf)R_z(L_x\varphi)(R_z\psi)\,dz$$

defines an element of $A(G)$. Now,

$$\begin{aligned}
(f*\check{S})(x)(\varphi*\check{\psi})(x) &= \left\langle \int_G (L_xf)R_z(L_x\varphi)(R_z\psi)\,dz, S\right\rangle \\
&= \int_G \langle (L_xf)R_z(L_x\varphi)(R_z\psi), S\rangle\,dz \\
&= \int_G \langle L_x(fR_z\varphi), (R_z\psi)\cdot S\rangle\,dz \\
&= \int_G [L_x(fR_z\varphi)*(R_z\psi\cdot S)](e)dz \\
&= \int_G [(fR_z\varphi)*(R_z\psi S)\check{}](x)dz.
\end{aligned}$$

□

Let N be a closed normal subgroup of G and let $S \in VN(G)$ with compact support. Let $w, w' \in A(G) \cap C_c(G)$ such that $w = 1$ and $w' = 1$ on a neighbourhood of $\operatorname{supp} S$, and choose $v \in A(G) \cap C_c(G)$ such that $v = 1$ on $\operatorname{supp} w \cup \operatorname{supp} w'$. Then, for every $u \in A(G/N)$,

$$\begin{aligned}\langle S, w(u \circ q)\rangle &= \langle S, (vw)(u \circ q)\rangle = \langle w \cdot S, v(u \circ q)\rangle \\ &= \langle w' \cdot S, v(u \circ q)\rangle = \langle S, w'(u \circ q)\rangle.\end{aligned}$$

Moreover,

$$|\langle S, w(u \circ q)\rangle| \leqslant \|S\| \cdot \|w\|_{A(G)} \|u \circ q\|_{B(G)} = \|S\| \cdot \|w\|_{A(G)} \cdot \|u\|_{A(G/N)}.$$

Thus, if $w = 1$ on a neighbourhood of $\operatorname{supp} S$, then $\langle S, w(u \circ q)\rangle$ does not depend on the choice of w and $u \to \langle S, w(u \circ q)\rangle$ defines a bounded linear operator on $A(G/N)$, denoted $\phi_N(S)$.

PROPOSITION 6.4.5. *Let G be an amenable locally compact group, N a closed normal subgroup of G and $q : G \to G/N$ the quotient homomorphism. Then there exists a (unique) contraction $\phi_N : UC(\widehat{G}) \to UC(\widehat{G/N})$ with the following property. If $T \in VN(G)$ has compact support and $w \in A(G) \cap C_c(G)$ is such that $w = 1$ on a neighbourhood of* $\operatorname{supp} T$, *then*

$$\langle \phi_N(T), u\rangle = \langle T, w(u \circ q)\rangle$$

for all $u \in A(G/N)$. Moreover ϕ_N satisfies:

(1) *if S has compact support, then* $\operatorname{supp} \phi_N(S) \subseteq q(\operatorname{supp} S)$;
(2) *for $f \in L^1(G)$, $\phi_N(\lambda_G(f)) = \lambda_{G/N}(T_N(f))$.*

PROOF. By the considerations above, we can define a linear map ϕ_N from $UC_c(\widehat{G})$ into $VN(G/N)$. Now, since G is amenable, for every compact subset K of G and $\epsilon > 0$, there exists $w \in A(G) \cap C_c(G)$ such that $w = 1$ on K and $\|w\| \leqslant 1 + \epsilon$. This implies that, for any $\epsilon > 0$,

$$\|\phi_N(T)\| = \sup\{|\langle T, w(u \circ q)\rangle| : \|u\|_{A(G/N)} \leqslant 1\} \leqslant (1 + \epsilon)\|T\|.$$

Thus ϕ_N is a contraction on $UC_c(\widehat{G})$ and hence extends uniquely to a contraction, also denoted ϕ_N, of $UC(\widehat{G})$ into $VN(G)$.

We next show (1) for $S \in UC_c(\widehat{G})$. To verify this, let $a \in G$ be such that $\dot{a} \in \operatorname{supp} \phi_N(S)$ and, towards a contradiction, assume that $aN \cap \operatorname{supp} S = \varnothing$. Choose a compact neighbourhood V of $\operatorname{supp} S$ such that $VN \cap aN = \varnothing$. Then VN is closed in G and $a \notin VN$, and hence there exists $u \in A(G) \cap C_c(G)$ such that $u \geqslant 0$, $u(a) \neq 0$ and $u = 0$ on VN. Then $T_N(u)(\dot{a}) > 0$ and $T_N(u)(\dot{x}) = 0$ for all $\dot{x} \in q(V)$. For every $v \in A(G/N) \cap C_c(G/N)$, it follows

$$\begin{aligned}\langle T_N(u) \cdot \phi_N(S), v\rangle &= \langle \phi_N(S), T_N(u)v\rangle = \langle S, w((T_N(u)v) \circ q)\rangle \\ &= \langle S, w(T_N(u) \circ q)(v \circ q)\rangle = 0,\end{aligned}$$

since $T_N(u) \circ q$ vanishes on the neighbourhood $q^{-1}(q(V))$ of $\operatorname{supp} S$. Thus $T_N(u) \circ \phi_N(S) = 0$ and hence $T_N(u)(\dot{a}) = 0$ by Proposition 2.5.3. This contradiction proves that $\operatorname{supp} \phi_N(S) \subseteq q(\operatorname{supp} S)$.

It remains to show (2). For $f \in L^1(G)$ and $v \in A(G/N)$, we have

$$\langle \lambda_{G/N}(T_N f), v\rangle = \int_{G/N} T_N f(\dot{x}) v(\dot{x}) d\dot{x} = \int_G f(t)(v \circ q)(t) dt.$$

On the other hand, if $f \in C_c(G)$ and $w = 1$ on a neighbourhood of $\operatorname{supp} f = \operatorname{supp} \lambda_G(f)$, then

$$\langle \phi_N(\lambda_G(f)), v\rangle = \langle \lambda_G(f), w(v \circ q)\rangle = \int_G f(t)(v \circ q)(t)dt.$$

Continuity now implies (2). □

THEOREM 6.4.6. *Let G be a locally compact group and N a closed normal subgroup of G. Let $q : G \to G/N$ denote the quotient homomorphism and let E be a closed subset of G/N. Then E is a local spectral set for $A(G/N)$ if and only if $q^{-1}(E)$ is a local spectral set for $A(G)$.*

PROOF. Suppose first that $q^{-1}(E)$ is a local spectral set for $A(G)$. Recall that we have to show $\langle S, u\rangle = 0$ for every $u \in I(E) \cap C_c(G/N)$ and $S \in VN(G/N)$ with compact support contained in E.

There exists $w \in A(G) \cap C_c(G)$ such that $\phi_N(w) = 1$ on a neighbourhood of $\operatorname{supp} S$. It follows from Lemma 6.4.1 that the assignment $f \to \langle S, \phi_N(wf)\rangle$, $f \in A(G)$, defines a bounded linear functional of $A(G)$. Thus there exists $T \in VN(G)$ such that $\langle T, f\rangle = \langle S, \phi_N(wf)\rangle$ for all $f \in A(G)$. Then T has compact support and $\operatorname{supp} T \subseteq q^{-1}(E)$. To verify this, assume first that $f \in A(G)$ vanishes in a neighbourhood of $\operatorname{supp} w$. Then $wf = 0$ and hence $\langle T, f\rangle = \langle S, \phi_N(wf)\rangle = 0$. This shows that $\operatorname{supp} T \subseteq \operatorname{supp} w$, which is compact. Now, let $f \in A(G) \cap C_c(G)$ be such that $f = 0$ in a neighbourhood of $q^{-1}(E)$. Then, since $\operatorname{supp} f \cap q^{-1}(E) = \varnothing$ and $\operatorname{supp} f$ is compact, there exists a neighbourhood W of e in G such that $\operatorname{supp} f \cap Wq^{-1}(E) = \varnothing$. It follows that $\phi_N(f)$ vanishes in a neighbourhood of E, an d this implies that $\langle T, f\rangle = \langle S, \phi_N(w)\phi_N(f)\rangle = 0$ since $\operatorname{supp} S \subseteq E$.

Now choose $v \in A(G) \cap C_c(G)$ such that $v = 1$ on a neighbourhood of $\operatorname{supp} T$. Then, since $v \cdot T = T$, $\phi_N(w) \cdot S = S$ and $q^{-1}(E)$ is a set of local synthesis,

$$\begin{aligned}
\langle S, u\rangle &= \langle \phi_N(w) \cdot S, u\rangle = \langle S, \phi_N(w)u\rangle \\
&= \langle S, \phi_N(w(u \circ q))\rangle = \langle T, u \circ q\rangle \\
&= \langle v \cdot T, u \circ q\rangle = \langle T, v(u \circ q)\rangle = 0,
\end{aligned}$$

as was to be shown.

For the converse, let E be a local spectral set for $A(G/N)$. We have to show that if $u \in I(q^{-1}(E)) \cap C_c(G)$ and $S \in UC(\widehat{G})$ such that $\operatorname{supp} S \subseteq q^{-1}(E)$, then $\langle S, u\rangle = 0$. Note first that we can assume $\operatorname{supp} S$ to be compact. In fact, let $v \in A(G) \cap C_c(G)$ be such that $v = 1$ on $\operatorname{supp} u$ and observe that $\langle v \cdot S, u\rangle = \langle S, vu\rangle = \langle S, u\rangle$ and $\operatorname{supp}(v \cdot S) \subseteq \operatorname{supp} v \cap \operatorname{supp} S$.

There exist nets $(w_\alpha)_\alpha$ and $(v_\alpha)_\alpha$ in $A(G) \cap C_c(G)$ such that the family of functions $(w_\alpha * \check{v_\alpha})\check{}|_N$ forms an approximate identity for $L^1(N)$. From Lemma 6.4.3 we get

$$\begin{aligned}
\int_N (u * \check{S})(n)(w_\alpha * \check{v_\alpha})(n)\, dn &= \int_N \left[\int_G (u \cdot R_z w_\alpha) * (u_\alpha \cdot L_z S)\check{}(n) dz\right] dn \\
&= \int_G \left[\int_N (u \cdot R_z w_\alpha) * (u_\alpha \cdot L_z S)\check{}(n) dn\right] dz.
\end{aligned}$$

Notice next that the last inner integral equals

$$\phi_N(u \cdot R_z w_\alpha) * \left[\Delta_{G/N}^{-1} \phi_N(\Delta_G^{-1} u_\alpha \cdot L_z S)\right]\check{}(\dot{e}).$$

In fact, for every $v \in A(G) \cap C_c(G)$ and $T \in UC(\widehat{G})$,

$$\begin{aligned}
\phi_N(v) * \left[\Delta_{G/N}^{-1}\phi_N(\Delta_G^{-1}T)\right]\check{}(\dot{e}) &= \langle \phi_N(v), \phi_N(T)\rangle \\
&= \left[(\phi_N(v)\cdot q) * \check{T}\right](\dot{e}) \\
&= \langle \phi_N(v)\cdot q, T\rangle \\
&= \left\langle \int_N L_n v\, dn, T\right\rangle \\
&= \int_N \langle L_n v, T\rangle dn \\
&= \int_N (v * \check{T})(n) dn.
\end{aligned}$$

Now $\phi_N(u\cdot R_z w_\alpha) \in A(G/N) \cap C_c(G/N)$ vanishes on E and the operator $\phi_N((\Delta_G^{-1}u_\alpha)\cdot L_z S) \in VN(G/N)$ has compact support contained in E. Since E is local spectral set, it follows that

$$\int_N (u\cdot R_z w_\alpha) * (u_\alpha \cdot L_z S)\check{}(n)\, dn = 0$$

for each α. Because $u * \check{S}$ is continuous on N and $((w_\alpha * \check{v_\alpha})\check{}|_N)_\alpha$ is an approximate identity for $L^1(N)$, we conclude

$$\langle u, S\rangle = (u * \check{S})(e) = \lim_\alpha \int_N (u * \check{S})(n)(w_\alpha * \check{v_\alpha})(n)\, dn = 0.$$

This completes the proof of the theorem. □

COROLLARY 6.4.7. *Let G be a locally compact group and N a closed normal subgroup of G. Suppose that both, $A(G)$ and $A(G/N)$, have approximate identities. Then a closed subset E of G/N is a set of synthesis for $A(G/N)$ if and only if $q^{-1}(E)$ is a set of synthesis for $A(G)$. The conclusion in particular holds if G is amenable.*

PROOF. The statement follows from Theorem 6.4.6 and Lemma 5.1.1. □

Let f, g, φ and ψ be functions in $L^2(G)$. Then it is straightforward to verify that

$$(f * \check{g})(x)(\varphi * \check{\psi})(x) = \int_G ((fR_z\varphi) * (R_z\psi * g)\check{})(x)\, dz$$

for all $x \in G$. The formula presented in the next lemma may be viewed as an analogue.

6.5. Bounded approximate identities I: Ideals

Let G be a locally compact abelian group. The closed ideals in the group algebra $L^1(G)$ have been completely described by Liu, van Rooij and Wang [**193**] in terms of the closed sets in the coset ring of the dual group $\widehat{G}$ of G. The main purpose of this section is to present a generalization to Fourier algebras $A(G)$ of amenable locally compact groups G. Amenability seems to be an indispensable hypothesis because if G is any locally compact group, then $A(G)$ has a bounded approximate identity if and only if G is amenable (Theorem 2.7.2). We are going to prove that a closed ideal I of $A(G)$ with bounded approximate identity is of the form $I = I(E)$ and characterize the closed subsets E of $G = \sigma(A(G))$ which determine ideals with bounded approximate identities (Theorem 6.5.11). Our first goal is to show that this class of subsets of G contains all closed subgroups of G (Theorem 6.5.5). This will be accomplished using the operator space structure of $A(G)$. We shall also address the problem of finding the best possible norm bound.

LEMMA 6.5.1. *Let H be an amenable locally compact group. Then there exists a completely contractive projection $P : \mathcal{B}(L^2(H)) \to VN(H)$.*

PROOF. First recall that $VN(H)$ is the commutant of the set $\{\rho_H(x) : x \in H\}$. Let m be a left invariant mean on $L^\infty(H)$. It will be convenient for us to regard m as a finitely additive measure on H. Now define $P : \mathcal{B}(L^2(H)) \to \mathcal{B}(L^2(H))$ to be the weak operator converging integral

$$P(T) = \int_H \rho_H(s) T \rho_H(s)^* dm(s)$$

for each each $T \in \mathcal{B}(L^2(H))$. That is,

$$\langle P(T)f, g\rangle = m(s \to \langle \rho_H(s) T \rho_H(s)^* f, g\rangle$$

for all $f, g \in L^2(H)$.

From the invariance of m, it is easy to see that for each $T \in \mathcal{B}(L^2(H))$,

$$\rho_H(t)P(T) = P(T)\rho_H(t)$$

for all $t \in H$. It follows that $P(T) \in VN(H)$. It is also clear that if $T \in VN(H)$, then $P(T) = T$. We have that P is completely positive, since each amplification

$$P^{(n)} : M_n(\mathcal{B}(L^2(H))) \to M_n(VN(H))$$

is given by the weak operator converging integral

$$P^{(n)}([T_{ij}]) = \int_H \mathrm{diag}(\rho_H(s))[T_{ij}]\mathrm{diag}(\rho_H(s))^* dm(s),$$

for every $[T_{ij}] \in M_n(\mathcal{B}(L^2(H)))$, where $\mathrm{diag}(\rho_H(s))$ denotes the $n \times n$ diagonal matrix with all diagonal entries equal to $\rho_H(s)$. Finally, since $P(I) = I$, we get that P is completely contractive. □

Assume that H is an amenable group. We claim that if $\mathcal{M}$ is a von Neumann algebra on a Hilbert space $\mathcal{E}$ such that $\mathcal{M}$ is $*$-isomorphic to $VN(H)$, then there exists a contractive completely positive expectation $\widetilde{P} : \mathcal{B}(\mathcal{E}) \to \mathcal{M}$. To see this, let $\Phi : VN(H) \to \mathcal{M}$ be a $*$-isomorphism. The Arveson-Wittstock Extension Theorem [**70**, Theorem 4.1.5] implies that $\Phi^{-1} : \mathcal{M} \to VN(H)$ admits a completely contractive extension $\Psi : \mathcal{B}(\mathcal{E}) \to \mathcal{B}(L^2(H))$. Moreover, since $\Psi(I) = I$, Ψ is completely positive [**70**, Corollary 5.1.2]. We now let $\widetilde{P} = \Phi \circ P \circ \Psi$, where P is the projection constructed in Lemma 6.5.1.

PROPOSITION 6.5.2. *Let G be a locally compact group and H an amenable closed subgroup of G. Then there exists a completely contractive projection from $VN(G)$ onto $I(H)^\perp$.*

PROOF. We recall that $VN_H(G)$ is $*$-isomorphic to $VN(H)$. It follows from Lemma 6.5.1 and the discussion preceding this proposition that there exists a completely contractive projection

$$\widetilde{P} : \mathcal{B}(L^2(G)) \to VN_H(G) = I(H)^\perp.$$

We now simply restrict $\widetilde{P}$ to $VN(G)$. □

LEMMA 6.5.3. *Let A be a commutative normed algebra with approximate identity of norm bound* 1, *and let $C > 0$. Then, for any closed ideal I of A, the following two conditions are equivalent.*

(i) *I has an approximate identity with norm bound C.*
(ii) *There exists a projection P from A^* onto $I^\perp$ commuting with the action of A on A^* such that $\|I - P\| \leqslant C$.*

PROOF. (i) $\Rightarrow$ (ii) Let $\mathcal{B}(A^*)$ denote the space of all bounded linear operators $T : A^* \to A^*$ equipped with the w^*-operator topology. Thus a net $(T_\alpha)_\alpha$ in $\mathcal{B}(A^*)$ converges to $T \in \mathcal{B}(A^*)$ if and only if $\langle T_\alpha(\varphi), x\rangle \to \langle T(\varphi), x\rangle$ for every $\varphi \in A^*$ and $x \in A$. Then, for any $r > 0$, the ball of radius r in $\mathcal{B}(A^*)$ is compact.

Now let $(u_\alpha)_\alpha$ be an approximate identity for I such that $\|u_\alpha\| \leqslant C$ for all α. For each α, define $P_\alpha : A^* \to A^*$ by

$$\langle P_\alpha(\varphi), x\rangle = \langle \varphi, x - u_\alpha x\rangle, \quad x \in A, \varphi \in A^*.$$

It is straightforward to check that, for each $\varphi \in A^*$,

$$\|P_\alpha\| \leqslant 1 + C \quad \text{and} \quad \|I - P_\alpha\| \leqslant C.$$

Thus, passing to a subnet if necessary, we can assume that $P_\alpha \to P$ for some $P \in \mathcal{B}(A^*)$ in the w^*-operator topology. Then $\|I - P\| \leqslant C$ and, for $x \in I$ and $\varphi \in A^*$,

$$\langle P(\varphi), x\rangle = \lim_\alpha \langle \varphi, x - u_\alpha x\rangle = 0,$$

so that $P(A^*) \subseteq I^\perp$, and conversely, if $\varphi \in I^\perp$, then $\langle P(\varphi), x\rangle = \langle \varphi, x\rangle$ for all $x \in A$, whence $P|_{I^\perp}$ is the identity. Moreover, $P^2 = P$ since, for any $\varphi \in A^*$ and $x \in A$,

$$\langle P^2(\varphi), x\rangle = \lim_\alpha \langle P(\varphi), x - u_\alpha x\rangle = \langle P(\varphi), x\rangle$$

because $u_\alpha x \in I$ and $P(\varphi) \in I^\perp$. Finally, P commutes with the action of A on A^*. Indeed,

$$\begin{aligned}
\langle P(a \cdot \varphi), x\rangle &= \lim_\alpha \langle a \cdot \varphi, x - u_\alpha x\rangle = \lim_\alpha \langle \varphi, ax - u_\alpha ax\rangle \\
&= \lim_\alpha \langle P_\alpha(\varphi), ax\rangle = \langle P(\varphi), ax\rangle \\
&= \langle a \cdot P(\varphi), x\rangle
\end{aligned}$$

for every $x \in A$ and $\varphi \in A^*$.

(ii) $\Rightarrow$ (i) It suffices to show that given $u \in I$ and $\epsilon > 0$, there exists $v \in I$ such that $\|v\| \leqslant C$ and $\|u - uv\| \leqslant \epsilon$. Fix u and ϵ and, for any finite nonempty subset F of A^*, let

$$I(F, \epsilon) = \{v \in I : \|v\| \leqslant C, |\langle \varphi, u\rangle - \langle \varphi, uv\rangle| \leqslant \epsilon \text{ for all } \varphi \in F\}.$$

We are going to show that $I(F, \epsilon) \neq \emptyset$ for each such F and $\epsilon > 0$. Observe first that there exists $w \in A$ with $\|w\| \leqslant 1$ and

$$\|u - uw\| \leqslant \frac{\epsilon}{2 \max\{\|\varphi\| : \varphi \in F\}}.$$

Now define $f : I^* \to \mathbb{C}$ by setting, for $\psi \in I^*$,

$$\langle f, \psi\rangle = \langle \varphi, w\rangle - \langle P(\varphi), w\rangle,$$

where $\varphi \in A^*$ is such that $\varphi|_I = \psi$. Note that f is well defined since P projects onto $I^\perp$. Moreover, $\|f\| \leqslant C$ since $\|w\| \leqslant 1$, $\|I - P\| \leqslant C$ and f or every $\psi \in I^*$

there exists $\varphi \in A^*$ such that $\varphi|_I = \psi$ and $\|\varphi\| = \|\psi\|$. Since the ball of radius C in I is w^*-dense in the ball of radius C in I^{**}, we find $v \in I$ such that $\|v\| \leqslant C$ and

$$|\langle (u \cdot \varphi)|_I, v\rangle - \langle f, (u \cdot \varphi)|_I\rangle| \leqslant \frac{\epsilon}{2}$$

for all $\varphi \in F$. By definition of f and since $P(u \cdot \varphi) = u \cdot P(\varphi)$, this means that

$$|\langle \varphi, uv\rangle - \langle \varphi, uw\rangle + \langle P(\varphi), uw\rangle| \leqslant \frac{\epsilon}{2}$$

for all $\varphi \in F$. By the choice of w and since $uw \in I$ and $P(\varphi) \in I^\perp$, it follows that, for each $\varphi_0 \in F$,

$$\begin{aligned} |\langle \varphi_0, u\rangle - \langle \varphi_0, uv\rangle| &\leqslant |\langle \varphi_0, u\rangle - \langle \varphi_0, uw\rangle| \\ &\quad + |\langle \varphi_0, uw\rangle - \langle \varphi_0, uv\rangle - \langle P(\varphi_0), uw\rangle| \\ &\leqslant \|\varphi_0\| \cdot \frac{\epsilon}{2\max\{\|\varphi\| : \varphi \in F\}} + \frac{\epsilon}{2} \leqslant \epsilon. \end{aligned}$$

This shows that $I(F, \epsilon) \neq \varnothing$.

We have seen so far that u lies in the w^*-closure of the set

$$J = \bigcup\{u \cdot I(F, \epsilon) : F \subseteq A^* \text{ finite, } \epsilon > 0\}.$$

It follows that u lies in the norm closure in A of the convex hull of J. Therefore, given $\epsilon > 0$, there exist $n \in \mathbb{N}$, finite subsets $F_1, \ldots, F_n$ of A^*, $c_1, \ldots, c_n > 0$, $\epsilon_1, \ldots, \epsilon_n > 0$ and $v_1, \ldots, v_n \in I$ such that

$$v_j \in I(F_j, \epsilon_j),\ 1 \leqslant j \leqslant n,\ \sum_{j=1}^n c_j = 1, \text{ and } \left\| u - \sum_{j=1}^n c_j u v_j \right\| \leqslant \epsilon.$$

Then the element $v = \sum_{j=1}^n c_j v_j$ of I satisfies $\|v\| \leqslant C$ and $\|u - uv\| \leqslant \epsilon$. Thus (i) holds. □

PROPOSITION 6.5.4. *Let G be an amenable locally compact group and H a closed subgroup of G. Then there exists a unital completely positive projection P from $VN(G)$ onto $VN_H(G) = I(H)^\perp$ such that $P(u \cdot T) = u \cdot P(T)$ for all $T \in VN(G)$ and $u \in A(G)$.*

PROOF. It follows from Proposition 6.5.2 that there exists a projection $Q : VN(G) \to I(H)^\perp$ with $\|Q\|_{cb} = \|Q(I)\| = \|I\| = 1$. Since G is amenable, the operator space projective tensor product $A(G) \widehat{\otimes} A(G)$ admits an approximate diagonal of norm bounded by 1. Now Lemma 6.5.3 applies taking $A = A(G)$, $A^* = VN(G)$ and $Y = VN_H(G) = I(H)^\perp$, now shows that there exists a completely contractive projection $P : VN(G) \to I(H)^\perp$ that commutes with the module action of $A(G)$ on $VN(G)$. Since $P(I) = I$, P is also completely positive. □

By Theorem 3.2.5, the preceding proposition applies to the Fourier algebra of any amenable locally compact group.

THEOREM 6.5.5. *Let G be an amenable locally compact group and H a closed subgroup of G. Then the ideal $I(H)$ has an approximate identity with norm bounded by 2.*

PROOF. By Proposition 6.5.4 there is a completely contractive projection P from $VN(G)$ onto $VN_H(G) = I(H)^\perp$, which commutes with the action of $A(G)$ on $VN(G)$. We can now apply Lemma 6.5.3 to conclude the existence of an approximate identity of $I(H)$ with norm bounded by 2. □

The following standard facts will be used in the sequel. Let I and J be closed ideals of a commutative Banach algebra A. Then

(1) If I and J both have bounded approximate identities, then so do $I \cap J$ and $\overline{I + J}$.

(2) If I and A/I have bounded approximate identities, then A has a bounded approximate identity.

LEMMA 6.5.6. *Let G be an amenable locally compact group and let H and K be closed subgroups of G such that $K \subseteq H$ and K is open in H. Then, for any $a, b \in G$, the ideal $I(aH \backslash bK)$ has a bounded approximate identity.*

PROOF. Note first that $aH \backslash bK = a(H \backslash a^{-1}bK)$ and $H \backslash a^{-1}bK \neq H$ only if $a^{-1}b \in H$, and in this case $H \backslash a^{-1}bK = a^{-1}b(H \backslash K)$. It therefore suffices to show that $I(H \backslash K)$ has a bounded approximate identity. Let

$$J = \{u \in A(H) : u(x) = 0 \text{ for all } x \in H \backslash K\}.$$

Since K is open in H, the map $u \to u|_K$ is an isometric isomorphism from J onto $A(K)$. As K is amenable, $A(K)$ has a bounded approximate identity, and hence so does J. Let $r : A(G) \to A(H)$ denote the restriction map. Then $I(H \backslash K) = r^{-1}(J)$ and therefore $I(H \backslash K)$ has a bounded approximate identity provided that $I(H)$ and $r^{-1}(J)/I(H)$ have bounded approximate identities. For $I(H)$ this is guaranteed by Theorem 6.5.5. Now the map $\phi : u + I(H) \to u|_H$ is an isometric isomorphism between $A(G)/I(H)$ and $A(H)$. Since J has a bounded approximate identity, so does $\phi^{-1}(J) = r^{-1}(J)/I(H)$, as required. □

PROPOSITION 6.5.7. *Let G be an amenable locally compact group and $E \in \mathcal{R}_c(G)$. Then*

(i) *E is a Ditkin set for $A(G)$.*

(ii) *The ideal $I(E)$ has a bounded approximate identity.*

PROOF. Note first that if F is a closed subset of G such that $I(F)$ has a bounded approximate identity and F is a set of synthesis, then $I(F)$ has an approximate identity consisting of elements in $j(F)$. Consequently, F is a Ditkin set.

Assume first that E is of the form $E = aH \backslash \bigcup_{j=1}^{m} c_j K_j$, where H is a closed subgroup of G, $m \in \mathbb{N}_0$, $a, c_j \in G$ and the K_j are open subgroups of H, $1 \leqslant j \leqslant m$. Then E is a set of synthesis. Indeed, this follows from Theorem 6.1.11 since aH is a set of synthesis (Corollary 6.1.10), $\varnothing$ is a Ditkin set and E is open and closed in aH. To see that $I(E)$ has a bounded approximate identity, consider the ideal

$$J = \overline{\sum_{j=1}^{m} I(aH \backslash c_j K_j)}$$

of $A(G)$. It follows from Lemma 6.5.6 and fact (1) mentioned prior to that lemma that J has a bounded approximate identity. Now

$$Z(J) = \bigcap_{j=1}^{m} (aH \backslash c_j K_j) = E,$$

and hence $J = I(E)$ since E is a set of synthesis. By the remark at the outset of the proof, E is a Ditkin set.

Next, let $E = \bigcup_{i=1}^{n} E_i$, where each E_i is of the above form. We prove by induction on n that E is a Ditkin set and $I(E)$ has a bounded approximate identity. The case $n = 1$ having been dealt with in the preceding paragraph, let $n \geqslant 2$ and set $F = \bigcup_{i=1}^{n-1} E_i$, and suppose that (i) and (ii) hold for F. Clearly, then $I(E) = I(F) \cap I(E_n)$ has a bounded approximate identity. Moreover, E_n and F

are both Ditkin sets. Since the union of two Ditkin sets is a Ditkin set (Theorem 6.1.12), $E = E_n \cup F$ is a Ditkin set as well. □

LEMMA 6.5.8. *Let J be a closed ideal of $A(G)$ and suppose that J has an approximate identity $(u_\alpha)_\alpha$ such that $\|u_\alpha\|_{A(G)} \leqslant C$ for all α. Then $1_{Z(J)}$ and $1_{G\setminus Z(J)}$ belong to $B(G_d)$. Moreover, $\|1_{G\setminus Z(J)}\|_{B(G_d)} \leqslant C$ and $\|1_{Z(J)}\|_{B(G_d)} \leqslant C+1$.*

PROOF. Let $u \in J$. Then

$$\|u_\alpha u - u\|_\infty \leqslant \|u_\alpha u - u\|_{A(G)} \to 0.$$

Thus $(u_\alpha)_\alpha$ converges to $1_{G\setminus Z(J)}$ pointwise. Since the set $\{v \in B(G_d) : \|v\| \leqslant C\}$ is closed in the topology of pointwise convergence,

$$1_{G\setminus Z(J)} \in B(G_d) \text{ and } \|1_{G\setminus Z(J)}\|_{B(G_d)} \leqslant C.$$

Consequently, $1_{Z(J)} = 1_G - 1_{G\setminus Z(J)} \in B(G_d)$ and $\|1_{Z(J)}\|_{B(G_d)} \leqslant C + 1$. □

COROLLARY 6.5.9. *Let J be a closed ideal of $A(G)$ with bounded approximate identity. Then $Z(J) \in \mathcal{R}_c(G)$.*

PROOF. By the preceding lemma, $1_{Z(J)} \in B(G_d)$ and hence $Z(J) \in \mathcal{R}(G_d)$ by Host's idempotent theorem. Since $Z(J)$ is closed in G, $Z(J) \in \mathcal{R}_c(G)$. □

REMARK 6.5.10. As shown in [**91**], the conclusion of Corollary 6.5.9 holds under a somewhat weaker hypothesis on the approximate identity of the ideal J. To explain this, we recall the following definition. Let A be a commutative Banach algebra with structure space $\Delta(A)$. A net $(e_\alpha)_\alpha$ is called a *bounded Δ-weak approximate identity* if $\langle \varphi, e_\alpha a - a\rangle \to 0$ for every $\varphi \in \Delta(A)$ and there exists a constant $C > 0$ such that $\|e_\alpha\| \leqslant C$ for all α.

Now suppose that J is a closed ideal of $A(G)$ possessing a bounded Δ-weak approximate identity $(u_\alpha)_\alpha$. Then the bounded net $(u_\alpha)_\alpha$ in $B(G_d)$ converges pointwise to the characteristic function $1_{G\setminus Z(J)}$. It follows from Corollary 3.2.3 that $1_{G\setminus Z(J)} \in B(G_d)$ and hence $1_{Z(J)} \in B(G_d)$. Then apply Cohen's theorem as in the proof of Corollary 6.5.9.

THEOREM 6.5.11. *Let G be an amenable locally compact group and let I be a closed ideal of $A(G)$. Then I has a bounded approximate identity if and only if $I = I(E)$ for some subset E of G of the form*

$$E = \bigcup_{i=1}^n \left(a_i H_i \setminus \bigcup_{j=1}^{m_i} b_{ij} K_{ij}\right),$$

where $a_i, b_{ij} \in G$, H_i is a closed subgroup of G and K_{ij} is an open subgroup of H_i ($n, m_i \in \mathbb{N}_0$, $1 \leqslant i \leqslant n$, $1 \leqslant j \leqslant m_i$).

PROOF. If E is of the indicated form (equivalently, $E \in \mathcal{R}_c(G)$), then $I(E)$ has a bounded approximate identity by Proposition. 6.5.7(ii). Conversely, let I be a closed ideal of $A(G)$ with bounded approximate identity and let $E = Z(I)$. Then $E \in \mathcal{R}_c(G)$ by Corollary 6.5.9 and E is a set of synthesis by Proposition 6.5.7(i). It follows that $I = I(E)$ and E is of the stated form. □

COROLLARY 6.5.12. *Let G be an amenable locally compact group and I a closed ideal of $A(G)$. Then I is operator amenable if and only if $I = I(E)$ for some $E \in \mathcal{R}_c(G)$.*

PROOF. Since $A(G)$ is an operator amenable completely contractive Banach algebra, a closed ideal I of $A(G)$ is operator amenable precisely when I has a bounded approximate identity. Thus the statement follows from Theorem 6.5.11. □

We now turn to the question of whether 2 is the best possible norm bound for an approximate identity of $I(H)$ when H is a closed subgroup of G.

LEMMA 6.5.13. *Let H be a nonopen closed subgroup of G. Let $P : VN(G) \to I(H)^\perp$ be a projection, and suppose that there exists a bounded net $(u_\alpha)_\alpha$ in $B(G)$ such that*

$$\langle u_\alpha \cdot T, u\rangle \to \langle P(T), u\rangle$$

for all $T \in VN(G)$ and all $u \in A(G)$. Then $P(T) = 0$ for all $T \in C^*_\lambda(G)$.

PROOF. To establish the lemma, it suffices to prove that $P(\lambda_G(g)) = 0$ for every $g \in C_c(G)$. Since H is nonopen, $|H| = 0$, and hence given $\epsilon > 0$, we find an open subset V of G containing $H \cap \operatorname{supp} g$ such that $|V| \leqslant \epsilon$. Let $f = g \cdot 1_{G\setminus V}$. Then

$$\begin{aligned}\|P(\lambda_G(g)) - P(\lambda_G(f))\| &\leqslant \|P\| \cdot \|\lambda_G(g-f)\|\\ &\leqslant \|P\| \cdot \|g-f\| \leqslant \|P\| \cdot |V| \cdot \|g\|_\infty\\ &\leqslant \epsilon \|g\|_\infty \|P\|.\end{aligned}$$

We now show that $P(\lambda_G(f)) = 0$. Let $K = \operatorname{supp} g \cap (G\setminus V)$, a compact subset of $G\setminus H$. Then

$$\int_K (fu)(x)u_\alpha(x)dx = \int_G (fu)(x)u_\alpha(x)dx = \langle u_\alpha \cdot \lambda_G(f), u\rangle \to \langle P(\lambda_G(f)), u\rangle$$

for all $u \in A(G)$. Since $A(G)$ is regular, there exists $v \in A(G)$ such that $v(x) = 0$ for all $x \in H$ and $v(x) = 1$ for all $x \in K$. Then, since $vu \in I(H)$ and $P(\lambda_G(f)) \in I(H)^\perp$,

$$\int_K (fu)(x)u_\alpha(x)dx = \int_K f(x)(vu)(x)u_\alpha(x)dx \to \langle P(\lambda_G(f)), vu\rangle = 0.$$

Thus $\langle P(\lambda_G(f)), u\rangle = 0$ for all $u \in A(G)$, as required. □

LEMMA 6.5.14. *Let $I_{L^2(G)}$ denote the identity of $\mathcal{B}(L^2(G))$. There exists $T \in C^*_\lambda(G)$ such that $\|T\| = 1$ and $\|I_{L^2(G)} - 2T\| = 1$.*

PROOF. Choose any $f \in C_c(G)$, $f \neq 0$, with $f = f^*$, and let A (respectively, B) denote the closed subalgebra of $\mathcal{B}(L^2(G))$ generated by $\lambda_G(f)$ and $I_{L^2(G)}$ (respectively, $\lambda_G(f)$). Then A and B are C^*-algebras, $\Delta(B) \subseteq \Delta(A)$ and the Gelfand homomorphisms are isometric isomorphisms between A and $C(\Delta(A))$ (respectively, B and $C_0(\Delta(B))$). Now, take any $g \in C(\Delta(A))$ such that $\|g\|_\infty = 1$, $\operatorname{supp} g \subseteq \Delta(B)$ and $0 \leqslant g \leqslant 1$. Let $T \in B \subseteq C^*_\lambda(G)$ such that $\widehat{T} = g$. Then $\|T\| = 1$ and $\|I_{L^2(G)} - 2T\| = 1$ since $(I_{L^2(G)} - 2T)\widehat{\ } = 1_{\Delta(A)} - 2g$ and $\|1_{\Delta(A)} - 2g\|_\infty = 1$. □

THEOREM 6.5.15. *Let G be a locally compact group and H a nonopen closed subgroup of G. Then 2 is the best possible norm bound for an approximate identity if the ideal $I(H)$.*

PROOF. Towards a contradiction, assume that there exists an approximate identity $(v_\alpha)_\alpha$ in $I(H)$ such that, for some constant $c < 2$, $\|v_\alpha\| \leqslant c$ for all α. After passing to a subnet if necessary, we may assume that the net $(v_\alpha)_\alpha$ converges in the w^*-topology of $VN(G)^*$. This gives rise to a projection P from $VN(G)$ onto $I(H)^\perp$ such that

$$\langle (1-v_\alpha)\cdot T, u\rangle \to \langle P(T), u\rangle$$

for all $T \in VN(G)$ and all $u \in A(G)$. Lemma 6.5.13 now yields that $P(C^*_\lambda(G)) = \{0\}$.

Let I denote the identity of $\mathcal{B}(VN(G))$. By Lemma 6.5.14, there exists $T \in C^*_\lambda(G)$ such that $\|T\| = 1$ and $\|I_{L^2(G)} - 2T\| = 1$. Since $P(T) = 0$, it follows that

$$(I-P)(I_{L^2(G)} - 2T) = -2T.$$

Since $\|T\| = 1$, we conclude that $\|I - P\| \geqslant 2$. Thus, since $c < 2$, there exist $S \in VN(G)$ and $u \in A(G)$ such that $\|S\| \leqslant 1$, $\|u\| \leqslant 1$ and

$$|\langle S - P(S), u\rangle| \geqslant 1 + \frac{c}{2}.$$

Then $|\langle v_\alpha \cdot S, u\rangle| > c$ for sufficiently large α since

$$\langle (1-v_\alpha)\cdot S, u\rangle \to \langle P(S), u\rangle.$$

On the other hand, we have

$$|\langle v_\alpha \cdot S, u\rangle| \leqslant \|v_\alpha\| \cdot \|S\| \cdot \|u\| \leqslant c.$$

This contradiction shows that 2 is indeed the best possible norm bound. □

We conclude this section by discussing the problem of when for a closed subset E of G, the ideal $I(E)$ has an approximate identity with norm bound 1. In view of the preceding result, this norm condition appears to be very restrictive. The following proposition identifies these sets E.

PROPOSITION 6.5.16. *Let G be any locally compact group and E a closed subset of G. Then the ideal $I(E)$ has an approximate identity with norm bound 1 if and only if $E = G\backslash aH$, where $a \in G$ and H is an amenable open subgroup of G.*

PROOF. Suppose first that E is of the indicated form. Then, since H is amenable, $A(H)$ has an approximate identity with norm bound 1. On the other hand, since H is open, the map $v \to \tilde{v}$, where for $v \in A(H)$, $\tilde{v}$ denotes the trivial extension of v to all of G, is an isometric isomorphism between $A(H)$ and $I(G\backslash H) = I(a^{-1}E)$. So $I(a^{-1}E)$ has an approximate identity of norm bound 1, and hence so does $I(E)$.

Conversely, assume that $I(E)$ has an approximate identity with norm bound 1. Then, as in the proof of Lemma 6.5.8, it follows that $1_{G\backslash E} \in B(G_d)$ and $\|1_{G\backslash E}\|_{B(G_d)} = 1$. Fix any $a \in G\backslash E$ and let $H = a^{-1}(G\backslash E)$. Then $e \in H$ and $\|1_H\|_{B(G_d)} = 1$, and these facts together imply that H is a subgroup of G. Indeed, this can be seen as follows. There exist a unitary representation π of G_d and $\xi, \eta \in \mathcal{H}(\pi)$ such that $1_H(x) = \langle \pi(x)\xi, \eta\rangle$ for all $x \in G_d$ and $\|1_H\|_{B(G_d)} = \|\xi\| \cdot \|\eta\|$. So $\langle \xi, \eta\rangle = \|\xi\| \cdot \|\eta\|$ and hence $\eta = \xi$. Consequently, 1_H is a positive definite function, and this readily implies that H is a subgroup of G. Since E is closed, H is open. Now, using again that $A(H)$ is isometrically isomorphic to $I(E)$, we conclude that $A(H)$ has a bounded approximate identity, whence H must be amenable. □

In case E is a closed subgroup of G, $E = H$ say, Proposition 6.5.16 shows that $I(H)$ has an approximate identity bounded by 1 if and only if H has index two in G and G is amenable.

6.6. Bounded approximate identities II

Let G be an amenable locally compact group and H a closed normal subgroup of G, and let $q : G \to G/H$ denote the quotient homomorphism. We establish the existence of a bijection $e : I \to e(I)$ between the set of closed ideals I of $A(G/H)$ and the set of all closed ideals of $A(G)$ which are left and right invariant under translation by elements of H. We start with the definition of $e(I)$. For a closed ideal I of $A(G/H)$, put

$$e(I) = \overline{\text{span}\{u(v \circ q) : u \in A(G), v \in I\}}.$$

Then $e(I)$ is a closed ideal of $A(G)$. Moreover, $e(I)$ is left and right H-translation invariant since $L_h(u(v \circ q)) = (L_h u)(v \circ q)$, and similarly for R_h, $h \in H$.

LEMMA 6.6.1. *Let I be a closed ideal of $A(G)$ such that $R_h u \in I$ for every $u \in I$ and $h \in H$. Then $\phi_H(v \cdot T) \in r(I)^\perp$ for every $v \in A(G)$ and $T \in I^\perp$.*

PROOF. It suffices to consider $v \in A(G) \cap C_c(G)$. Choose $w \in A(G) \cap C_c(G)$ such that $w = 1$ on a neighbourhood of $\operatorname{supp} v \supseteq \operatorname{supp}(v \cdot T)$. Then, for any $u \in I \cap C_c(G)$

$$\begin{aligned}\langle \phi_H(v \cdot T), T_H u\rangle &= \langle v \cdot T, w(T_H u \circ q)\rangle = \langle T, v(T_H u \circ q)\rangle \\ &= \langle T, v \cdot \int_H R_h u \, dh\rangle \\ &= \int_H \langle T, v \cdot R_h u\rangle \, dh = 0,\end{aligned}$$

since $T \in I^\perp$ and I is H-right translation invariant. Since $I \cap C_c(G)$ is dense in I and $r(I)$ is the closure of the set of all elements of the form $T_h u$, $u \in I \cap C_c(G)$, it follows that $\langle \phi_H(v \cdot T), w\rangle = 0$ for every $w \in r(I)$. □

REMARK 6.6.2. If H is a closed normal subgroup of G, $T_H(A(G) \cap C_c(G))$ is a dense ideal in $A(G/H)$. Indeed, since $v \cdot T_H u = T_H(u(v \circ q))$ for any $u \in A(G) \cap C_c(G)$ and $v \in A(G/H)$, $T_H(A(G) \cap C_c(G))$ is an ideal in $A(G/H)$. Clearly, for every $x \in G$ there exists $u \in A(G) \cap C_c(G)$ such that $T_H u(\dot{x}) \neq 0$. Thus the hull of $T_H(A(G) \cap C_c(G))$ is empty and hence this ideal is dense in $A(G/H)$ since $A(G/H)$ is Tauberian.

LEMMA 6.6.3. *Let I be a closed ideal of $A(G)$ and $T \in VN(G)$ such that $\phi_H(u \cdot T) \in r(I)^\perp$ for every $u \in A(G)$. Then $T \in I^\perp$.*

PROOF. Since $I \cap C_c(G)$ is dense in I, it suffices to show that $\langle T, u\rangle = 0$ for every $u \in I \cap C_c(G)$. Fix such a u and choose $w \in A(G) \cap C_c(G)$ such that $w = 1$ in a neighbourhood of $\operatorname{supp} u \supseteq \operatorname{supp}(u \cdot T)$. By hypothesis and the definition of ϕ_H, we then have

$$0 = \langle \phi_H(u \cdot T), T_h v\rangle = \langle u \cdot T, w(T_H v \circ q)\rangle$$

for all $v \in A(G) \cap C_c(G)$. Next, choose $w' \in A(G/H)$ such that $w' = 1$ on $q(\operatorname{supp} u)$. There exists a sequence $(v_n)_v$ in $A(G) \cap C_c(G)$ such that $T_H(v_n) \to w'$ in $A(G/H)$. Then $w' \circ q = 1$ on $\operatorname{supp} u$ and $T_H(v_n) \circ q \to w' \circ q$ in $B(G)$. It follows that

$$\begin{aligned}\langle u \cdot T, w(T_H(v_n) \circ q)\rangle &= \langle T, uw(T_H(v_n) \circ q)\rangle = \langle T, u(T_H(v_n) \circ q)\rangle \\ &\to \langle T, u(w' \circ q)\rangle = \langle T, u\rangle.\end{aligned}$$

This shows that $\langle T, u\rangle = 0$. □

For a closed ideal J of $A(G)$, let

$$r(J) = \overline{q(J \cap C_c(G))} \subseteq A(G/H).$$

Then $r(J)$ is a closed ideal of $A(G/H)$. In fact, if $u \in J \cap C_c(G)$ and $v \in A(G/H)$ and $w \in A(G)$ is such that $w = 1$ on $\operatorname{supp} u$, then $(v \circ q)u = (v \circ q_H)wu \in J \bigcap C_c(G)$ and $vq(u) = q((v \circ q)u)$.

THEOREM 6.6.4. *Let G be an amenable locally compact group and H a closed normal subgroup of G. Then the assignment $e : I \to e(I)$ is a bijection between the set of all closed ideals of $A(G/H)$ and the set of all H-translation invariant closed ideals of $A(G)$. More precisely:*

(i) *For every closed ideal I of $A(G/H)$, we have $I = r(e(I))$.*
(ii) *For every closed ideal J of $A(G)$ with $R_h u \in J$ for all $u \in J$ and $h \in H$, we have $J = e(r(J))$.*

PROOF. (i) For $u \in A(G) \cap C_c(G)$ and $v \in I$, we have $u(v \circ q) \in e(I) \cap C_c(G)$ and

$$T_H(u(v \circ q)) = vT_H(u) \in I.$$

Since the span of the elements on the left hand side is dense in $r(e(I))$, is follows that $r(e(I)) \subseteq I$. For the reverse inclusion, let $v \in I \cap C_c(G/H)$ and choose $w \in A(G/H)$ such that $w = 1$ on $\operatorname{supp} v$. There exists a sequence $(u_n)_n$ in $A(G) \cap C_c(G)$ such that $T_H(u_n) \to w$ in $A(G/H)$. Then

$$\begin{aligned} v &= vw = v \cdot \lim_{n\to\infty} T_H(u_n) = \lim_{n\to\infty} (vT_H(u_n)) \\ &= \lim_{n\to\infty} T_H(u_n(v \circ q)) \in r(e(I)). \end{aligned}$$

Since $I \cap C_c(G/H)$ is dense in I, we infer that $I \subseteq r(e(I))$.

(ii) Taking $r(I)$ in (i), we have $r(e(r(I))) = r(I)$. Moreover, I and $e(r(I))$ are both right H-translation invariant. Therefore, we only need to verify that if I and J are two closed, right H-translation invariant ideals of $A(G)$ such that $r(I) = r(J)$, then $I = J$. For that, it suffices to observe that the annihilators coincide. Let $T \in I^\perp$, then $\phi_H(u \cdot T) \in r(I)^\perp = r(J)^\perp$ by Lemma 6.6.1 for every $u \in A(G)$. Lemma 6.6.3 implies $T \in J^\perp$. Thus $I^\perp \subseteq J^\perp$. □

We are now turning to the second issue of this section. We are going to show that an ideal I of $A(G/H)$ has a bounded approximate identity (with bound $C > 0$) if and only if the ideal $e(I)$ of $A(G)$ has a bounded approximate identity (with bound C).

LEMMA 6.6.5. *Let G be an amenable locally compact group, H a closed normal subgroup of G and I a closed ideal of $A(G/H)$. If I has an approximate identity bounded by c, then $e(I)$ has an approximate identity of norm bounded by c.*

PROOF. Let $u \in e(I)$ and $\epsilon > 0$ be given. There exist $u_1, \ldots, u_n \in A(G)$ and $v_1, \ldots, v_n \in I$ such that

$$\left\| u - \sum_{j=1}^{n} u_j(v_j \circ q) \right\|_{A(G)} \leqslant \epsilon.$$

Since G is amenable, there exists $u' \in A(G)$ with $\|u'\|_{A(G)} \leqslant 1$ and

$$\left\| \sum_{j=1}^{n} u_j(v_j \circ q) - u' \cdot \sum_{j=1}^{n} u_j(v_j \circ q) \right\|_{A(G)} \leqslant \epsilon.$$

Since a bounded approximate identity is a multiple approximate identity, there exists $v' \in I$ such that $\|v'\|_{A(G/H)} \leqslant c$ and, for $1 \leqslant j \leqslant n$,

$$\|v_j - v_j v'\|_{A(G/H)} \leqslant \epsilon \left(1 + \sum_{i=1}^{n} \|u_i\|_{A(G)}\right)^{-1}.$$

Let $w = u'(v' \circ q_H) \in e(I)$. Then $\|w\|_{A(G)} \leqslant \|v' \circ q\|_{B(G)} \leqslant c$ and

$$\begin{aligned}
\|u - uw\|_{A(G)} \leqslant\ & \left\| u - \sum_{j=1}^{n} u_j(v_j \circ q) \right\|_{A(G)} \\
& + \left\| \sum_{j=1}^{n} u_j(v_j \circ q) - u' \sum_{j=1}^{n} u_j(v_j \circ q) \right\|_{A(G)} \\
& + \left\| (u' - w) \sum_{j=1}^{n} u_j(v_j \circ q) \right\|_{A(G)} \\
& + \left\| w(u - \sum_{j=1}^{n} u_j(v_j \circ q)) \right\|_{A(G)} \\
\leqslant\ & \epsilon(2 + c) + \left\| (u' - w) \sum_{j=1}^{n} u_j(v_j \circ q) \right\|_{A(G)} \\
\leqslant\ & \epsilon(2 + c) + \left\| u' \sum_{j=1}^{n} u_j[(v_j \circ q) - (v_j \circ q)(v' \circ q)] \right\|_{A(G)} \\
\leqslant\ & \epsilon(2 + c) + \sum_{j=1}^{n} \|u_j\| \cdot \|v_j - v_j v'\|_{A(G/N)} \\
\leqslant\ & \epsilon(3 + c).
\end{aligned}$$

Since $\epsilon > 0$ is arbitrary, the statement of the lemma follows. □

It is easy to verify that $h(e(I)) = q^{-1}(h(I))$ and $\mathcal{R}_c(G/H) \subseteq \mathcal{R}_c(G)$. Therefore, it follows from the much more difficult Theorem 6.5.11 that the ideal $e(I)$ has a bounded approximate identity whenever I has one. However, Theorem 6.5.11 does not provide any bound for the norm.

REMARK 6.6.6. (i) Let J be a closed ideal of $A(G)$ such that $L_h u \in J$ for every $u \in J$ and $h \in H$. Then $R_h u \in J$ for every $u \in J$ and $h \in H$. This follows from (ii) of the preceding theorem replacing J by the ideal $\widetilde{J} = \{\widetilde{u} : u \in J\}$. In fact, since

$$(R_h \widetilde{u})(x) = \overline{u(h^{-1}x^{-1})} = \overline{(L_{h^{-1}}u)(x^{-1})} = \widetilde{L_{h^{-1}}u}(x),$$

$\widetilde{J}$ is right H-translation invariant. Consequently, $\widetilde{J} = e(r(\widetilde{J}))$ is two-sided H-translation invariant.

(ii) Let E be a closed subset of G/H. Then

$$e(I(E)) = I(q^{-1}(E)) \quad \text{and} \quad r(I(q^{-1}(E))) = I(E).$$

The second equation follows from the first and Theorem 6.6.4(i). For the first equation, if $v \in I(E)$ then $u(v \circ q)$ vanishes on $q^{-1}(E)$ for every $u \in A(G)$, and hence $e(I(E)) \subseteq I(q^{-1}(E))$. Conversely, if $u \in I(q^{-1}(E)) \cap C_c(G)$, then $T_H u \in I(E)$. This shows that $r(I(q^{-1}(E)) \subseteq I(E)$, and since $I(q^{-1}(E))$ is right H-translation invariant, Theorem 6.6.4(i) yields

$$I(q^{-1}(E)) = e(r(I(q^{-1}(E))) \subseteq e(I).$$

COROLLARY 6.6.7. *Let G and H be as in Theorem 6.6.4 and let J be a closed ideal of $A(G)$. Then $e(r(J))$ coincides with the closed linear span of the set $\{R_h u : u \in J, h \in H\}$, which is the smallest H-translation invariant closed ideal of $A(G)$ containing J.*

PROOF. Let L denote the ideal in question. In view of Theorem 6.6.4, it suffices to show that $r(L) = r(J)$. For that, since $J \subseteq L$, it is enough to verify that $r(L) \subseteq r(J)$.

Let $v \in L \cap C_c(G)$ and $\epsilon > 0$ be given and choose a compact neighbourhood K of $\operatorname{supp} v$. By Lemma 6.4.3 there exists a constant $c > 0$ such that $\|q(u)\|_{A(G/H)} \leqslant c\|u\|_{A(G)}$ for every $u \in A(G)$ with $\operatorname{supp} u \subseteq K$. Since $A(G)$ is regular, we find $w \in A(G)$ with $w = 1$ on $\operatorname{supp} v$, $\operatorname{supp} w \subseteq K$ and $\|w\|_{A(G)} \leqslant 2$ Corollary 3.2.6. By definition of L there exist $u_1, \ldots, u_n \in J$ and $h_1, \ldots, h_n \in H$ such that

$$\left\| v - \sum_{j=1}^{n} R_{h_j}(u_j) \right\|_{A(G)} \leqslant \frac{\epsilon}{2c}.$$

For $j = 1, \ldots, n$, let $w_j = u_j R_{h_j^{-1}}(w)$. Then

$$\begin{aligned}
\left\| q(v) - \sum_{j=1}^{n} q(R_{h_j}(w)) \right\|_{A(G/H)} &= \left\| q\left(v - \sum_{j=1}^{n} R_{h_j}(w)) \right) \right\|_{A(G/H)} \\
&\leqslant c \cdot \left\| \left(v - \sum_{j=1}^{n} R_{h_j}(u_j) \right) w \right\|_{A(G)} \\
&\leqslant 2c \cdot \left\| v - \sum_{j=1}^{n} R_{h_j}(u_j) \right\|_{A(G)} \leqslant \epsilon.
\end{aligned}$$

Since $q(R_{h_j}(w_j)) \in r(J)$ and $\epsilon > 0$ was arbitrary, we conclude that $q(v) \in r(J)$. This shows that $r(L) \subseteq r(J)$. □

PROPOSITION 6.6.8. *For $\varphi \in UC(\widehat{G})^*$ and $T \in VN(G)$, there exists a unique element of $VN(G)$, denoted $\sigma_G(\varphi)(T)$, such that*

$$\langle \varphi, u \cdot T \rangle = \overline{\langle u, \sigma_G(\varphi)(T) \rangle}$$

for all $u \in A(G)$. The map $\sigma_G : \varphi \to \sigma_G(\varphi)$ is an isometric algebra isomorphism form $UC(\widehat{G})^*$ onto $Hom_{A(G)}(VN(G))$. Moreover, for every $\phi \in Hom_{A(G)}(VN(G))$ and $T \in UC(\widehat{G})$, we have

$$\langle \sigma_G^{-1}(\phi), T \rangle = \langle \phi(T), 1_G \rangle.$$

PROOF. It is clear that σ_G is a linear map and that $\|\sigma_G(\varphi)\| \leqslant \|\varphi\|$. To show that σ_G is actually an isometry and surjective, let $\phi \in \mathrm{Hom}_{A(G)}(VN(G))$ be given and define $\varphi \in UC(\widehat{G})^*$ by

$$\langle \varphi, T\rangle = \langle \phi(T), 1_G\rangle, \quad T \in UC(\widehat{G}).$$

Then $\|\varphi\| \leqslant \|\phi\|$ and, for $u \in A(G)$ and $T \in VN(G)$, since $u \cdot T \in UC(\widehat{G})$,

$$\begin{aligned} \overline{\langle u, \sigma_G(\varphi)(T)\rangle} &= \langle \varphi, u\cdot T\rangle = \langle \phi(u\cdot T), 1_G\rangle, \\ &= \langle u \cdot \phi(T), 1_G\rangle, = \langle \phi(T), u\rangle, \\ &= \overline{\langle u, \phi(T)\rangle}. \end{aligned}$$

This shows that $\sigma_G(\varphi) = \phi$ and $\|\sigma_G(\varphi)\| \geqslant \|\varphi\|$. Since σ_G is bijective, it also follows that $\sigma_G^{-1}(\phi)(T) = \langle \phi(T), 1_G\rangle$, for every $\phi \in Hom_{A(G)}(VN(G))$ and $T \in UC(\widehat{G})$.

To show that σ_G is multiplicative, let $\varphi, \psi \in UC(\widehat{G})^*$, $T \in VN(G)$ and $u \in A(G)$. Then, by definition of the product in $UC(\widehat{G})^*$ and since $\psi\cdot(u\cdot T) = u\cdot(\psi\cdot T)$,

$$\begin{aligned} \langle v, u\cdot \sigma_G(\psi)(T)\rangle &= \langle vu, \sigma_G(\psi)(T)\rangle = \overline{\langle \psi, (vu)\cdot T\rangle} \\ &= \langle vu, \psi\cdot T\rangle = \langle v, , u\cdot(\psi\cdot T)\rangle \end{aligned}$$

for all $v \in A(G)$, and hence

$$\begin{aligned} \overline{\langle u, \sigma_G(\varphi\cdot\psi)(T)\rangle} &= \langle \varphi\cdot\psi, u\cdot T\rangle = \langle \varphi, \psi\cdot(u\cdot T)\rangle \\ &= \langle \varphi, u\cdot(\psi\cdot T)\rangle = \langle \varphi, u\cdot(\sigma_G(\psi)(T))\rangle \\ &= \overline{\langle u, \sigma_G(\varphi)(\sigma_G(\psi)(T))\rangle}. \end{aligned}$$

Thus σ_G is an algebra homomorphism. This completes the proof. □

Note that, for $\varphi \in UC(\widehat{G})^*$ and $T \in UC(\widehat{G})$, $\sigma_G(\varphi)(T) = \varphi\cdot T$, by definition of $\varphi\cdot T$ and Proposition 6.6.8.

LEMMA 6.6.9. *The adjoint map* $\phi_H^* : UC(\widehat{G/H})^* \to UC(\widehat{G})^*$ *is an isometric algebra homomorphism.*

PROOF. To start with, let $T \in VN(G)$ with compact support, and choose $w \in A(G) \cap C_c(G)$ such that $w = 1$ in a neighbourhood of $\operatorname{supp} T$. Then, for all $u, v \in A(G/H)$,

$$\begin{aligned} \langle u\cdot\phi_H(T), v\rangle &= \langle T, w(uv\circ q)\rangle = \langle T, w^2(uv\circ q)\rangle \\ &= \langle w(u\circ q)\cdot T, w(v\circ q)\rangle \\ &= \langle \phi_H[w(u\circ q)\cdot T], v\rangle, \end{aligned}$$

since $\operatorname{supp}[w(u\circ q)\cdot T] \subseteq \operatorname{supp} T$. This implies, for every $\psi \in UC(\widehat{G/H})^*$ and $u \in A(G/H)$,

$$\begin{aligned} \langle u, \phi_H[\phi_H^*(\psi)\cdot T]\rangle &= \langle \phi_H^*(\psi)\cdot T, w(u\circ q)\rangle \\ &= \overline{\langle \phi_H^*(\psi), w(u\circ q)\cdot T\rangle} \\ &= \overline{\langle \psi, \phi_H[w(u\circ q)\cdot T]\rangle} \\ &= \overline{\langle \psi, u\cdot\phi_H(T)\rangle} \\ &= \langle u, \psi\cdot\phi_H(T)\rangle. \end{aligned}$$

For any two $\varphi, \psi \in UC(\widehat{G/H})^*$, we then obtain by the definition of multiplication in $UC(\widehat{G/H})^*$ and in $UC(\widehat{G})^*$,

$$\begin{aligned}
\langle \phi_H^*(\varphi) \cdot \phi_H^*(\psi), T \rangle &= \langle \phi_H^*(\varphi), \phi_H^*(\psi) \cdot T \rangle = \langle \varphi, \phi_H[\phi_H^*(\psi) \cdot T] \rangle \\
&= \langle \varphi, \psi \cdot \phi_H(T) \rangle = \langle \varphi \cdot \psi, \phi_H(T) \rangle \\
&= \langle \phi_H^*(\varphi \cdot \psi), T \rangle.
\end{aligned}$$

Since $UC_c(\widehat{G/H})$ is dense in $UC(\widehat{G/H})$, we conclude that ϕ_H^* is an algebra homomorphism. □

THEOREM 6.6.10. *Let I be a closed ideal of $A(G/H)$ and suppose that there exists a projection P from $VN(G/H)$ onto $I^\perp$ such that P commutes with the action of $A(G/H)$ on $VN(G/H)$. Then $Q = \sigma_G \circ \phi_H^* \circ \sigma_{G/H}^{-1}(P)$ is a projection of $VN(G)$ onto $e(I)^\perp$.*

PROOF. Let $T \in VN(G)$, $v \in A(G/H)$ and $u \in A(G) \cap C_c(G)$. Then

$$\begin{aligned}
\langle Q(T), u(v \circ q) \rangle &= \langle \sigma_G[(\phi_H^*(\sigma_{G/H}^{-1}(P)))(T)], u(v \circ q) \rangle \\
&= \overline{\langle \phi_H^*(\sigma_{G/H}^{-1}(P)), u(v \circ q) \cdot T \rangle} \\
&= \overline{\langle \sigma_{G/H}^{-1}(P), \phi_H[u(v \circ q) \cdot T] \rangle} \\
&= \overline{\langle \sigma_{G/H}^{-1}(P), (vT_H u) \cdot \phi_H(T) \rangle} \\
&= \overline{\langle P[(vT_H u) \cdot \phi_H(T)], 1_G \rangle}, \\
&= \overline{\langle (vT_H u) \cdot P(\phi_H(T)), 1_G \rangle}, \\
&= \langle P(\phi_H(T)), vT_H u \rangle.
\end{aligned}$$

This last term is zero if $v \in I$ since $vT_H u \in I$, $P \in \mathrm{Hom}_{A(G/H)}(VN(G/H))$ and P maps into $I^\perp$. Since $e(I)$ is the closed linear span of elements of the form $u(v \circ q)$, where $v \in I$ and $u \in A(G) \cap C_c(G)$, we conclude that $Q(T) \in e(I)^\perp$.

It remains to show that $Q(T) = T$ for every $T \in e(I)^\perp$. For such T and $u \in A(G)$ we have $\phi_H(u \cdot T) \in r(e(I))^\perp = I^\perp$ by Theorem 6.6.4 and Lemma 6.6.9 and hence $P(\phi_H(u \cdot T)) = \phi_H(u \cdot T)$. On the other hand, using $P \in \mathrm{Hom}_{A(G/H)}(VN(G/H))$, Propositions 6.6.8, 6.4.5 and choosing $w \in A(G) \cap C_c(G)$ such that $w = 1$ on a neighbourhood of $\operatorname{supp} u$, we have for any $v \in A(G/H)$,

$$\begin{aligned}
\langle u(v \circ q), Q(T) \rangle &= \langle \phi_H^*(\sigma_{G/H}^{-1}(P)), u(v \circ q) \cdot T \rangle \\
&= \langle \sigma_{G/H}^{-1}(P), \phi_H(u(v \circ q) \cdot T) \rangle \\
&= \overline{\langle P(\phi_H(u(v \circ q) \cdot T)), 1_{G/H} \rangle}, \\
&= \overline{P(v \cdot \phi_H(u \cdot T)), 1_{G/H} \rangle}, \\
&= \overline{\langle v \cdot P(\phi_H(u \cdot T)), 1_{G/H} \rangle}, \\
&= \overline{\langle v \cdot \phi_H(u \cdot T), 1_{G/H} \rangle}, \\
&= \overline{\langle \phi_H(u \cdot T), v \rangle} = \overline{\langle u \cdot T, w(v \circ q) \rangle} \\
&= \langle u(v \cdot q), T \rangle.
\end{aligned}$$

Since every $u \in A(G) \cap C_c(G)$ can be written as $u(v \circ q)$ for some $v \in A(G/H)$, it follows that $Q(T) = T$. □

6.7. Notes and references

Let G be a locally compact abelian group. The first example of a set of nonsynthesis in $\widehat{G} = \sigma(L^1(G))$ was given by L. Schwartz [**258**]. He proved that the unit sphere $S^2 \subseteq \mathbb{R}^3 = \sigma(L^1(\mathbb{R}^3))$ fails to be a set of synthesis by explicitly constructing a bounded linear functional on $L^1(\mathbb{R}^3)$ which annihilates $j(S^2)$, but does not annihilate $I(S^2)$. Actually, Schwartz's proof works for $S^{d-1} \subseteq \sigma(L^1(\mathbb{R}^d))$ for all $d \geqslant 3$. In contrast, Herz [**121**] has shown that the circle S^1 is a set of synthesis for $L^1(\mathbb{R}^2)$. On the other hand, spectral synthesis holds for $L^1(G)$ whenever G is a compact abelian group. That compactness of G is necessary for spectral synthesis to hold for $L^1(G)$, is the content of Malliavin's celebrated theorem [**207**]. A somewhat more accessible proof of Malliavin's theorem was given by Varopoulos [**278**] using tensor product methods. The ana logue of Malliavin's theorem for general locally compact groups, which was presented in Section 6.3, was shown in [**155**]. The proof given here follows [**155**] and utilizes structure theory of locally compact groups, Malliavin's theorem and a deep theorem of Zelmanov on the existence of infinite abelian groups in infinite compact groups [**293**]. It seems to be unknown whether there exist locally compact groups G at all, for which the condition $u \in \overline{uA(G)}$ is not satisfied.

The fact that a closed subgroup of a locally compact group G is a set of synthesis for $A(G)$ (Theorem 6.1.9), is due to Takesaki and Tatsuuma [**272**]. For amenable G, the same is true for the Figa-Talamanca-Herz algebras $A_p(G)$ [**123**].

Section 6.4, which treats Lohoué's projection theorem for local spectral sets, is an elaboration of the original article [**196**]. However, it should be pointed out that Lohoué accomplishes his result for Figà-Talamanca-Herz algebras $A_p(G)$, $1 < p < \infty$, rather than just Fourier algebras. In this general setting the proof turns out to be technically more complicated.

Several authors have studied the problem of which closed ideals of $A(G)$ have a bounded approximate identity ([**92**], [**193**], [**154**], [**79**], [**80**], [**45**] and [**46**]). The investigations departed from Leptin's theorem (Theorem 3.2.5) stating that $A(G)$ itself has a bounded approximate identity (actually, then one of norm bound 1) precisely when G is amenable. However, Proposition 6.5.16, which has been proved in [**79**], shows that proper closed ideals of $A(G)$ rarely have approximate identities of norm bound 1. As shown in Theorem 6.5.5, if G is amenable then the ideal $I(H)$ admits an approximate identity with norm bound 2 for any closed subgroup H of G. Moreover, if H is nonopen, 2 is the best possible such bound (Theorem 6.5.15). These results are due to Forrest and Spronk [**92**], and the first one even holds for Figà-Talamanca-Herz algebras. The most general earlier result stating the same ([**154**, Theorem 3.4]) required G to have the H-separation property (see Chapter 7 for the definition). As mentioned at the outset of Section 6.5, the closed ideals in the group algebra $L^1(G)$ of a locally compact abelian group G have been completely described by Liu, van Rooij and Wang [**193**] in terms of the closed sets in the coset ring of the dual group $\widehat{G}$ of G. Theorem 6.5.11, which gives the analogous description for ideals with bounded approximate identities in $A(G)$ for amenable groups G, is the main result of [**85**].

The material in Section 6.6 is taken from [**45**], and our presentation follows [**45**]. Delaporte and Derighetti have actually shown the results in the more general context of algebras $A_p(G)$, $1 < p < \infty$, of any amenable locally compact group G. However, proofs are less technical when $p = 2$. Although amenability of G is employed several times, it was conjectured in [**45**] that Theorem 6.6.4 should be true for nonamenable groups as well.

We conclude these notes by mentioning another line of investigation, which was extensively pursued in recent years. Generalizing the notion of spectral set, Warner [**286**] introduced the concept of a weak spectral set. If A is an arbitrary semisimple and regular commutative Banach algebra, a closed subset E of its Gelfand space $\sigma(A)$ in called a *weak spectral set* if there exists some $n \in \mathbb{N}$ such that $a^n \in \overline{j(E)}$ for every $a \in I(E)$. The smallest such n is then called the characteristic of E. Warner was motivated by the union problem, that is, the question of whether the union of two sets of synthesis for $L^1(G)$ is again a set of synthesis. This problem is still open, even for $\ell^1(\mathbb{Z})$. In contrast, Warner showed that the union of two weak spectral sets is a weak spectral set. Another motivation for studying weak spectral sets is Varopoulos' result [**277**] stating that the sphere S^{d-1} in $\mathbb{R}^d = \sigma(L^1(\mathbb{R}^d))$ is a weak spectral set with characteristic $\lfloor \frac{d+1}{2} \rfloor$. Subsequently, weak spectral sets have been investigated by several authors for commutative Banach algebra, especially Fourier algebras ([**150**], [**152**], [**153**] [**160**], [**225**], [**226**], [**228**]). We mention just one particular issue and otherwise refer to the literature. Modifying Varopoulos' proof [**278**] of Malliavin's theorem, it was shown in [**230**] that weak spectral synthesis fails for any nondiscrete abelian group. This result was extended to arbitrary locally compact groups in [**150**] and also in [**228**].

If A is an arbitrary semisimple and regular commutative Banach algebra and E_1 and E_2 are closed subsets of the Gelfand space $\sigma(A)$ of A such that $E_1 \cap E_2$ is a set of synthesis, then $E_1 \cup E_2$ is a set of synthesis if and only if both E_1 and E_2 are sets of synthesis [**151**, Theorem 5.2.5]. This general result covers the case $X = VN(G)$ of Theorem 6.1.12.

CHAPTER 7

Extension and Separation Properties of Positive Definite Functions

The Hahn-Banach extension theorem asserts that if E is a normed linear space and F is a closed linear subspace of E, then each continuous linear functional on F extends to a continuous linear functional on E. From this it follows that given $x \in E\backslash F$, there exists a continuous linear functional ϕ on E such that $\phi = 0$ on F and $\phi(x) \neq 0$ (the Hahn-Banach separation theorem). In this chapter we address the analogous properties for positive definite functions on locally compact groups.

Let G be an arbitrary locally compact group and H a closed subgroup of G. We have shown in Section 2.6 that the restriction map $u \to u|_H$ from $A(G)$ into $A(H)$ is surjective. The corresponding problem for Fourier-Stieltjes algebras is much more delicate. To fix terminology, let us say that G has the extension property if for every closed subgroup H, each $\varphi \in P(H)$ admits an extension $\phi \in P(G)$ (equivalently, $B(H) = B(G)|_H$). The largest class of locally compact groups sharing this extension property, is formed by the groups with small conjugation invariant neighbourhoods of the identity, the so-called SIN-groups. The converse implication is true for connected Lie groups and for compactly generated nilpotent groups. More precisely, a connected Lie group has the extension property only if it is a direct product of a vector group and a compact group. On the other hand, there exists a compactly generated 2-step solvable group which has the extension property, but fails to be a SIN-group. These results are developed in Sections 7.3 and 7.4.

In this situation, it is reasonable to pose the above problem also for a fixed subgroup of G. We establish various results concerning subgroups which are extending in the obvious sense and which prove helpful in identifying all the extending subgroups in specific examples. For instance, if G is a nilpotent locally compact group and H is a closed subgroup of G which is topologically isomorphic to either $\mathbb{R}$ or $\mathbb{Z}$, then H is extending in G. In addition, we present a theorem, due to Cowling and Rodway [**39**], providing a necessary and sufficient condition for a function $u \in B(N)$ to belong to $B(G)|_N$ when N is a closed normal subgroup of G (Section 7.2).

The second topic pursued in this chapter is a separation property of positive definite functions. A closed subgroup H of G is called a separating subgroup if for each $x \in G\backslash H$ there exists ϕ in

$$P_H(G) = \{\phi \in P(G) : \phi(h) = 1 \text{ for all } h \in H\}$$

with $\phi(x) \neq 1$. Accordingly, G is said to have the separation property if every closed subgroup of G is separating. The interest and importance of the separation property arise from the fact that it is useful in the study of the ideal theory of Fourier algebras. In Section 7.5 we first show that a neutral closed subgroup of any locally compact group is separating, and consequently every SIN-group has the separation property. The converse fails to be true. In fact, the so-called Fell group, which is 2-step solvable and totally disconnected, turns out to have the separation property, even though it is not an SIN-group. Moreover, we consider a number of illustrative examples, for each of which we identify all the separating subgroups. Similar to the extension property, it turns out that for G, an almost connected group or a compactly generated nilpotent group, validity of the separation property implies that G is an SIN-group. Actually, for both classes of locally compact groups, the separation property for cyclic subgroup already forces the group to have small invariant neighbourhoods. Concerning a single closed subgroup H of G, the most interesting result asserts that if G has an almost connected open normal subgroup, then H is separating in G (if and) only if H is neutral in G. When G is a connected nilpotent group, the only separating subgroups of G are the normal subgroups.

7.1. The extension property: Basic facts

We start by introducing the properties which will be the subject of the following four sections.

DEFINITION 7.1.1. Let G be a locally compact group and H a closed subgroup of G.

(i) We say that H *is extending in* G if for every $\varphi \in P(H)$ there exists $\phi \in P(G)$ such that $\phi|_H = \varphi$.

(ii) The group G is said to have the *extension property* if each closed subgroup of G is extending.

Since every function in $B(H)$ is a finite linear combination of continuous positive definite functions, the restriction map $u \to u|_H$ from $B(G)$ to $B(H)$ is surjective whenever H is extending in G. As will be seen in Lemma 7.1.11, the converse is also true.

In this section we collect a number of fundamental facts about extensibility which will be used subsequently. The first part of the following lemma is shown in [**212**, (32.43)].

LEMMA 7.1.2. *Let H be an open subgroup of the locally compact group G. Then H is extending in G. More precisely, given $\varphi \in B(H)$, there exists $\phi \in B(G)$ such that $\phi|_H = \varphi$ and $\|\phi\| = \|\varphi\|$.*

PROOF. Recall that ϕ can be defined by $\phi(x) = \varphi(x)$ for $x \in H$ and $\phi(x) = 0$ otherwise. Let now $\varphi(x) = \langle \pi(x)\xi, \eta \rangle$, $x \in G$, where π is a unitary representation of H and $\xi, \eta \in \mathcal{H}(\pi)$ are such that $\|\varphi\| = \|\xi\| \cdot \|\eta\|$. Then, realizing the induced representation $\mathrm{ind}_H^G \pi$ on the Hilbert space $\ell^2(G/H, \mathcal{H}(\pi))$, we have $\mathcal{H}(\pi) \subseteq \mathcal{H}(\mathrm{ind}_H^G \pi)$ and $\phi(x) = \langle \mathrm{ind}_H^G \pi(x)\xi, \eta \rangle$ for all $x \in G$. Thus $\|\phi\| \leqslant \|\xi\| \cdot \|\eta\| = \|\varphi\|$ and hence $\|\phi\| = \|\varphi\|$. □

LEMMA 7.1.3. *Let H and N be closed subgroups of G such that N is normal, $H \cap N = \{e\}$, HN is closed in G and H is topologically isomorphic to HN/N. If HN/N is extending in G/N, then H is extending in G.*

PROOF. Let $\varphi \in P(H)$ and define $\psi : HN/N \to \mathbb{C}$ by $\psi(hN) = \varphi(h)$, $h \in H$. Then ψ is positive definite, and since the map $h \to hN$ is a homeomorphism from H onto HN/N, ψ is continuous. By hypothesis, ψ extends to some $\rho \in P(G/N)$. Then $\phi : G \to \mathbb{C}$, defined by $\phi(x) = \rho(xN)$, is in $P(G)$ and extends φ. □

LEMMA 7.1.4. *Let H be a closed subgroup of G and let φ be a character of H. If φ extends to some continuous positive definite function on G, then there exists $\phi \in \mathrm{ex}(P(G))$ such that $\phi|_H = \varphi$.*

PROOF. Let $P_\varphi(G)$ denote the set of all functions in $P(G)$ which extend φ, and suppose first that the set $P_\varphi(G)$ has an extreme point ϕ. Then $\phi \in \mathrm{ex}(P(G))$. Indeed, if $\psi \in P(G)$ is such that $\psi \leqslant c\phi$ for some $c > 0$, then $\psi|_H \leqslant c\phi|_H = c\varphi$ and hence $\psi|_H = \varphi$ since φ is extremal.

Therefore it remains to show that $P_\varphi(G)$ has an extreme point. For that, let $K = \{\phi \in P(G) : \phi(e) \leqslant 1\}$ and consider the convex subset

$$K_\varphi = \{\phi \in K : \phi|_H = \phi(e)\varphi\}$$

of K. Recall that if $\rho \in P^1(G)$ and $h \in G$, then $\rho(xh) = \rho(x)\rho(h)$ for all $x \in G$ if and only if $|\rho(h)| = 1$ by Proposition 1.4.16(v). Applying this to $\rho = \frac{1}{\phi(e)}\phi$, where $\phi \in K$ and $\phi \neq 0$, we get that a function $\phi \in K$ belongs to K_φ if and only if $\phi(xh) = \phi(x)\phi(h)$ for all $x \in G$ and $h \in H$. This equation is equivalent to

$$\int_G \phi(xh)f(x)dx = \varphi(h)\int_G \phi(x)f(x)dx$$

for all $f \in L^1(G)$ and $h \in H$. This in turn implies that K_φ is closed in K in the w^*-topology. Since $K \subseteq P(G) \subseteq L^\infty(G)$ is w^*-compact, K_φ is w^*-compact and hence by the Krein-Milman theorem is the closed convex hull of its set of extreme points. Now, any nonzero extreme point ϕ of K_φ must satisfy $\phi(e) = 1$ and hence belongs to $P_\varphi(G)$. Thus $P_\varphi(G)$ has extreme points, and such extreme points are extreme points of $P(G)$ by the first paragraph of the proof. $\square$

We next observe that that extensibility of a positive definite function on a closed subgroup can be expressed in terms of representations.

LEMMA 7.1.5. *Let H be a closed subgroup of a locally compact group G and let $\varphi \in P^1(H)$. Then φ extends to some $\phi \in P(G)$ if and only if π_φ is a subrepresentation of $\pi|_H$ for some representation π of G.*

PROOF. If π_φ is a subrepresentation of $\pi|_H$, then there exists an isometric linear mapping U from $\mathcal{H}(\pi_\varphi)$ into $\mathcal{H}(\pi)$ such that $U\pi_\varphi(h) = \pi(h)U$ for all $h \in H$. It follows that, for some $\eta \in \mathcal{H}(\pi_\varphi)$ and all $h \in H$,

$$\varphi(h) = \langle \pi_\varphi(h)\eta, \eta\rangle = \langle U\pi_\varphi(h)\eta, U\eta\rangle = \langle \pi(h)U\eta, U\eta\rangle.$$

Thus ϕ, defined by $\phi(x) = \langle \pi(x)U\eta, U\eta\rangle$ for $x \in G$, is a continuous positive definite function extending φ.

Conversely, let $\phi \in P(G)$ be an extension of φ and let $\pi = \pi_\phi$. Then there exists $\xi \in \mathcal{H}(\pi)$ such that $\phi(x) = \langle \pi(x)\xi, \xi\rangle$ for all $x \in G$. On the other hand, $\varphi(h) = \langle \pi_\varphi(h)\eta, \eta\rangle$ for some cyclic vector η of π_φ and all $h \in H$. Let L denote the linear span of all $\pi_\varphi(h)\eta$, $h \in H$. For $\sum_{j=1}^n c_j\pi_\varphi(h_j)\eta \in L$, $h_j \in H, c_j \in \mathbb{C}$, we have

$$\begin{aligned}\Big\|\sum_{j=1}^n c_j\pi_\varphi(h_j)\eta\Big\|^2 &= \sum_{j,k=1}^n c_j\overline{c_k}\langle\pi_\varphi(h_k^{-1}h_j)\eta,\eta\rangle = \sum_{j,k=1}^n c_j\overline{c_k}\varphi(h_k^{-1}h_j)\\ &= \sum_{j,k=1}^n c_j\overline{c_k}\phi(h_k^{-1}h_j) = \sum_{j,k=1}^n c_j\overline{c_k}\langle\pi(h_j)\xi,\pi(h_k)\xi\rangle\\ &= \Big\|\sum_{j=1}^n c_j\pi(h_j)\xi\Big\|^2.\end{aligned}$$

Thus we can define a linear mapping from L into $\mathcal{H}(\pi)$ by

$$\sum_{j=1}^n c_j\pi_\varphi(h_j)\eta \to \sum_{j=1}^n c_j\pi(h_j)\xi,$$

$h_j \in H$, $c_j \in \mathbb{C}$, $n \in \mathbb{N}$, and this mapping is isometric. Since L is dense in $\mathcal{H}(\pi_\varphi)$, this mapping extends uniquely to an isometric linear mapping U from $\mathcal{H}(\pi_\varphi)$ into $\mathcal{H}(\pi)$. Finally, U satisfies

$$U\pi_\varphi(h_1)\pi_\varphi(h_2)\eta = \pi(h_1h_2)\xi = \pi(h_1)(U\pi_\varphi(h_2)\eta)$$

for all $h_1, h_2 \in H$. This implies that $U\pi_\varphi(h) = \pi(h)U$ for all $h \in H$, as required. □

COROLLARY 7.1.6. *Let H be a closed subgroup of G which is either compact or contained in the centre of G. Then H is extending in G.*

PROOF. Let $\varphi \in P(H)$ and let $\pi = \operatorname{ind}_H^G \pi_\varphi$, where π_φ is the cyclic representation associated with φ. In both cases it is known that $\pi|_H$ contains π_φ as a subrepresentation. Actually, if H is contained in the centre of G, then $\pi|_H$ is a multiple of π_φ. So the statement follows from Lemma 7.1.5. □

LEMMA 7.1.7. *Let H and N be closed subgroups of the locally compact group G such that N is normal and $H \supseteq N$. If H is extending in G, then H/N is extending in G/N.*

PROOF. Let $q : G \to G/N$ denote the quotient homomorphism and let $\varphi \in P(H/N)$. Then $\varphi\circ q \in P(H)$ and hence there exists $\phi \in P(G)$ such that $\phi|_H = \varphi\circ q$. Let $L = \{x \in G : \phi(x) = \phi(e)\}$. Then L is a closed subgroup of G, ϕ is constant on cosets of L, and L clearly contains N. Therefore ϕ is of the form $\phi = \psi \circ q$ for some $\psi \in P(G/N)$. For $h \in H$, we have

$$\psi(hN) = \phi(h) = \varphi \circ q(h) = \varphi(hN),$$

so that ψ extends φ. □

LEMMA 7.1.8. *Let G be a locally compact group and N a closed normal subgroup of finite index in G. If H is a closed subgroup of G such that $H \cap N$ is extending in N, then H is extending in G.*

PROOF. Let $\psi \in P(H)$, $K = H \cap N$ and $\varphi = \psi|_K$. By hypothesis, there exists $\phi \in P(N)$ such that $\phi|_K = \varphi$. We claim that the representation $\rho = \operatorname{ind}_N^G \pi_\phi$ has the property that $\rho|_H \geqslant \pi_\psi$.

To see this, note first that since H/K is finite,

$$\pi_\psi \leqslant \operatorname{ind}_K^H \pi_\varphi = \operatorname{ind}_K^H(\pi_{\phi|_K}) \leqslant \operatorname{ind}_K^H(\pi_\phi|_K).$$

On the other hand, for any representation τ of N,

$$\mathrm{ind}_K^H(\tau|_K) \leqslant \left(\mathrm{ind}_N^G \tau\right)|_H.$$

Indeed, since $K = H \cap N$ and G/N is finite, it is easily verified that the range of the map

$$T : \mathcal{H}(\mathrm{ind}_K^H(\tau|_K)) \to \{F : G \to \mathcal{H}(\tau)\},$$

defined by $T\xi(x) = 0$ if $x \notin HN$ and $T\xi(x) = \tau(y^{-1})\xi(h)$ if $x = hy$ with $h \in H$ and $y \in N$, is contained in

$$\mathcal{H}(\mathrm{ind}_N^G \tau) = \mathcal{H}((\mathrm{ind}_N^G \tau)|_H),$$

and that T provides an intertwining operator between the representations $\mathrm{ind}_K^H(\tau|_K)$ and $(\mathrm{ind}_N^G \tau)|_H$. Thus $\pi_\psi \leqslant (\mathrm{ind}_N^G \pi_\phi)|_H$, and an application of Lemma 7.1.5 gives that ψ extends to some element of $P(G)$. □

COROLLARY 7.1.9. *Let G_1 be a closed subgroup of finite index in the locally compact group G. Then G has the extension property if and only if G_1 has the extension property.*

PROOF. Because G_1 is extending in G, we only have to show the "if part". So suppose that G_1 has the extension property, and let N be the largest normal subgroup of G which is contained in G_1. Then N has finite index in G. If H is any closed subgroup of G, then by hypothesis, $H \cap N$ is extending in G_1 and hence in N. Now, Lemma 7.1.8 shows that H is extending in G. □

LEMMA 7.1.10. *Let H be a closed subgroup of G and let $\varphi, \psi \in P(H)$ be such that $\varphi - \psi \in P(H)$. If φ extends to some element of $P(G)$, then the same is true of ψ.*

PROOF. Write $\varphi(h) = \langle \pi_\varphi(h)\xi, \xi\rangle$, $h \in H$, for some cyclic vector $\xi \in \mathcal{H}(\pi_\varphi)$. By Lemma 7.1.5, there exists a representation π of G in a Hilbert space $\mathcal{H}$ containing $\mathcal{H}(\pi_\varphi)$ such that $\pi_\varphi(h)\eta = \pi(h)\eta$ for all $h \in H$ and $\eta \in \mathcal{H}(\pi_\varphi)$. Thus, for all $h \in H$, $\varphi(h) = \langle \pi(h)\xi, \xi\rangle$. Since $\varphi - \psi$ is positive definite on H, there exists a positive operator T on $\mathcal{H}(\pi_\varphi)$, commuting with all operators $\pi_\varphi(h)$, $h \in H$, such that $\psi(h) = \langle T\pi(h)\xi, \xi\rangle$. Let S be the positive square root of T. Then S commutes with all $\pi_\varphi(h)$ and we have

$$\begin{aligned}\psi(h) &= \langle T\pi_\varphi(h)\xi, \xi\rangle = \langle S\pi_\varphi(h)S\xi, \xi\rangle = \langle \pi_\varphi(h)S\xi, S\xi\rangle \\ &= \langle \pi(h)S\xi, S\xi\rangle.\end{aligned}$$

So the positive definite function $x \to \langle \pi(x)S\xi, S\xi\rangle$ on G extends ψ. □

With the aid of the preceding lemma we can now show that if the restriction map $B(G) \to B(H)$ is surjective, then H is extending in G.

LEMMA 7.1.11. *Let H be a closed subgroup of G and let $\varphi \in P(H)$. If φ extends to some function in $B(G)$, then φ also extends to some function in $P(G)$.*

PROOF. Let first φ be an arbitrary hermitian function in $B(H)$, that is, $\varphi(h) = \overline{\varphi(h^{-1})}$ for all $h \in H$, and let $\psi \in B(G)$ be an extension of φ. Write $\psi = \psi_1 + i\psi_2$, where ψ_1 and ψ_2 are hermitian functions in $B(G)$. For $h \in H$, we then have $\varphi(h) = \psi_1(h) + i\psi_2(h)$ and φ and ψ_1 are hermitian on H, and hence $i\psi_2$ has to be hermitian on H. Thus, since ψ_2 is hermitian,

$$i\psi_2(h) = \overline{i\psi_2(h^{-1})} = -i\overline{\psi_2(h^{-1})} = -i\psi_2(h)$$

for all $h \in H$. So $\psi_2 = 0$ on H and the hermitian function ψ_1 is itself an extension of φ.

Now suppose that φ is positive definite. Since ψ_1 is hermitian, it can be written as $\psi_1 = \psi^+ - \psi^-$, where ψ^+ and ψ^- are in $P(G)$. For $h \in H$, we then have $\psi^+(h) = \varphi(h) + \psi^-(h)$. Since $\psi^+|_H - \varphi \in P(H)$ and $\psi^+|_H$ has the positive definite extension ψ^+, it follows from Lemma 7.1.10 that φ has an extension in $P(G)$. □

7.2. Extending from normal subgroups

In Section 7.1 we have identified a few simple cases of extending closed subgroups H of a locally compact group G. For arbitrary H, the problem appears to be untractable. However, the situation improves when H is normal in G. The theorem of Cowling and Rodway, mentioned in the introduction to this chapter, is presented below (Theorem 7.2.3). We start with a special and considerably simpler case, namely that of a character of an abelian normal subgroup, which will actually turn out to be much easier applicable.

PROPOSITION 7.2.1. [Douady's observation] *Let N be an abelian closed normal subgroup of G and χ a character of N. Suppose that χ extends to some $\phi \in P(G)$. Then the stability group*

$$G_\chi = \{x \in G : \chi(x^{-1}yx) = \chi(y) \text{ for all } y \in N\}$$

of χ is open in G.

PROOF. There exist a representation π of G and a vector $\xi \in \mathcal{H}(\pi)$ such that $\phi(x) = \langle \pi(x)\xi, \xi\rangle$ for all $x \in G$. Then $\|\xi\| = \phi(e)^{1/2} = \chi(e)^{1/2} = 1$ and

$$1 = |\chi(y)| = |\langle \pi(y)\xi, \xi\rangle| \leqslant \|\pi(y)\xi\| \cdot \|\xi\| = 1$$

for all $y \in N$. It follows that $\pi(y)\xi$ and ξ are linearly dependent, and then $\pi(y)\xi = \chi(y)\xi$. For $x \in G$, put $\xi_x = \pi(x)\xi$. Then

$$\pi(y)\xi_x = \pi(x)[\pi(x^{-1}yx)\xi] = \pi(x)[\chi(x^{-1}yx)\xi] = \chi(x^{-1}yx)\xi_x.$$

This means that ξ_x is an eigenvector of $\pi(y)$ with eigenvalue $\chi(x^{-1}yx)$.

Towards a contradiction, assume now that G_χ is not open in G. Since π is strongly continuous, there exists a neighbourhood U of the identity in G such that $\|\pi(x)\xi - \xi\| < \sqrt{2}$ for all $x \in U$. By assumption we find $x \in U$ and $y \in N$ such that $\chi(x^{-1}yx) \neq \chi(y)$. Then ξ_x and ξ are both eigenvectors of $\pi(y)$ with distinct eigenvalues and therefore they must be orthogonal. Since both are unit vectors, it follows that $\|\xi_x - \xi\| = \sqrt{2}$. This contradiction completes the proof. □

Since the characters of an abelian locally compact group separate the elements of the group, Proposition 7.2.1 implies:

COROLLARY 7.2.2. *Let N be an abelian closed normal subgroup of G. If N is extending in G and G/N is connected, then N is contained in the centre of G.*

Let N be a closed normal subgroup of the locally compact group G. Recall that G acts on N by inner automorphisms and hence on spaces of functions on G and on representations in the obvious manner. For instance, if $u \in B(N)$ and $x \in G$, then $x \cdot u \in B(N)$ is defined by $x \cdot u(n) = u(x^{-1}nx)$ for $n \in N$. It is clear that $\|x \cdot u\|_{B(N)} = \|u\|_{B(N)}$. The following theorem now provides a necessary and sufficient condition for a function u in $B(N)$ to be extensible to some element of $B(G)$.

THEOREM 7.2.3. *Let N be a closed normal subgroup of the locally compact group G and let $u \in B(N)$. Then u extends to some function in $B(G)$ if and only if*

$$\|x \cdot u - u\| \to 0 \text{ as } x \to e,$$

and in this case

$$\|u\| = \inf\{\|v\| : v|_N = u\}.$$

PROOF. Note first that if $v \in B(G)$, then $v|_N \in B(N)$ and $\|v|_N\| \leqslant \|v\|$. Moreover, if $v(x) = \langle \pi(x)\xi, \eta \rangle$, where $\|v\|_{B(G)} = \|\xi\| \cdot \|\eta\|$, and $x_\alpha \to e$, then

$$x_\alpha \cdot v(x) v(x) = \langle \pi(x)\pi(x_\alpha)\xi, \pi(x_\alpha)\eta \rangle - \langle \pi(x)\xi, \eta \rangle$$

and hence

$$\|x_\alpha \cdot v - v\| \leqslant \|\pi(x_\alpha)\xi - \xi\| \cdot \|\pi(x_\alpha)\eta\| + \|\pi(x_\alpha)\eta - \eta\| \cdot \|\xi\| \to 0$$

and therefore

$$\|x_\alpha \cdot (v|_N) - v|_N\|_{B(N)} \leqslant \|x_\alpha \cdot v - v\| \to 0.$$

The proof of the converse is much more complicated. It suffices to show that for any $u \in B(N)$ satisfying $\|x \cdot u - u\| \to 0$ as $x \to e$ and any $\epsilon > 0$, there exists $v \in B(G)$ such that $\|v\|_{B(G)} = \|u\|$ and $\|v|_N - u\| \leqslant \epsilon$. In fact, this can be seen as follows. Choose $v_1 \in B(G)$ with $\|v_1\| = \|u\|$ and $\|v_1|_N - u\| \leqslant \epsilon/2$, and then construct inductively a sequence $(v_n)_n$ in $B(G)$ such that, for $n \geqslant 2$,

$$\left\| v_n|_N - \left(u - \sum_{j=1}^{n-1} v_j|_N \right) \right\| \leqslant \frac{\epsilon}{2^n}$$

and

$$\|v_n\| = \left\| u - \sum_{j=1}^{n-1} v_j|_N \right\|.$$

Then $\|v_n\| \leqslant 1/2^{n-1}$ for $n \geqslant 2$ and hence the series $v = \sum_{n=1}^{\infty} v_n$ converges in $B(G)$ and satisfies $v|_N = u$ and

$$\|v\| \leqslant \|u\| + \sum_{n=2}^{\infty} \|v_n\| = \|u\| + \sum_{n=2}^{\infty} \frac{\epsilon}{2^{n-1}} = \|u\| + \epsilon.$$

Given u and ϵ, we find a neighbourhood U of e in G such that

$$\|x \cdot u - u\| < \frac{\epsilon}{2} \tag{7.1}$$

for all $x \in U$ and

$$\|L_n u - u\| < \frac{\epsilon}{2} \tag{7.2}$$

for all $n \in U \cap N$. Choose a compact symmetric neighbourhood V of e in G such that $V^2 \subseteq U$. Let Haar measures on G, N and G/N be normalized so that Weil's formula holds. Take a nonnegative continuous functions f on G with supp $f \subseteq V$

such that

$$\begin{aligned}
1 &= \int_{G/N} \left(\int_N f(xn)dn \right)^2 d\dot{x} \\
&= \int_{G/N} \left(\int_N f(xn)dn \right) \left(\int_N f(xm)dm \right) d\dot{x} \\
&= \int_{G/N} \left(\int_N f(xn)dn \right) \left(\int_N f(xnm)dm \right) d\dot{x} \\
&= \int_G \left(\int_N f(x)f(xn)dn \right) dx. \qquad (7.3)
\end{aligned}$$

We now define a function v on G by

$$v(y) = \int_G \left(\int_N f(yx)f(xn)u(n)dn \right) dx, \ y \in G.$$

We claim that $v \in B(G)$. To verify this, we write u in the form $u(x) = \langle \pi(x)\xi, \eta \rangle$, where π is a unitary representation of N and $\xi, \eta \in \mathcal{H}(\pi)$ are such that $\|u\| = \|\xi\| \cdot \|\eta\|$. Then, for any $y \in G$,

$$\begin{aligned}
v(y) &= \int_G \left(\int_N f(x)f(y^{-1}xn)u(n)dn \right) dx \\
&= \int_{G/N} \left(\int_N \int_N f(xm)f(y^{-1}xmn)\langle \pi(n)\xi, \eta \rangle dndm \right) d\dot{x} \\
&= \int_{G/N} \left(\int_N \int_N f(y^{-1}xn)f(xm)\langle \pi(n)\xi, \pi(m)\eta \rangle dndm \right) d\dot{x} \\
&= \int_{G/N} \left\langle \int_N f(y^{-1}xn)\pi(n)\xi dn, \int_N f(xm)\pi(m)\eta dm \right\rangle d\dot{x}.
\end{aligned}$$

This formula shows that v is a coordinate function of the unitary representation of G induced from π. Moreover,

$$\|v\| \leqslant \left\| \int_N f(xn)\pi(n)\xi dn \right\| \cdot \left\| \int_N f(xm)\pi(m)\eta dm \right\|$$

and, for any $\omega \in \mathcal{H}(\pi)$, using (3),

$$\begin{aligned}
\left\| \int_N f(xn)\pi(n)\omega dn \right\| &= \left(\int_{G/N} \left\| \int_N f(xn)\pi(n)\omega dn \right\|^2 d\dot{x} \right)^{1/2} \\
&\leqslant \|\omega\| \left(\int_{G/N} \left(\int_N f(xn)dn \right)^2 d\dot{x} \right)^{1/2} \\
&= \|\omega\|.
\end{aligned}$$

So $\|v\| \leqslant \|\xi\| \cdot \|\eta\| = \|u\|$.

To finish the proof, it remains to show that

$$\|v|_N - u\| \leqslant \epsilon.$$

To that end, we first obtain an expression for v as follows. For any $n \in N$,

$$
\begin{aligned}
v(n) &= \int_G \int_N f(x) f(n^{-1}xm) u(m) dm dx \\
&= \int_G \int_N f(x) f(x(x^{-1}n^{-1}x)m) u(m) dm dx \\
&= \int_G \int_N f(x) f(xm) u(x^{-1}nxm) dm dx \\
&= \int_G \int_N f(x) f(xm) [(xm) \cdot (L_{m^{-1}} u)](n) dm dx. \qquad (7.4)
\end{aligned}
$$

The map from $G \times N$ to $B(N)$ defined by

$$(x, m) \to f(x) f(xm)(xm) \cdot (L_{m^{-1}} u)$$

is continuous and has compact support. Therefore, the vector-valued integral

$$\int_G \int_N f(x) f(xm)[(xm) \cdot (L_{m^{-1}} u)] dm dx$$

exists in $B(N)$ and equals $v|_N$ by the pointwise equality (7.4). From (7.4) we get

$$u = \int_G \int_N f(x) f(xn) u dn dx,$$

and together with (7.4), this yields

$$
\begin{aligned}
\|v|_N - u\| &\leqslant \int_G \int_N f(x) f(xn) \|(xn) \cdot (L_{n^{-1}} u) - u\| dn dx \\
&\leqslant \int_G \int_N f(x) f(xn) \|(xn) \cdot (L_{n^{-1}} u) - (xn) \cdot u\| dn dx \\
&\quad + \int_G \int_N f(x) f(xn) \|(xn) \cdot u - u\| dn dx \\
&\leqslant \int_G \int_N f(x) f(xn) \|L_{n^{-1}} u - u\| dn dx \\
&\quad + \int_G \int_N f(x) f(xn) \|(xn) \cdot u - u\| dn dx. \qquad (7.5)
\end{aligned}
$$

If $f(x) f(xn) \neq 0$, then both x and xn are in V and hence $xn \in U$ and $n^{-1} \in V^{-1}V \cap N \subseteq U \cap N$. Thus, by (7.1) and (7.2),

$$\|(xn) \cdot u - u\| < \epsilon/2 \text{ and } \|L_{n^{-1}} u - u\| < \epsilon/2.$$

Using (7.5), it follows that

$$
\begin{aligned}
\|v|_N - u\| &\leqslant \sup \{\|(xn) \cdot [L_{n^{-1}} u - u]\| : n \in U \cap N, xn \in U\} \\
&\quad + \sup \{\|(xn) \cdot u - u\| : n \in U \cap N, xn \in U\} \\
&\leqslant \epsilon/2 + \epsilon/2 = \epsilon.
\end{aligned}
$$

This completes the proof of the theorem. □

We conclude this section with a result which at first glance might not be surprising in view of the fact that open subgroups are always extending, but the proof turns out to be fairly involved and requires a clever inductive procedure.

THEOREM 7.2.4. *For an arbitrary locally compact group G, the connected component G_0 of the identity is extending in G.*

PROOF. Since G/G_0 is totally disconnected, we can choose an open subgroup H of G containing G_0 such that H/G_0 is compact. As H is extending in G, we can henceforth assume that G/G_0 is compact. Then G is a projective limit of Lie groups G/K_α (Section 1.2) and, with $q_\alpha : G \to G/K_\alpha$ denoting the quotient homomorphism, $B(G/K_\alpha) \circ q_\alpha \subseteq B(G)$. For each α, let μ_α be normalized Haar measure on the compact normal subgroup $G_0 \cap K_\alpha$.

Now, let $\varphi \in P(G_0)$ and let π be the associated cyclic representation. Since φ is continuous at e, we have

$$|\varphi(e) - (\varphi * \mu_\alpha)(e)| < \frac{1}{2}\|\varphi\|$$

for sufficiently large α. Fix such an α and let $K = K_\alpha$, $\mu = \mu_\alpha$ and $q = q_\alpha$. Since $K \cap G_0$ is a compact normal subgroup of G_0, π decomposes into a direct sum $\pi = \pi_1 \oplus \pi_2$, where $\pi_1|_{K\cap G_0}$ is a multiple of the trivial 1-dimensional representation of $K \cap G_0$ and $\pi_2|_{K\cap G_0}$ is disjoint from $\pi_1|_{K\cap G_0}$. Accordingly, φ can be written as $\varphi = \varphi_1 + \varphi_2$, where φ_j is associated to π_j, $j = 1, 2$. Then $\varphi * \mu = \varphi_1$ and hence $\varphi_2 = \varphi - \varphi * \mu$ is a positive definite function of norm $< \frac{1}{2}\|\varphi\|$. Now $\varphi_1 = \varphi * \mu$ is constant on $(K \cap G_0)$-cosets in G_0 and hence defines an element ψ_1 of $P(G_0K/K)$ by

$$\psi_1(xK) = \varphi_1(x), \ x \in G_0.$$

As G_0K/K is open in G/K, ψ_1 extends to some $\phi_1' \in P(G/K)$ such that

$$\|\phi_1'\| = \|\psi_1\| = \|\varphi_1\|.$$

Let $\phi_1 = \phi_1' \circ q \in P(G)$. Then $\phi_1|_{G_0} = \varphi_1$ and $\|\phi_1\| = \|\varphi_1\|$.

Having constructed φ_1 and ϕ_1, we now apply the same procedure to $\varphi - \varphi_1$ instead of φ to obtain $\varphi_2 \in P(G_0)$ such that $(\varphi - \varphi_1) - \varphi_2 \in P(G_0)$ and

$$\|(\varphi - \varphi_1) - \varphi_2\| < \frac{1}{2}\|\varphi - \varphi_1\| < \frac{1}{4}\|\varphi\|$$

and $\phi_2 \in P(G)$ such that $\phi_2|_{G_0} = \varphi_2$ and $\|\phi_2\| = \|\varphi_2\|$.

Continuing this process, we construct inductively sequences $(\varphi_n)_n$ in $P(G_0)$ and $(\phi_n)_n$ in $P(G)$ satisfying $\phi_n|_{G_0} = \varphi_n$, $\|\phi_n\| = \|\varphi_n\|$ and

$$\|\varphi - (\varphi_1 + \ldots + \varphi_n)\| < 2^{-n}\|\varphi\|$$

for all $n \in \mathbb{N}$. It follows that the series $\sum_{n=1}^\infty \varphi_n$ and $\sum_{n=1}^\infty \phi_n$ converge in $B(G_0)$ and $B(G)$, respectively. Then $\sum_{n=1}^\infty \varphi_n = \varphi$ and if we put $\phi = \sum_{n=1}^\infty \phi_n$, then clearly ϕ is continuous and positive definite and $\phi|_{G_0} = \varphi$. □

Alternatively, the preceding theorem can also be shown by employing Theorem 7.2.3. However, this would require projective limit arguments.

7.3. Connected groups and SIN-groups

So far, the only locally compact groups which we know to have the extension property, are either discrete, or compact or locally compact abelian. There exists a natural class of groups comprising all these, which we now introduce and each member of which will turn out to have the extension property.

DEFINITION 7.3.1. Let H be a closed subgroup of the locally compact group G. Then G is said to be an $[\mathrm{SIN}]_H$-group (a group with *small H-invariant neighbourhoods*) if G has a neighbourhood basis $\mathcal{V}$ of the identity such that $h^{-1}Vh = V$ for all $V \in \mathcal{V}$ and all $h \in H$. $[\mathrm{SIN}]_G$-groups are simply referred to as SIN-*groups*.

THEOREM 7.3.2. *Let G be a locally compact group and H a closed subgroup of G such that G is an* $[\mathrm{SIN}]_H$*-group. Then $B(G)|_H = B(H)$ and, if $u \in B(H)$, then*

$$\|u\|_{B(H)} = \inf\{\|v\|_{B(G)} : v|_H = u\}.$$

PROOF. As in the proof of Theorem 7.2.3, it suffices to show that for any $u \in B(H)$ and $\epsilon > 0$, there exists $v \in B(G)$ such that $\|v\|_{B(G)} \leqslant \|u\|_{B(H)}$ and $\|v|_H - u\|_{B(H)} < \epsilon$. Let C be a compact H-invariant neighbourhood of e in G. Then

$$|C| = \int_G 1_C(x)dx = \Delta_G(h)\int_G 1_C(h^{-1}xh)dx = \Delta_G(h)|C|$$

for every $h \in H$. So $\Delta_G|_H = 1$ and also $\Delta_H = 1$. Therefore, there exists an H-invariant measure on the quotient space G/H (see Section 1.3). We assume that Haar measures on G and H are adjusted so that Weil's formula

$$\int_G f(x)dx = \int_{G/H}\left(\int_H f(xh)dh\right)d\dot{x}$$

holds for all $f \in C_c(G)$. Choose a compact neighbourhood V of e in G such that

$$\|L_{h^{-1}}u - u\| < \epsilon \text{ for all } h \in V^{-1}V \cap H. \tag{7.6}$$

Let f be a nonnegative continuous function on G such that $\operatorname{supp} f \subseteq V$, $f(x) = f(h^{-1}xh)$ for all $x \in G$ and $h \in H$, and

$$\int_{G/H}\left(\int_H f(xh)^2 dh\right)d\dot{x} = 1. \tag{7.7}$$

We define a function v on G by

$$v(y) = \int_G\left(\int_H f(yx)f(xh)u(h)dh\right)dx.$$

Then, using that f is H-invariant and H is unimodular,

$$\begin{aligned} v(h_1) &= \int_G\left(\int_H f(x)f(h_1^{-1}xh)u(h)dh\right)dx \\ &= \int_G\left(\int_H f(x)f(xhh_1^{-1})u(h)h\right)dx \\ &= \int_G\left(\int_H f(x)f(xh)L_{h^{-1}}u(h_1)dh\right)dx \end{aligned}$$

for every $h_1 \in H$. This pointwise equality is analogous to equation (7.3) in the proof of Theorem 7.2.3. Repeating the arguments in the proof of Theorem 7.2.3 and using (7.6) and (7.7) and the fact that $\operatorname{supp} f \subseteq V$, it is straightforward to deduce that $\|v|_H - u\| < \epsilon$. Moreover, the proof that $v \in B(G)$ and $\|v\| \leqslant \|u\|$ is identical to the corresponding part of the proof of Theorem 7.2.3. □

As an immediate consequence of Theorem 7.3.2 we state

COROLLARY 7.3.3. *Every* SIN*-group has the extension property.*

A locally compact group is called an IN-*group* if it possesses at least one compact conjugation invariant neighbourhood of the identity. It is worthwhile to mention that IN-groups need not have the extension property. As an example, let H denote the 3-dimensional Heisenberg group and let D be a central discrete subgroup isomorphic to $\mathbb{Z}$. Then $G = H/D$ has a compact invariant neighbourhood of the identity, but it follows from Proposition 7.2.1 that G has a normal abelian subgroup of the form $\mathbb{R} \times \mathbb{T}$, which is not extending. We now investigate almost connected locally compact groups.

PROPOSITION 7.3.4. *Let G be a connected group and suppose that G possesses a descending sequence*

$$R = R_n \supseteq R_{n-1} \supseteq \ldots \supseteq R_1 \supseteq R_0 = \{e\}$$

of closed normal subgroups such that all the subquotients R_j/R_{j-1}, $j = 1, \ldots, n$, are abelian. If every closed normal subgroup of G is extending, then R is contained in the centre of G.

PROOF. The proof essentially consists of repeatedly applying Proposition 7.2.1. We proceed by induction on k to show that R_k is contained in the centre $Z(G)$ of G.

To start with, observe first that $R_1 \subseteq Z(G)$ since R_1 is an extending abelian closed normal subgroup of G and G/R_1 is connected. Similarly, for each $1 \leqslant k \leqslant n$, $R_k/R_{k-1} \subseteq Z(G/R_{k-1})$. Now assume that $R_k \subseteq Z(G)$ for some $1 \leqslant k \leqslant n-1$. We show that then R_{k+1} is contained in the centre of G. To that end, take any $a \in R_{k+1} \backslash R_k$ and let H denote the closed subgroup of G generated by a and R_k. Then H is abelian since $R_k \subseteq Z(G)$, and H is normal in G since $R_{k+1}/R_k \subseteq Z(G/R_k)$. By hypothesis, H is extending in G and therefore $H \subseteq Z(G)$. As $a \in R_{k+1} \backslash R_k$ is arbitrary, it follows that $R_{k+1} \subseteq Z(G)$. This finishes the inductive step and thus completes the proof. □

We remind the reader that the radical of a connected Lie group G is the maximal connected solvable normal subgroup of G.

COROLLARY 7.3.5. *Let G be a connected Lie group and suppose that every closed normal subgroup of G is extending. Then the radical of G is contained in the centre of G.*

PROOF. The statement follows from Proposition 7.3.4, taking for R the radical of G and for the descending sequence of normal subgroups the commutator series $R \supseteq [R, R] \supseteq [[R, R], [R, R]] \supseteq \ldots$ of R. It is clear that this sequence satisfies the hypothesis of Proposition 7.3.4. □

THEOREM 7.3.6. *For a connected Lie group G, the following properties are equivalent.*

(i) *G has the extension property.*
(ii) *G is of the form $G = V \times K$, where V is a vector group and K is a compact group.*

PROOF. (ii) $\Rightarrow$ (i) A group of the form $V \times K$ is a SIN-group and therefore has the extension property by Corollary 7.3.3.

(i) $\Rightarrow$ (ii) If (i) holds, then by Corollary 7.3.5 the radical R of G is contained in the centre of G, and G/R is a semisimple Lie group. If G/R is compact, then by the structure theorem for connected central groups (see [**110**] or [**60**]), G is in fact a direct product of a vector group and a compact group.

So suppose that G/R is noncompact. Then G/R admits an Iwasawa decomposition $G/R = KAN$, where K is compact, A is abelian and N is nilpotent. The group A normalizes N, but does not centralize N. Let $q : G \to G/R$ denote the quotient homomorphism. The group $q^{-1}(N)$ is extending in G and hence also in $q^{-1}(AN)$. But $q^{-1}(N)$ is normal in $q^{-1}(AN)$ and connected and solvable and therefore contained in the radical of $q^{-1}(AN)$. Proposition 7.3.4 shows that $q^{-1}(N) \subseteq Z((q^{-1}(AN))$. It follows that $N = q^{-1}(N)/R$ is contained in the centre of $AN = q^{-1}(AN)/R$. This is a contradiction, and hence G/R has to be compact. □

In the proof of the following corollary and later in this chapter, we shall use the next lemma.

LEMMA 7.3.7. *Let G be a projective limit of* SIN*-groups. Then G itself is an* SIN*-group.*

PROOF. Let U be an open neighbourhood of the identity e of G. Then, by hypothesis, there exists a compact normal subgroup K of G such that $K \subseteq U$ and G/K is an SIN-group. Since K is compact and U is open, there exists an open neighbourhood W of e in G such that $KW \subseteq U$. Thus, replacing U by KW, we can assume that $U = q^{-1}(q(U))$, where $q : G \to G/K$ denotes the quotient homomorphism. Now, as G/K is SIN, we find a conjugation invariant neighbourhood V of $\{K\}$ in G/K contained in $q(U)$. Then $q^{-1}(V) \subseteq q^{-1}(q(U)) = U$ and $q^{-1}(V)$ is conjugation invariant. □

COROLLARY 7.3.8. *Let G be an almost connected locally compact group. Then G has the extension property (if and) only if G is an* SIN*-group.*

PROOF. We only need to show that if G has the extension property, then G must be an SIN-group. Since G/G_0 is compact, G is a projective limit of Lie groups G/K_α. Then, for each α, $(G/K_\alpha)_0$ is open in G/K_α and hence has finite index in G/K_α. Since G/K_α has the extension property, by Theorem 7.3.6 the connected Lie group $(G/K_\alpha)_0$ is an SIN-group, and hence so is G/K_α because $(G/K_\alpha)_0$ has finite index in G/K_α. Finally, the preceding lemma shows that G is an SIN-group. □

The results of this section raise the question of whether an arbitrary locally compact group which has the extension property must be an SIN-group. It turns out that this is not the case. In fact, there exists an example, which we present below, of a compactly generated, 2-step solvable locally compact group which has the extension property and nevertheless fails to be an SIN-group.

EXAMPLE 7.3.9. Let $G = \mathbb{R} \rtimes \mathbb{Z}$, where $n \in \mathbb{Z}$ acts on $\mathbb{R}$ by $t \to 2^n t$. Clearly, G has no compact invariant neighbourhood of the identity. We show that nevertheless G has the extension property. For this, we determine all the closed subgroups of G.

We identify $\mathbb{R}$ and $\mathbb{Z}$ with the subgroups $\mathbb{R} \times \{0\}$ and $\{0\} \times \mathbb{Z}$, respectively. Let H be a closed subgroup of G which is not contained in $\mathbb{R}$. Then there exists $s \in \mathbb{R}$, $s \neq 0$, such that $sHs^{-1} \cap \mathbb{Z} \neq \{0\}$. So we may assume that $H \cap \mathbb{Z} = m\mathbb{Z}$ for some $m \in \mathbb{N}$. Suppose that $H \neq m\mathbb{Z}$ and take $t \in \mathbb{R}$, $t \neq 0$, and $k \in \mathbb{Z}$ such that

$x = tk \in H$. Then, as is easily verified,

$$x^m = \left(\sum_{j=0}^{m-1} 2^{jk}\right) t(mk).$$

Since $m\mathbb{Z} \subseteq H$, it follows that $r = \left(\sum_{j=0}^{m-1} 2^{jk}\right) t \in H$. Again, since $m\mathbb{Z} \subseteq H$, we get that

$$(2^{mn}r, 0) = (0, mn)(r, 0)(0, mn)^{-1} \in H \cap \mathbb{R}$$

for all $n \in \mathbb{Z}$. Since $r \neq 0$ and $H \cap \mathbb{R}$ is a closed subgroup of $\mathbb{R}$, it follows that $\mathbb{R} \subseteq H$. Thus we have seen that every closed subgroup H of G either is contained in $\mathbb{R}$ or contains $\mathbb{R}$ or is conjugate to a subgroup of $\mathbb{Z}$. In the first two cases H is extending in G since $\mathbb{R}$ is open in G, and in the third case it follows from Lemma 7.1.2 that H is extending.

7.4. Nilpotent groups and 2-step solvable examples

In the preceding section we have seen that if G is a connected Lie group satisfying the extension property, then G has to be an SIN-group. The main theme of this section is to show that the same conclusion holds for compactly generated nilpotent locally compact groups. On the other hand, if H is a closed subgroup of a nilpotent group G and H is topologically isomorphic to either $\mathbb{R}$ or $\mathbb{Z}$, then H is extending in G. In addition, we determine all the extending subgroups of several examples of 2-step solvable groups.

Later in this section we shall exploit to some extent the structure of general nilpotent locally compact groups, as developed in [**99**]. For any locally compact group G, let G^c denote the set of all compact elements of G (that is, elements which generate a relatively compact subgroup). Then G is said to be *compact-free* if $G^c = \{e\}$. Suppose that G is nilpotent. Then G^c is a closed (normal) subgroup of G [**99**, Corollary 3.5.1 and Lemma 3.8], and G/G^c is compact-free and a Lie group. If, in addition, G is compactly generated, then G^c is compact [**99**, Theorem 9.7] and the subgroup G_0G^c is open in G. We remind the reader that when G is discrete, G^c is simply the set of all elements of finite order, which is usually denoted G^t and called the *torsion subgroup* of G.

We shall need below the following result on positive definite functions.

LEMMA 7.4.1. *Let $\varphi \in P^1(G)$. Then, for all $x, y \in G$,*

$$\begin{aligned}\frac{1}{2}|1 - \varphi([x, y])| &\leqslant 1 - |\varphi(x)| + 1 - |\varphi(y)| \\ &\quad +(1 - |\varphi(x)|)(1 - |\varphi(y)|) \\ &\quad +(1 - |\varphi(x)|)^{1/2} + (1 - |\varphi(y)|)^{1/2}.\end{aligned}$$

PROOF. Recall first that

$$|\varphi(a)\varphi(b) - \varphi(ab)|^2 \leqslant (1 - |\varphi(a)|^2)(1 - \varphi(b)|^2)$$

for all $a, b \in G$ (Proposition 1.4.16(iv)). Using several times this inequality as well as $|\varphi(a)| \leqslant 1$ and $\varphi(a^{-1}) = \overline{\varphi(a)}$ for all $a \in G$, we obtain, for $x, y \in G$,

$$\begin{aligned}
|1-\varphi([x,y])| &\leqslant 1-|\varphi(x)|^2+|\varphi(x)|^2(1-|\varphi(y)|^2)\\
&\quad + |\varphi(x)\varphi(y)|^2-\varphi(x)\varphi(y)\varphi(x^{-1}y^{-1})|\\
&\quad + |\varphi(x)\varphi(y)\varphi(x^{-1}y^{-1})-\varphi(x)\varphi(yx^{-1}y^{-1})|\\
&\quad + |\varphi(x)\varphi(yx^{-1}y^{-1})-\varphi([x,y])|\\
&\leqslant 2(1-|\varphi(x)|)+2(1-|\varphi(y)|)+|\varphi(x)\varphi(y)-\varphi(yx)|\\
&\quad + |\varphi(y)\varphi(x^{-1}y^{-1})-\varphi(yx^{-1}y^{-1})|\\
&\quad + |\varphi(x)\varphi(yx^{-1}y^{-1})-\varphi([x,y])|\\
&\leqslant 2(1-|\varphi(x)|)+2(1-|\varphi(y)|)\\
&\quad +(1-|\varphi(x)|^2)^{1/2}(1-|\varphi(y)|^2)^{1/2}\\
&\quad +(1-|\varphi(y)|^2)^{1/2}(1-|\varphi(x^{-1}y^{-1}|^2)^{1/2}\\
&\quad +(1-|\varphi(x)|^2)^{1/2}(1-|\varphi(yx^{-1}y^{-1})|^2)^{1/2}\\
&\leqslant 2(1-|\varphi(x)|)+2(1-|\varphi(y)|)\\
&\quad +(1-|\varphi(x)|^2)^{1/2}(1-|\varphi(y)|^2)^{1/2}\\
&\quad +2(1-|\varphi(x)|)^{1/2}+2(1-|\varphi(y)|)^{1/2},
\end{aligned}$$

which implies the desired inequality. □

Before proceeding, we remind the reader that

$$\{e\} = Z_0(G) \subseteq Z_1(G) = Z(G) \subseteq \ldots \subseteq Z_m(G) \subseteq \ldots$$

denotes the ascending central series of G.

LEMMA 7.4.2. *Let H be a closed subgroup of the locally compact group G. Suppose that every character of H extends to some function in $P(G)$. Then*

$$H \cap [H, Z_2(G)_0] \subseteq [H, H].$$

In particular, if G is connected and 2-step nilpotent, then $H \cap [H, G] \subseteq [H, H]$.

PROOF. Towards a contradiction, assume there exists $h_0 \in H \cap [H, Z_2(G)_0]$ such that $h_0 \notin [H, H]$. There exists a character χ of H with $\chi(h_0) \neq 1$. By hypothesis, we find some $\phi \in P(G)$ so that $\phi|_H = \chi$, and by Lemma 7.1.4 ϕ can be chosen in $\text{ex}(P^1(G))$. Then $\phi|_{Z(G)}$ is a character of $Z(G)$. Indeed, the representation π_ϕ is irreducible and hence $\pi_\phi|_{Z(G)}$ is a multiple of a one-dimensional representation. Since $|\phi(h)| = 1$ for all $h \in H$, we have by Lemma 7.4.1,

$$\frac{1}{2}|1-\phi([h,x])| \leqslant 1-|\phi(x)|+(1-|\phi(x)|)^{1/2}$$

for all $h \in H$ and $x \in G$. Thus there exists an open neighbourhood V of the identity in G such that $|1 - \phi([h, x])| < \sqrt{3}$ for all $x \in V$ and $h \in H$.

Now let $x \in V \cap Z_2(G)$. Then the map $h \to [h, x]$ is a homomorphism from H into $Z(G)$. Since $\phi|_{Z(G)}$ is a character, it follows that the set

$$\Gamma_x = \{\phi([h, x]) : h \in H\}$$

is a subgroup of $\mathbb{T}$ with the property that $|z - 1| < \sqrt{3}$ for all $z \in \Gamma_x$. As is well-known, this implies $\Gamma_x = \{1\}$. Finally, since for each $h \in H$ the map $x \to [h, x]$ is a homomorphism from $Z_2(G)$ into $Z(G)$, we conclude that $\phi([h, x]) = 1$ for all $h \in H$

and all x in the open subgroup of $Z_2(G)$ generated by $V \cap Z_2(G)$. Consequently, $\phi(y) = 1$ for all $y \in [H, Z_2(G)_0]$. Since $h_0 \in [H, Z_2(G)_0]$, we have reached the contradiction $1 = \phi(h_0) = \chi(h_0)$. □

COROLLARY 7.4.3. *Let G be a connected 2-step nilpotent locally compact group and let H be an abelian closed subgroup of G. Suppose that H is σ-compact and that $H[H,N]$ is closed in G. Then H is extending in G if and only if $H \cap [H,G] = \{e\}$.*

PROOF. The necessity of the condition follows from Lemma 7.4.2. Conversely, let $N = [H,G]$ and suppose that $H \cap N = \{e\}$. By hypothesis, HN is closed in G and, since G is 2-step nilpotent, HN/N is contained in the centre of G/N. So HN/N is extending in G/N and hence, since $H \cap N = \{e\}$, H is extending in G by Lemma 7.1.3. □

LEMMA 7.4.4. *Let G be a semidirect product $G = N \rtimes A$, where N is abelian, connected and contained in $Z_m(G)$ for some $m \in \mathbb{N}$. If G has the extension property, then N is contained in the centre of G.*

PROOF. We prove the lemma by induction on m. Thus, let $N \subseteq Z_{m+1}(G)$ for some m and suppose that if $H = M \rtimes A$ is such that H has the extension property and M is abelian and connected and contained in $Z_m(H)$, then $M \subseteq Z(H)$.

Let $M = [A,N]$ and $H = M \rtimes A$. Then H has the extension property, M is abelian and connected and $M \subseteq Z_m(H)$. So $M \subseteq Z(H)$, and since N is abelian, M is even central in G. Now, $H = AM$ is an extending subgroup of G. Since N is connected and $N \subseteq Z_2(G)$, Lemma 7.4.2 implies

$$[A,N] = AM \cap [A,N] \subseteq H \cap [H, Z_2(G)_0] \subseteq [H,H] = [A,A].$$

Since $A \cap N = \{e\}$, it follows that $[A,N] = \{e\}$. Consequently, N is contained in the centre of G since N is abelian. □

COROLLARY 7.4.5. *Let G be a compact-free nilpotent locally compact group. Then G has the extension property if and only if the centre of G is open in G.*

PROOF. We only have to show that if G has the extension property then the centre of G is open. The connected component G_0 is open and G/G_0 is torsion-free. Since G_0 has the extension property, $G_0 = V \times K$, where V is a vector group and K is a compact group (Theorem 7.3.6). Now K, being a connected and nilpotent compact group, K is abelian. Take any element $a \in G\backslash G_0$ and let A be the subgroup of G generated by a. Since G_0 is open and G/G_0 is torsion-free, $A \cap G_0 = \{e\}$ and hence $G_0A = G_0 \rtimes A$. Since $G_0 \rtimes A$ has the extension property and G_0 is abelian and G is nilpotent, Lemma 7.4.4 applies and shows that $G_0 \subseteq Z(G_0A)$. Since $a \in G\backslash G_0$ is arbitrary, it follows that G_0 is contained in the centre of G. This finishes the proof since G_0 is open. □

Using the preceding preliminary results, we are now ready to deduce the analogue of Theorem 7.3.6 for compactly generated nilpotent groups.

THEOREM 7.4.6. *Let G be a compactly generated nilpotent locally compact group. Then G has the extension property if and only if G is an* SIN*-group.*

PROOF. We only have to show that if G has the extension property then G is an SIN-group. We use the fact that a compactly generated nilpotent group is a projective limit of Lie groups [**127**, Theorem 9]. Since each quotient group of G

has the extension property and a projective limit of SIN-groups is a SIN-group, we can assume that G is a Lie group. Then G_0 is open in G and G/G_0 is a finitely generated nilpotent group. Let $N \subseteq G$ be the pullback of the torsion subgroup $(G/G_0)^t$ of G/G_0. Then $N/G_0 = (G/G_0)^t$ is finite, and G/N is torsion-free.

Since G_0, being open in G, has the extension property, $G_0 = V \times K$, where V is a vector group and K is a compact group (Theorem 7.3.6). Being a connected, nilpotent, compact group, K is abelian.

Now consider any $a \in G\backslash N$ and let A denote the cyclic subgroup generated by a. Then, since G/N is torsion-free, G_0A is actually a semidirect product $G_0 \rtimes A$. Since G_0A has the extension property, an application of Lemma 7.4.4 shows that a commutes with all elements of G_0. As N/G_0 is finite and G_0 is abelian, it follows that G acts as a finite group of inner automorphisms on G_0. Thus $G_0 \in [\mathrm{SIN}]_G$, and then G is a SIN-group since G_0 is open in G. □

Example 7.3.9 shows that Theorem 7.4.6 does not hold for compactly generated 2-step solvable groups.

We now turn to the second aim of this section, the identification of certain extending subgroups of general nilpotent groups and of all the extending subgroups of a number of 2-step solvable groups.

LEMMA 7.4.7. *Let G be a connected and simply connected nilpotent Lie group and H a closed subgroup which is topologically isomorphic to either $\mathbb{R}$ or $\mathbb{Z}$. Then H is extending in G.*

PROOF. Let $\mathfrak{g}$ denote the Lie algebra of G, $\exp : \mathfrak{g} \to G$ the exponential map and $\mathfrak{z}_0 = \{0\} \subseteq \mathfrak{z}_1 \dots$ the ascending central series of $\mathfrak{g}$. Suppose first that H is isomorphic to $\mathbb{R}$ and let $H = \exp \mathfrak{h}$ and $Z_j = \exp \mathfrak{z}_j$, $j = 0, 1, \dots$. Since $\mathfrak{h}$ is 1-dimensional, $\mathfrak{h} \cap \mathfrak{z}_j \neq \{0\}$ implies that $\mathfrak{h} \subseteq \mathfrak{z}_j$. Let j be minimal such that $\mathfrak{h} \subseteq \mathfrak{z}_{j+1}$. Then $\mathfrak{h} \cap \mathfrak{z}_j = \{0\}$ and hence $H \cap Z_j = \{e\}$. Moreover, HZ_j is closed in G, $HZ_j/Z_j \subseteq Z_{j+1}/Z_j$ and, since G is σ-compact, H is topologically isomorphic to HZ_j/Z_j. Lemma 7.1.3 now shows that H is extending in G.

Finally, if H is isomorphic to $\mathbb{Z}$, then there exists a closed subgroup L of G which contains H and is isomorphic to $\mathbb{R}$. Since H is open in L and L is extending, H is extending as well. □

Applying Lemmas 7.4.7, 7.1.3 and 7.1.7 and structure theory of nilpotent groups, it is possible to generalize Lemma 7.4.7 to arbitrary nilpotent locally compact groups as follows.

THEOREM 7.4.8. *Let G be a nilpotent locally compact group and let H be a closed subgroup which is topologically isomorphic to either $\mathbb{R}$ or $\mathbb{Z}$. Then H is extending in G.*

PROOF. Suppose first that H is topologically isomorphic to $\mathbb{R}$. Then $H \subseteq G_0$, and since G_0 is extending in G by Theorem 7.2.4, we only have to observe that H is extending in G_0. To verify this, let K be the maximal compact normal subgroup of G_0. Then G_0/K is simply connected and HK/K is topologically isomorphic to $\mathbb{R}$ since $H \cap K = \{e\}$. Therefore HK/K is extending in G_0/K by Lemma 7.4.7 and hence H is extending in G_0 by Lemma 7.1.3.

Now let H be an infinite cyclic subgroup of G. Since it suffices to show that H is extending in some compactly generated open subgroup of G containing H, we can assume that G is compactly generated. Let $L = G_0G^c$, which is open in G

and almost connected since G^c is compact. We now distinguish the two cases that $H \cap L = \{e\}$ and $H \cap L \neq \{e\}$.

In the first case, H is extending in G by Lemma 7.1.3 since G/L is discrete. In the second case, let $M = HL$. Then L has finite index in M and hence M is almost connected. Since M is a projective limit of Lie groups, there exists a compact normal subgroup K of M such that M/K is a Lie group. Then $H \cap K = \{e\}$ since H is infinite cyclic, and hence H is topologically isomorphic to HK/K. Thus, applying Lemma 7.1.3 again and passing to M/K, we can henceforth assume that M is an almost connected nilpotent Lie group. So M_0 has finite index in M, and by Lemma 7.1.8 it suffices to show that $H \cap M_0$ is extending in M_0. Consequently, we can assume that M is a connected nilpotent Lie group. Then M^c is compact and M/M^c is simply connected. Since $H \cap M^c = \{e\}$, we can argue as before and apply Lemma 7.1.3 and Lemma 7.4.7 to conclude that H is extending in M, and hence in G. □

Applying Proposition 7.2.1, Lemma 7.4.7 and Lemma 7.4.2, we can now determine the extending subgroups of the 3-dimensional Heisenberg group.

EXAMPLE 7.4.9. We identify the 3-dimensional Heisenberg group G, as a set, with $\mathbb{R}^3$. Multiplication of G is then given by

$$(x_1, y_1, z_1)(x_2, y_2, z_2) = (x_1 + x_2, y_1 + y_2, z_1 + z_2 + x_1 y_2),$$

$(x_j, y_j, z_j) \in \mathbb{R}^3$, $j = 1, 2$. Then $Z(G) = \{(0,0,z) : z \in \mathbb{R}\} = [G, G]$ and $N = \{(0, y, z) : y, z \in \mathbb{R}\}$ is an abelian normal subgroup of G isomorphic to $\mathbb{R}^2$. Let H be a proper closed subgroup of G. We are going to show that H is extending in G if and only if H is isomorphic to either $\mathbb{R}$ or $\mathbb{Z}$.

Assume first that H is isomorphic to $\mathbb{R}$ or $\mathbb{Z}$. Then H is extending by Lemma 7.4.7.

Conversely, suppose that H is extending and note that $H \cap [H, G] = [H, H]$ by Lemma 7.4.2. Since $Z(G) = \mathbb{R}$, we can assume that H is not contained in $Z(G)$. Then $[H, G]$ is a nontrivial connected subgroup of $Z(G)$, hence equal to $Z(G)$, and therefore $H \cap Z(G) = [H, H]$. We now distinguish the two cases $H \subseteq N$ and $H \not\subseteq N$. If $H \subseteq N$, then $H \cap Z(G) = \{e\}$ and so H is topologically isomorphic to $HZ(G)/Z(G) = N/Z(G) = \mathbb{R}$. This means that H is isomorphic to $\mathbb{R}$ or $\mathbb{Z}$.

Finally, let $H \not\subseteq N$ and, towards a contradiction, assume that H is isomorphic to neither $\mathbb{R}$ nor $\mathbb{Z}$. It then follows as before that $H \cap N \neq \{e\}$. We now further split the discussion into the two cases $H_0 \subseteq N$ and $H_0 \not\subseteq N$. If $H_0 \not\subseteq N$, so that $H_0 N = G$, then for any $a \in H \cap N$, $[H_0, a]$ is a non-trivial connected subgroup of $Z(G)$. Thus $[H, H] \supseteq Z(G)$ and hence H_0 is normal in G. Also, $\dim H_0 = 2$ since $H_0 \not\subseteq N$ and $H_0 \neq G$. Now a 2-dimensional connected Lie group is abelian. Since H_0 is open in H, H_0 is extending in G. This contradicts Proposition 7.2.1. It remains to consider the case that $H \not\subseteq N$, but $H_0 \subseteq N$. Let $\widetilde{H}$ denote the unique minimal connected subgroup of G containing H. Then, as before, $\dim \widetilde{H} = 2$ and hence H is abelian. Also $H \cap Z(G) \neq \{e\}$ since H is not isomorphic to $\mathbb{R}$ or $\mathbb{Z}$. On the other hand, by Lemma 7.4.2

$$H \cap Z(G) = H \cap [H, G] = [H, H] = \{e\}.$$

This is contradiction.

We continue with two lemmas which appear to be of interest in their own and which will be used to determine the extending subgroups of the $ax+b$-group and the motion group of the plane.

LEMMA 7.4.10. *Suppose that G has a closed normal subgroup N such that, with $q : G \to G/N$ denoting the quotient homomorphism,*

$$B(G) \subseteq B(G/N) \circ q + C_0(G).$$

Let H be a noncompact closed subgroup of G. If H is extending, then $H \cap N \subseteq [H, H]$.

PROOF. Towards a contradiction, suppose there exists $h \in H \cap N$ with $h \notin [H, H]$. Since $h[H, H]$ is not the identity of the abelian group $H/[H, H]$, we can choose a character χ of H such that $\chi(h) \neq 1$. By the hypothesis, χ extends to some $\phi \in P(G)$ and $\phi = \phi_1 + \phi_2$, where ϕ_1 is constant on cosets of N and ϕ_2 vanishes at infinity. Then, since $h \in N$,

$$\phi(xh) = \phi_1(x) + \phi_2(xh)$$

for all $x \in G$ and, since $|\phi(h)| = 1$,

$$\phi(xh) = \phi(x)\phi(h) = \chi(h)(\phi_1(x) + \phi_2(x)).$$

It follows that

$$(1 - \chi(h))\phi_1(x) = \chi(h)\phi_2(x) - \phi_2(x)$$

for all $x \in G$. Now, the function $x \to \chi(h)\phi_2(x) - \phi_2(xh)$ vanishes at infinity, and hence so does ϕ_1 since $\chi(h) \neq 1$. Therefore $\phi = \phi_1 + \phi_2$ vanishes at infinity, whence $\chi = \phi|_H$ vanishes at infinity on H. Since χ is a character, this forces H to be compact, which is a contradiction. □

LEMMA 7.4.11. *Let G be a locally compact group and let N be a noncompact closed normal subgroup of G such that G/N is compact and second countable. Suppose that*

$$B(G) \subseteq B(G/N) \circ q + C_0(G).$$

If H is an abelian closed subgroup of G and H is extending, then H must be compact.

PROOF. Let χ be a character of H and let $\phi \in P(G)$ be an extension of χ. Write $\phi = \phi_1 + \phi_2$, where $\phi_1 \in B(G/N) \circ q$ and $\phi_2 \in C_0(G)$. As in the proof of Lemma 7.4.10,

$$\phi_1(xh) - \chi(h)\phi_1(x) = \chi(h)\phi_2(x) - \phi_2(xh)$$

for all $x \in G$ and $h \in H$. Since the function $x \to \phi_1(xh) - \chi(h)\phi_1(x)$ is constant on N, whereas $x \to \chi(h)\phi_2(x) - \phi_2(xh)$ vanishes at infinity, and since N is noncompact, it follows that

$$\phi_1(xh) = \chi(h)\phi_1(x) \quad \text{and} \quad \phi_2(xh) = \chi(h)\phi_2(x)$$

for all $x \in G$ and $h \in H$.

Now assume that H is noncompact. Since the function $h \to \phi_2(xh)$ vanishes at infinity on H and its absolute value equals $|\phi_2(x)|$, we conclude that $\phi_2(x) = 0$ for every $x \in G$. Thus $\phi = \phi_1$ and so $\phi \in P(G/N) \circ q$. Let $L = \overline{HN}$. Then L/N is abelian since HN/N is abelian and dense in L/N. By what we have shown, for each $\chi \in \widehat{H}$ there exists $\phi \in P^1(L/N) \circ q$ such that $\phi|_H = \chi$. By Lemma 7.1.4, we can assume that $\phi \in \mathrm{ex}(P^1(L))$ and hence, since L/N is abelian,

$$\phi \in \mathrm{ex}(P^1(L/N)) \circ q = \widehat{L/N} \circ q.$$

Thus we have seen that $\widehat{H} \subseteq (\widehat{L/N} \circ q)|_H$. But $\widehat{L/N}$ is countable since L/N is compact and second countable. Consequently, $\widehat{H}$ is countable. This contradicts the noncompactness of H. □

EXAMPLE 7.4.12. We find out all the extending subgroups of the $ax+b$-group $G = \mathbb{R} \rtimes \mathbb{R}_+^\times$. We identify $\mathbb{R}$ with the normal subgroup $\{1\} \times \mathbb{R}$ and $\mathbb{R}_+^\times$ with the subgroup $\mathbb{R}_+^\times \times \{0\}$. Lemma 7.4.10 applies to G with $N = \mathbb{R}$. It is known that the normal subgroup $\mathbb{R}$ is not extending. Let H be a nontrivial extending subgroup of G and assume first that H is connected. Since $[H, H] \subseteq \mathbb{R}$, by Lemma 7.4.10, $H \cap \mathbb{R} = [H, H]$. Thus, since $[H, H]$ is connected, either $H \cap \mathbb{R} = \mathbb{R}$ or $H \cap \mathbb{R} = \{0\}$. In the first case, $H = \mathbb{R}$ or $H = G$ as H is connected. In the second case, with $q : G \to \mathbb{R}_+^\times$ denoting the quotient homomorphism, $q(H) = \mathbb{R}_+^\times$ since $q(H)$ is connected. Thus $G = \mathbb{R} \ltimes H$, whence H is extending.

Now, let H be an arbitrary extending subgroup of G. Since H_0 is open in H, H_0 is extending as well. By what we already know, either $H_0 = \{0\}$, $H_0 = G$ or G is the semidirect product of $\mathbb{R}$ with H_0. In the last case, $H_0 = H$. Indeed, given any $a \in \mathbb{R}_+^\times$, there is a unique $x_a \in \mathbb{R}$ such that $(a, x_a) \in H_0$. Now

$$(1, y)(a, x_a)(1, y)^{-1} = (a, x_a + y - ay),$$

so that $y = ay$ for all $a \in \mathbb{R}_+^\times$ and hence $y = 0$ whenever $(1, y)$ normalizes H_0. Thus, it remains to consider the case that H is discrete. Then $H \cap N$ is extending and hence $H \cap N \subseteq [H \cap N, H \cap N] = \{0\}$ by Lemma 7.4.10. Consequently, if H is any proper extending subgroup of G, then $H \cap \mathbb{R} = \{0\}$.

We claim that conversely each closed subgroup H with $H \cap \mathbb{R} = \{0\}$ is extending. To that end, we observe that HN is closed in G. In fact, let $(t_n, a_n) \in H$, $n \in \mathbb{N}$, be such that $a_n \to a \in \mathbb{R}_+^\times$, and fix some $(s, b) \in H$ with $b \neq 1$. Then, since H is abelian,

$$(s + bt_n, ba_n) = (s, b)(t_n, a_n) = (t_n, a_n)(s, b) = (t_n + a_n s, a_n b)$$

and hence $(b-1)t_n = (a_n - 1)s$. It follows that

$$(t_n, a_n) \to \left(\frac{a-1}{b-1} s, a\right) \in H.$$

Since HN is closed and $H \cap N = \{0\}$, H is topologically isomorphic to HN/N. By Lemma 7.1.3, this implies that H is extending since G/N is abelian.

Summarizing, we have seen that a proper closed subgroup H of G is extending if and only if $H \cap \mathbb{R} = \{0\}$.

EXAMPLE 7.4.13. Let $G = SO(2) \ltimes \mathbb{R}^2$ where $SO(2)$ acts on $\mathbb{R}^2$ by rotation, and identify $\mathbb{R}^2$ with the normal subgroup $N = \{(E, x) : x \in \mathbb{R}^2\}$, where E is the unit matrix. We claim that a proper closed subgroup of G is extending only if it is compact. For that, notice first that no non-trivial subgroup of $\mathbb{R}^2$ can be extending since $B(G) \subseteq B(G/N) \circ q + C_0(G)$.

Now let H be a proper extending subgroup of G. Since H_0 is open in H, H_0 is extending as well. By Lemma 7.4.10, $H_0 \cap \mathbb{R}^2 = [H_0, H_0]$ which is connected. So $H_0 \cap \mathbb{R}^2$ is either equal to $\{0\}$ or $\mathbb{R}^2$ or $\mathbb{R}v$ for some $v \in \mathbb{R}^2, v \neq 0$. If $H \supseteq \mathbb{R}^2$ then, since H is proper, $H_0 = \mathbb{R}^2$ which fails to be extending. Next, let $H_0 \cap \mathbb{R}^2 = \mathbb{R}v$. Then $H_0 = \mathbb{R}v$ since E and $-E$ are the only elements of $SO(2)$ mapping $\mathbb{R}v$ to itself. However, $\mathbb{R}v$ is not extending. Finally, let $H_0 \cap \mathbb{R}^2 = \{0\}$. Then either $H_0\mathbb{R}^2 = G$ or H_0 is trivial. In the first case, H_0 is compact and from $H_0 \cap \mathbb{R}^2 = \{0\}$

and $H_0\mathbb{R}^2 = G$ it is easily deduced that H_0 is its own normalizer. In particular, $H_0 = H$. In the second case H is discrete and hence $H \cap \mathbb{R}^2$ is extending. Thus $H \cap \mathbb{R}^2 = \{0\}$ and hence H is abelian. Now Lemma 7.4.11 shows that H is finite. This finishes the proof of the above claim. Thus a proper closed subgroup of G is extending if and only if it either finite or conjugate to $SO(2)$.

7.5. The separation property: Basic facts and examples

In the following three sections we study the separation property of positive definite functions exposed at the outset of this chapter. To start with, we introduce the basic definitions.

DEFINITION 7.5.1. Let G be a locally compact group. For any closed subgroup H of G, let

$$P_H(G) = \{\phi \in P(G) : \phi(h) = 1 \text{ for all } h \in H\}.$$

(1) We say that G has the *H-separation property* or *H is separating in G* if for every $x \in G\backslash H$, there exists $\phi \in P_H(G)$ such that $\phi(x) \neq 1$. When G has the H-separation property for every closed subgroup H of G, we refer to G as a group with the *separation property.*

(2) We say that G has the *separation property for cyclic subgroups* if every closed subgroup of G which is isomorphic to $\mathbb{Z}$ is separating.

REMARK 7.5.2. (1) If H is either compact, or open, or normal, then G has the H-separation property.

(2) The separation property for cyclic subgroups, which is apparently weaker than the separation property, is of interest because for special classes of locally compact groups, such as compactly generated nilpotent groups and almost connected groups, it will be seen to actually enforce the separation property (see Theorems 7.7.10 and 7.6.7). Note, however, that the separation property for cyclic subgroups is a local property in the sense that a locally compact group has this property if (and only if) every compactly generated open subgroup does so.

(3) Recall that for any $\phi \in P(G)$ we have

$$|\phi(xy) - \phi(x)\phi(y)|^2 \leqslant (1 - |\phi(x)|^2)(1 - |\phi(y)|^2)$$

for all $x, y \in G$ (Proposition 1.4.16(iv))). In particular, if H is a closed subgroup of G and $\phi \in P_H(G)$, then

$$\phi(h_1 x h_2) = \phi(x)$$

for all $x \in G$ and $h_1, h_2 \in H$. This simple property will be used frequently.

We now determine the separating subgroups for several classical groups. Some of these examples will be used later when dealing with solvable connected Lie groups.

EXAMPLE 7.5.3. Let G be the 3-dimensional Heisenberg group. Thus $G = \mathbb{R}^3$ with multiplication

$$(x_1, x_2, x_3)(y_1, y_2, y_3) = (x_1 + y_1, x_2 + y_2, x_3 + y_3 + x_1 y_2),$$

$x_i, y_i \in \mathbb{R}$, $i = 1, 2, 3$. Let H be a closed subgroup of G. Then G has the H-separation property (if and) only if H is normal in G. To see this, consider elements $a = (x_1, x_2, x_3) \in H$ and $b = (y_1, y_2, y_3) \in G$. It is straightforward to check that

$$[a, b] = aba^{-1}b^{-1} = (0, 0, x_1 y_2 - x_2 y_1).$$

For any $n \in \mathbb{N}$, let

$$a_n = (nx_1, nx_2, nx_3) \text{ and } b_n = (y_1/n, y_2/n, y_3/n).$$

Then $a_n \in H$ and $[a_n, b_n] = [a, b]$ for all n. Now, let $\varphi \in \mathrm{ex}(P_H(G)) \subseteq \mathrm{ex}(P^1(G))$. Then $\varphi(xz) = \varphi(x)\varphi(z)$ for all $x \in G$ and $z \in Z(G)$, and hence

$$\begin{aligned}\varphi(b_n) &= \varphi(a_n b_b a_n^{-1}) = \varphi([a_n, b_n]b_n) \\ &= \varphi([a_n, b_n])\varphi(b_n) = \varphi([a, b])\varphi(b_n).\end{aligned}$$

Since $b_n \to e$ as $n \to \infty$, $\varphi(b_n) \to 1$, and this implies that $\varphi([a, b]) = 1$. Since this holds for every $\varphi \in \mathrm{ex}(P_H(G))$, we conclude that $\phi([a, b]) = 1$ for all $\phi \in P_H(G)$. Finally, because G has the H-separation property, it follows that $[a, b] \in H$, and since $a \in H$ and $b \in G$ were arbitrary, it follows that H is normal in G.

The next lemma might be viewed as an analogue of Lemmas 7.4.10 and 7.4.11.

LEMMA 7.5.4. *Let G be a locally compact group and suppose that G has a closed normal subgroup N such that, with $q : G \to G/N$ denoting the quotient homomorphism,*

$$B(G) \subseteq B(G/N) \circ q + C_0(G).$$

If H is a closed subgroup of G and G has the H-separation property, then either $N \subseteq H$ or H is compact.

PROOF. Assume that H does not contain N and choose $a \in N$ with $a \notin H$. There exists $\phi \in P_H(G)$ such that $\phi(a) \neq 1$. By hypothesis, $\phi = \phi_1 + \phi_2$, where $\phi_2 \in C_0(G)$ and $\phi_1 = \varphi \circ q$ for some $\varphi \in B(G/N)$. For all $h \in H$ we have

$$\phi_1(ha) = \phi(ha) - \phi_2(ha) = \phi(a) - \phi_2(ha)$$

and also, since $a \in N$,

$$\phi_1(ha) = \phi_1(h) = \phi(h) - \phi_2(h) = 1 - \phi_2(h).$$

Combining these two equations, we get

$$\phi_2(h) - \phi_2(ha) = 1 - \phi(a) \neq 0.$$

Since the function $h \to \phi_2(h) - \phi_2(ha)$ vanishes at infinity on H, we conclude that H must be compact. □

EXAMPLE 7.5.5. Let G be the $ax + b$-group. Then the closed subgroups H of G such that G has the H-separation property are precisely the normal ones. To see this, we only have to observe that if H is nontrivial and G has the H-separation property, then H contains the normal subgroup $\mathbb{R} \times \{1\}$. This follows from Lemma 7.5.4 since $B(G) = B(\mathbb{R}_+^\times) \circ q + A(G)$, where $q(b, a) = a$ for $(b, a) \in G$.

EXAMPLE 7.5.6. Let G be the motion group of $\mathbb{R}^d$ ($d \geqslant 2$), that is, $G = \mathbb{R}^d \rtimes SO(d)$ where the special orthogonal group $SO(d)$ acts on $\mathbb{R}^d$ by rotation. Let H be a closed subgroup of G such that G has the H-separation property. We claim that H is either compact (and hence conjugate to some closed subgroup of $SO(d)$) or H contains $\mathbb{R}^d$ (and hence $H = \mathbb{R}^d \rtimes K$ for some closed subgroup K of $SO(d)$). It is clear that conversely, for every such subgroup H, G has the H-separation property.

Let E denote the $d \times d$-unit matrix. We shall use the following fact. Given $0 < r < s$ and $v \in \mathbb{R}^d$ such that $\|v\| \geqslant sd^{1/2}$, there exists $D \in SO(d)$ such that $\|E - D\| \leqslant r/s$ and $\|(E - D)v\| = r$. This can be seen as follows. There exists an orthonormal basis $\{e_1, \ldots, e_d\}$ of $\mathbb{R}^d$ such that, with $v = \sum_{j=1}^d \alpha_j e_j$, $|\alpha_1| \geqslant s$. For $\varphi \in \mathbb{R}$, let

$$w = (\alpha_1 \cos \varphi - \alpha_2 \sin \varphi)e_1 + (\alpha_1 \sin \varphi + \alpha_2 \cos \varphi)e_2 + \sum_{j=3}^{d} \alpha_j e_j.$$

Then $w = D_\varphi v$, where $D_\varphi \in SO(d)$ leaves $e_3, \ldots, e_d$ fixed and is the rotation associated to φ in the (e_1, e_2)-plane. It follows that $\|E - D_\varphi\| = |1 - e^{i\varphi}|$ and

$$\|(E - D_\varphi)v\| = \|v - w\| = |1 - e^{i\varphi}|(\alpha_1^2 + \alpha_2^2)^{1/2}.$$

Since the range of the continuous function $\varphi \to |1 - e^{i\varphi}|(\alpha_1^2 + \alpha_2^2)^{1/2}$ contains the interval $[0, 2s]$, there exists φ such that $\|(E - D_\varphi)v\|_r$. For this φ, it follows that

$$r = |1 - e^{i\varphi}|(\alpha_1^2 + \alpha_2^2)^{1/2} \geqslant |1 - e^{i\varphi}|s,$$

and therefore $\|E - D_\varphi\| \leqslant r/s$.

Now, let H be a noncompact subgroup of G and suppose that G has the H-separation property. We first show that given $r > 0$, there exists $y \in \mathbb{R}^d$ such that $\|y\| = r$ and $(y, E) \in H$. Since H is noncompact, there exists a sequence $(x_n, A_n)_n$ in H with $\|x_n\| \geqslant nd^{1/2}$ for all n. Then, for $C \in SO(d)$ and $\phi \in P_H(G)$,

$$\begin{aligned} \phi((0, C)) &= \phi((x_n, A_n)^{-1}(0, C)(x_n, A_n)) \\ &= \phi((A_n^{-1}(Cx_n - x_n), A_n^{-1}CA_n)). \end{aligned}$$

By what we seen above, for each n large enough, there exists $C_n \in SO(d)$ such that $\|E - C_n\| \leqslant r/n$ and $\|C_n x_n - x_n\| = r$. Passing to a subsequence if necessary, we can assume that $A_n^{-1}(C_n x_n - x_n) \to y$ for some $y \in \mathbb{R}^d$. Then $\|y\| = r$ and, for every $\phi \in P_H(G)$,

$$\begin{aligned} 1 &= \lim_{n\to\infty} \phi((0, C_n)) = \lim_{n\to\infty} \phi((A_n^{-1}(Cx_n - x_n), A_n^{-1}CA_n)) \\ &= \phi((y, E)). \end{aligned}$$

Since G has the H-separation property, it follows that $(y, E) \in H$. In particular, as $r > 0$ is arbitrary, we conclude that $H \cap \mathbb{R}^d$ is a nondiscrete subgroup of $\mathbb{R}^d$. Let V denote the connected component of $H \cap \mathbb{R}^d$. Then V is a vector space of positive dimension. Towards a contradiction, suppose that $\dim V < d$. Choose a linear subspace W of $\mathbb{R}^d$ containing V such that $\dim(W/V) = 1$, and define closed subgroups $S(W)$ and $S_V(W)$ of $SO(d)$ by

$$S(W) = \{A \in SO(d) : A(W) = W\}$$

and

$$S_V(W) = \{A \in S(W) : A(V) = V\}.$$

Then, for any $A \in S(W) \backslash S_V(W)$, we have

$$W = \{x + Ay : x, y \in V\}.$$

Now, for $x, y \in V$, $A \in SO(d)$ and $\phi \in P_H(G)$,

$$\phi((0, A)) = \phi((x, E)(0, A)(y, E)) = \phi((x + Ay, A)).$$

There exists a sequence $(A_n)_n$ in $S(W) \backslash S_V(W)$ such that $A_n \to E$ in $SO(d)$. It follows that $\phi((0, A_n)) = \phi((w, A_n))$ and hence

$$1 = \lim_{n\to\infty} \phi((0, A_n)) = \lim_{n\to\infty} \phi((w, A_n)) = \phi((w, E))$$

for every $\phi \in P_H(G)$ and $w \in W$. Thus $W \times \{E\} \subseteq H$. This contradiction proves that $\dim V = d$, whence $H \supseteq \mathbb{R}^d$, as was to be shown.

Example 7.5.6 could alternatively be treated by appealing to Lemma 7.5.4 and using that for the motion group $G = \mathbb{R}^d \rtimes SO(d)$ one has $B(G) \subseteq B(G/\mathbb{R}^d) + C_0(G)$. This latter fact can either be deduced from [**8**, Corollary of Theorem 4] or from a result of Chou [**31**, Theorem 3.3] stating that $W(G) = AP(G) + C_0(G)$, where $AP(G)$ and $W(G)$ denotes the space of almost periodic and weakly almost periodic functions of G, respectively. However, the proof of [**31**, Theorem 3.3] is fairly intricate and involves Grothendieck's criterion for weak almost periodicity of continuous functions.

We now introduce a concept which turns out to be of considerable relevance as it describes precisely the separating subgroups of locally compact groups having an almost connected open normal subgroup (see Theorem 7.5.9).

DEFINITION 7.5.7. Let H be a closed subgroup of G. Then H is called *neutral* in G if given any neighbourhood U of the identity, there exists a neighbourhood V of the identity such that $VH \subseteq HU$.

Such subgroups were first considered in [**239**] in the context of invariant measures on homogeneous spaces and have later been studied extensively in [**190**] and [**244**].

REMARK 7.5.8. (1) A subgroup H of G, which is either open, or compact, or closed and normal, is obviously neutral. More generally, if $G \in [\mathrm{SIN}]_H$ then H is neutral in G. In fact, given U as above, there exists a neighbourhood V of e such that $V \subseteq U$ and $hVh^{-1} = V$ for all $h \in H$, and this implies that $HV = VH \subseteq UH$.

(2) If H is a neutral subgroup of G, then there exists a neighbourhood basis $\mathcal{V}$ of the identity such that $VH = HV$ for all $V \in \mathcal{V}$. This can be seen as follows. Replacing U by $U \cap U^{-1}$, we can assume that U is symmetric. Then choose a symmetric neighbourhood W of the identity with $HW \subseteq UH$ and let $V = U \cap HWH$. It is easily verified that V satisfies $(VH)^{-1} \subseteq VH$ and hence $(VH)^{-1} = VH$. Consequently, $HV = (VH)^{-1} = VH$.

THEOREM 7.5.9. *Let G be a locally compact group and H a closed neutral subgroup of G. Then, given any compact subset C of G with $C \cap H = \varnothing$, there exists $u \in P_H(G)$ such that $u(x) = 0$ for all $x \in C$. In particular, G has the H-separation property.*

PROOF. Note first that since H is neutral in G, there exists a quasi-invariant measure μ on the left coset space G/H (equivalently, the modular functions of G and H agree on H). So Weil's formula

$$\int_G f(x)dx = \int_{G/H} \left(\int_H f(xh)dh \right) d\mu(xH)$$

holds for all $f \in L^1(G)$ (Section 1.3). As before, let T_H denote the homomorphism from $L^1(G)$ onto $L^1(G/H, \mu)$.

We choose a symmetric neighbourhood U of the identity in G such that $UCU \cap H = \varnothing$. Since H is neutral in G, there exists a compact symmetric neighbourhood V of the identity such that $V \subseteq U$ and $HV = VH$. Let $q : G \to G/H$ denote the quotient map and let v be a non-negative function in $L^1(G)$ such that $T_H v$ is equal to $\mu(q(V))^{-1/2}$ on $q(V)$ and $T_H v$ vanishes on $G/H \backslash q(V)$. Then $T_H v$ has norm one in $L^2(G/H, \mu)$.

Now define a function u on G by

$$u(x) = \int_G v(y) T_H v(x^{-1} y H) dy, \ x \in G.$$

Using the choice of U, V and v, it is then easily verified that $u(x) = 0$ for all $x \in C$. We claim that $u(h) = 1$ for all $h \in H$. To see this, observe that if $y \in V$, then there exist $k \in H$ and $z \in V$ such that $hy = zk$. Thus

$$\int_H v(hyt) dt = \int_H v(zs) ds = \mu(q(V))^{-1/2}.$$

Since μ is H-invariant, it follows that

$$\begin{aligned} u(h) &= \int_{G/H} \left(T_H v(h^{-1} y H) \int_H v(yt) dt \right) d\mu(yH) \\ &= \int_{q(V)} \left(T_H v(yH) \int_H v(hyt) dt \right) d\mu(yH) \\ &= 1. \end{aligned}$$

Finally, denoting by π the unitary representation of G induced from the trivial one-dimensional representation of H, the formula defining u can be rewritten as

$$u(x) = \int_{G/H} T_H v(x^{-1} y H) \overline{T_H v(yH)} d\mu(yH) = \langle \pi(x) T_H v, T_H v \rangle.$$

This shows that u is positive definite and completes the proof. $\square$

Theorem 7.5.9 especially shows that every SIN-group has the separation property. It also raises the question of whether conversely a closed subgroup H has to be neutral whenever G has the H-separation property. This will turn out to be true for locally compact groups G containing an almost connected open normal subgroup (Theorem 7.7.2). The proof of Theorem 7.7.2 will build on the following sequence of lemmas.

LEMMA 7.5.10. *Let H be a closed subgroup of G and let $\mathcal{U}$ be a neighbourhood basis of the identity of G. If G has the H-separation property, then*

$$H = \bigcap \{ \overline{HUH} : U \in \mathcal{U} \}.$$

PROOF. Let $x \in \bigcap \{ \overline{HUH} : U \in \mathcal{U} \}$ and, towards a contradiction, assume that $x \notin H$. Then, since G has the H-separation property, there exists $\phi \in P_H(G)$ such that $\phi(x) \neq 1$. Let $\epsilon = |1 - \phi(x)| > 0$. There exists $U \in \mathcal{U}$ such that $|\phi(z) - \phi(y)| < \frac{\epsilon}{2}$ for all $z, y \in G$ with $yz^{-1} \in U$.

Now, since $x \in \overline{HUH}$, there exist $a, b \in H$ and $u \in U$ such that $aub \in Ux$. Since $\phi(aub) = \phi(u)$, it follows that

$$|1 - \phi(x)| \leqslant |1 - \phi(u)| + |\phi(aub) - \phi(x)| < \epsilon.$$

This contradiction shows that $x \in H$. $\square$

LEMMA 7.5.11. *Let H be a closed subgroup of G such that, for some neighbourhood basis $\mathcal{U}$ of e in G, $H = \bigcap \{ \overline{HUH} : U \in \mathcal{U} \}$. Let K be a compact normal subgroup of G and $q : G \to G/K$ the quotient homomorphism. Then*

$$q(H) = \bigcap \{ \overline{q(H) V q(H)} : V \in \mathcal{V} \}$$

for some neighbourhood basis $\mathcal{V}$ of $q(e)$ in G/K.

PROOF. Note first that $q(H)$ is closed in G/K since $q^{-1}(q(H)) = HK$ is closed in G. Let $x \in G$ such that $q(x) \notin q(H)$, that is, $xK \cap H = \varnothing$. By hypothesis, for each $k \in K$ there exists $U_k \in \mathcal{U}$ such that $xk \notin \overline{HU_kH}$. This implies that $xV_k \cap \overline{HU_kH} = \varnothing$ for some neighbourhood V_k of k in G. Since K is compact, $K \subseteq \bigcup_{j=1}^{n} V_{k_j}$ for certain $k_1, \ldots, k_n \in K$. Let $U = \cap_{j=1}^{n} U_{k_j}$. Then

$$xK \cap \overline{HUH} \subseteq x\left(\bigcup_{j=1}^{n} V_{k_j}\right) \cap \overline{HUH} \subseteq \bigcup_{j=1}^{n} \left(xV_{k_j} \cap \overline{HU_{k_j}H}\right) = \varnothing,$$

and hence $q(x) \notin q(\overline{HUH})$. Now, since $q(\overline{HUH})$ is closed in G/K,

$$\overline{q(H)q(U)q(H)} \subseteq \overline{q(HUH)} \subseteq q(\overline{HUH}),$$

and therefore $q(x) \notin \overline{q(H)q(U)q(H)}$. Taking for $\mathcal{V}$ the set of all finite intersections of sets $q(U)$, $U \in \mathcal{U}$, the conclusion of the lemma follows. □

LEMMA 7.5.12. *Let G be a locally connected locally compact group, and let H be a closed subgroup of G such that $H = \bigcap\{\overline{HUH} : U \in \mathcal{U}\}$, where $\mathcal{U}$ is a neighbourhood basis of e. Then H is neutral in G.*

PROOF. Let $q : G \to G/H$ denote the map $x \to xH$, and equip G/H with the quotient topology. Let $\mathcal{U}$ be the collection of all open neighbourhoods of e in G. It suffices to show that if $U \in \mathcal{U}$ is relatively compact, then there exists $V \in \mathcal{U}$ such that $HV \subseteq UH$. To that end, observe first that there exists $V \in \mathcal{U}$ such that $q(HV) \cap q(\overline{U}) \subseteq q(U)$. In fact, $q(\overline{U})$ is compact, $q(U)$ is an open neighbourhood of $q(e)$ in G/H and

$$\bigcap\{q(\overline{HVH}) \cap \overline{q(U)} : V \in \mathcal{U}\} = \{q(e)\},$$

and this implies that $q(HVH) \cap \overline{q(U)} \subseteq q(U)$ for some V since the sets $q(HVH)$, $V \in \mathcal{U}$, have the finite intersection property.

Since G is locally connected, we can assume that V is connected. Then $q(HV)$ is connected since, for each $h \in H$, $q(hV)$ is connected and $q(e) = q(h) \in q(hV)$. Now $q(HV) \cap \overline{q(U)} \subseteq q(U)$ means that $q(HV) \cap \overline{q(U)}$ is open and closed in $q(HV)$, and this in turn implies that $q(HV) \subseteq q(U)$ since $q(HV)$ is connected and $q(HV) \cap q(U) \neq \varnothing$. It follows that $HV \subseteq UH$. □

LEMMA 7.5.13. *Let G be a locally compact group and C a compact normal subgroup of G such that G/C is locally connected. If G has the separation property for cyclic subgroups, then G/C has the corresponding property.*

PROOF. Let B be a closed subgroup of G/C which is isomorphic to $\mathbb{Z}$ and let $q : G \to G/C$ denote the quotient homomorphism. Choose $a \in G$ such that $q(a)$ generates B and let A be the closed subgroup of G generated by a. Then A cannot be compact and hence is isomorphic to $\mathbb{Z}$ [**125**, Theorem (9.1)]. Moreover, $q(A) = B$. Let $\mathcal{U}$ be a neighbourhood basis of the identity in G. Since A is separating in G, $A = \cap\{\overline{AUA} : U \in \mathcal{U}\}$ (Lemma 7.5.10). By Lemma 7.5.11, it follows that

$$B = \bigcap\{\overline{BVB} : V \in \mathcal{V}\}$$

for some neighbourhood basis $\mathcal{V}$ of $\{C\}$ in G/C. Since G/C is locally connected, this implies that B is neutral in G/C (Lemma 7.5.11). Finally, B being neutral, it is separating in G/C by Theorem 7.5.9. □

LEMMA 7.5.14. *Let G be a projective limit of groups G/K_α, $\alpha \in A$, and let H be a closed subgroup of G. If HK_α/K_α is neutral in G/K_α for all α, then H is neutral in G.*

PROOF. Let W be a neighbourhood of e in G, and choose $K = K_\alpha$ such that $K \subseteq W$. Since K is compact, there exists a neighbourhood U of e such that $UK \subseteq W$. Denote by $q : G \to G/K$ the quotient homomorphism. By hypothesis, there exists a neighbourhood V of $\{K\}$ in G/K such that $q(H)V \subseteq q(U)q(H)$. Then

$$Hq^{-1}(V) \subseteq q^{-1}(q(U)q(H)) = UKHK = UKH \subseteq WH.$$

As W was an arbitrary neighbourhood of e, it follows that H is neutral in G. □

COROLLARY 7.5.15. *Let G be a Lie-projective locally compact group and H a closed subgroup of G. If $H = \bigcap\{\overline{HUH} : U \in \mathcal{U}\}$, where $\mathcal{U}$ is a neighbourhood basis of the identity, then H is neutral in G.*

PROOF. Since every Lie group is locally connected, the statement of the corollary follows from Lemma 7.5.11 and Lemma 7.5.13. □

We now present an example of a 2-step solvable, totally disconnected group which fails to be an SIN-group and nevertheless has the separation property.

EXAMPLE 7.5.16. Let p be a prime number and let N be the additive group of the p-adic number fields Ω_p. Let K be the multiplicative group of p-adic numbers of valuation one and let $G = N \rtimes K$, where K acts on N by multiplication. We are going to show that G has the separation property.

For that we first observe that if H is a noncompact separating subgroup of G, then $H \supseteq N$. Indeed, this follows from Lemma 7.5.4 since $B(G) = B(K) \circ q + A(G)$, where $q : G \to K$ is the quotient homomorphism.

We now show that an arbitrary closed subgroup of G either is compact or contains N. Let H be a closed subgroup of G such that $N \nsubseteq H$. We have to show that H is compact. Then $H \cap N$ is either trivial or a proper closed subgroup of N and as such is compact. In fact, every nontrivial proper closed subgroup of N is of the form $p^k\Delta_p$, where Δ_p is the subring of p-adic integers and $k \in \mathbb{Z}$. Moreover, $H/H \cap N$ is abelian since the homomorphism $H/H \cap N \to G/N = K$ is injective. We claim that $q(H)$ is closed in K. To see this, let $(x_n, a_n) \in H$, $n \in \mathbb{N}$, such that $a_n \to a$ for some $a \in K$. Assume that $q(H) \neq \{1\}$ and fix $(y, b) \in H$ with $b \neq 1$. Since $H/H \cap N$ is abelian, there exists a sequence $(z_n)_n$ in $H \cap N$ such that

$$\begin{aligned}(x_n + a_n y, a_n b) &= (x_n, a_n)(y, b) = (z_n, 1)(y, b)(x_n, a_n) \\ &= (z_n + y + bx_n, ba_n),\end{aligned}$$

and consequently, for all n,

$$x_n + a_n y = z_n + y + bx_n.$$

Moreover, since $H \cap N$ is compact, we can assume that $z_n \to z$ for some $z \in H \cap N$. It follows that

$$(1 - b)x_n = z_n + (1 - a_n)y \to z + (1 - a)y.$$

Thus $(x_n)_n$ converges to $x = (1 - b)^{-1}(z + (1 - a)y) \in N$. Hence $(x_n, a_n) \to (x, a)$ in H, whence $a \in q(H)$, as required.

Finally, since H is σ-compact and $q(H)$ is closed in K, the continuous isomorphism from $H/H \cap N$ onto $HN/N = q(H)$ is a homeomorphism. So H is compact since both $H \cap N$ and $H/H \cap N$ are compact.

We conclude this section by briefly pointing out the connection of the separation property with the so-called Mautner phenomenon (see also Section 7.8).

DEFINITION 7.5.17. Let G be a locally compact group and $(\widetilde{H}, H)$ a pair of closed subgroups of G with $\widetilde{H} \supseteq H$. Then the *Mautner phenomenon* holds for $(\widetilde{H}, H)$ if for every unitary representation π of G, each unit vector $\xi \in \mathcal{H}(\pi)$ which is invariant under H is also invariant under $\widetilde{H}$, that is, if $\pi(h)\xi = \xi$ for all $h \in H$, then $\pi(x)\xi = \xi$ for all $x \in \widetilde{H}$.

REMARK 7.5.18. (1) Let H be a closed subgroup of G and $\varphi \in P(G)$. There exist a unitary representation π of G and $\xi \in \mathcal{H}(\pi)$ such that $\varphi(x) = \langle \pi(x)\xi, \xi \rangle$ for all $x \in G$. Then it is easily verified that $\varphi \in P_H(G)$ if and only if $\|\xi\| = 1$ and $\phi(h)\xi = \xi$ for all $h \in H$. Thus G has the H-separation property precisely when H is the only closed subgroup $\widetilde{H}$ of G containing H such that the Mautner phenomenon holds for $(\widetilde{H}, H)$.

(2) Let $\varphi \in P^1(G)$. Then $\varphi^{-1}(1)$ is a closed subgroup of G. For any closed subgroup H of G, define H_{sep} by

$$H_{\text{sep}} = \bigcap \{\varphi^{-1}(1) : \varphi \in P_H(G)\}.$$

Then H_{sep} is the smallest closed subgroup of G containing H for which G has the separation property. In particular, G has the H-separation property if and only if $H_{\text{sep}} = H$. Furthermore, the Mautner phenomenon holds for (H_{sep}, H), and H_{sep} is the largest subgroup of G with this property.

7.6. The separation property: Nilpotent Groups

In the remainder of this chapter our goal is twofold. On the one hand we aim at a criterion for a given closed subgroup to be separating. On the other hand we want to derive a structure theorem for groups having the separation property. In this section we treat nilpotent groups, whereas the next section is devoted to groups which possess an almost connected open normal subgroup.

LEMMA 7.6.1. *Let G be a locally compact group and H a closed subgroup of G. Let*

$$N = \{x \in G : |\varphi(x)| = 1 \text{ for all } \varphi \in \text{ex}(P_H(G))\}.$$

Then N is a closed subgroup of G containing the centre of G, and G has the N-separation property.

PROOF. Clearly, N is a closed subgroup of G. To see that N contains the centre of G, let $\varphi \in \text{ex}(P_H(G))$, and let π denote the Gelfand-Naimark-Segal representation of G associated with φ. Since $\text{ex}(P_H(G)) \subseteq \text{ex}(P^1(G))$, π is irreducible, and hence $\pi|_{Z(G)}$ is a multiple of some character $\chi \in \widehat{Z(G)}$. Let $\xi \in \mathcal{H}(\pi)$ such that $\varphi(x) = \langle \pi(x)\xi, \xi \rangle$ for all $x \in G$. Then

$$|\varphi(z)| = |\chi(z)| \cdot \|\xi\|^2 = 1$$

for all $z \in Z(G)$. Now, let

$$F = \{|\varphi|^2 : \varphi \in \text{ex}(P_H(G))\} \subseteq P_N(G).$$

If $x \in G \backslash N$, then $|\varphi(x)| < 1$ for some $\varphi \in \text{ex}(P_H(G))$ and hence $\psi(x) \neq 1$ for some $\psi \in F$. This shows that G has the N-separation property. □

LEMMA 7.6.2. *Let G, H and N be as in Lemma 7.6.1. Suppose that G has the H-separation property and that N is normalized by G_0, the connected component of the identity. Then G_0 normalizes H.*

PROOF. Let $\varphi \in \text{ex}(P_H(G))$ and put $V_\varphi = \{x \in G : \varphi(x) \neq 0\}$. Then V_φ is an open set containing N. For $x \in G_0$ and $h \in H$ we have $[h, x] \in N$ since $H \subseteq N$ and G_0 normalizes N. So, since φ is of absolute value one on N,

$$\varphi(x) = \varphi([h, x]x) = \varphi([h, x])\varphi(x).$$

Hence, if $x \in V_\varphi \cap G_0$,

$$\varphi(xhx^{-1}) = \varphi([h, x]) = 1$$

for all $h \in H$. Now, let $x \in V_\varphi \cap G_0$, $y \in G_0$ and $h \in H$. Then, since $zHz^{-1} \subseteq N$ for each $z \in G_0$,

$$\begin{aligned} \varphi(x)\varphi(yhy^{-1}) &= \varphi(xyhy^{-1}) = \varphi((xy)(h(xy)^{-1}x) \\ &= \varphi((xy)h(xy)^{-1})\varphi(x). \end{aligned}$$

Thus $\varphi(xyH(xy)^{-1}) = \varphi(yHy^{-1})$ since $\varphi(x) \neq 0$. This equation shows that the set of all elements y in G_0 such that $\varphi(yHy^{-1}) = \{1\}$ contains the subgroup generated by $V_\varphi \cap G_0$. Since G_0 is connected, it follows that $\varphi(xHx^{-1}) = \{1\}$ for all $x \in G_0$ and all $\varphi \in \text{ex}(P_H(G))$. Define a closed subgroup K of G by

$$K = \{y \in G : \varphi(y) = 1 \text{ for all } \varphi \in \text{ex}(P_H(G))\}.$$

We claim that $K \subseteq H$. For that, notice first that if C is a closed subgroup of G and $(\phi_\alpha)_\alpha$ is a net in $P_C(G)$ converging to some $\phi \in P(G)$ in the w^*-topology, then $\phi(cx) = \phi(x)$ for all $x \in G$ and $c \in C$. Indeed, for any $c \in C$, $L_c\phi_\alpha \to L_c\phi$ in the w^*-topology and $L_c\phi_\alpha = \phi_\alpha$, whence $\phi = L_c\phi$. It follows that $[0, 1]P_C(G) = \{\lambda\varphi : \lambda \in [0, 1], \varphi \in P_C(G)\}$ is w^*-closed in $L^\infty(G)$. By the Krein-Milman theorem, this implies that

$$[0, 1]P_H(G) = \overline{\text{co}(\text{ex}([0, 1]P_H(G)))} = \overline{\text{co}(\{0\} \cup \text{ex}(P_H(G)))},$$

where the closures are taken in the w^*-topology. Now, by definition of K, $\text{ex}(P_H(G)) \subseteq P_K(G)$. Thus $[0, 1]P_H(G) \subseteq [0, 1]P_K(G)$ and hence $P_H(G) \subseteq P_K(G)$. Since G has the H-separation property, we conclude that $K \subseteq H$. We have seen above that $xHx^{-1} \subseteq K$ for all $x \in G_0$. This proves that G_0 normalizes H. □

LEMMA 7.6.3. *Let G be a locally compact group such that $G_0 \subseteq Z_m(G)$ for some $m \in \mathbb{N}$. Let H be a closed subgroup of G and suppose that G has the H-separation property. Then G_0 normalizes H.*

PROOF. Notice first that if M is a closed normal subgroup of G which is contained in G_0, then $(G/M)_0 = G_0/M$. To prove the statement of the lemma, we apply the preceding two lemmas and induction on m. Let $M = G_0 \cap Z_1(G)$, and let N be as in Lemma 7.6.1. Then G has the N-separation property and hence G/M has the N/M-separation property. Since $(G/M)_0 = G_0/M \subseteq Z_{m-1}(G/M)$, by the inductive hypothesis G_0/M normalizes N/M and hence G_0 normalizes N. Then, by Lemma 7.6.2, G_0 normalizes H since G has the H-separation property. □

The following theorem is the first main result of this section. It in particular shows that if G is a connected nilpotent group and H is a closed subgroup of G, then G has the H-separation property (if and) only if H is normal in G. Moreover, for such a group G, the neutral subgroups are precisely the normal subgroups.

THEOREM 7.6.4. *Let G be a locally compact group such that G_0 is open in G and $G_0 \subseteq Z_m(G)$ for some $m \in \mathbb{N}$. For a closed subgroup H of G, the following conditions are equivalent.*

(i) *G has the H-separation property.*
(ii) *G_0 normalizes H.*
(iii) *H is neutral in G.*

PROOF. (i) $\Rightarrow$ (ii) is a consequence of Lemma 7.6.3. The implication (ii) $\Rightarrow$ (iii) is trivial since G_0 is open in G, and (iii) $\Rightarrow$ (i) holds for any locally compact group by Theorem 7.5.9. □

We now investigate the problem of when a compactly generated nilpotent has the separation property.

COROLLARY 7.6.5. *Let G be a connected nilpotent Lie group which has the separation property for cyclic subgroups. Then G is abelian.*

PROOF. Let a be an arbitrary element of G and let A be the closed subgroup generated by a. Then either A is compact or isomorphic to $\mathbb{Z}$. In any case, A is separating in G, and hence normal in G by Theorem 7.6.4. If now $A = \mathbb{Z}$, then $[G, a] = \{e\}$ because $[G, a]$ is connected and contained in A. If A is compact, then consider the action of G on the discrete dual group of A. Since G is connected, this action is trivial and hence so is the action of G on A itself. In either case a is contained in the centre of G. □

LEMMA 7.6.6. *Let G be a nilpotent Lie group and suppose that G has the separation property for cyclic subgroups. Then G_0 is contained in the centre of G.*

PROOF. Since G_0 has the separation property for cyclic subgroups, $G_0 = V \times K$ where V is a vector group and K is a compact connected Lie group (Corollary 7.6.5). Moreover, because K is nilpotent, $K = \mathbb{T}^d$ for some $d \in \mathbb{N}_0$.

Let a be an arbitrary element of G and let A denote the closed subgroup of G generated by a. Since A is either compact or isomorphic to $\mathbb{Z}$ and G has the separation property for cyclic subgroups, it follows from Lemma 7.6.3 that G_0 normalizes A. We now distinguish the two cases that a has finite order or infinite order modulo the open normal subgroup G_0.

Assume first that $a^q \in G_0$ for some $q \in \mathbb{N}$. Using the facts that G_0 and A are abelian and that G_0 normalizes A, it is easily verified by induction that, for all $y \in G_0$ and $n \in \mathbb{N}$,

$$[y^n, a] = [y, a]^n = [y, a^n].$$

Let now x be an arbitrary element of G_0. Since both V and $\mathbb{T}^d$ are divisible, there exists $y \in G_0$ such that $y^q = x$. Then

$$[x, a] = [y^q, a] = [y, a^q] = e.$$

This shows that $[G_0, a] = \{e\}$. Now let a have infinite order modulo G_0. Then $A \cap G_0 = \{e\}$ and therefore $[G_0, a] \subseteq A \cap G_0 = \{e\}$. Summing up we have seen that G_0 is contained in the centre of G. □

We have seen in Section 7.5 that every SIN-group has the separation property. Now we are able to prove the converse for compactly generated nilpotent groups. Actually, the implication (ii) $\Rightarrow$ (iv) of the following theorem constitutes an even stronger result.

THEOREM 7.6.7. *Let G be a compactly generated nilpotent locally compact group. Then the following conditions are equivalent.*

(i) *G has the separation property.*
(ii) *G has the separation property for cyclic subgroups.*
(iii) *G is a projective limit of groups each of which has an open centre.*
(iv) *G is an* SIN*-group.*

PROOF. Since a projective limit of SIN-groups is again an SIN-group and every SIN-group has the separation property, it only remains to prove the implication (ii) $\Rightarrow$ (iii). As G is a compactly generated nilpotent group, it is a projective limit of Lie groups G/C_α [**127**, 9, Theorem]. By Lemma 7.5.13, the separation property for cyclic subgroups passes to the quotient groups G/C_α. Since G/C_α is a Lie group, the connected component $(G/C_\alpha)_0$ is open in G/C_α. Lemma 7.6.6 now shows that $(G/C_\alpha)_0$ is contained in the centre of G/C_α. This proves (iii). □

We remind the reader that a locally compact group G is called *locally nilpotent* if every compactly generated (open) subgroup of G is nilpotent. Examples of such groups can be constructed as follows. Let A be an infinite index set, and for each $\alpha \in A$ let G_α be a locally compact nilpotent group of nilpotency length $\ell(G_\alpha)$. Suppose that there is a finite subset B of A such that for all $\alpha \in A\backslash B$, G_α possesses a compact open subgroup K_α such that $\sup\{\ell(K_\alpha) : \alpha \in A\backslash B\} < \infty$. Form the restricted direct product

$$G = \{(x_\alpha)_{\alpha\in A} \in \textstyle\prod_{\alpha\in A} G_\alpha : x_\alpha \in K_\alpha \text{ for almost all } \alpha \in A\}.$$

Then, equipped with the product topology, G is a locally compact group which is locally nilpotent, but not nilpotent whenever the set of numbers $\ell(G_\alpha)$ is unbounded.

REMARK 7.6.8. In the situation of Theorem 7.6.7 condition (iii) is equivalent to G_0 being contained in the centre of G. In fact, if G is the projective limit of groups G/C_α and each G/C_α has an open centre, then $G_0C_\alpha/C_\alpha = (G/C_\alpha)_0 \subseteq Z(G/C_\alpha)$ for all α, and this implies that $G_0 \subseteq Z(G)$. Conversely, if $G_0 \subseteq Z(G)$ and G is a projective limit of Lie groups G/C_α, then $(G/C_\alpha)_0 = G_0C_\alpha/C_\alpha \subseteq Z(G/C_\alpha)$ and $(G/C_\alpha)_0$ is open in G/C_α for every α, so that (iii) holds.

COROLLARY 7.6.9. *Let G be a locally nilpotent locally compact group. Then G has the separation property for cyclic subgroups if and only if G_0 is contained in the centre of G.*

PROOF. If G_0 is contained in the centre of G, then by Theorem 7.6.7 and Remark 7.6.8, every compactly generated open subgroup of G has the separation property for cyclic subgroups. Hence the same is true of G.

Conversely, suppose that G has the separation property for cyclic subgroups and let a be an arbitrary element of G. Choose a compactly generated open subgroup H of G containing a. Then H is nilpotent and has the separation property for cyclic subgroups. Therefore, Theorem 7.6.7 applies to H and, using Remark 7.6.8 again, we conclude that $G_0 = H_0$ is contained in the centre of H. Thus $[G_0, a] = \{e\}$, and since $a \in G$ was arbitrary, it follows that G_0 is contained in the centre of G. □

7.7. The separation property: Almost connected groups

In this section we consider a class of locally compact groups for which the results are considerably more complete than for nilpotent groups. The first main achievement is Theorem 7.7.2 below which, for a locally compact group having an almost connected open normal subgroup, identifies the separating subgroups as precisely the neutral subgroups.

PROPOSITION 7.7.1. *Let G be a locally compact group containing an almost connected open normal subgroup N. Let H be a closed subgroup of G such that $G = HN$ and G has the H-separation property. Then there exists a compact normal subgroup C of G such that $C \subseteq H$ and G/C is a projective limit of Lie groups.*

PROOF. Note first that N, being almost connected, possesses a maximal compact normal subgroup K. Then N/K is a Lie group, and since K is characteristic in N and N is normal in G, K is normal in G.

For any $\tau \in \widehat{K}$, let χ_τ denote the normalized trace of τ, that is, the function $\chi_\tau(x) = d_\tau^{-1}\mathrm{tr}(\tau(x))$, $x \in G$. Here d_τ denotes the dimension of $\mathcal{H}(\tau)$ and, for any operator T in a Hilbert space, $\mathrm{tr}(T)$ denotes the trace of T. Let $E(K) = \{\chi_\tau : \tau \in \widehat{K}\} \subseteq L^1(K)$. If π is a unitary representation of G, then $\mathcal{H}(\pi) = \mathcal{H}(\pi|_K)$ admits an orthogonal decomposition

$$\mathcal{H}(\pi) = \bigoplus\{\pi(\chi)\mathcal{H}(\pi) : \chi \in E(K)\}.$$

Consider $\xi \in \mathcal{H}(\pi)$ satisfying $\pi(h)\xi = \xi$ for all $h \in H$. Then, for each $\chi \in E(K)$,

$$\begin{aligned}\pi(h)\pi(\chi)\xi &= \pi(h)\pi(\chi)\pi(h^{-1})\xi = \int_K \chi(x)\pi(hxh^{-1})\xi\, dx \\ &= \int_K \chi(h^{-1}xh)\pi(x)\xi\, dx = \pi(h\cdot\chi)\xi\end{aligned}$$

and hence $\|\pi(\chi)\xi\| = \|\pi(h\cdot\chi)\xi\|$ for all $h \in H$. This implies that ξ belongs to the closed linear span of those subspaces $\pi(\chi)\mathcal{H}(\pi)$ for which the H-orbit $H\cdot\chi$ of χ is finite. Now, let

$$C = \bigcap\nolimits_{x\in G} x(H\cap K)x^{-1} = K \cap \left(\bigcap\nolimits_{x\in G} xHx^{-1}\right),$$

and consider any $y \in K\backslash C$. Then there exists $x \in G$ such that $z = xyx^{-1} \notin H$. Since G has the H-separation property, there exist a unitary representation π of G and $\xi \in \mathcal{H}(\pi)$ such that $\pi(z)\xi \neq \xi$ and $\pi(h)\xi = \xi$ for all $h \in H$. By what we have observed above about such ξ, we can replace ξ by $\sum_{\chi'\in H\cdot\chi}\pi(\chi')\xi$, where $\chi \in E(K)$ is such that $H\cdot\chi$ is finite. Notice next that if a connected group acts on K, then the orbits in the discrete dual space $\widehat{K}$ are trivial. Since N is almost connected, it follo ws that the N-orbits in $\widehat{K}$ and hence the N-orbits in $E(K)$ are all finite. Now, by hypothesis, $G = HN = NH$. This implies that the G-orbit $G\cdot\chi$ of χ is finite.

Let $\tau \in \widehat{K}$ such that $\chi = \chi_\tau$, and define a closed subgroup L of K by

$$L = \{k \in K : x\cdot\tau(k) = I \text{ for all } x \in G\}.$$

Then L is normal in G, and L is the kernel of σ, the direct sum of the representations in $G\cdot\tau$. Since $G\cdot\tau$ is finite, σ defines a faithful finite dimensional representation of K/L. In particular, K/L is a Lie group, and since N/K is a Lie group and N is open in G, it follows that G/L is a Lie group.

Summarizing, for each $y \in K\backslash C$ we have found a closed normal subgroup L_y of G such that $y \notin L_y$, $L_y \subseteq K$ and G/L_y is a Lie group. Since then $L_y \cap V_y = \varnothing$ for some neighbourhood V_y of y in K, it follows that given any compact subset Y of $K\backslash C$, there exist $y_1, \dots, y_n \in K\backslash C$ such that $Y \cap (\bigcap_{j=1}^n L_{y_j}) = \varnothing$. Since $G/\cap_{j=1}^n L_{y_j}$ is a Lie group, we conclude that given any neighbourhood V of e in G, there exists a closed normal subgroup L_V of G such that $C \subseteq L_V \subseteq VC$ and G/L_V is a Lie group. In other words, G/C is a projective limit of Lie groups. This finishes the proof of the proposition. □

Apart from results obtained in Section 6.5, the preceding proposition is the main tool needed to establish the following theorem.

THEOREM 7.7.2. *Let G be a locally compact group which has an open almost connected normal subgroup. Then, for a closed subgroup H of G, the following are equivalent.*

(i) *G has the H-separation property.*
(ii) *H is neutral in G.*

PROOF. (ii) $\Rightarrow$ (i) follows from Theorem 7.5.9

Conversely, suppose that (i) holds and let N be an open almost connected normal subgroup of G and put $\widetilde{G} = HN$. If H is shown to be neutral in $\widetilde{G}$, then it is neutral in G because $\widetilde{G}$ is open in G. Therefore we can assume that $G = HN$. Then, by Proposition 7.7.1, there exists a compact normal subgroup C of G such that $C \subseteq H$ and G/C is a projective limit of Lie groups. Of course, it suffices to show that H/C is neutral in G/C. Consequently, we can further assume that $C = \{e\}$, so that G is a projective limit of Lie groups. Using once more that G has the H-separation property, we have $H = \bigcap\{\overline{HUH} : U \in \mathcal{U}\}$, where $\mathcal{U}$ is any neighbourhood basis of e in G (Lemma 7.5.10). Since G is locally connected, Corollary 7.5.15 now shows that H is neutral in G. □

In view of Theorem 7.7.2 it is surprising that it requires much more effort to derive a criterion for an almost connected locally compact group to have the separation property. A first step towards such a criterion is the treatment of solvable connected Lie groups (Lemma 7.7.4) through reducing to small dimensional groups.

EXAMPLE 7.7.3. With the notation of [**14**, p. 180-182], let $G = G_{3,4}(\alpha)$, $\alpha \in \mathbb{R}$. Then G is a semidirect product of $\mathbb{R}^2$ with $\mathbb{R}$ and can be realized as $\mathbb{R}^3$ with multiplication given by

$$\begin{aligned}(x_1, y_1, t_1)(x_2, y_2, t_2) &= (x_1 + e^{\alpha t_1}(x_2 \cos t_1 - y_2 \sin t_1),\\ &\quad y_1 + e^{\alpha t_1}(x_2 \sin t_1 + y_2 \cos t_1), t_1 + t_2),\end{aligned}$$

$x_j, y_j, t_j \in \mathbb{R}$, $j = 1, 2$. We now give an example of an infinite cyclic subgroup H of G which is not separating.

First, let $\alpha = 0$. Then, for all $x, y, t \in \mathbb{R}$,

$$(x, 0, 0)(0, 0, t)(y, 0, 0) = (x + y \cos t, y \sin t, t).$$

Let $H = \{(m, 0, 0) : m \in \mathbb{Z}\}$ and choose $t_n \in (0, \pi/2)$ and $x_n, y_n \in \mathbb{R}$, $n \in \mathbb{N}$, such that $\sin t_n = 1/n$, $y_n = n$ and $x_n = -\lfloor n \cos t_n \rfloor$, where for any $t \in \mathbb{R}$, $\lfloor t \rfloor$ denotes the largest integer $\leqslant t$. Then, for any $\phi \in P_H(G)$,

$$\phi(0,0,t_n) = \phi(n\cos t_n - \lfloor n\cos t_n\rfloor, 1, t_n).$$

Since $t_n \to 0$ as $n \to \infty$ and some subsequence of the sequence $n \to n\cos t_n - \lfloor n\cos t_n\rfloor$ converges to some $s \in [0,1]$, it follows that $1 = \phi(s,1,0)$. Thus H is not separating.

Now let $\alpha > 0$, the case $n < 0$ being treated similarly. Let $H = \{(0,0,2\pi n) : n \in \mathbb{Z}\}$ and $\phi \in P_H(G)$. Since, for all $x \in \mathbb{R}$ and $n \in \mathbb{Z}$,

$$(0,0,2\pi n)(x,0,0)(0,0,2\pi n) = (e^{2\pi\alpha n}x, 0, 0),$$

taking $x_n = e^{-2\pi\alpha n}$, we get

$$1 = \lim_{n\to\infty} \phi(x_n,0,0) = \lim_{n\to\infty} \phi((0,0,2\pi n)(x_n,0,0)(0,0,-2\pi n)) = \phi(1,0,0).$$

This shows that G does not have the H-separation property.

LEMMA 7.7.4. *Let G be a solvable simply connected Lie group, and suppose that G has the separation property for cyclic subgroups. Then G is abelian.*

PROOF. Towards a contradiction, assume that there exists a nonabelian, simply connected, solvable Lie group which has the separation property for cyclic subgroups, and let G be such a group of minimal dimension. We claim that G contains a nonabelian, simply connected subgroup H of dimension 2 or 3.

To that end, let V be a nontrivial normal vector subgroup of G of minimal dimension, and let $q : G \to G/V$ denote the quotient homomorphism. It is well-known that V is of dimension 1 or 2 since G is solvable. Since G/V has the separation property, by the minimality of G, G/V must be abelian and hence a vector group. Now, if $\dim V = 1$, then choose noncommuting elements x and y in G and let $H = q^{-1}(\mathbb{R}q(x) + \mathbb{R}q(y))$. If $\dim V = 2$, due to the minimality of V, V cannot be contained in the centre of G. Then choose $x \in V$ and $y \in G$ such that $[x,y] \neq e$ and let $H = q^{-1}(\mathbb{R}q(y))$.

Thus it suffices to show that no nonabelian, simply connected, solvable Lie group G of dimension $\leqslant 3$ does have the separation property for cyclic subgroups. Now, the noncommutative, solvable, real Lie algebras $\mathfrak{g}$ of dimension $\leqslant 3$ are classified in [**14**, p. 180-182]. Retaining the notation of [**14**], any such $\mathfrak{g}$, except for the Heisenberg Lie algebra and the algebras $\mathfrak{g}_{3,4}(\alpha)$, $\alpha \in \mathbb{R}$, contain the Lie algebra of the $ax+b$-group as a subalgebra. Since the separation property for cyclic subgroups is inherited by closed subgroups, it therefore remains to observe that none of the Heisenberg group, the $ax+b$-group and $G_{3,4}(\alpha)$, the simply connected Lie group corresponding to $\mathfrak{g}_{3,4}(\alpha)$, does have the separation property for cyclic subgroups. However, in the first two cases the only separating subgroups are the normal ones (Examples 7.5.3 and 7.5.6), whereas for $G_{3,4}(\alpha)$ we have exhibited an example of a nonseparating subgroup in the preceding example. □

The next step in our approach to Theorem 7.7.10 below consists of generalizing Lemma 7.7.4 to the extent that an arbitrary connected Lie group, which has the separation property for cyclic subgroups, has to be the direct product of a vector group and a compact group (Proposition 7.7.9).

LEMMA 7.7.5. *Let G be a connected Lie group and let R be the radical of G. If G has the separation property for cyclic subgroups, then G/R is compact.*

PROOF. Suppose that the semisimple Lie group G/R is noncompact. Then G/R has an Iwasawa decomposition $G = KAN$, where K is a compact group, A is a nontrivial vector group and N is nontrivial connected nilpotent group. The group A normalizes, but does not centralize N. Then, with $q : G \to G/R$ the quotient homomorphism, $q^{-1}(AN)$ is a connected solvable Lie group which has the separation property for cyclic subgroups. It follows from Lemma 7.7.4 that $q^{-1}(AN)$ is abelian. In particular, AN is abelian, which is a contradiction. Thus G/R is compact. □

LEMMA 7.7.6. *Let G be a semidirect product $G = V \rtimes A$, where V is a vector group and A is a connected and locally connected group. If G has the separation property for cyclic subgroups, then G is the direct product of the subgroups V and A.*

PROOF. Let $a \to \alpha_a$ denote the homomorphism from A into $GL(V)$ defining this semidirect product, and let e and e_A denote the identity of G and A, respectively.

We have to show that $\alpha_a(y) = y$ for all $y \in V$ and $a \in A$. Suppose that there exist $y \in V$ and $a \in A$ such that $\alpha_a(y) \neq y$, and let $H = \{(ny, e_A) : n \in \mathbb{Z}\}$. Since G has the H-separation property, Lemma 7.5.10 implies that $H = \cap\{\overline{HUH} : U \in \mathcal{U}\}$, where $\mathcal{U}$ is any neighbourhood basis of e in G. This in turn implies that H is neutral in G because G is locally connected (Lemma 7.5.11). Since H is discrete and G is locally connected, we find a symmetric, compact, connected neighbourhood W of e such that $HW = WH$ and $W^2 \cap H = \{e\}$. The latter condition and $W = W^{-1}$ imply that the sets Wh, $h \in H$, are pairwise disjoint. Now the compact set $(y, e_A)W$ is a finite union of disjoint closed sets of the form $(y, e_A)W \cap Wh$, $h \in H$. Since W is connected, it follows that $(y, e_A)W \subseteq W(y, e_A)^m$ for some $m \in \mathbb{Z}$. But $(y, e_A) = v(y, e_A)^m$ for some $v \in W$ and $m \in \mathbb{Z}$ with $m \neq 1$ implies that $H \cap W \neq \{e\}$. This contradiction shows that $(y, e_A)W \subseteq W(y, e_A)$. The same argument gives $W(y, e_A) \subseteq (y, e_A)W$. We now have, for all $x \in V$ and $a \in A$ such that $(0, a) \in W$,

$$(x - \alpha_a(x), a) = (x, e_A)(0, a)(x, e_A)^{-1} \in W.$$

There exists $a \in A$ such that $(0, a) \in W$ and $\alpha_a(y) \neq y$. Indeed, otherwise we have $\alpha_a(y) = y$ for all $a \in A$ since A is connected. Since every continuous automorphism of a vector group is linear, we conclude that

$$(n(y - \alpha_a(y)), a) = (ny - \alpha_a(ny), a) \in W$$

for all $n \in \mathbb{N}$. This is impossible since W is compact. Consequently $\alpha_a(y) = y$ for all $a \in A$ and $y \in V$ and hence G is the direct product of V and A. □

COROLLARY 7.7.7. *Let G be a locally compact group containing a closed normal vector subgroup V such that G/V is a compact connected Lie group. If G has the separation property for cyclic subgroups, then there exists a compact subgroup K of G such that G is the direct product of V and K.*

PROOF. Since V is a vector group and G/V is compact, there exists a compact subgroup K of G such that G is the semidirect product of V and K [**128**, Theorem VIII]. Now apply Lemma 7.7.6. □

In the proof of the next proposition we shall use the statement of the following lemma. This appears to be folklore. However, being unaware of a reference, we include the proof for the readers convenience.

LEMMA 7.7.8. *Let N be a compact and solvable normal subgroup of a connected locally compact group G. Then N is contained in the centre $Z(G)$ of G.*

PROOF. Assume first that N is abelian, and let $(x, \chi) \to x \cdot \chi$ denote the continuous action of G on the dual group $\widehat{N}$ of N given by $x \cdot \chi(s) = \chi(x^{-1}sx)$ for $\chi \in \widehat{N}$, $x \in G$ and $y \in N$. Then, since G is connected and $\widehat{N}$ is discrete, $x \cdot \chi = \chi$ for all χ and x. Since the characters of N separate the points of N, it follows that $N \subseteq Z(G)$.

The statement of the lemma now follows by induction on the length $\ell(N)$ of the descending commutator series of N. Suppose that every solvable compact normal subgroup H of G with $\ell(H) < \ell(N)$ is contained in the centre of G. Then take $H = [N, N]$, the closed commutator subgroup of N. Then $[N, N] \subseteq Z(G)$ by what we have already seen. To conclude that $N \subseteq Z(G)$, assume that there exists some $x \in N \backslash Z(G)$ and let M denote the closed subgroup of G generated by x and N. Then M is abelian, and also normal in G because $N/[N, N]$ is an abelian compact normal subgroup of the connected group $G/[N, N]$. Applying the first paragraph again, we get $M \subseteq Z(G)$. This contradicts $x \notin Z(G)$ and completes the proof. □

PROPOSITION 7.7.9. *Let G be a connected Lie group which has the separation property for cyclic subgroups. Then G is the direct product of a vector group and a compact group.*

PROOF. To start with, suppose that G is solvable and let C be the maximal compact connected normal subgroup of G. Then G/C is simply connected and has the separation property for cyclic subgroups by Lemma 7.5.13. Lemma 7.7.8 implies that G/C is abelian and hence a vector group. Moreover, since C is a compact normal solvable subgroup of the connected Lie group G, C is contained in the centre of G and $C = \mathbb{T}^d$ for some $d \in \mathbb{N}_0$. Now Corollary 7.6.5 applies and yields that G is a direct product $G = W \times \mathbb{T}^d$, where W is a vector group.

Now let G be an arbitrary connected Lie group having the separation property for cyclic subgroups, and let R be the radical of G. Then, by the first paragraph, $R = W \times \mathbb{T}^d$ as above. By Lemma 7.7.5, G/R is compact.

Finally, consider the action of G on R through inner automorphisms. Since $\mathbb{T}^d$ is characteristic in R, it is normal in G. Since G/R is connected, the induced action of G on the discrete dual group of $\mathbb{T}^d$ is trivial. Thus $\mathbb{T}^d$ is contained in the centre of G. Since G/R is compact, the stable decomposition theorem [**110**, Theorem 1.1] asserts the existence of a topological automorphism γ of R such that $V = \gamma(W)$ is normal in G. Consequently, V is a normal vector subgroup of G such that the quotient group G/V, which is an extension of G/R by $\mathbb{T}^d$, is compact and connected. An application of Corollary 7.7.7 gives that $G = V \times K$, where K is a compact group. This completes the proof of the proposition. □

Combining the preceding proposition with results obtained in Section 7.5, we are now ready to deduce the second main result of this section.

THEOREM 7.7.10. *Let G be an almost connected locally compact group. Then the following four conditions are equivalent.*

(i) *G has the separation property.*
(ii) *G has the separation property for cyclic subgroups.*
(iii) *G contains an open normal subgroup N of finite index such that N is a direct product of a compact group and a vector group.*
(iv) *G is an* SIN*-group.*

PROOF. The implication (i) $\Rightarrow$ (ii) is trivial and (iv) $\Rightarrow$ (i) was shown in Theorem 7.5.9. Notice next that if G is as in (iii) (and not necessarily almost connected), then G is a projective limit of SIN-groups and hence an SIN-group itself. It therefore only remains to prove that (ii) implies (iii). Since G is almost connected, it is a projective limit of Lie groups $G_\alpha = G/C_\alpha$ [**214**, Theorem 4.6]. Then each G_α has the separation property for cyclic subgroups because G_α is locally connected (Lemma 7.5.13). By Proposition 7.7.8, the connected component of the identity of G_α, which has finite index in G_α, is a direct product of a compact group and a vector group. In particular, each G_α is an SIN-group, and hence so is G. Thus, by Theorem 2.13 of [**110**], G has an open normal subgroup N such that N is the direct product of a vector group and a compact group. Finally, since G is almost connected, N must have finite index in G. Consequently, G is an SIN-group. □

We have seen in Theorem 7.7.2 that if G is a locally compact group having an almost connected open normal subgroup, then the separating subgroups of G are precisely the neutral subgroups. It is therefore tempting to try to establish Theorem 7.7.10 by showing that an almost connected locally compact group G has to be an SIN-group provided that every closed cyclic subgroup of G is neutral in G. However, there appears to be no more direct and simpler proof than the one presented above.

EXAMPLE 7.7.11. Let G be a simple Lie group with finite centre. Veech [**279**, Theorem 1.4] has shown that $W(G) = \mathbb{C} + C_0(G)$, where $W(G)$ denotes the space of all weakly almost periodic functions on G. Therefore, Lemma 7.5.4 applies with $N = G$ and shows that the compact subgroups are the only separating subgroups of G. We remark that, alternatively, the fact that $B(G) \subseteq \mathbb{C} + C_0(G)$ could be derived from vanishing at infinity theorems for matrix coefficients of unitary representations of G (see [**294**, Sections 2.2 and 2.4]).

7.8. Notes and references

Let G be a locally compact and let H be a closed subgroup of G. As observed in Section 7.1, it is easy to see that H is extending in G if H is either open, or compact, or contained in the centre of G. The last two cases are covered by Theorem 7.3.2, which states that H is extending whenever G has small H-invariant neighbourhoods and which was independently shown, using induced representations, by Henrichs [**118**] and Cowling and Rodway [**39**]. Of course, this result was raising the question of whether conversely a locally compact group, which has the extension property, must be an SIN-group. That this is true for connected groups was shown in [**39**] and can also be deduced directly from [**120**, Theorem 2]. Our proof that a connected Lie group which has the extension property, has to be a direct product of a vector group and a compact group (Theorem 7.3.6), is taken from [**39**]. That the answer to the above question is also affirmative for compactly generated nilpotent groups (Theorem 7.4.6). However, according to Example 7.3.9, there exist locally compact groups which have the extension property, but nevertheless fail to be SIN-groups.

In connection with Theorem 7.4.6, it is worth pointing out that Example 7.3.9, which was given in [**119**], is a 2-step solvable and compactly generated group.

The problem of when a single given closed subgroup is extending was studied by several authors (see [**28**], [**39**], [**118**], [**119**], [**156**], [**157**], [**194**] and [**212**]) and naturally is a more difficult question. The first substantial result in this direction was Douady's observation that if N is an abelian closed normal subgroup of G and χ is a character of N, then χ extends to a continuous positive definite function on G only if the stabilizer of χ is open in G (Proposition 7.2.1). Theorem 7.2.3, which extends Proposition 7.2.1 considerably and which, for a normal subgroup N of G and $\varphi \in P^1(N)$, gives a necessary and sufficient condition for φ to be extendible to some element of $P(G)$, was established by Cowling and Rodway [**39**, Theorem 1]. The exposition in Section 7.2 follows closely that in [**39**]. Since, as indicated in Section 7.1, the extension property can be formulated in terms of representation theory, it is not surprising that the extension of φ constructed in the proof is realized as a coordinate function of the representation induced by π_φ. We feel that the full strength of [**39**, Theorem 1] has not yet been exploited. Theorem 7.2.4 saying that the connected component of the identity is always extending, is due to Liukkonen and Mislove [**194**], who were not primarily interested in the extension property, but used Theorem 7.2.4 in their study of symmetry of the Fourier-Stieltjes algebra. The various examples presented in Section 7.4 as well as the useful Lemmas 7.4.10 and 7.4.11 are taken from [**157**]. The same applies to several other results accumulated in Section 7.1 to 7.4.

The separation property for positive definite functions first appeared in [**177**] in connection with the invariant complementation problem for w^*-closed invariant subalgebras of $VN(G)$. The first general fact that a closed subgroup H of G is separating whenever $G \in [\mathrm{SIN}]_H$ was obtained in [**81**]. In [**81**] as well as in [**154**] the interest in the separation property arose from applications in the ideal theory of the Fourier algebra. Theorem 7.5.9, which covers the $[\mathrm{SIN}]_H$-group case and which was the first to indicate the importance of the neutral subgroup notion, is an adaptation of the $[\mathrm{SIN}]_H$-group result.

It was expected for some time that a locally compact group having the separation property, must be an SIN-group. This turns out to be true for compactly generated nilpotent groups on the one hand and almost connected groups on the other hand. Moreover, for both classes of locally compact groups the separation property for cyclic subgroups already ensures the small invariant neighbourhood property. These results, Theorem 7.6.7 and Theorem 7.7.10, were established in [**154**] and [**156**], respectively. The proofs of both theorems involve structure theory of the groups in question, some Lie group theory and the treatment of several special cases. For instance, Examples 7.5.3, 7.5.5 and 7.7.3 are used to prove Theorem 7.7.10 for solvable connected Lie groups.

However, there exists an example of a 2-step solvable, totally disconnected and compactly generated group which has the separation property, but nevertheless fails to be an SIN-group. The example was given in [**157**]. This group, which is often referred to as the Fell group, has served as a counterexample to other conjectures as well. It has, for instance, been the first example of a noncompact group which has a countable dual space.

Of course, the reader will have observed that there are many similarities between the extension and the separation properties for the classes of nilpotent locally compact groups and almost connected groups. Yet, there are marked differences as soon as single subgroups or locally compact groups which are neither nilpotent nor almost connected are considered. A striking example is provided by Douady's observation: A normal subgroup is always separating, whereas an abelian normal subgroup of a connected group is extending only if it is contained in the centre.

In [**156**] several partial results have been obtained towards when a given closed subgroup H of G is separating in G (see Theorems 1.3 and 2.4 of [**156**]). In particular, a normal subgroup of a connected nilpotent group is separating only if it is normal. The most satisfying result in this direction, however, is due to Losert [**202**, Theorem 2]. The proof presented here is a streamlined version of the one given in [**202**].

There is one aspect of the separation property which the authors of [**154**], [**156**] and [**157**] have not been aware of and which was brought to their attention through [**202**]. This is the relation with the so-called Mautner phenomenon explained in Definition 7.5.17 and Remark 7.5.18. The Mautner phenomenon describes just the opposite behaviour to the separation property. The classical case, done by Mautner [**208**], was the $ax+b$-group. A very deep study, culminating in [**216**], of the Mautner phenomenon for one-parameter subgroups of connected Lie groups was carried out by C.C. Moore. Subsequently, S.P. Wang investigated the same problem for cyclic subgroups of connected Lie groups [**283**] and for p-adic groups [**284**], with applications to Kazhdan's property (T). Moore and Wang established, using Mackey's theory of induced unitary representations, fairly explicit descriptions of the group H_{sep} (compare Definition 7.5.17) in terms of compactness conditions of the group of inner automorphisms defined by the elements of H.

As pointed out above, Theorem 7.7.2 was shown in [**202**, Theorem 2]. In addition, using the results of Moore and Wang, Losert has also shown that in the situation of 7.7.2, the neutral subgroups H of G can be built up from normal subgroups and $[\mathrm{SIN}]_H$-groups. More precisely (see (iii) of [**202**, Theorem 2]), H is neutral in G if and only if H contains a closed subgroup N which is normal in G_0H such that G/N is a $[\mathrm{SIN}]_H$-group. We felt that this characterization, the proof of which is rather complicated, is far beyond the scope of this monograph. The methods developed in [**202**] also allow to handle the examples presented in Sections 7.5 and 7.7 and to give an alternative, though definitely not shorter, approach to, e.g., Theorems 7.7.10 and 7.6.4. Moreover, [**202**] also contains, for closed subgroups H of G, a comparison of neutrality, local neutrality and the $[\mathrm{SIN}]_H$-property. Finally, very recently, Losert [**203**] has presented a description of the subgroup H_{sep} in the situation when G contains an open, almost connected, H-invariant subgroup.

We have seen in Example 7.5.16 that in general the separation property does not force a group to be an SIN-group. Heavily exploiting constructions due to Ol'shanskii [**222**], Losert has produced a very striking example demonstrating this phenomenon; he constructed a locally compact group which is not an SIN-group, but has a compact open normal subgroup (thus, in particular, is a so-called IN-group) and such that every proper closed subgroup is compact. Note that such a group also has the extension property.

We conclude these comments and historical remarks by pointing out some open problems which arise naturally from the results and discussion in this chapter.

1. Determine all the extending subgroups of the motion groups $\mathbb{R}^d \rtimes SO(d)$, $d \geqslant 2$ (compare Example 7.4.12 for $d=2$).

2. Let N be a closed normal subgroup of a connected nilpotent group G. Find necessary and sufficient conditions for N to be extending in G. A necessary condition is that every irreducible representation of N is fixed under the action of G on the dual space of N. Moreover, it follows by induction from Douady's observation that the descending central series of N is obtained by intersecting the members of the descending central series of G with N.

3. In view of Losert's example [**203**], how restrictive for a locally compact group G is the hypothesis that G satisfies both, the extension and the separation properties?

4. Nothing is known concerning both properties for nondiscrete, totally disconnected, nonabelian, noncompact groups.

5. Is, for general locally compact groups, the separation property (like the one for cyclic groups) a local property in the sense of Remark 7.5.2(2)?

6. Let G be a simple Lie group with finite centre. Then the proper separating subgroups are exactly the compact subgroups (Example 7.7.11). In contrast, it does not seem to be known which subgroups are extending.

7. If G is the 3-dimensional Heisenberg group, then a proper closed subgroup of G is extending if and only if it is either cyclic or isomorphic to $\mathbb{R}$. However, for general 2-step nilpotent simply connected Lie groups the question is open.

8. When does a not necessarily compactly generated, nilpotent group have the extension or the separation property? Does there exist a group theoretic characterization of extending or separating subgroups?

9. Let G be any locally compact group and H a closed neutral subgroup of G. Must H be separating? Of course, by Theorem 7.7.2 the answer is affirmative when G contains an open, almost connected, normal subgroup.

APPENDIX A

A.1. The closed coset ring

For any (discrete) group H, let $\mathcal{R}(H)$ denote the Boolean ring generated by the left cosets of subgroups of H. $\mathcal{R}(H)$ is called the *coset ring* of H. If G is a locally compact group and G_d denotes the group G with the discrete topology, then the *closed coset ring* $\mathcal{R}_c(G)$ is defined by

$$\mathcal{R}_c(G) = \{E \in \mathcal{R}(G_d) : E \text{ is closed in } G\}.$$

Gilbert [**98**] and Schreiber [**258**] have independently determined explicitly the structure of sets in $\mathcal{R}_c(G)$ when G is a locally compact abelian group. In [**80**], Forrest gave the description of sets in $\mathcal{R}_c(G)$ for arbitrary locally compact groups saying that Gilbert's long and complicated proof can be carried over line by line. However, a very elegant proof for abelian groups was found by Saeki [**255**]. This proof, which is based on an idea of Cohen [**34**], can easily be adapted to work for general locally compact groups and in what follows we present the details.

The first lemma is due to Cohen [**34**] for abelian groups. The generalization to arbitrary groups can be found in [**81**].

LEMMA A.1.1. *A group G cannot be the union of a finite number of left cosets of subgroups, each of which has infinite index in G.*

PROOF. Suppose that there exist subgroups $H_1, \ldots, H_n$ of G and finite subsets $F_1, \ldots, F_n$ of G such that $G = \bigcup_{i=1}^n F_i H_i$, but none of the H_i has finite index in G. Of course, we can assume that the family $\{H_1, \ldots, H_n\}$ is minimal in this respect. Then $n > 1$, and because H_n has infinite index, we find some $x \in G$ such that $xH_n \cap F_n H_n = \varnothing$. Then $xH_n \subseteq \bigcup_{i=1}^{n-1} F_i H_i$ and hence

$$F_n H_n = (F_n x^{-1}) x H_n \subseteq \bigcup_{i=1}^{n-1} (F_n x^{-1} F_i) H_i.$$

This implies that

$$G = \bigcup_{i=1}^n F_i H_i = \bigcup_{i=1}^{n-1} (F_i \cup F_n x^{-1} F_i) H_i,$$

which is impossible by the minimality of the set $\{H_1, \ldots, H_n\}$. □

For any family $\mathcal{J}$ of subsets of a group G, let $\mathcal{R}(\mathcal{J})$ denote the smallest Boolean algebra that contains $\mathcal{J}$ and is invariant under left translation. In the following proposition [**34**, Lemma p. 223], which is an essential step towards proving Theorem A.1.5 below, $f(E, \cdot)$ will denote the characteristic function of a set E.

PROPOSITION A.1.2. *Let H_i be finitely many subgroups of a group G, and for each i, let K_{ij} be finitely many left cosets of H_i. Let $c_{ij} \in \mathbb{C}$ and*

$$f(x) = \sum_{i,j} c_{ij} f(K_{ij}, x), \quad x \in G.$$

Let B_k denote the pairwise disjoint subsets of G on which f attains its finite number of values. Then there are finitely many subgroups L_i of G such that

$$\mathcal{R}(\{B_k\}_k) = \mathcal{R}(\{L_i\}_i).$$

PROOF. As in the proof of Lemma A.1.1, we can assume that for all i and j, the index $[H_i : (H_i \cap H_j)]$ is either one or infinite. Under this assumption, we shall prove that if the function

$$f_i(x) = \sum_j c_{ij} f(K_{ij}, x)$$

attains its finitely many values on sets C_k, each of which is of course a finite union of left cosets H_i, then each C_k is a finite union of left cosets of some subgroup L_i, which in turn lies in the Boolean algebra generated by the sets B_k and their left translates. Moreover, H_i has finite index in L_i.

We now proceed by induction on the number of distinct subgroups H_i which occur. Suppose first that there is only one subgroup $H_i = H$. It suffices to prove that the statement of the proposition in the case where the c_{ij} are all one. Let L be the subgroup of all elements $a \in G$ such that $f(xa) = f(x)$ for all $x \in G$. Then $L \supseteq H$ and

$$f(x) = \sum_{j=1}^{r} f(a_j L, x)$$

for certain distinct elements $a_1, \ldots, a_r$ of G. By translating f from the left, we can assume that $a_1 = e$. We prove by induction on r that L is in the Boolean algebra generated by the support of f and its left translates. The case $r = 1$ being trivial, assume that $r \geqslant 2$. The union of cosets $a_j L$ does not form a group, because otherwise this group would have to coincide with L, contradicting $r \geqslant 2$. Therefore, for some j, the function $f(a_j^{-1}x) - f(x)$ is not identically zero. But

$$f(a_j^{-1}x) - f(x) = g_1(x) - g_2(x),$$

where g_1 and g_2 are characteristi c functions of unions of less than r left cosets of L. By the induction hypothesis, it follows that the group L_1, which leaves the support of g_1 invariant under right translation is in the wanted Boolean algebra. Also, by considering $f - g_1$, the subgroup L_2 leaving it invariant is also in that Boolean algebra. Since L is the intersection of L_1 and L_2, the statement follows in this case.

Now assume that there are more than just one H_i. Then at least one of them, H_1 say, has the property that it is not contained in H_i for any $i \neq 1$. Then $[H_1 : (H_1 \cap H_i)] = \infty$ for all $i \neq 1$ and hence H_1 is not contained in the union of finitely many left cosets of H_i (Lemma A.1.1). Therefore, we find $a_1, a_2 \in H_1$ such that the three cosets $K_{2j}, a_1 K_{2j}$ and $a_2 K_{2j}$ are all distinct. Now the two functions

$$f(x) - f(a_1^{-1}x) \quad \text{and} \quad f(x) - f(a_2^{-1}x)$$

both don't involve H_1 and so the functions

$$\sum_j c_{2j}(f(K_{2j}, x) - f(a_1 K_{2j}, x) \quad \text{and} \quad \sum_j c_{2j}(f(K_{2j}, x) - f(a_a K_{2j}, x))$$

take their different values on unions of left cosets of subgroups M_2 and N_2, respectively, such that $H_2 \subseteq M_2, N_2$ and M_2 and N_2 lie in the Boolean algebra generated by the sets B_k and their left translates. Since the cosets $K_{2j}, a_1 K_{2j}$ and $a_2 K_{2j}$ are distinct, it follows th at the function $f_2 = \sum_j c_{2j} f(K_{2j}, \cdot)$ takes its different values on unions of left cosets of $L_2 = M_2 \cap N_2$. Similarly, for all $i \neq 1$. This finishes the proof of the proposition. □

Let $\mathcal{C}$ and $\mathcal{S}$ denote the family of all closed subsets and all subgroups of G, respectively.

LEMMA A.1.3. *Let $E \in \mathcal{R}(\mathcal{C})$ be a coset in G. Then E is closed in G.*

PROOF. Let $E = xH$, where H is a subgroup of G. Then $H \in \mathcal{R}(\mathcal{C})$ and it suffices to show that $H = \overline{H}$. Therefore, replacing G by $\overline{H}$, we can assume that H is dense in G. As $H \in \mathcal{R}(\mathcal{C})$, there exist $C_1, \ldots, C_n \in \mathcal{C}$ and open subsets $U_1, \ldots, U_n$ of G such that

$$H = \bigcup_{j=1}^{n} (C_j \cap U_j).$$

Since H is dense in G, at least one of the sets $\overline{C_j \cap U_j}$ has nonempty interior. Fix such j and let $U \neq \varnothing$ be open and contained in $\overline{C_j \cap U_j}$. Then $U \cap (C_j \cap U_j) \neq \varnothing$ and hence

$$\varnothing \neq U \cap U_j \subseteq C_j \cap U_j \subseteq H.$$

It follows that H is open and hence closed. □

COROLLARY A.1.4. *$\mathcal{R}(\mathcal{C} \cap \mathcal{S}) = \mathcal{R}(\mathcal{C}) \cap \mathcal{R}(G_d)$.*

PROOF. It is clear that $\mathcal{R}(\mathcal{C} \cap \mathcal{S}) \subseteq \mathcal{R}(\mathcal{C}) \cap \mathcal{R}(G_d)$. For the reverse inclusion, it suffices to show that every closed set E in $\mathcal{R}(G_d)$ belongs to $\mathcal{R}(\mathcal{C} \cap \mathcal{S})$. By Lemma A.1.1, there are finitely many subgroups $H_1, \ldots, H_n$ of G such that $\mathcal{R}(\{E\}) \subseteq \mathcal{R}(\{H_1, \ldots, H_n\})$. Since $E \in \mathcal{R}(\mathcal{C})$, it follows that $H_i \in \mathcal{R}(\mathcal{C})$, $1 \leqslant i \leqslant n$. Now, by Lemma A.1.3, every H_i is closed in G and hence $E \in \mathcal{R}(\{H_1, \ldots, H_n\}) \subseteq \mathcal{R}(\mathcal{C} \cap \mathcal{S})$. □

THEOREM A.1.5. *Let G be any locally compact group and E a subset of G. Then E belongs to $\mathcal{R}_c(G)$ if and only if E is of the form*

$$E = \bigcup_{i=1}^{n} \left(E_i \setminus \bigcup_{j=1}^{m_i} E_{ij} \right), \tag{A.1}$$

where E_i, $1 \leqslant i \leqslant n$ is a left coset of some closed subgroup H_i of G and, for each $1 \leqslant i \leqslant n$ and $1 \leqslant j \leqslant m_i$, E_{ij} is either empty or a left coset of some open subgroup of H_i.

PROOF. It is clear that each such set belongs to $\mathcal{R}_c(G)$. Conversely, let $E \in \mathcal{R}_c(G_d)$; then $E \in \mathcal{R}(\mathcal{C}) \cap \mathcal{R}(G_d)$ by Corollary A.1.4 and hence

$$E = \bigcup_{i=1}^{n} \left(E_i \setminus \bigcup_{j=1}^{m_i} E_{ij} \right), \quad n, m_i \in \mathbb{N}, 1 \leqslant i \leqslant n,$$

where the E_i and E_{ij} are closed cosets in G and $E_{ij} \subseteq E_i$ (or $E_{ij} = \varnothing$). Since E is closed,

$$E = \bigcup_{i=1}^{n} \overline{C_i \setminus \bigcup_{j=1}^{m_i} C_{ij}}.$$

We claim that if C is any closed coset and $C_1, \ldots, C_m$ are closed cosets contained in C, then

$$\overline{C \setminus \bigcup_{j=1}^{m} C_j} = C \setminus \bigcup \{C_j : 1 \leqslant j \leqslant m, C_j \text{ open in } G\}.$$

To see this, let I be the set of all indices j such that C_j is open in C (equivalently, C_j has nonempty interior relative C). Then $\bigcup_{k\notin I} C_k$ has empty interior in C and hence $\overline{C\setminus\bigcup_{k\notin I} C_k} = C$. Since $C\setminus\bigcup_{j\in I} C_j$ is open and closed in C, it follows that

$$\begin{aligned}\overline{C\setminus\bigcup_{j\in I} C_j} &= \overline{\left(C\setminus\bigcup_{j\in I} C_j\right)\cap\left(C\setminus\bigcup_{k\notin I} C_k\right)}\\ &= \left(C\setminus\bigcup_{j\in I} C_j\right)\cap\overline{\left(C\setminus\bigcup_{k\notin I} C_k\right)}\\ &= C\setminus\bigcup_{j\in I} C_j.\end{aligned}$$

This completes the proof of the theorem. □

EXAMPLE A.1.6. (1) Every coset of the integer group $\mathbb{Z}$ is of the form $n\mathbb{Z}+m$, $n, m \in \mathbb{Z}$. Thus, if $n_1, \dots, n_r, m_1, \dots, m_r \in \mathbb{Z}$ and F is any finite subset of $\mathbb{Z}$, then $E = \bigcup_{i=1}^n (n_i\mathbb{Z}+m_i)\setminus F \in \mathcal{R}(\mathbb{Z})$. Conversely, every $E \in \mathcal{R}(\mathbb{Z})$ is of this form. Indeed, this follows from the description of sets in $\mathcal{R}(G)$ and the fact that if H and K are subgroups of $\mathbb{Z}$ with $\varnothing \neq K \subseteq H$, then $K = mH$ for some $m \in \mathbb{N}$ and hence $H\setminus K = \bigcup_{k=1}^m (k+K)$.

(2) Every nonempty set $E \in \mathcal{R}_c(\mathbb{R})$ is of the form

$$E = \bigcup_{i=1}^n \left(a_i + \left(H_i\setminus\bigcup_{j=1}^{m_i}(b_{ij}+K_{ij})\right)\right), \quad a_i, b_{ij} \in \mathbb{R},$$

where H_i and K_{ij} are closed subgroups of G and K_{ij} is open in H_i. To see this, note first that if one of the subgroups H_i equals $\mathbb{R}$, then $E = \mathbb{R}$. Since every proper closed subgroup of $\mathbb{R}$ is of the form $\alpha\mathbb{Z}$, $\alpha \in \mathbb{R}$, using the description of $\mathcal{R}(\mathbb{Z})$ in (1), we conclude that the sets $E \in \mathcal{R}_c(\mathbb{R})$ are precisely the sets of the form

$$E = \bigcup_{i=1}^n (\alpha_i\mathbb{Z}+\beta_i)\setminus F, \quad \alpha_i, \beta_i \in \mathbb{R},$$

where F is a finite (possibly empty) subset of $\mathbb{R}$.

(3) Since every proper closed subgroup of the circle group $\mathbb{T}$ is finite, $\mathcal{R}_c(\mathbb{T})$ consists of $\mathbb{T}$ and all the finite subsets of $\mathbb{T}$.

A.2. Amenability and weak amenability of Banach algebras

If A is a Banach algebra, then the *diagonal operator* is defined through

$$\Delta_A : A\widehat{\otimes} A \to A, \quad a\otimes b \to ab.$$

$A\widehat{\otimes} A$ becomes a Banach A-bimodule through

$$a\cdot(b\otimes c) = ab\otimes c \quad \text{and} \quad (b\otimes c)\cdot a = b\otimes ca, \quad a, b, c \in A.$$

DEFINITION A.2.1. A bounded net $(u_\alpha)_\alpha$ in $\Delta_A : A\widehat{\otimes} A$ is called an *approximate diagonal* for A if, for all $a \in A$,

$$a\cdot u_\alpha - u_\alpha\cdot a \to 0 \quad \text{and} \quad a\Delta_A(u_\alpha) \to a.$$

Let A be a Banach algebra, and let E be a Banach A-bimodule. A *derivation* $D : A \to E$ is a bounded linear map satisfying

$$D(ab) = a\cdot D(b) + D(a)\cdot b, a, b \in A.$$

A derivation is called *inner* if there exists $x \in E$ such that

$$D(a) = a\cdot x - x\cdot a$$

for all $a \in A$. The dual space E^* is a Banach A-bimodule through

$$\langle x, a\cdot\varphi\rangle = \langle x\cdot a, \varphi\rangle \quad \text{and} \quad \langle x, \varphi\cdot a\rangle = \langle a\cdot x, \varphi\rangle$$

for $a \in A$, $x \in E$ and $\varphi \in E^*$. Modules of this kind are called dual Banach A-bimodules.

DEFINITION A.2.2. A is called *amenable* if every derivation from A into a dual Banach A-bimodule is inner.

THEOREM A.2.3. *Let G be a locally compact group. Then $L^1(G)$ is amenable if and only if G is amenable.*

LEMMA A.2.4. *Let A be a Banach algebra with bounded approximate identity. and let E be a Banach A-bimodule such that $A \cdot E = \{0\}$. Then every bounded derivation $D : A \to E^*$ is inner.*

PROOF. Note that $E^* \cdot A = \{0\}$ since $\langle \varphi \cdot a, u \rangle = \langle \varphi, a \cdot u \rangle$ for $\varphi \in E^*$, $a \in A$ and $u \in E$. Let $(e_\alpha)_\alpha$ be a bounded approximate identity for A, and let $\varphi \in E^*$ be a w^*-cluster point of the net $(D(e_\alpha))_\alpha$. Passing to a subnet if necessary, we may assume that $\varphi = w^*\text{-}\lim_\alpha D(e_\alpha)$. Then, for any $a \in A$,

$$D(a) = \lim_\alpha D(ae_\alpha) = lim_\alpha a \cdot D(e_\alpha) = a \cdot \varphi = a \cdot \varphi - \varphi \cdot a.$$

Hence D is an inner derivation. □

DEFINITION A.2.5. Let A be a Banach algebra. A Banach A-bimodule E is called *pseudo-unital* if $E = \{a \cdot u \cdot b : a, b \in A, u \in E\}$.

LEMMA A.2.6. *Let A be a Banach algebra with bounded approximate identity. If for each pseudo-unital Banach A-bimodule E every bounded derivation from A into E^* is inner, then A is amenable.*

PROOF. Let E be an arbitrary Banach A-bimodule and $D : A \to E^*$ a bounded derivation. Let

$$E_0 = \{a \cdot u \cdot b : a, b \in A, u \in E\}.$$

By Cohen-Hewitt factorization theorem, E_0 is a closed A-submodule of E. Let $r : E^* \to E_0^*$ denote the restriction map $\varphi \to \varphi|_{E_0}$. Then r is a module homomorphism and hence $r \circ D$ is a bounded derivation of A into E_0^*. By hypothesis, there exists $\varphi_0 \in E_0^*$ such that $(r \circ D)(a) = a \cdot \varphi_0 - \varphi_0 \cdot a$ for all $a \in A$. It is easily checked that then $\widetilde{D} = D - \mathrm{ad}\, \varphi_0$ defines a bounded derivation from A into $E_0^\perp \subseteq E^*$. □

PROPOSITION A.2.7. *Let A be an amenable Banach algebra, B a Banach algebra and $\phi : A \to B$ a continuous homomorphism with dense range. Then B is amenable. In particular, if I is a closed ideal of A, then A/I is amenable.*

THEOREM A.2.8. *Let A be an amenable Banach algebra and I a closed ideal of A. Then the following are equivalent.*

(i) *I is amenable.*
(ii) *I has a bounded approximate identity.*
(iii) *I is weakly complemented.*

PROPOSITION A.2.9. *Let I be a closed ideal of the Banach algebra A. If both I and A/I are amenable, then A is amenable.*

PROPOSITION A.2.10. *Let A and B be commutative Banach algebras and $\phi : A \to B$ a continuous homomorphism with dense range. If A is weakly amenable, then so is B.*

If $\varphi \in A^*$, the map $\Delta_\varphi : a \mapsto a \cdot \varphi - \varphi \cdot a$ is a derivation. Derivations Δ_φ are called *inner*. A Banach algebra A is *weakly amenable* if every continuous derivation from A to A^* is inner. As shown by B. Johnson [**142**], the group algebra $L^1(G)$ of a locally compact graph is always weakly amenable (see [**58**] for a beautiful short proof).

PROPOSITION A.2.11. *Let A be a weakly amenable Banach algebra and I a closed ideal of A. Then I is weakly amenable if and only if I equals the closed linear span of the set $\{ab : a, b \in I\}$.*

The following theorem was shown in [**108**, Theorem 3.2].

THEOREM A.2.12. *Let A be a commutative Banach algebra and let $m : A_e \hat{\otimes} A_e \to A_e$ be the multiplication map. Then the following are equivalent.*

(i) *A is weakly amenable.*
(ii) $\ker m = \overline{\operatorname{span}(\ker m)^2}$.

If A has a bounded approximate identity, then these are equivalent to

(iii) $\ker(m|_{A \hat{\otimes} A}) = \overline{\operatorname{span}(\ker(m|_{A \hat{\otimes} A})^2)}$.

THEOREM A.2.13. *(i) Let G be a locally compact group. Then $L^1(G)$ is weakly amenable.*

(ii) Every C^*-algebra is weakly amenable.

A.3. Operator spaces

Let H be a Hilbert space and $\mathcal{B}(H)$ the C^*-algebra of all bounded linear operators on H. A *(concrete) operator space* is a subspace of $\mathcal{B}(H)$. For $n \in \mathbb{N}$, let E be a linear subspace of $B(H)$. We define

$$M_n(E) = \{[x_{ij}] = (x_{ij})_{i,j=1}^n : x_{ij} \in E\},$$

the space of $n \times n$-matrices with entries in E. Let $\ell_2^n(H)$ denote the Hilbert space direct sum of n copies of H. Then there is a natural identification of $M_n(\mathcal{B}(H))$ with $\mathcal{B}(\ell_2^n(H))$ given by $[T_{ij}] \to T$, where T is defined to be

$$\left(T\left(\sum_{j=1}^n \xi_j\right)\right)_i = \sum_{j=1}^n T_{ij}\xi_j, \quad 1 \leqslant i \leqslant n,$$

for $\xi_1, \ldots, \xi_n \in H$. Equip $M_n(\mathcal{B}(H))$ and its subspace $M_n(E)$ with the norm induced from $\mathcal{B}(\ell_2^n(H))$, denoted $\|\cdot\|_n$.

Let H and K be Hilbert spaces and E and F be closed subspaces of $\mathcal{B}(H)$ and $\mathcal{B}(K)$, respectively. Given a linear map $\phi : E \to F$, the nth *amplification*, ϕ_n, $n \in \mathbb{N}$, of ϕ is the map $\phi_n : M_n(E) \to M_n(F)$ defined by $\phi_n[x_{ij}] = [\phi(x_{ij})]$, $[x_{ij}] \in M_e(E)$. Then ϕ is said to be *completely contractive* if $\|\phi\| \leqslant 1$. Moreover, a linear map $\phi : V \to W$ is called *completely positive* if $\phi_n \geqslant 0$ for all $n \in \mathbb{N}$. Let $\|\phi_n\|$ denote the norm of ϕ_n. Then the map ϕ is called *completely bounded* if $\sup\{\|\phi_n\| : n \in \mathbb{N}\} < \infty$. In this case the *completely bounded norm* $\|\phi\|_{cb}$ of ϕ is defined to be $\|\phi\|_{cb} = \sup\{\|\phi_n\| : n \in \mathbb{N}\}$. Let $CB(E, F)$ denote the set of all completely bounded maps from E into F. Then $(CB(E, F), \|\cdot\|_{cb})$ is a normed linear space, and $CB(E, F)$ is complete whenever F is a Banach space.

THEOREM A.3.1. *Let H be a Hilbert space and let X be a subspace of $\mathcal{B}(H)$ and $n, m \in \mathbb{N}$.*

(i) *For each* $A \in M_n(X)$ *and* $B \in M_m(X)$,

$$\left\| \begin{bmatrix} A & 0 \\ 0 & B \end{bmatrix} \right\|_{n+m} = \max\{\|A\|_n, \|B\|_m\}.$$

(ii) *For* $A \in M_m(X)$, $[a_{ij}] \in M_{n,m}(\mathbb{C})$ *and* $[b_{ij}] \in M_{m,n}(\mathbb{C})$,

$$\|[a_{ij}]A\,[b_{ij}]\|_n \leqslant \|[a_{ij}]\| \cdot \|A\|_m \cdot \|[b_{ij}]\|.$$

Let E be an arbitrary vector space over $\mathbb{C}$. A *matricial norm* on E is a sequence $(\|\cdot\|_n)_{n\in\mathbb{N}}$, where $\|\cdot\|_n$ is a norm on $M_n(E)$ such that conditions (i) and (ii) of Theorem A.3.1 hold for all $n, m \in \mathbb{N}$. A linear space equipped with a matricial norm is called an *(abstract) operator space.* The following theorem, known as *Ruan's representation theorem*, shows that every abstract operator space is actually isomorphic in the appropriate sense to a concrete operator space.

THEOREM A.3.2. *Let* V *be an abstract operator space. Then there exist a Hilbert space* H*, a concrete operator space* $E \subseteq \mathcal{B}(H)$ *and a complete isometry* ϕ *of* V *onto* E*, that is, for each* $n \in \mathbb{N}$*, the map* $\phi_n : M_n(V) \to M_n(E)$*, defined by* $\phi_n[x_{ij}] = [\phi(x_{ij})]$*, is an isometry.*

LEMMA A.3.3. *Let* E *be an operator space and* F *a closed subspace of* E*. Then* F *and* E/F *are also operator spaces.*

PROPOSITION A.3.4. *Let* E *be an operator space and* $F : E \to M_n(\mathbb{C})$*,* $n \in \mathbb{N}$*, a bounded linear map. Then* F *is completely bounded and* $\|F\|_{cb} = \|F\|$*. In particular, every* $f \in E^*$ *is completely bounded and* $\|f\|_{cb} = \|f\|$*.*

The following theorem, which is one of the fundamental results in the theory of operator spaces, is usually referred to as the *Arveson-Wittstock extension theorem.*

THEOREM A.3.5. *Let* V *be a subspace of an operator space* W*, and let* H *be a Hilbert space. Then every complete contraction* $\Phi : V \to \mathcal{B}(H)$ *admits a completely contractive extension* $\widetilde{\Phi} : W \to \mathcal{B}(H)$*.*

Given an operator space E, we can now define the operator space dual E^* of E. Each $f = [f_{ij}] \in M_n(E^*)$ determines a linear map $E \to M_n(\mathbb{C})$, $x \to [\langle f_{ij}, x\rangle]$. This gives a linear isomorphism $M_n(E^*) \cong CB(E, M_n(\mathbb{C})$, which allows us to transfer the completely bounded norm on $CB(E, M_n(\mathbb{C}))$ to $M_n(E^*)$. Thus, using Proposition A.3.4, for any $f \in M_n(E^*)$,

$$\begin{aligned} \|f\|_n &= \|f\|_{cb} = \|f_n\| \\ &= \sup\{|\langle f_n, x\rangle| : x \in M_n(E), \|x\| \leqslant 1\}. \end{aligned}$$

We call $(E^*, \|\cdot\|_n$ as defined above the *operator space dual* of E.

Let V and W be operator spaces. An element u of $M_n(V \otimes W)$, the space of $n \times n$-matrices with entries in the tensor product of V and W, has a representation $u = \alpha(v \otimes w)\beta$, where $v \in M_p(V)$, $w \in M_q(W)$, $\alpha \in M_{n,p\times q}$ and $\beta \in M_{p\times q,n}$, $p, q \in \mathbb{N}$. We define

$$\|u\|_n = \inf\{\|\alpha\| \cdot \|v\| \cdot \|w\| \cdot \|\beta\| : u = \alpha(v \otimes w)\beta\},$$

where the infimum is taken over all such representations of u.

The assignment $\|\cdot\| : n \to \|\cdot\|_n$ defines an operator space matrix norm on $V \otimes W$, and the completion of $V \otimes W$ with respect to $\|\cdot\|$ is called the *operator space projective tensor product* of V and W and denoted $V \widehat{\otimes}_{op} W$.

Proposition A.3.6. *For any operator space E, the canonical embedding $E \to E^{**}$ is a complete isometry.*

More generally, if E and F are operator spaces, then $CB(E,F)$ carries the operator space structure given by the identification

$$M_n(CB(E,F)) \cong CB(E, M_n(F)), \quad [\phi_{ij}] \to (x \to [\phi_{ij}(x)],$$

$\phi_{ij} \in CB(E,F)$, $1 \leqslant i,j \leqslant n$, $x \in E$.

Proposition A.3.7. *Given operator spaces E and F and a completely bounded linear map $\phi : E \to F$, the dual map $\phi^* : F^* \to E^*$ is also completely bounded and satisfies $\|(\phi^*)_n\| = \|\phi_n\|$ for all $n \in \mathbb{N}$. In particular, $\|\phi^*\|_{cb} = \|\phi\|_{cb}$.*

We write $E \cong F$ if there exists a linear complete isometry from E onto F.

The following proposition was shown in [**27**, Lemma 1.5].

Proposition A.3.8. *Let $\mathcal{M}$ and $\mathcal{N}$ be von Neumann algebras, and let $\phi : \mathcal{M} \to \mathcal{M}$ be a σ-weakly continuous, completely bounded map. Then there exists a (unique) σ-weakly continuous map $\widetilde{\phi} : \mathcal{M}\overline{\otimes}\mathcal{N} \to \mathcal{M}\overline{\otimes}\mathcal{N}$ such that*

$$\widetilde{\phi}(x \otimes y) = \phi(x) \otimes y, \quad x \in \mathcal{M}, y \in \mathcal{N},$$

where $\mathcal{M}\overline{\otimes}\mathcal{N}$ is the usual von Neumann tensor product of $\mathcal{M}$ and $\mathcal{N}$.

Moreover, $\|\widetilde{\phi}\| \leqslant \|\phi\|_{cb}$.

A.4. Operator amenability

A Banach algebra A is called *completely contractive* if A is a (complete) operator space and the multiplication $m : A \times A \to A$ is a completely contractive mapping, i.e. m determines a completely contractive bilinear map from $A \widehat{\otimes} A$ into A. Equivalently, if $a = [a_{ij}] \in M_m(A)$ and $b \in [b_{kl}] \in M_n(A)$, $m, n \in \mathbb{N}$, then

$$\|[a_{ij}, b_{kl}]\| \leqslant \|a\| \cdot \|b\|.$$

Let A be a completely contractive Banach algebra and X an A-bimodule. Then X is called ab *operator A-bimodule* if X is also an operator space and the module operations

$$(a,x) \to a \cdot a \quad \text{and} \quad (a,x) \to x \cdot a$$

are completely bounded. Equivalently, there is a constant $c > 0$ such that

$$\|[a_{ij} \cdot x_{kl}\| \leqslant c\|a\| \cdot \|x\| \text{ and } \|[x_{kl} \cdot a_{ij}\| \leqslant c\|a\| \cdot \|x\|$$

for all $a = [a_{ij}] \in M_m(A)$ and $x = [x_{kl}] \in M_n(X)$, $m, n \in \mathbb{N}$. In this case X^* is also an operator A-bimodule with the module actions given by

$$\langle a \cdot f, x\rangle = \langle f, x \cdot a\rangle \quad \text{and} \quad \langle f \cdot a, x\rangle = \langle f, a \cdot x\rangle$$

for $a \in A, x \in X$ and $f \in X^*$. Moreover, the bimodule operations are completely bounded.

A completely bounded Banach algebra A is called *operator amenable Banach algebra* if for every operator A-bimodule X, every completely bounded derivation from A into X^* is inner.

Proposition A.4.1. *Let A be a completely contractive Banach algebra. If A is operator amenable, then A has a bounded approximate identity.*

PROOF. We first show that A has a bounded right approximate identity. Recall that the left A-module action of A on A^{**} is given by $\langle a\cdot\varphi, f\rangle = \langle\varphi, f\cdot a\rangle$ for $a \in A$, $f \in A^*$ and $\varphi \in A^{**}$. We define a right module operation of A on A^{**} by $\varphi\cdot a = 0$ for all $a \in A$ and $\varphi \in A^{**}$. Then clearly A^{**} is a dual operator A-bimodule. The canonical embedding $D : A \to A^{**}$ is a completely bounded derivation since

$$\begin{aligned}\langle D(ab), f\rangle &= \langle f, ab\rangle = \langle f\cdot a, b\rangle = \langle D(b), f\cdot a\rangle = \langle a\cdot D(b), f\rangle \\ &= \langle D(a)\cdot b, f\rangle + \langle a\cdot D(b), f\rangle\end{aligned}$$

for all $a, b \in A$ and $f \in A^*$.

Since A is operator amenable, there exists $\varphi \in A^{**}$ such that $D(a) = a\cdot\varphi - \varphi\cdot a = a\cdot\varphi$ for all $a \in A$. So φ is a right identity for A in A^{**}. By Goldstine's theorem there exists a net $(e_\alpha)_\alpha$ in A such that $\|e_\alpha\| \leqslant \|\varphi\|$ for all α and $e_\alpha \to \varphi$ in the w^*-topology $\sigma(A^{**}, A^*)$. In particular, $ae_\alpha \to a$ in the weak topology $\sigma(A, A^*)$ for each $a \in A$.

Now let $C = \{x \in A : \|x\| \leqslant \|\varphi\|\}$ and let A^A denote the product topological vector space $(A, \|\cdot\|)^A$ with the product topology. Then the weak topology on A^A is the product of the weak topologies $\sigma(A, A^*)$. Define $\phi : C \to A^A$ by

$$\phi(x)(a) = ax - a, \quad x \in C, a \in A.$$

Then $\phi(e_\alpha) \to 0$ in the weak topology of A^A. Since on bounded convex sets the weak topology coincides with the norm topology, replacing e_α by a convex combination if necessary, we may assume that $\|ae_\alpha - a\| \to 0$ for all $a \in A$. This shows that A has a bounded right approximate identity.

Similarly, by considering the usual right A-module operation of A on A^{**} and the zero left A-module operation on A^{**}, one can show that A has a bounded left approximate identity, $(f_\beta)_\beta$ say. Finally, let

$$u_{(\alpha,\beta)} = e_\alpha + f_\beta - e_\alpha f_\beta.$$

Then the $u_{(\alpha,\beta)}$ form a bounded net, and for each $a \in A$,

$$\begin{aligned}\|au_{(\alpha,\beta)} - a\| &= \|(ae_\alpha - a) + (a - ae_\alpha)f_\beta\| \\ &\leqslant \|ae_\alpha - a\|(1 + \|f_\beta\|) \to 0\end{aligned}$$

and $\|u_{(\alpha,\beta)}a - a\| \leqslant \|f_\beta a - a\| + \|e_\alpha\|\cdot\|a - f_\beta a\| \to 0$. Thus $(u_{(\alpha,\beta)})_{(\alpha,\beta)}$ is a bounded approximate identity for A. □

Let A be a Banach algebra.

(1) A bounded net $(u_\alpha)_\alpha$ in $A\widehat{\otimes}A$ is called a *bounded approximate diagonal* for A if

$$\|a\cdot u_\alpha - u_\alpha\cdot a\| \to 0 \text{ and } \|m(u_\alpha)a - a\| \to 0$$

for each $a \in A$.

(2) An element M of $(A\check{\otimes}A)^*$ is called a *virtual diagonal* if

$$a\cdot M - M\cdot a \text{ and } a\cdot m^{**}(M) = a$$

for all $a \in A$.

PROPOSITION A.4.2. *Let A be a completely contractive Banach algebra. If A has a bounded approximate diagonal, then A is operator amenable.*

PROOF. Let $(u_\alpha)_\alpha$ be a bounded approximate diagonal in $A\hat{\otimes}A$. Let X be an operator A-bimodule and $D : A \to X^*$ a completely bounded derivation. Then $(m(u_\alpha))_\alpha$ is a bounded approximate identity for A.

For $x \in X$, we define $F_x \in (A\hat{\otimes}A)^*$ by

$$\langle F_x, a \otimes b\rangle = \langle x, a \cdot D(b)\rangle, \quad a, b \in A.$$

Then $\|F_x\| \leqslant \|x\| \cdot \|D\|_{cb}$.

Let M be a w^*-cluster point of the net $(m(u_\alpha))_\alpha$ in A^{**}. Then $a \cdot M = M \cdot a$ for all $a \in A$. In fact, for all $f \in A^*$,

$$\begin{aligned} \langle a \cdot M, f\rangle &= \langle M, f \cdot a\rangle = \lim_\alpha \langle m(u_\alpha), f \cdot a\rangle \\ &= \lim_\alpha \langle am(u_\alpha), f\rangle = \lim_\alpha \langle m(au_\alpha), f\rangle \\ &= \lim_\alpha \langle m(u_\alpha a), f\rangle = \langle M, a \cdot f\rangle = \langle M \cdot a, f\rangle. \end{aligned}$$

Now define $f \in X^*$ by

$$\langle f, x\rangle = \langle M, , F_x\rangle, \quad x \in X.$$

We are going to show that $D = D_f$. To see this, let $a, b \in A$ and $x \in X$. Then

$$\begin{aligned} \langle b \otimes c, F_x \cdot a - a \cdot F_x\rangle &= \langle a \cdot (b \otimes c) - (b \otimes c) \cdot a, F_x\rangle \\ &= \langle ab \otimes c - b \otimes ca, F_x\rangle \\ &= \langle x, (ab) \cdot D(c) - b \cdot D(ca)\rangle \\ &= \langle x, (ab) \cdot D(c) - b \cdot D(c) \cdot a\rangle - \langle x, (bc) \cdot D(a)\rangle \\ &= \langle x \cdot a - a \cdot x, b \cdot D(c)\rangle - \langle x, (bc) \cdot D(a)\rangle \\ &= \langle b \otimes c, F_{x \cdot a - a \cdot a}\rangle - \langle x, (bc) \cdot D(a)\rangle. \end{aligned}$$

This implies that, for each α,

$$\langle u_\alpha, F_x \cdot a - a \cdot F_x\rangle = \langle u_\alpha, F_{x \cdot a - a \cdot x}\rangle - \langle x, m(u_\alpha) \cdot D(a)\rangle$$

and therefore, since $a \cdot M = M \cdot a$,

$$\begin{aligned} \langle x, a \cdot f - f \cdot a\rangle &= \langle x \cdot a - a \cdot x, f\rangle = \langle F_{x \cdot a - a \cdot x}, M\rangle \\ &= \langle F_x \cdot a - a \cdot F_x, M\rangle + \lim_\alpha \langle x, m(u_\alpha) \cdot D(a)\rangle \\ &= \langle F_x, a \cdot M - M \cdot a\rangle + \lim_\alpha \langle x, m(u_\alpha) \cdot D(a)\rangle \\ &= \lim_\alpha \langle x, m(u_\alpha) \cdot D(a)\rangle \end{aligned}$$

for all $x \in X$ and $a \in A$. Now x is of the form $x = b \cdot y \cdot c$ for certain $y \in X$ and $b, c \in A$. Since $(m(u_\alpha))_\alpha$ is an approximate identity for A, it follows that

$$\begin{aligned} \langle x, a \cdot f - f \cdot a\rangle &= \lim_\alpha \langle b \cdot y \cdot (c\, m(u_\alpha)), D(a)\rangle \\ &= \langle b \cdot y \cdot c, D(a)\rangle = \langle x, D(a)\rangle. \end{aligned}$$

This shows that $D(a) = a \cdot f - f \cdot a$ for all $a \in A$, as required. □

THEOREM A.4.3. *For a completely contractive Banach algebra A, the following are equivalent.*

(i) *A is operator amenable.*
(ii) *A has a virtual diagonal.*
(iii) *A has a bounded approximate diagonal.*

THEOREM A.4.4. *Let A be an operator amenable completely contractive Banach algebra and let I be a closed ideal of A. Then I is operator amenable if and only if I has a bounded approximate identity.*

PROOF. Since I is completely contractive and since we already know from Proposition A.4.1 that every operator amenable completely contractive Banach algebra has a bounded approximate identity, we only need to show that if conversely I has bounded approximate identity, then I is operator amenable. □

THEOREM A.4.5. *Let A be a Banach algebra which is an operator space, and let I be a closed ideal of A such that both I and A/I are operator amenable. Then A is operator amenable.*

A.5. Operator weak amenability

Let A be a completely contractive Banach algebra. Then A is said to be *operator weakly amenable* if every completely bounded derivation from A into A^* is an inner derivation.

LEMMA A.5.1. *Let A be an operator weakly amenable Banach algebra. Then A^2 is dense in A.*

PROOF. Towards a contradiction, assume that A^2 is not dense in A. Then there exists $f \in A^*$ such that $f \neq 0$ and $f(A^2) = \{0\}$. Define $D : A \to A^*$ by $D(a) = f(a)f$. Then D is a derivation since $\langle f, xy\rangle = 0$ for all $x, y \in A$. By hypothesis, there exists $g \in A^*$ such that $D(a) = a \cdot g - g \cdot a$ for all $a \in A$. Choose $b \in A$ with $\langle f, b\rangle \neq 0$. Then $\langle D(b), b\rangle = \langle f, b\rangle \neq 0$, but on the other hand $\langle D(b), b\rangle = \langle b \cdot g - g \cdot b, b\rangle = 0$. This contradiction shows that A^2 must be dense in A. □

PROPOSITION A.5.2. *Let A be a commutative completely contractive Banach algebra. Then A is operator weakly amenable if and only if every completely bounded derivation from A into a symmetric operator A-bimodule is trivial.*

PROOF. Since A is commutative, A^* is symmetric. Hence the "if" direction is trivial.

Conversely, suppose that A is operator weakly amenable and let X be a symmetric operator A-bimodule and $D : A \to X$ a completely bounded derivation. By Lemma A.5.1, A^2 is dense in A. Assume that D is nonzero and choose $a \in A$ and $f \in X^*$ such that $D(a^2) \neq 0$ and $\langle f, D(a^2)\rangle \neq 0$. For $x \in X$, define $f_x \in A^*$ by $\langle f_x, b\rangle = \langle f, b \cdot x\rangle$. Then the map $F : x \to f_x$ is completely bounded. Indeed, for each $n \in \mathbb{N}$,

$$\begin{aligned}
\|F_n\| &= \sup\{\|F_n([x_{ij}])\| : [x_{ij}] \in M_n(X), \|[x_{ij}]\| \leqslant 1\} \\
&= \sup\{\|[f(x_{ij})]\| : [x_{ij}] \in M_n(X), \|[x_{ij}]\| \leqslant 1\} \\
&= \sup\{\|[\langle f_{x_{ij}}, a\rangle]\| : a \in A, \|a\| \leqslant 1, [x_{ij}] \in M_n(X), \|[x_{ij}]\| \leqslant 1\} \\
&= \sup\{\|\langle F_n, [a \cdot x_{ij}]\rangle\| : a \in A, \|a\| \leqslant 1, [x_{ij}] \in M_n(X), \|[x_{ij}]\| \leqslant 1\} \\
&\leqslant \|F\|_{cb} = \|F\|.
\end{aligned}$$

Now define $\widetilde{D} : A \to A^*$ by $\widetilde{D}(a) = F(D(a))$. It is straightforward to check that $\widetilde{D}$ is a completely bounded derivation. Finally, since X is symmetric,

$$\langle \widetilde{D}(a), a\rangle = \langle f_{D(a)}, a\rangle = \langle f, a \cdot D(a)\rangle = \frac{1}{2}\langle f, D(a^2)\rangle \neq 0.$$

Since A is operator weakly amenable, $\widetilde{D}(a)$ is inner. So $\widetilde{D}(a) = a\cdot g - g\cdot a$ for some $g \in A^*$ and hence $\langle \widetilde{D}(a), a\rangle = 0$. This contradiction finishes the proof. □

COROLLARY A.5.3. *Let A and B be commutative completely contractive Banach algebras and let $\phi : A \to B$ be a completely bounded homomorphism with dense range. If A is operator weakly amenable, then so is B.*

PROOF. We may assume that $\|\phi\|_{cb} = 1$. Observe that B^* is an operator A-bimodule with the action given by

$$(a, f) \to \phi(a)\cdot f = f\cdot \phi(a), \quad a \in A, f \in B^*.$$

Let $D : B \to B^*$ be a completely bounded derivation and define $\widetilde{D} : A \to B^*$ by $\widetilde{D}(a) = D(\phi(a))$. Then $\widetilde{D}$ is a completely bounded derivation and hence, by Proposition A.5.2, $\widetilde{D} = 0$ since A is operator weakly amenable. Since ϕ has dense range, it follows that $D = 0$. □

THEOREM A.5.4. *Let A be a commutative operator weakly amenable Banach algebra and let I be a closed ideal of A. Then I is operator weakly amenable if and only if I^2 is dense in I.*

PROOF. If I is operator weakly amenable, then I^2 is dense in I by Lemma A.5.1.

Conversely, suppose that I^2 is dense in I and let $D : I \to I^*$ be a completely bounded derivation. Let $\mathrm{Hom}_{cb}(I, I^*)$ denote the set of all completely bounded left I-module homomorphisms, i..e. of all completely bounded linear maps $T : I \to I^*$ such that $T(a\cdot f) = a\cdot T(f)$ for all $a \in A$ and $f \in I^*$.

For $a \in A$, we now define $D_a : A \to I^*$ by

$$D_a(b) = D(ab) - b\cdot D(a), \quad b \in A.$$

Then, since A is commutative and D is a derivation, for all $x, y \in I$ and $a, b \in A$,

$$\begin{aligned}
D_{xy}(ab) &= D(xyab) - (ab)\cdot D(xy)\\
&= (ax)\cdot D(yb) + (yb)\cdot D(ax) - (ab)\cdot D(xy)\\
&= a\cdot[x\cdot D(yb) - (bx)\cdot D(y)] + b\cdot[(ax)\cdot D(y) + y\cdot D(ax) - a\cdot D(xy)]\\
&= a\cdot[D(xyb) - b\cdot D(xy)] + b\cdot[D(xya) - a\cdot D(xy)]\\
&= a\cdot D_{xy}(b) + b\cdot D_{xy}(a).
\end{aligned}$$

This shows that, for every $z \in I^2$, $a \to D_z(a)$ is a derivation from A into I^*. Being the composition of completely bounded maps, D_z is completely bounded. Since A is an operator space, so is I and hence I^*. Moreover, I^* is a symmetric A-bimodule since A is commutative. As A is operator weakly amenable, it follows from Proposition A.5.2 that $D_z = 0$ for all $z \in I^2$. Now, for $x, a \in I$,

$$D_x(a) = D(xa) - a\cdot D(x) = x\cdot D(a) + D(x)\cdot a - a\cdot D(x) = x\cdot D(a).$$

So, if $x, y, a \in I$, then $(xy)\cdot D(a) = D_{xy}(a) = 0$ and hence $\langle I^2\rangle\cdot D(I) = \{0\}$. As $\langle I^2\rangle$ is dense in I, $x\cdot D(I) = D(I)\cdot x = \{0\}$ for all $x \in I$ and so $D(I^2) = \{0\}$. Again, using that $\overline{\langle I^2\rangle} = I$, we conclude that $D(I) = \{0\}$. □

LEMMA A.5.5. *If $S \in CB(A, X^*) \cong (A_e + X)^*$ is a derivation, then S annihilates $[K; A]$.*

PROOF. First note that

$$K = \{u - e \otimes m_l(u) : u \in A_e \hat{\otimes} X\} = \overline{\mathrm{span}\{a \otimes c - e \otimes a \cdot x : a \in A_e, x \in X\}}.$$

Since S is a derivation, for any $a, b \in A_e$ and $x \in X$, we have

$$\begin{aligned}\langle b \otimes x - e \otimes b \cdot x\rangle, S \cdot a - a \cdot S\rangle &= \langle x, S(ab) - a \cdot S(b)\rangle \\ &\quad -\langle b \cdot x, S(a) - a \cdot S(e)\rangle \\ &= \langle x, S(ab) - a \cdot S(b)\rangle - \langle x, S(a) \cdot b\rangle \\ &= \langle x, S(ab) - a \cdot S(b) - S(a) \cdot b\rangle = 0.\end{aligned}$$

Thus $S \cdot a - a \cdot S \in K^{\perp}$ for all $a \in K$. This implies

$$\langle a \cdot u - u \cdot a, S\rangle = \langle u, S \cdot a - a \cdot S\rangle = 0$$

for all $u \in K$ and $a \in A_e$, and hence S annihilates $[K; A]$. □

PROPOSITION A.5.6. *Let A be a commutative completely contractive Banach algebra and $m : A_e \hat{\otimes} A_e \to A_e$ the multiplication map. Let $K_1 = \ker m$ and $K_0 = K_1 \cap (A \hat{\otimes} A)$. If $\overline{A^2} = A$ and $\overline{K_0^2} = \overline{(A \hat{\otimes} A)K_1}$, then A_e and hence A is operator weakly amenable.*

PROOF. Let $D : A_e \to A_e^*$ be a completely bounded derivation and view D as an element of $(A_e \otimes A_e)^*$. Then D vanishes on $\overline{K_1^2}$. Indeed, with $u_j = a_j \otimes b_j - e \otimes a_j b_j$, $a_j, b_j \in A_e$, $j = 1, 2$, we have

$$\begin{aligned}D(u_1 u_2) &= \langle a_1 a_2 \otimes b_b b_2, D\rangle - \langle a_1 \otimes a_2 b_1 b_2, D\rangle \\ &\quad -\langle a_2 \otimes a_1 b_1 b_2, D\rangle + \langle e \otimes a_1 a_2 b_1 b_2, D\rangle \\ &= \langle b_1 b_2, D(a_1 a_2)\rangle - \langle a_2 b_1 b_2, D(a_1)\rangle \\ &\quad -\langle a_2 b_1 b_2, D(a_2)\rangle - \langle a_1 a_2 b_1 b_2, .D(e)\rangle = 0.\end{aligned}$$

Since K_1 is the closed linear span of elements of the form $a \otimes b - e \otimes ab$, $a, b \in A_e$, it follows that D vanishes on $\overline{K_1^2}$. Since $K_0 \subseteq K_1$ and $\overline{K_0^2} = \overline{(A \hat{\otimes} A)K_1}$, D vanishes on $\overline{K_1(A \hat{\otimes} A)}$. In particular, D vanishes on elements of the form

$$(e \otimes c) - c \otimes e)(a \otimes b) = a \otimes bc - ac \otimes b, \quad a, b, c \in A.$$

D being a derivation, we conclude that

$$\langle ab, D(c)\rangle = \langle b, D(ac)\rangle - \langle bc, D(a)\rangle = \langle ac \otimes b - a \otimes bc, D\rangle = 0$$

for all $a, b, c \in A$. Because A^2 is dense in A, it follows that D vanishes on A and hence also on A_e since $D(e) = 0$. Thus A_e is operator weakly amenable and hence so is the ideal A by Theorem A.5.4. □

Bibliography

[1] Charles A. Akemann and Martin E. Walter, *The Riemann-Lebesgue property for arbitrary locally compact groups*, Duke Math. J. **43** (1976), no. 2, 225–236. MR0399757

[2] Alberto Alesina and Leonede De Michele, *A dichotomy for a class of positive definite functions*, Pacific J. Math. **103** (1982), no. 2, 251–257. MR705226

[3] A. S. Amitsur and J. Levitzki, *Minimal identities for algebras*, Proc. Amer. Math. Soc. **1** (1950), 449–463. MR0036751

[4] Wolfgang Arendt and Jean De Cannière, *Order isomorphisms of Fourier algebras*, J. Funct. Anal. **50** (1983), no. 1, 1–7. MR689996

[5] Gilbert Arsac, *Sur l'espace de Banach engendré par les coefficients d'une représentation unitaire* (French), Publ. Dép. Math. (Lyon) **13** (1976), no. 2, 1–101. MR0444833

[6] W. G. Bade, P. C. Curtis Jr., and H. G. Dales, *Amenability and weak amenability for Beurling and Lipschitz algebras*, Proc. London Math. Soc. (3) **55** (1987), no. 2, 359–377. MR896225

[7] Larry Baggett, *A separable group having a discrete dual space is compact*, J. Functional Analysis **10** (1972), 131–148. MR0346090

[8] Larry Baggett and Keith F. Taylor, *Riemann-Lebesgue subsets of* $\mathbf{R}^n$ *and representations which vanish at infinity*, J. Functional Analysis **28** (1978), no. 2, 168–181. MR0476911

[9] Mohammed E. B. Bekka, *A characterization of locally compact amenable groups by means of tensor products*, Arch. Math. (Basel) **52** (1989), no. 5, 424–427. MR998613

[10] M. B. Bekka, E. Kaniuth, A. T. Lau, and G. Schlichting, *Weak*-closedness of subspaces of Fourier-Stieltjes algebras and weak*-continuity of the restriction map*, Trans. Amer. Math. Soc. **350** (1998), no. 6, 2277–2296. MR1401762

[11] M. E. B. Bekka, A. T. Lau, and G. Schlichting, *On invariant subalgebras of the Fourier-Stieltjes algebra of a locally compact group*, Math. Ann. **294** (1992), no. 3, 513–522. MR1188135

[12] John J. Benedetto, *Spectral synthesis*, B. G. Teubner, Stuttgart, 1975. Mathematische Leitfäden. MR0622037

[13] G. Bennett, *Schur multipliers*, Duke Math. J. **44** (1977), no. 3, 603–639. MR0493490

[14] P. Bernat, N. Conze, M. Duflo, M. Lévy-Nahas, M. Raïs, P. Renouard, and M. Vergne, *Représentations des groupes de Lie résolubles*, Dunod, Paris, 1972. Monographies de la Société Mathématique de France, No. 4. MR0444836

[15] Robert J. Blattner, *On a theorem of G. W. Mackey*, Bull. Amer. Math. Soc. **68** (1962), 585–587. MR0141738

[16] David P. Blecher, *The standard dual of an operator space*, Pacific J. Math. **153** (1992), no. 1, 15–30. MR1145913

[17] David P. Blecher and Vern I. Paulsen, *Tensor products of operator spaces*, J. Funct. Anal. **99** (1991), no. 2, 262–292. MR1121615

[18] Frank F. Bonsall and John Duncan, *Complete normed algebras*, Springer-Verlag, New York-Heidelberg, 1973. Ergebnisse der Mathematik und ihrer Grenzgebiete, Band 80. MR0423029

[19] Marek Bożejko, *Remark on Herz-Schur multipliers on free groups*, Math. Ann. **258** (1981/82), no. 1, 11–15. MR641665

[20] Marek Bożejko, *Positive definite bounded matrices and a characterization of amenable groups*, Proc. Amer. Math. Soc. **95** (1985), no. 3, 357–360. MR806070

[21] Marek Bożejko and Gero Fendler, *Herz-Schur multipliers and completely bounded multipliers of the Fourier algebra of a locally compact group* (English, with Italian summary), Boll. Un. Mat. Ital. A (6) **3** (1984), no. 2, 297–302. MR753889

[22] Marek Bożejko and Gero Fendler, *Herz-Schur multipliers and uniformly bounded representations of discrete groups*, Arch. Math. (Basel) **57** (1991), no. 3, 290–298. MR1119903

[23] Marek Bożejko and Massimo A. Picardello, *Weakly amenable groups and amalgamated products*, Proc. Amer. Math. Soc. **117** (1993), no. 4, 1039–1046. MR1119263

[24] Michael Brannan and Brian Forrest, *Extending multipliers of the Fourier algebra from a subgroup*, Proc. Amer. Math. Soc. **142** (2014), no. 4, 1181–1191. MR3162241

[25] Gavin Brown and William Moran, *Point derivations on $M(G)$*, Bull. London Math. Soc. **8** (1976), no. 1, 57–64. MR0417695

[26] Daniel Bump, *Lie groups*, Graduate Texts in Mathematics, vol. 225, Springer-Verlag, New York, 2004. MR2062813

[27] Jean De Cannière and Uffe Haagerup, *Multipliers of the Fourier algebras of some simple Lie groups and their discrete subgroups*, Amer. J. Math. **107** (1985), no. 2, 455–500. MR784292

[28] L. N. Carling, *On the restriction map of the Fourier-Stieltjes algebra $B(G)$ and $B_p(G)$*, J. Functional Analysis **25** (1977), no. 3, 236–243. MR0458051

[29] Yemon Choi and Mahya Ghandehari, *Weak and cyclic amenability for Fourier algebras of connected Lie groups*, J. Funct. Anal. **266** (2014), no. 11, 6501–6530. MR3192460

[30] Yemon Choi and Mahya Ghandehari, *Weak amenability for Fourier algebras of 1-connected nilpotent Lie groups*, J. Funct. Anal. **268** (2015), no. 8, 2440–2463. MR3318655

[31] Ching Chou, *Minimally weakly almost periodic groups*, J. Funct. Anal. **36** (1980), no. 1, 1–17. MR568972

[32] Ching Chou and Guangwu Xu, *The weak closure of the set of left translation operators*, Proc. Amer. Math. Soc. **127** (1999), no. 2, 465–471. MR1468187

[33] Paul J. Cohen, *On a conjecture of Littlewood and idempotent measures*, Amer. J. Math. **82** (1960), 191–212. MR0133397

[34] Paul J. Cohen, *On homomorphisms of group algebras*, Amer. J. Math. **82** (1960), 213–226. MR0133398

[35] Michael Cowling, *The Kunze-Stein phenomenon*, Ann. Math. (2) **107** (1978), no. 2, 209–234. MR0507240

[36] Michael Cowling, *The Fourier-Stieltjes algebra of a semisimple group*, Colloq. Math. **41** (1979), no. 1, 89–94. MR550633

[37] Michael Cowling, *Sur les coefficients des représentations unitaires des groupes de Lie simples* (French), Analyse harmonique sur les groupes de Lie (Sém., Nancy-Strasbourg 1976–1978), II, Lecture Notes in Math., vol. 739, Springer, Berlin, 1979, pp. 132–178. MR560837

[38] Michael Cowling and Uffe Haagerup, *Completely bounded multipliers of the Fourier algebra of a simple Lie group of real rank one*, Invent. Math. **96** (1989), no. 3, 507–549. MR996553

[39] Michael Cowling and Paul Rodway, *Restrictions of certain function spaces to closed subgroups of locally compact groups*, Pacific J. Math. **80** (1979), no. 1, 91–104. MR534697

[40] Michael Cowling, Brian Dorofaeff, Andreas Seeger, and James Wright, *A family of singular oscillatory integral operators and failure of weak amenability*, Duke Math. J. **127** (2005), no. 3, 429–486. MR2132866

[41] H. G. Dales, *Banach algebras and automatic continuity*, London Mathematical Society Monographs. New Series, vol. 24, The Clarendon Press, Oxford University Press, New York, 2000. Oxford Science Publications. MR1816726

[42] H. G. Dales, F. Ghahramani, and A. Ya. Helemskii, *The amenability of measure algebras*, J. London Math. Soc. (2) **66** (2002), no. 1, 213–226. MR1911870

[43] Matthew Daws, *p-operator spaces and Figà-Talamanca–Herz algebras*, J. Operator Theory **63** (2010), no. 1, 47–83. MR2606882

[44] Mahlon M. Day, *Fixed-point theorems for compact convex sets*, Illinois J. Math. **5** (1961), 585–590. MR0138100

[45] Jacques Delaporte and Antoine Derighetti, *On ideals of A_p with bounded approximate units and certain conditional expectations*, J. London Math. Soc. (2) **47** (1993), no. 3, 497–506. MR1214911

[46] Jacques Delaporte and Antoine Derighetti, *p-pseudomeasures and closed subgroups*, Monatsh. Math. **119** (1995), no. 1-2, 37–47. MR1315682

[47] Jacques Delaporte and Antoine Derighetti, *Best bounds for the approximate units for certain ideals of $L^1(G)$ and of $A_p(G)$*, Proc. Amer. Math. Soc. **124** (1996), no. 4, 1159–1169. MR1301019

[48] Jacques Delaporte and Antoine Derighetti, *Invariant projections and convolution operators*, Proc. Amer. Math. Soc. **129** (2001), no. 5, 1427–1435. MR1814169

[49] Leonede De-Michele and Alessandro Figà-Talamanca, *Positive definite functions on free groups*, Amer. J. Math. **102** (1980), no. 3, 503–509. MR573099

[50] A. Derighetti, *Some results on the Fourier-Stieltjes algebra of a locally compact group*, Comm. Math. Helv. **46** (1973), 328-339.

[51] Antoine Derighetti, *Relations entre les convoluteurs d'un groupe localement compact et ceux d'un sous-groupe fermé* (French, with English summary), Bull. Sci. Math. (2) **106** (1982), no. 1, 69–84. MR654425

[52] Antoine Derighetti, *À propos des convoluteurs d'un groupe quotient* (French, with English summary), Bull. Sci. Math. (2) **107** (1983), no. 1, 3–23. MR699988

[53] Antoine Derighetti, *Quelques observations concernant les ensembles de Ditkin d'un groupe localement compact* (French, with English summary), Monatsh. Math. **101** (1986), no. 2, 95–113. MR843295

[54] Antoine Derighetti, *On Herz's projection theorem*, Illinois J. Math. **48** (2004), no. 2, 463–476. MR2085420

[55] Antoine Derighetti, *Closed subgroups as Ditkin sets*, J. Funct. Anal. **266** (2014), no. 3, 1702–1715. MR3146833

[56] Antoine Derighetti, *Convolution operators on groups*, Lecture Notes of the Unione Matematica Italiana, vol. 11, Springer, Heidelberg; UMI, Bologna, 2011. MR2809956

[57] Antoine Derighetti, *Injection theorem for local Ditkin sets*, Ann. Funct. Anal. **7** (2016), no. 1, 96–101. MR3449342

[58] M. Despić and F. Ghahramani, *Weak amenability of group algebras of locally compact groups*, Canad. Math. Bull. **37** (1994), no. 2, 165–167. MR1275699

[59] Jacques Dixmier, *Les algèbres d'opérateurs dans l'espace hilbertien (algèbres de von Neumann)* (French), Gauthier-Villars Éditeur, Paris, 1969. Deuxième édition, revue et augmentée; Cahiers Scientifiques, Fasc. XXV. MR0352996

[60] Jacques Dixmier, *C^*-algebras*, North-Holland Publishing Co., Amsterdam-New York-Oxford, 1977. Translated from the French by Francis Jellett; North-Holland Mathematical Library, Vol. 15. MR0458185

[61] Robert S. Doran and Josef Wichmann, *Approximate identities and factorization in Banach modules*, Lecture Notes in Mathematics, vol. 768, Springer-Verlag, Berlin-New York, 1979. MR555240

[62] Brian Dorofaeff, *The Fourier algebra of* $\mathrm{SL}(2,\mathbf{R}) \rtimes \mathbf{R}^n$*, $n \geqslant 2$, has no multiplier bounded approximate unit*, Math. Ann. **297** (1993), no. 4, 707–724. MR1245415

[63] Brian Dorofaeff, *Weak amenability and semidirect products in simple Lie groups*, Math. Ann. **306** (1996), no. 4, 737–742. MR1418350

[64] N. Dunford and J.T. Schwartz, *Linear operators. I*, Wiley, 1988.

[65] Charles F. Dunkl and Donald E. Ramirez, *C^*-algebras generated by Fourier-Stieltjes transforms*, Trans. Amer. Math. Soc. **164** (1972), 435–441. MR0310548

[66] Charles F. Dunkl and Donald E. Ramirez, *Weakly almost periodic functionals on the Fourier algebra*, Trans. Amer. Math. Soc. **185** (1973), 501–514. MR0372531

[67] W. F. Eberlein, *Characterizations of Fourier-Stieltjes transforms*, Duke Math. J. **22** (1955), 465–468. MR0072424

[68] Edward G. Effros, *Order ideals in a C^*-algebra and its dual*, Duke Math. J. **30** (1963), 391–411. MR0151864

[69] Edward G. Effros and Zhong-Jin Ruan, *On the abstract characterization of operator spaces*, Proc. Amer. Math. Soc. **119** (1993), no. 2, 579–584. MR1163332

[70] Edward G. Effros and Zhong-Jin Ruan, *Operator spaces*, London Mathematical Society Monographs. New Series, vol. 23, The Clarendon Press, Oxford University Press, New York, 2000. MR1793753

[71] L. Ehrenpreis and F. I. Mautner, *Uniformly bounded representations of groups*, Proc. Nat. Acad. Sci. U. S. A. **41** (1955), 231–233. MR0071434

[72] John Ernest, *A new group algebra for locally compact groups*, Amer. J. Math. **86** (1964), 467–492. MR0167566

[73] Pierre Eymard, *L'algèbre de Fourier d'un groupe localement compact* (French), Bull. Soc. Math. France **92** (1964), 181–236. MR0228628

[74] J. M. G. Fell, *Weak containment and induced representations of groups. II*, Trans. Amer. Math. Soc. **110** (1964), 424–447. MR0159898

[75] Alessandro Figà-Talamanca, *Positive definite functions which vanish at infinity*, Pacific J. Math. **69** (1977), no. 2, 355–363. MR0493175

[76] Alessandro Figà-Talamanca and Massimo Picardello, *Multiplicateurs de $A(G)$ qui ne sont pas dans $B(G)$* (French), C. R. Acad. Sci. Paris Sér. A-B **277** (1973), A117–A119. MR0333597

[77] Christian Fiorillo, *An extension property for the Figà-Talamanca Herz algebra*, Proc. Amer. Math. Soc. **137** (2009), no. 3, 1001–1011. MR2457440

[78] Gerald B. Folland, *A course in abstract harmonic analysis*, Studies in Advanced Mathematics, CRC Press, Boca Raton, FL, 1995. MR1397028

[79] Brian Forrest, *Amenability and bounded approximate identities in ideals of $A(G)$*, Illinois J. Math. **34** (1990), no. 1, 1–25. MR1031879

[80] Brian Forrest, *Amenability and ideals in $A(G)$*, J. Austral. Math. Soc. Ser. A **53** (1992), no. 2, 143–155. MR1175708

[81] Brian Forrest, *Complemented ideals in the Fourier algebra and the Radon Nikodým property*, Trans. Amer. Math. Soc. **333** (1992), no. 2, 689–700. MR1112546

[82] Brian Forrest, *Amenability and the structure of the algebras $A_p(G)$*, Trans. Amer. Math. Soc. **343** (1994), no. 1, 233–243. MR1181182

[83] Brian Forrest, *Fourier analysis on coset spaces*, Rocky Mountain J. Math. **28** (1998), no. 1, 173–190. MR1639849

[84] Brian Forrest, *Completely bounded multipliers and ideals in $A(G)$ vanishing on closed subgroups*, Banach algebras and their applications, Contemp. Math., vol. 363, Amer. Math. Soc., Providence, RI, 2004, pp. 89–94. MR2097953

[85] B. Forrest, E. Kaniuth, A. T. Lau, and N. Spronk, *Ideals with bounded approximate identities in Fourier algebras*, J. Funct. Anal. **203** (2003), no. 1, 286–304. MR1996874

[86] Brian E. Forrest and Volker Runde, *Amenability and weak amenability of the Fourier algebra*, Math. Z. **250** (2005), no. 4, 731–744. MR2180372

[87] Brian E. Forrest and Volker Runde, *Norm one idempotent cb-multipliers with applications to the Fourier algebra in the cb-multiplier norm*, Canad. Math. Bull. **54** (2011), no. 4, 654–662. MR2894515

[88] Brian E. Forrest, Volker Runde, and Nico Spronk, *Operator amenability of the Fourier algebra in the cb-multiplier norm*, Canad. J. Math. **59** (2007), no. 5, 966–980. MR2354398

[89] Brian E. Forrest, Ebrahim Samei, and Nico Spronk, *Weak amenability of Fourier algebras on compact groups*, Indiana Univ. Math. J. **58** (2009), no. 3, 1379–1393. MR2542091

[90] Brian E. Forrest, Ebrahim Samei, and Nico Spronk, *Convolutions on compact groups and Fourier algebras of coset spaces*, Studia Math. **196** (2010), no. 3, 223–249. MR2587297

[91] Brian Forrest and Mahatheva Skantharajah, *A note on a type of approximate identity in the Fourier algebra*, Proc. Amer. Math. Soc. **120** (1994), no. 2, 651–652. MR1166356

[92] Brian Forrest and Nicolaas Spronk, *Best bounds for approximate identities in ideals of the Fourier algebra vanishing on subgroups*, Proc. Amer. Math. Soc. **134** (2006), no. 1, 111–116. MR2170550

[93] Brian Forrest and Peter Wood, *Cohomology and the operator space structure of the Fourier algebra and its second dual*, Indiana Univ. Math. J. **50** (2001), no. 3, 1217–1240. MR1871354

[94] Brian E. Forrest, Ebrahim Samei, and Nico Spronk, *Weak amenability of Fourier algebras on compact groups*, Indiana Univ. Math. J. **58** (2009), no. 3, 1379–1393. MR2542091

[95] Steven A. Gaal, *Linear analysis and representation theory*, Springer-Verlag, New York-Heidelberg, 1973. Die Grundlehren der mathematischen Wissenschaften, Band 198. MR0447465

[96] Fereidoun Ghahramani and Ross Stokke, *Approximate and pseudo-amenability of the Fourier algebra*, Indiana Univ. Math. J. **56** (2007), no. 2, 909–930. MR2317550

[97] Mahya Ghandehari, *Amenability properties of Rajchman algebras*, Indiana Univ. Math. J. **61** (2012), no. 3, 1369–1392. MR3071703

[98] John E. Gilbert, *On projections of $L^\infty(G)$ onto translation-invariant subspaces*, Proc. London Math. Soc. (3) **19** (1969), 69–88. MR0244705

[99] V. M. Gluškov, *Locally nilpotent locally bicompact groups* (Russian), Trudy Moskov. Mat. Obšč. **4** (1955), 291–332. MR0072422

[100] Roger Godement, *Les fonctions de type positif et la théorie des groupes* (French), Trans. Amer. Math. Soc. **63** (1948), 1–84. MR0023243

[101] Colin C. Graham and O. Carruth McGehee, *Essays in commutative harmonic analysis*, Grundlehren der Mathematischen Wissenschaften [Fundamental Principles of Mathematical Science], vol. 238, Springer-Verlag, New York-Berlin, 1979. MR550606

[102] Edmond E. Granirer, *Weakly almost periodic and uniformly continuous functionals on the Fourier algebra of any locally compact group*, Trans. Amer. Math. Soc. **189** (1974), 371–382. MR0336241

[103] Edmond E. Granirer, *On some properties of the Banach algebras $A_p(G)$ for locally compact groups*, Proc. Amer. Math. Soc. **95** (1985), no. 3, 375–381. MR806074

[104] Edmond E. Granirer, *Some results on $A_p(G)$ submodules of* $\mathrm{PM}_p(G)$, Colloq. Math. **51** (1987), 155–163. MR891283

[105] E. E. Granirer and M. Leinert, *On some topologies which coincide on the unit sphere of the Fourier-Stieltjes algebra $B(G)$ and of the measure algebra $M(G)$*, Rocky Mountain J. Math. **11** (1981), no. 3, 459–472. MR722579

[106] Frederick P. Greenleaf, *Norm decreasing homomorphisms of group algebras*, Pacific J. Math. **15** (1965), 1187–1219. MR0194911

[107] F.P. Greenleaf, *Invariant means on topological groups*, van Nostrand, New York, 1969.

[108] Niels Groenbaek, *A characterization of weakly amenable Banach algebras*, Studia Math. **94** (1989), no. 2, 149–162. MR1025743

[109] Niels Grønbæk, *Commutative Banach algebras, module derivations, and semigroups*, J. London Math. Soc. (2) **40** (1989), no. 1, 137–157. MR1028919

[110] Siegfried Grosser and Martin Moskowitz, *Compactness conditions in topological groups*, J. Reine Angew. Math. **246** (1971), 1–40. MR0284541

[111] Uffe Haagerup, *An example of a nonnuclear C^*-algebra, which has the metric approximation property*, Invent. Math. **50** (1978/79), no. 3, 279–293. MR520930

[112] Uffe Haagerup, *Solution of the similarity problem for cyclic representations of C^*-algebras*, Ann. of Math. (2) **118** (1983), no. 2, 215–240. MR717823

[113] U. Haagerup, *Group C^*-agebras without the completely bounded approximation property*, unpublished manuscript, 1986.

[114] Uffe Haagerup and Jon Kraus, *Approximation properties for group C^*-algebras and group von Neumann algebras*, Trans. Amer. Math. Soc. **344** (1994), no. 2, 667–699. MR1220905

[115] Mogens Lemvig Hansen, *Weak amenability of the universal covering group of* $\mathrm{SU}(1,n)$, Math. Ann. **288** (1990), no. 3, 445–472. MR1079871

[116] Hun Hee Lee, Jean Ludwig, Ebrahim Samei, and Nico Spronk, *Weak amenability of Fourier algebras and local synthesis of the anti-diagonal*, Adv. Math. **292** (2016), 11–41. MR3464018

[117] A. Ya. Helemskii, *The homology of Banach and topological algebras*, Mathematics and its Applications (Soviet Series), vol. 41, Kluwer Academic Publishers Group, Dordrecht, 1989. Translated from the Russian by Alan West. MR1093462

[118] Rolf Wim Henrichs, *Über Fortsetzung positiv definiter Funktionen* (German), Math. Ann. **232** (1978), no. 2, 131–150. MR0481931

[119] Rolf Wim Henrichs, *On characters of subgroups*, Nederl. Akad. Wetensch. Indag. Math. **41** (1979), no. 3, 273–281. MR546368

[120] Rolf Wim Henrichs, *On one-sided harmonic analysis*, Proc. Amer. Math. Soc. **80** (1980), no. 4, 627–630. MR587940

[121] C. S. Herz, *Spectral synthesis for the circle*, Ann. of Math. (2) **68** (1958), 709–712. MR0100758

[122] Carl Herz, *The theory of p-spaces with an application to convolution operators.*, Trans. Amer. Math. Soc. **154** (1971), 69–82. MR0272952

[123] Carl Herz, *Harmonic synthesis for subgroups* (English, with French summary), Ann. Inst. Fourier (Grenoble) **23** (1973), no. 3, 91–123. MR0355482

[124] Carl Herz, *Une généralisation de la notion de transformée de Fourier-Stieltjes* (French, with English summary), Ann. Inst. Fourier (Grenoble) **24** (1974), no. 3, xiii, 145–157. MR0425511

[125] Edwin Hewitt and Kenneth A. Ross, *Abstract harmonic analysis. Vol. I: Structure of topological groups. Integration theory, group representations*, Die Grundlehren der mathematischen Wissenschaften, Bd. 115, Academic Press, Inc., Publishers, New York; Springer-Verlag, Berlin-Göttingen-Heidelberg, 1963. MR0156915

[126] Edwin Hewitt and Kenneth A. Ross, *Abstract harmonic analysis. Vol. II: Structure and analysis for compact groups. Analysis on locally compact Abelian groups*, Die Grundlehren der mathematischen Wissenschaften, Band 152, Springer-Verlag, New York-Berlin, 1970. MR0262773

[127] Karl H. Hofmann, John R. Liukkonen, and Michael W. Mislove, *Compact extensions of compactly generated nilpotent groups are pro-Lie*, Proc. Amer. Math. Soc. **84** (1982), no. 3, 443–448. MR640250

[128] K. H. Hofmann and Paul Mostert, *Splitting in topological groups*, Mem. Amer. Math. Soc. No. **43** (1963), 75. MR0151544

[129] B. Host, *Le théorème des idempotents dans* $B(G)$ (French, with English summary), Bull. Soc. Math. France **114** (1986), no. 2, 215–223. MR860817

[130] Zhiguo Hu, *Extreme non-Arens regularity of quotients of the Fourier algebra* $A(G)$, Colloq. Math. **72** (1997), no. 2, 237–249. MR1426699

[131] Zhiguo Hu, *Open subgroups of* G *and almost periodic functionals on* $A(G)$, Proc. Amer. Math. Soc. **128** (2000), no. 8, 2473–2478. MR1662249

[132] Zhiguo Hu, *Inductive extreme non-Arens regularity of the Fourier algebra* $A(G)$, Studia Math. **151** (2002), no. 3, 247–264. MR1917836

[133] Zhiguo Hu, *Open subgroups and the centre problem for the Fourier algebra*, Proc. Amer. Math. Soc. **134** (2006), no. 10, 3085–3095. MR2231636

[134] A. Hulanicki, *Means and Følner condition on locally compact groups*, Studia Math. **27** (1966), 87–104. MR0195982

[135] Monica Ilie, *On Fourier algebra homomorphisms*, J. Funct. Anal. **213** (2004), no. 1, 88–110. MR2069782

[136] Monica Ilie and Nico Spronk, *Completely bounded homomorphisms of the Fourier algebras*, J. Funct. Anal. **225** (2005), no. 2, 480–499. MR2152508

[137] Monica Ilie and Nico Spronk, *The algebra generated by idempotents in a Fourier-Stieltjes algebra*, Houston J. Math. **33** (2007), no. 4, 1131–1145. MR2350085

[138] Monica Ilie and Ross Stokke, *Weak*-continuous homomorphisms of Fourier-Stieltjes algebras*, Math. Proc. Cambridge Philos. Soc. **145** (2008), no. 1, 107–120. MR2431642

[139] Monica Ilie and Ross Stokke, *Extension of Fourier algebra homomorphisms to duals of algebras of uniformly continuous functionals*, J. Funct. Anal. **256** (2009), no. 8, 2518–2541. MR2502524

[140] B. E. Johnson, *Isometric isomorphisms of measure algebras*, Proc. Amer. Math. Soc. **15** (1964), 186–188. MR0160846

[141] Barry Edward Johnson, *Cohomology in Banach algebras*, American Mathematical Society, Providence, R.I., 1972. Memoirs of the American Mathematical Society, No. 127. MR0374934

[142] B. E. Johnson, *Weak amenability of group algebras*, Bull. London Math. Soc. **23** (1991), no. 3, 281–284. MR1123339

[143] B. E. Johnson, *Non-amenability of the Fourier algebra of a compact group*, J. London Math. Soc. (2) **50** (1994), no. 2, 361–374. MR1291743

[144] Paul Jolissaint, *A characterization of completely bounded multipliers of Fourier algebras*, Colloq. Math. **63** (1992), no. 2, 311–313. MR1180643

[145] Richard V. Kadison, *Isometries of operator algebras*, Ann. Of Math. (2) **54** (1951), 325–338. MR0043392

[146] Richard V. Kadison, *A generalized Schwarz inequality and algebraic invariants for operator algebras*, Ann. of Math. (2) **56** (1952), 494–503. MR0051442

[147] Richard V. Kadison and John R. Ringrose, *Fundamentals of the theory of operator algebras. Vol. I*, Pure and Applied Mathematics, vol. 100, Academic Press, Inc. [Harcourt Brace Jovanovich, Publishers], New York, 1983. Elementary theory. MR719020

[148] Shizuo Kakutani and Kunihiko Kodaira, *Über das Haarsche Mass in der lokal bikompakten Gruppe* (German), Proc. Imp. Acad. Tokyo **20** (1944), 444–450. MR0014401

[149] Eberhard Kaniuth, *Extension of positive definite functions from subgroups of nilpotent locally compact groups*, Proc. Amer. Math. Soc. **132** (2004), no. 3, 865–874. MR2019967

[150] Eberhard Kaniuth, *Weak spectral synthesis in commutative Banach algebras*, J. Funct. Anal. **254** (2008), no. 4, 987–1002. MR2381201

[151] Eberhard Kaniuth, *A course in commutative Banach algebras*, Graduate Texts in Mathematics, vol. 246, Springer, New York, 2009. MR2458901

[152] Eberhard Kaniuth, *Weak spectral synthesis in Fourier algebras of coset spaces*, Studia Math. **197** (2010), no. 3, 229–246. MR2607490

[153] Eberhard Kaniuth, *Weak spectral synthesis in commutative Banach algebras. II*, J. Funct. Anal. **259** (2010), no. 2, 524–544. MR2644112

[154] Eberhard Kaniuth and Anthony T. Lau, *A separation property of positive definite functions on locally compact groups and applications to Fourier algebras*, J. Funct. Anal. **175** (2000), no. 1, 89–110. MR1774852

[155] Eberhard Kaniuth and Anthony T. Lau, *Spectral synthesis for $A(G)$ and subspaces of $VN(G)$*, Proc. Amer. Math. Soc. **129** (2001), no. 11, 3253–3263. MR1845000

[156] Eberhard Kaniuth and Anthony T. Lau, *On a separation property of positive definite functions on locally compact groups*, Math. Z. **243** (2003), no. 1, 161–177. MR1953055

[157] Eberhard Kaniuth and Anthony T. Lau, *Extension and separation properties of positive definite functions on locally compact groups*, Trans. Amer. Math. Soc. **359** (2007), no. 1, 447–463. MR2247899

[158] Eberhard Kaniuth and Keith F. Taylor, *Induced representations of locally compact groups*, Cambridge Tracts in Mathematics, vol. 197, Cambridge University Press, Cambridge, 2013. MR3012851

[159] Eberhard Kaniuth and Ali Ülger, *The Bochner-Schoenberg-Eberlein property for commutative Banach algebras, especially Fourier and Fourier-Stieltjes algebras*, Trans. Amer. Math. Soc. **362** (2010), no. 8, 4331–4356. MR2608409

[160] Eberhard Kaniuth and Ali Ülger, *The structure of power bounded elements in Fourier-Stieltjes algebras of locally compact groups*, Bull. Sci. Math. **137** (2013), no. 1, 45–62. MR3007099

[161] Eberhard Kaniuth and Ali Ülger, *Weak spectral synthesis in commutative Banach algebras. III*, J. Funct. Anal. **268** (2015), no. 8, 2142–2170. MR3318645

[162] Eberhard Kaniuth and Ali Ülger, *Strong Ditkin sets in the spectrum of a commutative Banach algebra and the Fourier algebra*, Arch. Math. (Basel) **105** (2015), no. 1, 67–77. MR3365576

[163] E. Kaniuth, A. T. Lau, and G. Schlichting, *Lebesgue type decomposition of subspaces of Fourier-Stieltjes algebras*, Trans. Amer. Math. Soc. **355** (2003), no. 4, 1467–1490. MR1946400

[164] E. Kaniuth, A. T. Lau, and A. Ülger, *Multipliers of commutative Banach algebras, power boundedness and Fourier-Stieltjes algebras*, J. Lond. Math. Soc. (2) **81** (2010), no. 1, 255–275. MR2580464

[165] E. Kaniuth, A. T. Lau, and A. Ülger, *Power boundedness in Fourier and Fourier-Stieltjes algebras and other commutative Banach algebras*, J. Funct. Anal. **260** (2011), no. 8, 2366–2386. MR2772374

[166] E. Kaniuth, A. T. Lau, and A. Ülger, *The Rajchman algebra $B_0(G)$ of a locally compact group G*, Bull. Sci. Math. **140** (2016), no. 3, 273–302. MR3473390

[167] Idriss Khalil, *Sur l'analyse harmonique du groupe affine de la droite* (French), Studia Math. **51** (1974), 139–167. MR0350330

[168] Søren Knudby, *Fourier algebras of parabolic subgroups*, Math. Scand. **120** (2017), no. 2, 272–290. MR3657416

[169] R. A. Kunze and E. M. Stein, *Uniformly bounded representations and harmonic analysis of the 2×2 real unimodular group*, Amer. J. Math. **82** (1960), 1–62. MR0163988

[170] K. Kuratowski, *Topology. Vol. I*, New edition, revised and augmented. Translated from the French by J. Jaworowski, Academic Press, New York-London; Państwowe Wydawnictwo Naukowe, Warsaw, 1966. MR0217751

[171] Anselm Lambert, Matthias Neufang, and Volker Runde, *Operator space structure and amenability for Figà-Talamanca-Herz algebras*, J. Funct. Anal. **211** (2004), no. 1, 245–269. MR2054624

[172] Ronald Larsen, *An introduction to the theory of multipliers*, Springer-Verlag, New York-Heidelberg, 1971. Die Grundlehren der mathematischen Wissenschaften, Band 175. MR0435738

[173] Anthony To Ming Lau, *The Fourier-Stieltjes algebra of a topological semigroup with involution*, Pacific J. Math. **77** (1978), no. 1, 165–181. MR507628

[174] Anthony To Ming Lau, *Uniformly continuous functionals on the Fourier algebra of any locally compact group*, Trans. Amer. Math. Soc. **251** (1979), 39–59. MR531968

[175] Anthony To Ming Lau, *Invariantly complemented subspaces of $L_\infty(G)$ and amenable locally compact groups*, Illinois J. Math. **26** (1982), no. 2, 226–235. MR650390

[176] Anthony To-Ming Lau and Michael Leinert, *Fixed point property and the Fourier algebra of a locally compact group*, Trans. Amer. Math. Soc. **360** (2008), no. 12, 6389–6402. MR2434292

[177] Anthony To Ming Lau and Viktor Losert, *Weak*-closed complemented invariant subspaces of $L_\infty(G)$ and amenable locally compact groups*, Pacific J. Math. **123** (1986), no. 1, 149–159. MR834144

[178] Anthony To Ming Lau and Viktor Losert, *Complementation of certain subspaces of $L_\infty(G)$ of a locally compact group*, Pacific J. Math. **141** (1990), no. 2, 295–310. MR1035445

[179] Anthony To Ming Lau and Viktor Losert, *The C^*-algebra generated by operators with compact support on a locally compact group*, J. Funct. Anal. **112** (1993), no. 1, 1–30. MR1207935

[180] A. T.-M. Lau and R. J. Loy, *Weak amenability of Banach algebras on locally compact groups*, J. Funct. Anal. **145** (1997), no. 1, 175–204. MR1442165

[181] Anthony To-Ming Lau and J. Ludwig, *Fourier-Stieltjes algebra of a topological group*, Adv. Math. **229** (2012), no. 3, 2000–2023. MR2871165

[182] Anthony To-Ming Lau and Jean Ludwig, *Unitary closure and Fourier algebra of a topological group*, Studia Math. **231** (2015), no. 1, 1–28. MR3460624

[183] Anthony To Ming Lau and Peter F. Mah, *Normal structure in dual Banach spaces associated with a locally compact group*, Trans. Amer. Math. Soc. **310** (1988), no. 1, 341–353. MR937247

[184] A. T.-M. Lau, R. J. Loy, and G. A. Willis, *Amenability of Banach and C^*-algebras on locally compact groups*, Studia Math. **119** (1996), no. 2, 161–178. MR1391474

[185] Anthony To Ming Lau and Ali Ülger, *Some geometric properties on the Fourier and Fourier-Stieltjes algebras of locally compact groups, Arens regularity and related problems*, Trans. Amer. Math. Soc. **337** (1993), no. 1, 321–359. MR1147402

[186] Hun Hee Lee and Ebrahim Samei, *Beurling-Fourier algebras, operator amenability and Arens regularity*, J. Funct. Anal. **262** (2012), no. 1, 167–209. MR2852259

[187] Marcel Lefranc, *Sur certaines algèbres de fonctions sur un groupe* (French), C. R. Acad. Sci. Paris Sér. A-B **274** (1972), A1882–A1883. MR0316975

[188] Michael Leinert, *Faltungsoperatoren auf gewissen diskreten Gruppen* (German), Studia Math. **52** (1974), 149–158. MR0355480

[189] Michael Leinert, *Abschätzung von Normen gewisser Matrizen und eine Anwendung* (German), Math. Ann. **240** (1979), no. 1, 13–19. MR523997

[190] Martin Leischner and Walter Roelcke, *On neutral subgroups of topological groups*, Math. Ann. **282** (1988), no. 1, 113–129. MR960836

[191] Horst Leptin, *Sur l'algèbre de Fourier d'un groupe localement compact* (French), C. R. Acad. Sci. Paris Sér. A-B **266** (1968), A1180–A1182. MR0239002

[192] H. Leptin, *On locally compact groups with invariant means*, Proc. Amer. Math. Soc. **19** (1968), 489–494. MR0239001

[193] Teng-Sun Liu, Arnoud van Rooij, and Ju Kwei Wang, *Projections and approximate identities for ideals in group algebras*, Trans. Amer. Math. Soc. **175** (1973), 469–482. MR0318781

[194] John R. Liukkonen and Michael W. Mislove, *Symmetry in Fourier-Stieltjes algebras*, Math. Ann. **217** (1975), no. 2, 97–112. MR0420148

[195] Noël Lohoué, *Sur certains ensembles de synthèse dans les algèbres $A_p(G)$* (French), C. R. Acad. Sci. Paris Sér. A-B **270** (1970), A589–A591. MR0264337

[196] N. Lohoué, *Remarques sur les ensembles de synthèse des algèbres de groupe localement compact* (French), J. Functional Analysis **13** (1973), 185–194. MR0361624

[197] Noël Lohoué, *Sur les convoluteurs de $L^p(G)$* (French), C. R. Acad. Sci. Paris Sér. A **278** (1974), 1543–1545. MR0346426

[198] Lynn H. Loomis, *An introduction to abstract harmonic analysis*, D. Van Nostrand Company, Inc., Toronto-New York-London, 1953. MR0054173

[199] V. Losert, *Some properties of groups without the property P_1*, Comment. Math. Helv. **54** (1979), no. 1, 133–139. MR522036

[200] Viktor Losert, *On tensor products of Fourier algebras*, Arch. Math. (Basel) **43** (1984), no. 4, 370–372. MR802314

[201] Viktor Losert, *Properties of the Fourier algebra that are equivalent to amenability*, Proc. Amer. Math. Soc. **92** (1984), no. 3, 347–354. MR759651

[202] Viktor Losert, *Separation property, Mautner phenomenon, and neutral subgroups*, Banach algebras and their applications, Contemp. Math., vol. 363, Amer. Math. Soc., Providence, RI, 2004, pp. 223–234. MR2097963

[203] Viktor Losert, *Separation property, Mautner phenomenon, and neutral subgroups*, Banach algebras and their applications, Contemp. Math., vol. 363, Amer. Math. Soc., Providence, RI, 2004, pp. 223–234. MR2097963

[204] Jean Ludwig and Lyudmila Turowska, *On the connection between sets of operator synthesis and sets of spectral synthesis for locally compact groups*, J. Funct. Anal. **233** (2006), no. 1, 206–227. MR2204679

[205] George W. Mackey, *On induced representations of groups*, Amer. J. Math. **73** (1951), 576–592. MR0042420

[206] George W. Mackey, *Induced representations of locally compact groups. I*, Ann. of Math. (2) **55** (1952), 101–139. MR0044536

[207] Paul Malliavin, *Impossibilité de la synthèse spectrale sur les groupes abéliens non compacts* (French), Inst. Hautes Études Sci. Publ. Math. **1959** (1959), 85–92. MR0107129

[208] Giancarlo Mauceri, *Square integrable representations and the Fourier algebra of a unimodular group*, Pacific J. Math. **73** (1977), no. 1, 143–154. MR0486289

[209] Giancarlo Mauceri and Massimo A. Picardello, *Noncompact unimodular groups with purely atomic Plancherel measures*, Proc. Amer. Math. Soc. **78** (1980), no. 1, 77–84. MR548088

[210] F. I. Mautner, *Geodesic flows on symmetric Riemann spaces*, Ann. of Math. (2) **65** (1957), 416–431. MR0084823

[211] Kelly McKennon, *Multipliers, positive functionals, positive-definite functions, and Fourier-Stieltjes transforms*, American Mathematical Society, Providence, R. I., 1971. Memoirs of the American Mathematical Society, No. 111. MR0440299

[212] John R. McMullen, *Extensions of positive-definite functions*, American Mathematical Society, Providence, R. I., 1972. Memoirs of the American Mathematical Society, No. 117. MR0445229

[213] Tianxuan Miao, *Decomposition of $B(G)$*, Trans. Amer. Math. Soc. **351** (1999), no. 11, 4675–4692. MR1608490

[214] Deane Montgomery and Leo Zippin, *Topological transformation groups*, Interscience Publishers, New York-London, 1955. MR0073104

[215] Calvin C. Moore, *Groups with finite dimensional irreducible representations*, Trans. Amer. Math. Soc. **166** (1972), 401–410. MR0302817

[216] Calvin C. Moore, *The Mautner phenomenon for general unitary representations*, Pacific J. Math. **86** (1980), no. 1, 155–169. MR586875

[217] T. K. Muraleedharan, *Weak Wiener-Ditkin sets*, Acta Math. Hungar. **133** (2011), no. 4, 359–364. MR2851138

[218] Varadharajan Muruganandam, *Fourier algebra of a hypergroup. I*, J. Aust. Math. Soc. **82** (2007), no. 1, 59–83. MR2301971

[219] Claudio Nebbia, *Multipliers and asymptotic behaviour of the Fourier algebra of nonamenable groups*, Proc. Amer. Math. Soc. **84** (1982), no. 4, 549–554. MR643747

[220] Matthias Neufang, Zhong-Jin Ruan, and Nico Spronk, *Completely isometric representations of $M_{\mathrm{cb}}A(G)$ and $\mathrm{UCB}(\hat{G})$*, Trans. Amer. Math. Soc. **360** (2008), no. 3, 1133–1161. MR2357691

[221] Paul Nouyrigat, *Sur le prédual de l'algèbre de von Neumann associée à une représentation unitaire d'un groupe localement compact* (French), Publ. Dép. Math. (Lyon) **9** (1973), no. 2, 31–59. MR0346089

[222] A.Yu. Ol'shanskii, *Geometry of defining relations in groups*, Kluwer Academic Press, Dordrecht, 1991.

[223] Theodore W. Palmer, *Banach algebras and the general theory of $*$-algebras. Vol. I*, Encyclopedia of Mathematics and its Applications, vol. 49, Cambridge University Press, Cambridge, 1994. Algebras and Banach algebras. MR1270014

[224] K. Parthasarathy and N. Shravan Kumar, *Fourier algebras on homogeneous spaces*, Bull. Sci. Math. **135** (2011), no. 2, 187–205. MR2773397

[225] K. Parthasarathy and R. Prakash, *Spectral synthesis and operator synthesis*, Studia Math. **177** (2006), no. 2, 173–181. MR2285239

[226] Krishnan Parthasarathy and Rajendran Prakash, *Spectral synthesis in the Fourier algebra and the Varopoulos algebra*, Tohoku Math. J. (2) **59** (2007), no. 3, 441–454. MR2365350

[227] K. Parthasarathy and R. Prakash, *Spectral subspaces for the Fourier algebra*, Colloq. Math. **108** (2007), no. 2, 179–182. MR2291631

[228] K. Parthasarathy and R. Prakash, *Malliavin's theorem for weak synthesis on nonabelian groups*, Bull. Sci. Math. **134** (2010), no. 6, 561–578. MR2679529

[229] Krishnan Parthasarathy and Nageswaran Shravan Kumar, *Ditkin sets in homogeneous spaces*, Studia Math. **203** (2011), no. 3, 291–307. MR2786168

[230] K. Parthasarathy and Sujatha Varma, *On weak spectral synthesis*, Bull. Austral. Math. Soc. **43** (1991), no. 2, 279–282. MR1097069

[231] Alan L. T. Paterson, *Amenability*, Mathematical Surveys and Monographs, vol. 29, American Mathematical Society, Providence, RI, 1988. MR961261

[232] Alan L. T. Paterson, *The Fourier algebra for locally compact groupoids*, Canad. J. Math. **56** (2004), no. 6, 1259–1289. MR2102633

[233] Vern I. Paulsen, *Completely bounded maps and dilations*, Pitman Research Notes in Mathematics Series, vol. 146, Longman Scientific & Technical, Harlow; John Wiley & Sons, Inc., New York, 1986. MR868472

[234] Hung Le Pham, *Contractive homomorphisms of the Fourier algebras*, Bull. Lond. Math. Soc. **42** (2010), no. 5, 937–947. MR2728697

[235] Massimo Angelo Picardello, *Lacunary sets in discrete noncommutative groups* (English, with Italian summary), Boll. Un. Mat. Ital. (4) **8** (1973), 494–508. MR0344804

[236] Jean-Paul Pier, *Amenable locally compact groups*, Pure and Applied Mathematics (New York), John Wiley & Sons, Inc., New York, 1984. A Wiley-Interscience Publication. MR767264

[237] Gilles Pisier, *Introduction to operator space theory*, London Mathematical Society Lecture Note Series, vol. 294, Cambridge University Press, Cambridge, 2003. MR2006539

[238] R. Plymen, *Fourier algebra of a compact Lie group*, arXiv. math. FA/0104018.

[239] Jean Poncet, *Une classe d'espaces homogènes possédant une mesure invariante* (French), C. R. Acad. Sci. Paris **238** (1954), 553–554. MR0059289

[240] Arlan Ramsay and Martin E. Walter, *Fourier-Stieltjes algebras of locally compact groupoids*, J. Funct. Anal. **148** (1997), no. 2, 314–367. MR1469344

[241] Hans Reiter and Jan D. Stegeman, *Classical harmonic analysis and locally compact groups*, 2nd ed., London Mathematical Society Monographs. New Series, vol. 22, The Clarendon Press, Oxford University Press, New York, 2000. MR1802924

[242] P. F. Renaud, *Centralizers of Fourier algebra of an amenable group*, Proc. Amer. Math. Soc. **32** (1972), 539–542. MR0293019

[243] Charles E. Rickart, *General theory of Banach algebras*, The University Series in Higher Mathematics, D. van Nostrand Co., Inc., Princeton, N.J.-Toronto-London-New York, 1960. MR0115101

[244] Walter Roelcke and Susanne Dierolf, *Uniform structures on topological groups and their quotients*, McGraw-Hill International Book Co., New York, 1981. Advanced Book Program. MR644485

[245] Zhong-Jin Ruan, *The operator amenability of $A(G)$*, Amer. J. Math. **117** (1995), no. 6, 1449–1474. MR1363075

[246] Walter Rudin, *On isomorphisms of group algebras*, Bull. Amer. Math. Soc. **64** (1958), 167–169. MR0096981

[247] Walter Rudin, *Fourier analysis on groups*, Interscience Tracts in Pure and Applied Mathematics, No. 12, Interscience Publishers (a division of John Wiley and Sons), New York-London, 1962. MR0152834

[248] Walter Rudin, *Spaces of type $H^\infty + C$* (English, with French summary), Ann. Inst. Fourier (Grenoble) **25** (1975), no. 1, vi, 99–125. MR0377520

[249] Volker Runde, *Lectures on amenability*, Lecture Notes in Mathematics, vol. 1774, Springer-Verlag, Berlin, 2002. MR1874893

[250] Volker Runde, *Applications of operator spaces to abstract harmonic analysis*, Expo. Math. **22** (2004), no. 4, 317–363. MR2075743

[251] Volker Runde, *The amenability constant of the Fourier algebra*, Proc. Amer. Math. Soc. **134** (2006), no. 5, 1473–1481. MR2199195

[252] Volker Runde, *Cohen-Host type idempotent theorems for representations on Banach spaces and applications to Figà-Talamanca-Herz algebras*, J. Math. Anal. Appl. **329** (2007), no. 1, 736–751. MR2306837

[253] Volker Runde and Nico Spronk, *Operator amenability of Fourier-Stieltjes algebras*, Math. Proc. Cambridge Philos. Soc. **136** (2004), no. 3, 675–686. MR2055055

[254] Volker Runde and Nico Spronk, *Operator amenability of Fourier-Stieltjes algebras. II*, Bull. Lond. Math. Soc. **39** (2007), no. 2, 194–202. MR2323448

[255] Sadahiro Saeki, *On strong Ditkin sets*, Ark. Mat. **10** (1972), 1–7. MR0310554

[256] Shôichirô Sakai, *C^*-algebras and W^*-algebras*, Springer-Verlag, New York-Heidelberg, 1971. Ergebnisse der Mathematik und ihrer Grenzgebiete, Band 60. MR0442701

[257] Ebrahim Samei, *Hyper-Tauberian algebras and weak amenability of Figà-Talamanca-Herz algebras*, J. Funct. Anal. **231** (2006), no. 1, 195–220. MR2190169

[258] Bert M. Schreiber, *On the coset ring and strong Ditkin sets*, Pacific J. Math. **32** (1970), 805–812. MR0259502

[259] Laurent Schwartz, *Sur une propriété de synthèse spectrale dans les groupes non compacts* (French), C. R. Acad. Sci. Paris **227** (1948), 424–426. MR0027096

[260] N. Shravan Kumar, *Ideals with bounded approximate identities in the Fourier algebras on homogeneous spaces*, Indag. Math. (N.S.) **24** (2013), no. 1, 1–14. MR2997748

[261] Martha Smith, *Group algebras*, J. Algebra **18** (1971), 477–499. MR0277625

[262] Nico Spronk, *Operator weak amenability of the Fourier algebra*, Proc. Amer. Math. Soc. **130** (2002), no. 12, 3609–3617. MR1920041

[263] Nico Spronk, *Measurable Schur multipliers and completely bounded multipliers of the Fourier algebras*, Proc. London Math. Soc. (3) **89** (2004), no. 1, 161–192. MR2063663

[264] Nico Spronk, *Amenability properties of Fourier algebras and Fourier-Stieltjes algebras: a survey*, Banach algebras 2009, Banach Center Publ., vol. 91, Polish Acad. Sci. Inst. Math., Warsaw, 2010, pp. 365–383. MR2777488

[265] Nico Spronk and Lyudmila Turowska, *Spectral synthesis and operator synthesis for compact groups*, J. London Math. Soc. (2) **66** (2002), no. 2, 361–376. MR1920408

[266] Ross Stokke, *Amenable representations and coefficient subspaces of Fourier-Stieltjes algebras*, Math. Scand. **98** (2006), no. 2, 182–200. MR2243701

[267] Ross Stokke, *Contractive homomorphisms of measure algebras and Fourier algebras*, Studia Math. **209** (2012), no. 2, 135–150. MR2943839

[268] Ryszard Szwarc, *Groups acting on trees and approximation properties of the Fourier algebra*, J. Funct. Anal. **95** (1991), no. 2, 320–343. MR1092129

[269] Masamichi Takesaki, *A characterization of group algebras as a converse of Tannaka-Stinespring-Tatsuuma duality theorem*, Amer. J. Math. **91** (1969), 529–564. MR0244437

[270] Masamichi Takesaki, *Theory of operator algebras. I*, Springer-Verlag, New York-Heidelberg, 1979. MR548728

[271] Masamichi Takesaki and Nobuhiko Tatsuuma, *Duality and subgroups*, Ann. of Math. (2) **93** (1971), 344–364. MR0281843

[272] Masamichi Takesaki and Nobuhiko Tatsuuma, *Duality and subgroups. II*, J. Functional Analysis **11** (1972), 184–190. MR0384995

[273] Nobuhiko Tatsuuma, *A duality theorem for locally compact groups*, J. Math. Kyoto Univ. **6** (1967), 187–293. MR0217222

[274] Joseph L. Taylor, *Measure algebras*, Published for the Conference Board of the Mathematical Sciences by the American Mathematical Society, Providence, R. I., 1973. Expository lectures from the CBMS Regional Conference held at the University of Montana, Missoula, Mont., June 1972; Conference Board of the Mathematical Sciences Regional Conference Series in Mathematics, No. 16. MR0427949

[275] Keith F. Taylor, *Geometry of the Fourier algebras and locally compact groups with atomic unitary representations*, Math. Ann. **262** (1983), no. 2, 183–190. MR690194

[276] A. Ülger, *Some results about the spectrum of commutative Banach algebras under the weak topology and applications*, Monatsh. Math. **121** (1996), no. 4, 353–379. MR1389676

[277] N. Th. Varopoulos, *Spectral synthesis on spheres*, Proc. Cambridge Philos. Soc. **62** (1966), 379–387. MR0201908

[278] N. Th. Varopoulos, *Tensor algebras and harmonic analysis*, Acta Math. **119** (1967), 51–112. MR0240564

[279] William A. Veech, *Weakly almost periodic functions on semisimple Lie groups*, Monatsh. Math. **88** (1979), no. 1, 55–68. MR550072

[280] Martin E. Walter, *W^*-algebras and nonabelian harmonic analysis*, J. Functional Analysis **11** (1972), 17–38. MR0352879

[281] Martin E. Walter, *On the structure of the Fourier-Stieltjes algebra*, Pacific J. Math. **58** (1975), no. 1, 267–281. MR0425008

[282] Martin E. Walter, *On a theorem of Figà-Talamanca*, Proc. Amer. Math. Soc. **60** (1976), 72–74 (1977). MR0420149

[283] S. P. Wang, *On the Mautner phenomenon and groups with property* (T), Amer. J. Math. **104** (1982), no. 6, 1191–1210. MR681733

[284] John S. P. Wang, *The Mautner phenomenon for p-adic Lie groups*, Math. Z. **185** (1984), no. 3, 403–412. MR731685

[285] P. S. Wang, *On isolated points in the dual spaces of locally compact groups*, Math. Ann. **218** (1975), no. 1, 19–34. MR0384993

[286] C. Robert Warner, *A class of spectral sets*, Proc. Amer. Math. Soc. **57** (1976), no. 1, 99–102. MR0410275

[287] C. Robert Warner, *Spectral synthesis in the Mirkil algebra*, J. Math. Anal. Appl. **167** (1992), no. 1, 176–181. MR1165267

[288] J. G. Wendel, *On isometric isomorphism of group algebras*, Pacific J. Math. **1** (1951), 305–311. MR0049910

[289] J. G. Wendel, *Left centralizers and isomorphisms of group algebras*, Pacific J. Math. **2** (1952), 251–261. MR0049911

[290] Matthew Wiersma, *L^p-Fourier and Fourier-Stieltjes algebras for locally compact groups*, J. Funct. Anal. **269** (2015), no. 12, 3928–3951. MR3418075

[291] Peter J. Wood, *Complemented ideals in the Fourier algebra of a locally compact group*, Proc. Amer. Math. Soc. **128** (2000), no. 2, 445–451. MR1616589

[292] Peter J. Wood, *The operator biprojectivity of the Fourier algebra*, Canad. J. Math. **54** (2002), no. 5, 1100–1120. MR1924715

[293] E. I. Zel′manov, *On periodic compact groups*, Israel J. Math. **77** (1992), no. 1-2, 83–95. MR1194787

[294] Robert J. Zimmer, *Ergodic theory and semisimple groups*, Monographs in Mathematics, vol. 81, Birkhäuser Verlag, Basel, 1984. MR776417

[295] A. Zygmund, *Trigonometric series. Vol. I, II*, 3rd ed., Cambridge Mathematical Library, Cambridge University Press, Cambridge, 2002. With a foreword by Robert A. Fefferman. MR1963498

Index

Selected Published Titles in This Series

231 **Eberhard Kaniuth and Anthony To-Ming Lau,** Fourier and Fourier-Stieltjes Algebras on Locally Compact Groups, 2018

228 **Zhenbo Qin,** Hilbert Schemes of Points and Infinite Dimensional Lie Algebras, 2018

227 **Roberto Frigerio,** Bounded Cohomology of Discrete Groups, 2017

226 **Marcelo Aguiar and Swapneel Mahajan,** Topics in Hyperplane Arrangements, 2017

225 **Mario Bonk and Daniel Meyer,** Expanding Thurston Maps, 2017

224 **Ruy Exel,** Partial Dynamical Systems, Fell Bundles and Applications, 2017

223 **Guillaume Aubrun and Stanisław J. Szarek,** Alice and Bob Meet Banach, 2017

222 **Alexandru Buium,** Foundations of Arithmetic Differential Geometry, 2017

221 **Dennis Gaitsgory and Nick Rozenblyum,** A Study in Derived Algebraic Geometry, 2017

220 **A. Shen, V. A. Uspensky, and N. Vereshchagin,** Kolmogorov Complexity and Algorithmic Randomness, 2017

219 **Richard Evan Schwartz,** The Projective Heat Map, 2017

218 **Tushar Das, David Simmons, and Mariusz Urbański,** Geometry and Dynamics in Gromov Hyperbolic Metric Spaces, 2017

217 **Benoit Fresse,** Homotopy of Operads and Grothendieck–Teichmüller Groups, 2017

216 **Frederick W. Gehring, Gaven J. Martin, and Bruce P. Palka,** An Introduction to the Theory of Higher-Dimensional Quasiconformal Mappings, 2017

215 **Robert Bieri and Ralph Strebel,** On Groups of PL-homeomorphisms of the Real Line, 2016

214 **Jared Speck,** Shock Formation in Small-Data Solutions to 3D Quasilinear Wave Equations, 2016

213 **Harold G. Diamond and Wen-Bin Zhang (Cheung Man Ping),** Beurling Generalized Numbers, 2016

212 **Pandelis Dodos and Vassilis Kanellopoulos,** Ramsey Theory for Product Spaces, 2016

211 **Charlotte Hardouin, Jacques Sauloy, and Michael F. Singer,** Galois Theories of Linear Difference Equations: An Introduction, 2016

210 **Jason P. Bell, Dragos Ghioca, and Thomas J. Tucker,** The Dynamical Mordell–Lang Conjecture, 2016

209 **Steve Y. Oudot,** Persistence Theory: From Quiver Representations to Data Analysis, 2015

208 **Peter S. Ozsváth, András I. Stipsicz, and Zoltán Szabó,** Grid Homology for Knots and Links, 2015

207 **Vladimir I. Bogachev, Nicolai V. Krylov, Michael Röckner, and Stanislav V. Shaposhnikov,** Fokker–Planck–Kolmogorov Equations, 2015

206 **Bennett Chow, Sun-Chin Chu, David Glickenstein, Christine Guenther, James Isenberg, Tom Ivey, Dan Knopf, Peng Lu, Feng Luo, and Lei Ni,** The Ricci Flow: Techniques and Applications: Part IV: Long-Time Solutions and Related Topics, 2015

205 **Pavel Etingof, Shlomo Gelaki, Dmitri Nikshych, and Victor Ostrik,** Tensor Categories, 2015

204 **Victor M. Buchstaber and Taras E. Panov,** Toric Topology, 2015

203 **Donald Yau and Mark W. Johnson,** A Foundation for PROPs, Algebras, and Modules, 2015

202 **Shiri Artstein-Avidan, Apostolos Giannopoulos, and Vitali D. Milman,** Asymptotic Geometric Analysis, Part I, 2015

For a complete list of titles in this series, visit the
AMS Bookstore at **www.ams.org/bookstore/survseries/**.